NON-EQUILIBRIUM SOFT MATTER PHYSICS

SERIES IN SOFT CONDENSED MATTER **ISSN: 1793-737X**

Published:

Vol. 1 Polymer Thin Films
edited by Ophelia K. C. Tsui and Thomas P. Russell

Vol. 2 Polymers, Liquids and Colloids in Electric Fields:
Interfacial Instabilities, Orientation and Phase Transitions
edited by Yoav Tsori and Ullrich Steiner

Vol. 3 Understanding Soft Condensed Matter via Modeling and Computation
edited by Wenbing Hu and An-Chang Shi

Vol. 4 Non-Equilibrium Soft Matter Physics
edited by Shigeyuki Komura and Takao Ohta

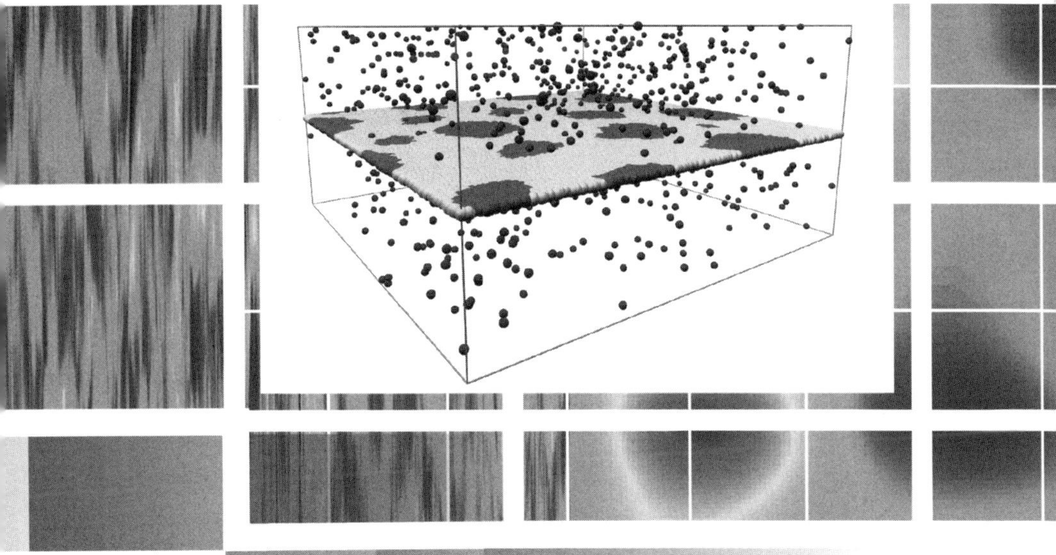

Series in Soft Condensed Matter **Vol. 4**

NON-EQUILIBRIUM SOFT MATTER PHYSICS

Editors

Shigeyuki Komura
Tokyo Metropolitan University, Japan

Takao Ohta
Kyoto University, Japan

World Scientific

NEW JERSEY • LONDON • SINGAPORE • BEIJING • SHANGHAI • HONG KONG • TAIPEI • CHENNAI

Published by

World Scientific Publishing Co. Pte. Ltd.
5 Toh Tuck Link, Singapore 596224
USA office: 27 Warren Street, Suite 401-402, Hackensack, NJ 07601
UK office: 57 Shelton Street, Covent Garden, London WC2H 9HE

British Library Cataloguing-in-Publication Data
A catalogue record for this book is available from the British Library.

Series in Soft Condensed Matter — Vol. 4
NON-EQUILIBRIUM SOFT MATTER PHYSICS

ISBN-13 978-981-4360-62-3
ISBN-10 981-4360-62-7

Printed in Singapore by World Scientific Printers.

Foreword

The study of *Soft Condensed Matter* has stimulated fruitful interactions between physicists, chemists, and engineers, and is now reaching out to biologists. A broad interdisciplinary community involving all these areas of science has emerged over the last 30 years or so, and with it our knowledge of *Soft Condensed Matter* has grown considerably with the active investigations of polymers, supramolecular assemblies of designed organic molecules, liquid crystals, colloids, lyotropic systems, emulsions, biopolymers, and biomembranes, among others. Taking into account that research in *Soft Condensed Matter* involves ideas coming physics, chemistry, materials science as well as biology, this series may form a bridge between all these disciplines with the aim to provide a comprehensive and substantial understanding a broad spectrum of phenomena relevant to *Soft Condensed Matter*.

The present Book Series, initiated by the late Pierre-Gilles de Gennes, comprises independent book volumes that touch on a wide and diverse range of topics of current interest and importance, covering a large number of diverse aspects, both theoretical and experimental, in all areas of *Soft Condensed Matter*. These volumes will be edited books on advanced topics with contributions by various authors and monographs in a lighter style, written by experts in the corresponding areas. The Book Series mainly addresses graduate students and junior researchers as an introduction to new fields, but it should also be useful to experienced people who want to obtain a general idea on a certain topic or may consider a change of their field of research. This Book Series aims to provide a comprehensive and instructive overview of all *Soft Condensed Matter* phenomena.

The present volume of this Book Series, edited by Shigeyuki Komura and Takao Ohta, takes as its central aim an exploration of dynamical and non-equilibrium effects of *Soft Condensed Matter*. Among others, these include rheology of polymers and liquid crystals, dynamical properties of Langmuir monolayers at the air/water interface, hydrodynamics of membranes and twisted filaments as well as dynamics of deformable self-propelled particles and migration of biological cells. Complex non-equilibrium phenomena are explored from microscopic to macroscopic scales. The added complexity of soft matter systems manifests itself by intrinsic multi-scale and hierarchical features producing a wealth of challenging dynamical effects. The seven self-contained and clearly written chapters of this volume can serve both as an introduction to students and novice researchers as well as a useful reference to researchers as it covers state of the art topics in the field of non-equilibrium soft matter.

Within the next years, our Series on Soft Condensed Matter will continuously grow and eventually cover the whole spectrum of phenomena in *Soft Condensed Matter*. We hope that many interested colleagues and scientists will profit from this book series.

David Andelman and Günter Reiter
Series Editors

Preface

One of the fundamental characteristic features of soft matter is that it exhibits various mesoscopic structures originating from a large number of internal degrees of freedom of each molecule. Due to such intermediate structures, soft matter can easily be brought into non-equilibrium states and cause non-linear responses by imposing external fields such as an electric field, a mechanical stress or a shear flow. As Volume 4 of the series in Soft Condensed Matter, this book focuses on the non-linear and non-equilibrium properties of soft matter.

In the past years, the disciplines of non-equilibrium physics and soft matter physics have developed somewhat independently. When the systematic research of non-equilibrium systems started in the early 1970's, the main subjects were macroscopic non-linear dissipative systems. Rayleigh–Benard convection, Belousov–Zhabotinsky reaction and crystal growth are typical examples. In the 1990's, new experimental techniques to manipulate meso/nanoscopic structures and to obtain real space-time information were developed such as using atomic force microscopy, laser tweezers and three-dimensional tomography. These achievements produced a new era of non-equilibrium physics which deals with out-of-equilibrium situations at a small scale. In fact, soft matter offers a rich variety of experimental systems that exhibit unique out-of-equilibrium phenomena. In turn, they have motivated further development of non-equilibrium statistical mechanics. Such an academic phase in the last decade promoted the momentum toward studying these two research fields in an interactive manner. This is the central concept of non-equilibrium soft matter physics; a new scientific frontier in the twenty-first century.

Studying non-equilibrium soft matter includes several aspects. (i) In dense polymer systems, for instance, due to a huge internal freedom and a high susceptibility, characteristic structures are formed under external fields. The polymer dynamics such as gel formation is strongly influenced by the microscopic structure itself. This coupling effect in different space-time scales is one of the key issues in modern polymer science. (ii) Soft matter often exhibits structural transitions by applying external fields such as an electric field or a shear flow. These morphological transitions lead to a drastic change in the physical properties of the material. It is of great interest and importance to elucidate the underlying kinetic pathways of these different non-equilibrium transitions. (iii) Soft matter driven far-from-equilibrium manifests striking nonlinear properties as a result of its high sensitivity to external fields. In those circumstances, non-equilibrium fluctuations and structural transitions at a mesoscopic level strongly affect macroscopic spatio-temporal structures. In order to explore these non-equilibrium phenomena, new experimental and theoretical tools from the statistical mechanics point of view have recently been developed.

Yet another exciting and important direction of non-equilibrium soft matter physics is to aim at biology. Living matter consists of various soft materials (hence soft matter complex) and can be regarded as one of the ultimate non-equilibrium systems. Generally speaking, biological systems offer a rich variety of fascinating non-equilibrium phenomena that can be fundamentally understood by means of appropriate physical modeling. In order to describe them by using the language of physics, various methods and accumulated knowledge in soft condensed matter physics are indispensable. We believe that the progression of non-equilibrium soft matter physics will provide us with a promising outcome in the future of biophysics.

Although many independent researchers in the world were consciously aware of the necessity of non-equilibrium soft matter physics, it has been strongly pursued in Japan during the last decade as an intensive scientific stream in physics and chemistry (and some biology). Such a trend was supported by the Grant-in-Aid for Scientific Research on Priority Areas "Creation of non-equilibrium soft matter physics: structure and dynamics of mesoscopic systems" (2006–2010) from the Ministry of Education, Culture, Sports, Science and Technology of Japan. This national project was headed by one of the editors of this book (TO), and involved more than 100 Japanese researchers. Almost 1,000 papers have been published in the last five years out of this project. More importantly, some new concepts such as soft matter composites, active soft matter, or driven soft matter began to emerge. We now believe that the importance of this new field is shared not only by Japanese scientists but also by many soft matter scientists around the world. In fact, several international conferences have been held in the recent years on this subject: "International Symposium on Non-Equilibrium Soft Matter 2010" (August 17–20, 2010, Nara), "Mini-

Symposium on Non-Equilibrium Soft Matter" (July 5, 2011, Berlin), "Soft Matter Far from Equilibrium" (Gordon Research Conference, August 14–19, 2011, New London).

As the editors, our aim in this book is to present some of the recent achievements of non-equilibrium soft matter physics in a comprehensive manner. Although the topics discussed are widespread ranging from microscopic molecular dynamics of polymers to macroscopic behavior of structured fluids or migration of cells, their common features are the formation of mesoscopic and hierarchical structures under non-equilibrium situations. The main purpose of this book is to deliver an overview on the exciting and rapidly growing area of research. We required that each independent chapter should be self-contained. It is our great pleasure if students, professors and researchers find this book a useful guide and resource to the field of non-equilibrium soft matter physics.

This book consists of seven chapters. In Chapter 1, Doi describes Onsager's variational principle which gives a unified framework in formulating various dynamics of soft matter. In Chapter 2, the rheodielectric behavior of polymers, liquid crystals, and soft matter composites is discussed by Watanabe *et al.* In Chapter 3, Orihara presents experimental study of morphology and rheology of immiscible polymer blends under both electric and shear flow fields using a new system combining a rheometer and a confocal scanning laser microscope. In Chapter 4, using a photosensitive two-dimensional monolayer prepared from an azobenzene surfactant, Sagués *et al.* discuss dynamical properties of the forced Langmuir monolayer. In Chapter 5, Komura *et al.* describe hydrodynamic effects in multicomponent fluid membranes. The presence of the outer bulk solvent is shown to play an essential role for the membrane dynamics. In Chapter 6, Wada *et al.* study non-equilibrium twist dynamics of rotationally driven semi-flexible polymers and filaments in a viscous fluid by means of analytical and numerical approaches. In the final chapter, Sano *et al.* present numerical and analytical studies on the dynamics of deformable self-propelled particles. Some experiments revealing the correlation of the shape and the stress distribution in a migrated cell are also discussed. It is important to notice, however, that these topics still do not cover all the aspects of non-equilibrium soft matter physics. Those who are further interested in this field may refer such as to the special section of "Non-equilibrium soft matter" published in Journal of Physics: Condensed Matter, Volume 23, 2011.

We are grateful to the contributors of this book who have taken enormous time and effort in completing chapters. Special thanks are due to Prof. David Andelman and Prof. Günter Reiter (series editors) who initially gave us the opportunity to edit and contribute to this book. We appreciate the anonymous reviewers who had kindly spent their time and given constructive suggestions for improvements of this book. Finally, this book would not have been published without the help of Mr. Alvin Chong from World Scientific Publishing who has been continuously supporting us in our endeavors.

Shigeyuki Komura
Takao Ohta

Contents

Foreword v

Preface vii

1. Onsager's Variational Principle in Soft Matter Dynamics 1
M. Doi

2. Rheo-Dielectric Behavior of Soft Matters 37
H. Watanabe, Y. Matsumiya, K. Horio, Y. Masubuchi and T. Uneyama

3. Morphology and Rheology of Immiscible Polymer Blends in Electric and Shear Flow Fields 89
H. Orihara

4. Dynamical Aspects of Two-Dimensional Soft Matter 145
F. Sagués, J. Claret and J. Ignés-Mullol

5. Hydrodynamic Effects in Multicomponent Fluid Membranes 197
S. Komura, S. Ramachandran and M. Imai

6. Actively Twisted Polymers and Filaments in Biology 275
H. Wada and R. R. Netz

7. Dynamics of Deformable Self-Propelled Particles: Relations with Cell Migration 365

M. Sano, M. Y. Matsuo and T. Ohta

Index 417

Chapter 1

Onsager's Variational Principle in Soft Matter Dynamics

Masao Doi*

Department of Applied Physics, University of Tokyo, Hongo, Tokyo, Japan

It is shown that Onsager's variational principle gives us a unified framework in discussing various dynamics of soft matter such as diffusion, rheology and their coupling. The variational principle gives kinetic equations which satisfy Onsager's reciprocal relation. With many examples, it is shown that the kinetic equations are usually written as non-linear partial differential equations for state variables and can describe various non-linear and non-equilibrium phenomena in soft matter. The physics underlying the variational principle is discussed.

1. Introduction

Soft matter is characterized by strongly non-linear and non-equilibrium responses to external fields. For example, polymer solutions show various non-linear viscoelastic behaviors such as shear thinning, stress relaxation and Weissenberg effect etc.[1–3] Liquid crystals show dramatic change of their optical properties under weak external fields (of mechanical, electrical or magnetic nature).[4] Gels shows various spatial patterns due to the coupled effects of solvent permeation and gel deformation.[5]

Various equations have been proposed to describe such phenomena. Examples are the Smoluchowskii equation in the dynamics of polymer solutions,[1–3] Leslie-Ericksen equation in the dynamics of liquid crystals.[4] and the gel-dynamics equations.[6] These equations have been, so far, proposed for each problems. In this paper, I will show that these equations are derived from a common framework, the Onsager's variational principle.

The variational principle was proposed by Onsager in his celebrated paper on the reciprocal relation in the kinetic equations for irreversible

*doi@rheo.t.u-tokyo.ac.jp

processes.[8] As Onsager stated in the paper, the variational principle is an extension of Rayleigh's least energy dissipation principle.[9]

The least energy dissipation principle is known as the principle in determining the steady state in linear systems such as viscous flow in Newtonian fluid, and electric current in Ohmic devices.[10] The Onsager's variational principle which will be discussed in this paper, however, refers to a principle slightly different from this (although they are connected). It is a principle to derive the time evolution equation in irreversible processes. The equations derived from the variational principle is equivalent to Onsager's kinetic equation.

Although the variational principle, or the kinetic equations, of Onsager represents a linear relation between forces and fluxes, the derived equations are usually non-linear time evolution equations and can describe non-linear phenomena.[10–12] It will be shown that Onsager's variational principle gives us various merits, not only to simplify calculations, but also to give us new insight to understand the problem. The purpose of this paper is to explain and to demonstrate these merits. Earlier version of this work was published in[13–15]

The construction of this paper is as follows. First, the least energy dissipation principle in hydrodynamics is discussed since I think that it is the earlier version of Onsager's principle and includes the essential physics underlying Onsager's theory. Second, Onsager's variational principle is discussed in a general (and somewhat abstract) form. After this, various applications of Onsager's principle are discussed. These sections will demonstrate the convenience and the usefulness of Onsager's variational principle.

2. Particle Motion in Viscous Fluid

2.1. *Stokesian hydrodynamics*

Consider particles moving in viscous fluid subject to some potential forces, for example, particles sedimenting in a gravitational field, or particles attracting each other via interparticle potential For small particle of colloidal dimension, one can assume that the velocity field $\boldsymbol{v}(\boldsymbol{r})$ of the fluid surrounding the particle obeys the Stokes equation:[16,17]

$$\eta \nabla^2 \boldsymbol{v} = \nabla p, \qquad \nabla \cdot \boldsymbol{v} = 0 \tag{1}$$

where η is the viscosity of the surrounding fluid and p is the pressure. The Stokes equation is the limit of zero Reynolds number of Navier Stokes equation, and represents the situtation that the fluid flow is in a steady state for given boundary condition. The first equation of the Stokes equation represents the force balance for each fluid element, and the second equation represents the incompressible condition for the fluid.

In the Stokes limit, the hydrodynamic drag acting on particles is proportional to the particle velocity. This gives a variational principle which we shall refer to as the hydrodynamic variational principle.

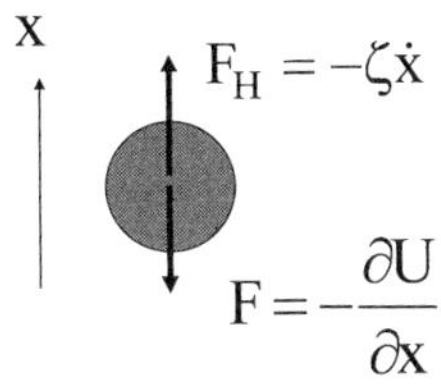

Fig. 1. Force balance of a sphere moving in a potential field $U(x)$.

Let us consider a very simple situation, i.e., a one-dimensional motion of a spherical particle in a viscous fluid driven by some potential field $U(x)$ (see Fig.1), where x stands for the position of the particle. Since the inertia effect is assumed to be negligibly small in the Stokesian hydrodynamics, the particle motion is determined by the condition that the total force acting on the particle is zero. In the present problem, two forces are acting on the particle. One is the potential force given by

$$F(x) = -\frac{\partial U}{\partial x} \tag{2}$$

and the other is the frictional force which is proportional to the particle volocity $\dot{x}$

$$F_H = -\zeta\dot{x} \tag{3}$$

The friction coefficient ζ can be calculated by the Stokes equation, and is given by $\zeta = 6\pi\eta a$ where a is the radius of the particle.

The force balance equation is then written as

$$\zeta\dot{x} = -\frac{\partial U}{\partial x} \tag{4}$$

This equation describes the time evolution of the particle position $x(t)$:

$$\frac{dx}{dt} = -\frac{1}{\zeta}\frac{\partial U}{\partial x} \tag{5}$$

Equation (4) can be cast in the form of a variational principle. If one defines a function $R(\dot{x})$ by

$$R(\dot{x}) = \frac{1}{2}\zeta\dot{x}^2 + \frac{\partial U}{\partial x}\dot{x} \tag{6}$$

Equation (4) is equivalent to the condition that R be minimized with respect to $\dot{x}$, i.e., $\partial R/\partial \dot{x} = 0$. The function R is called Rayleighian.

The Rayleighian consists of two terms. The first term

$$\Phi(\dot{x}) = \frac{1}{2}\zeta\dot{x}^2 \tag{7}$$

is called the energy dissipation function. When a particle moves with velocity $\dot{x}$ in a viscous fluid, it does a work to the fluid per unit time $-F_H\dot{x} = \zeta\dot{x}^2$, which is immediately dissipated into heat. Therefore the energy dissipation function $\Phi(\dot{x})$ stands for the half of the energy dissipation rate (the energy dissipated in the viscous fluid per unit time) when the particle moves with velocity $\dot{x}$. (The coefficient 1/2 is put by historical reason.)

The second term in the Rayleighian represents the rate of the potential energy change when the particle moves with velocity $\dot{x}$. We shall write this term as $\dot{U}(\dot{x})$:

$$\dot{U}(\dot{x}) = \frac{\partial U}{\partial x}\dot{x} \tag{8}$$

Given Φ and $\dot{U}$, the Rayleighian is written as

$$R(\dot{x}) = \Phi(\dot{x}) + \dot{U}(\dot{x}) \tag{9}$$

and the actual velocity $\dot{x}$ is determined by the condition $\partial R/\partial \dot{x} = 0$.

2.2. *Hydrodynamic reciprocal relation*

The above rewriting of the force balance equation into the variational principle is trivial. However, in more complex systems, the variational principle represents a non-trivial law in Stokesian hydrodynamics.

Consider a system of particles having many degrees of freedom, and let x_i $(i = 1, 2, ...f)$ be the set of variables describing the state of the sytem. For example, the state of a rigid particle is specified by six variables, three to specify the position, and the other three to specify the orientation. For N particle systems, $6N$ variables are needed.

Suppose that the state of the particles changes with rate $\dot{x}_i$, then there will be a frictional force exerted on the particles by the surrounding fluid.

Let F_{Hi} be the frictional force conjugate to $\dot{x}_i$ i.e., the work done to the fluid per unit time is written as

$$2\Phi = -\sum_i F_{Hi}\dot{x}_i \tag{10}$$

In the Stokesian hydrodynamics, the frictional force is always written as a linear function of $\dot{x}_i$:[16,17]

$$F_{Hi} = -\sum_j \zeta_{ij}\dot{x}_j \tag{11}$$

The coefficients ζ_{ij} are called friction coefficients, and can be calculated by solving the Stokes equation (see Appendix A).

The friction coefficients ζ_{ij} have an important property : they are symmetric, and positive definite, i.e.,

$$\zeta_{ij} = \zeta_{ji}, \tag{12}$$

and

$$\sum_{ij} \zeta_{ij}\dot{x}_i\dot{x}_j \geq 0 \qquad \text{for any } \dot{x}_i \tag{13}$$

The proof of these relations is given in Appendix A.

The reciprocal relation (12) is not a trivial relation. For example, consider the motion of a rod-like particle caused by a force $\boldsymbol{F}$ applied at the center of the particle. For rod-like particle, the velocity $\boldsymbol{V}$ is in general not parallel to the force $\boldsymbol{F}$. As it is shown in Fig.2(a), the force F_x in x direction causes a velocity component V_y in y direction, and as shown in Fig.2(b), the force F_y causes a velocity component V_x. The reciprocal relation (12) indicates that V_y/F_x is equal to V_x/F_y. This relation can be proven by symmetry argument for rod, but for a particle of general shape, validity of the relation $V_y/F_x = V_x/F_y$ is not obvious. For more complex shaped particles, the reciprocal relation gives more surprising result. As it is shown in Fig.2(c), a force F acting on a helical particle causes a rotational angular velocity ω of the particle, and as shown in Fig.2(d), a torque T causes a tranlational velocity V The reciprocal relation indicates the equality $\omega/F = V/T$, which is quite non-trivial.

2.3. *Hydrodynamic variational principle*

Now consider that the motion of the particles is driven by a potential $U(x)$, then the potential force conjugate to x_i is given by $F_i = -\partial U/\partial x_i$. There-

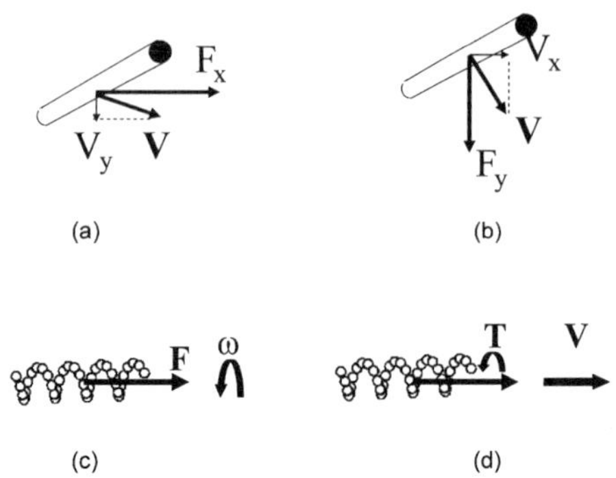

Fig. 2. Hydrodynamic reciprocal relation. Suppose that a force F_x in x-direction causes the y component velocity V_y as in (a), then a force F_y in y direction causes a x component velocity given by $V_x = (V_y/F_x)F_y$ as in (b). Suppose that a force F causes a rotation of helical particle with angular velocity ω as in (c), then a torque T causes a translational velocity $V = (\omega/F)T$ as in (d).

fore the force balance equation is written as

$$\sum_j \zeta_{ij}\dot{x}_j = -\frac{\partial U}{\partial x_i} \tag{14}$$

Let $(\zeta^{-1})_{ij}$ be the inverse of the matrix ζ_{ij}, then Eq. (14) gives the following time evolution equation for x_i:

$$\frac{dx_i}{dt} = -\sum_j (\zeta^{-1})_{ij}\frac{\partial U}{\partial x_j} \tag{15}$$

Notice that ζ_{ij}, $(\zeta^{-1})_{ij}$, and U are functions of x_i. Therefore, Eq. (15) is, in general, rather complex non-linear equations for x_i.

Using the reciprocal relation (12), the time evolution equation (14) can be cast into a variational principle. Let Φ and $\dot{U}$ be defined by

$$\Phi = \frac{1}{2}\sum_{i,j}\zeta_{ij}\dot{x}_i\dot{x}_j \tag{16}$$

$$\dot{U} = \sum_i \frac{\partial U}{\partial x_i}\dot{x}_i \tag{17}$$

and the function R be defined by

$$R = \Phi + \dot{U} = \frac{1}{2}\sum_{i,j}\zeta_{ij}\dot{x}_i\dot{x}_j + \sum_i \frac{\partial U}{\partial x_i}\dot{x}_i \tag{18}$$

then the force balance equation (14) is equivalent to the condition $\partial R/\partial \dot{x}_i = 0$, i.e., the velocity $\dot{x}_i$ is determined by the condition that R be minimum with respect to $\dot{x}_i$. This is the general form of the hydrodynamic variational principle.

It should be noted that the rewriting of the force balance equation (14) to the variational principle is justified only when the reciprocal relation ($\zeta_{ij} = \zeta_{ji}$) is satisfied.

3. Onsager's Variational Principle

3.1. *Onsager's kinetic equation*

Onsager showed that the above variational principle holds for general irreversible processes. Let $x = (x_1, x_2, ...)$ be the set of variables describing the non-equilibrium state of the system, and let us assume that the evolution law of the state can be written in the form

$$\frac{dx_i}{dt} = \sum_j L_{ij} \frac{\partial S(x)}{\partial x_j} \tag{19}$$

where $S(x)$ is the entropy of the system, and L_{ij} are the phenomenological kinetic coefficients. Equation (19) is called Onsager's kinetic equation. Many equations known in physics and chemistry can be written in this form as it will be shown later. At this point, let us proceed assuming the validity of Eq. (19).

Using the time reversal symmetry in the fluctuation at equilibrium state, Onsager proved that the coefficeints L_{ij} in Eq. (19) must be symmetric†

$$L_{ij} = L_{ji} \tag{20}$$

This reciprocal relation allows us to write the kinetic equation (19) in the form of a variational principle:the time evolution of the system is determined by minimizing

$$\tilde{R} = \frac{1}{2} \sum_{i,j} (L^{-1})_{ij} \dot{x}_i \dot{x}_j - \frac{\partial S}{\partial x_i} \dot{x}_i \tag{21}$$

This principle is called Onsager's variational principle. The second term in Eq. (21) stands for the reversible entropy change i.e., the entropy change caused by infinitely slow process (the qusi-equilibrium processes). On the

†Here we are assuming that the state variables x_i are invariant under time reversal transformation, and that there is no magnetic field.

other hand, the first term stands for the half of of irreversible entropy change caused by a process having finite speed.

Onsager's variational principle has a structure very similar to hydrodynamic variational principle. In fact, if the temperature of the system is constant, themodynamic force F_i conjugate to the variable x_i is given by

$$F_i = -\frac{\partial A}{\partial x_i} = T\frac{\partial S}{\partial x_i} \tag{22}$$

where A is the free energy of the system. and Onsager's variational principle is written in the same form as the hydrodynamic variational principle.

$$R = \frac{1}{2}\sum_j \zeta_{ij}\dot{x}_i\dot{x}_j + \frac{\partial A}{\partial x_i}\dot{x}_i \tag{23}$$

The difference between Eq. (23) and (18) is that x_i in Eq. (23) stands for a general coordinate specifying the non-equilibrium state, and that the potential energy U is replaced by the free energy A. The relation between the Onsager's variational principle and the hydrodynamic variational principle will be discussed in more detail in the next section.

In the following discussion, we shall assume that the temperature of the system is constant, and use Onsager's variational principle in the form of Eq.(23). The kinetic equation of Onsager is then given by

$$\frac{dx_i}{dt} = -\sum_j (\zeta^{-1})_{ij}\frac{\partial A}{\partial x_j} \tag{24}$$

3.2. *Validity of the variational principle*

As we have seen, Onsager's variational principle is equilvalent to Onsager's kinetic equation with symmetric coefficient ζ_{ij}. Therefore the validity of Onsager's variational principle entirely depends on the validity of the kinetic equation. Let us therefore spend some time to discuss the base of this equation.

Usual argument to derive the kinetic equation (19) is as follows. If the system is at equilibrium, the entropy takes the maximum value and the state does not change, i.e., both $\partial S(x)/\partial x_i$ and dx_i/dt are equal to zero. If the system is not at equilibrium, both $\partial S(x)/\partial x_i$ and dx_i/dt are not equal to zero. If the deviation from the equilibrium state is small, one can assume a linear relation between $\partial S(x)/\partial x_i$ and dx_i/dt, which is equation (19).

This argument presumes that the deviation of x_i from the equilibrium value $x_{i,eq}$ is small, and that the kinetic equation (19) is a linear equation for $x_i - x_{i,eq}$. In fact, Onsager used this assumption to prove the reciprocal relation.

On the other hand, it is known that in many systems, the time evolution equations for the state variables are written in the form of Eq.(19) even when the equations become non-linear equations for $x_i - x_{i,eq}$, yet the reciprocal relation (20) are satisfied. We have seen this in the case of particle motion in viscous fluid. Other example is the usual diffusion equation for the particle concentration $n(x;t)$

$$\frac{\partial n}{\partial t} = \frac{\partial n}{\partial x}\left(D(n)\frac{\partial n}{\partial x}\right) \tag{25}$$

which comes from the Fuch's law for the diffusion flux $j = -D(\partial n/\partial x)$. Equation (25) is in general a non-linear partial differential equation for n since the diffusion coefficient D can be a function of n. It will be shown later (see Sec. 4), that Eq.(25) can be written in the form of the kinetic equation (19) with symmetric coefficients ($\zeta_{ij} = \zeta_{ji}$).

It is now generally accepted[10–12] that Onsager's theory is valid even when the kinetic equation becomes a non-linear equation for x_i. The non-linearity comes from two sources; one is that the forces $\partial S(x)/\partial x_i$ (or $\partial A(x)/\partial x_i$) can be non-linear functions of x, and the other is that the kinetic coefficients L_{ij} (or ζ_{ij}) can be a function of x.

How can we justify the reciprocal relation in the situation that the time evolution equation becomes a non-linear equation for the state variable $x = (x_1, x_2, ...x_f)$? This question can be cast in a more specific question: is the hydrodynamic reciprocal relation (12) a special case of Onsager's reciprocal relation? The latter question has been discussed by hydrodynamicists[16] who raised several questions for the assertion that the hydrodynamic reciprocal relation is a special case of Onsager's reciprocal relation.

One can justify this assertion in the following way. Although the friction coefficients ζ_{ij} depend on the particle configuration x, they are independent of the potential force which drives the particle motion. Therefore, in proving the reciprocal relation, one can generally assume that there are some hypothetical forces acting on the particles, and that the state x is close to equilibrium (under the given hypothetical forces). Therefore it is justified to use Onsager's argument. to prove the reciprocal relation for the friction coefficients. In this sense, one can regard the hydrodynamic reciprocal relation as a special case of Onsager's reciprocal relation. In general, as far as the kinetic coefficients $L_{ij}(x)$ is independent of the driving force $\partial S/\partial x_i$, one can justify Onsager's argument to prove the the reciprocal relation.

The underlying assumption for the kinetic equations (24) is that there is a set of slow variables $x = (x_1, x_2, ...x_f)$, the relaxation time of which are dintinctively longer than those of other fast variables. The free energy $A(x)$ can be defined only for such systems:in fact $A(x)$ can be calculated by statistical mechanics assuming that the fast variables are in equilibrium for given values of slow variables. If the slow variables change with the rate $\dot{x}$, the fast variables will not be in equilibrium, but as long as the deviation of *fast variables* from their equilibrium state is small, one can assume a linear relation between the rate of state change and the driving force, and can use Onsager's argument to justify the reciprocal relation. This is the reason why we can use Onsager's variational principle for the system whose time evolution equation becomes a non-linear equation for the *slow variables.*

3.3. *Merit of the variational principle*

The variational principle we have discussed is a simple rewriting of the kinetic equation for the state variables. However, formulating the evolution law in the variational principle has several advantages.

(a) The variational principle gives kinetic equations which automatically satisfy Onsager's reciprocal relation. In the usual fomulation of irreversible thermodynamics, careful considerations are needed to identify proper set of variables to ensure the reciprocal relation. This task becomes even more complicated if there are certain constraints such as the incompressible condition or conservation law. If one uses the variational principle, proper kinetic equations are obtained easily. It will be shown (Sec. 3.4) that the equations obtained by the variational principle generally satisfy the reciprocal relation. Examples of this merit will be demonstrated in the problem of diffusion (Sec. 4) and nematodynamics (Sec. 8).

(b) The variational principle allows us flexibility in choosing state variables and gives us new approaches for the problems. Any set of variables x can be chosen for state variables as long as their time derivatives $\dot{x}$ are uniquely determined by the state variables. Examples of this merit will be seen in the discussion of diffusion (Sec. 4) and rotational Browinian motion (Sec.5).

(c) The variational principle gives us a convenient formula to find out the forces needed to cause certain controll parameters such as the system volume, or shear strain. The mathematical base for this usage is explained in Sec. 3.5, and its application will be shown in Sec. 4.3 and Sec. 5.5.

In the following, these points are explained in more detail.

3.4. *Reciprocal relation in the kinetic equation*

First, a general reason is given for why the reciprocal relation is guranteed for the equations obtained by the variational principle.

If both $\Phi(\dot{x})$ and $\dot{A}(\dot{x})$ are given explicitly as a function of $\dot{x} = (\dot{x}_1, \dot{x}_2, ...)$ as in Eq.(23), it is obvious that the variational principle $\partial R/\partial \dot{x}_i = 0$ give equations which satisfy the reciprocal relation. In many situations, however, although $\dot{A}(\dot{x})$ is obtained easily, $\Phi(\dot{x})$ is not easily found. It quite often happens that the energy dissipation function $\Phi(\dot{x})$ can be expressed as a function of other variables $\dot{y}$ which determine $\dot{x}$, i.e., the dissipation function is expressed as

$$\Phi = \frac{1}{2}\sum_{ij} \zeta_{ij}\dot{y}_i\dot{y}_j \tag{26}$$

and $\dot{x}_i$ is determined by

$$\dot{x}_i = \sum_j a_{ij}\dot{y}_j \tag{27}$$

In Eq.(27), a_{ij} are certain coefficients (which can be a function of x). The matrix (a_{ij}) needs not be a square matrix: the number of variables $\dot{y}_i$ can be different from the number of variables $\dot{x}_i$.

An example of such situation is seen in the problem of diffusion. The non-equilibrium state of the system is described by the concentration profile $n(x)$. The free energy A is espressed easily as a functional of $n(x)$, while the energy dissipation function Φ cannot be expressed by $\dot{n}(x)$: Φ is expressed as a functional of the flux $j(x)$ which determines $\dot{n}(x)$.

It is easy to prove the reciprocal relation for the kinetic equations derived from the variational principle. The Rayleighian is given by

$$R = \Phi + \sum_i \frac{\partial A}{\partial x_i}\dot{x}_i = \frac{1}{2}\sum_{ij}\zeta_{ij}\dot{y}_i\dot{y}_j + \sum_{ij}\frac{\partial A}{\partial x_i}a_{ij}\dot{y}_j \tag{28}$$

Minimization of R gives

$$\dot{y}_i = \sum_{jk}(\zeta^{-1})_{ij}\frac{\partial A}{\partial x_k}a_{kj} \tag{29}$$

This gives the following kinetic equation

$$\dot{x}_i = \sum_j L_{ij} \frac{\partial A}{\partial x_j} \tag{30}$$

with

$$L_{ij} = \sum_{l,k} a_{ik} a_{jl} (\zeta^{-1})_{kl} \tag{31}$$

which satisfies the reciprocal relation.

The reciprocal relation can also be proven in the situation that $\dot{y}_i$ are not independent of each other, but are subject to a set of constraints

$$\sum_i b_i^{(p)} \dot{y}_i = 0, \qquad p = 1, 2, .. \tag{32}$$

With such constraints, the Rayleighian becomes

$$R = \frac{1}{2} \sum_{ij} \zeta_{ij} \dot{y}_i \dot{y}_j + \sum_{ij} \frac{\partial A}{\partial x_i} a_{ij} \dot{y}_j - \sum_p \lambda^{(p)} \sum_i b_i^{(p)} \dot{y}_i \tag{33}$$

where $\lambda^{(p)}$ $(p = 1, 2, ...)$ are the Lagrange multipliers for the constraints. Minimization of Eq.(33) gives the following kinetic coefficients:

$$L_{ij} = \sum_{l,k} a_{ik} a_{jl} (\zeta^{-1})_{kl} - \sum_{p,q} c_i^{(p)} c_j^q (A^{-1})_{pq} \tag{34}$$

where

$$c_i^{(p)} = \sum_{i,j,k} a_{ij} (\zeta^{-1})_{jk} b_k^{(p)} F \tag{35}$$

and $(A^{-1})_{pq}$ is the inverse of the matrix

$$A_{pq} = \sum_{i,j} b_i^{(p)} b_j^{(p)} (\zeta^{-1})_{ij} \tag{36}$$

It is easy to verify that L_{ij} is a symmetric matrix.

3.5. *Forces needed to controll the state variables*

Certain state variables in the Rayleighian can be changed externally. For example, the particle position shown in Fig. 1 can be changed by applying a certain force on the particle. It can be shown that the force $F_i^{(e)}$ needed to change the parameter x_i with the rate $\dot{x}_i$ is given by

$$F_i^{(e)} = \frac{\partial R}{\partial \dot{x}_i} \tag{37}$$

For example, if one moves the particle with velocity $\dot{x}$, one has to apply an extra force on the particle

$$F^{(e)} = \zeta \dot{x} + \frac{\partial U}{\partial x} \tag{38}$$

which is equal to $\partial R/\partial \dot{x}$.

Equation (37) can be shown as follows. Suppose that the variables x_i $(i = 1, 2, ...f)$ are the internal variables (the time evolution of which is given by $\partial R/\partial \dot{x}_i = 0$), and the other variables x_i, $(i = f+1, f+2, ...g)$ are the externally controllable variables. For such system, the force balance equations are written as

$$\sum_{j=1}^{g} \zeta_{ij} \dot{x}_j = -\frac{\partial A}{\partial x_i}, \qquad i = 1, 2, ...f \tag{39}$$

$$\sum_{j=1}^{g} \zeta_{ij} \dot{x}_j = -\frac{\partial A}{\partial x_i} + F_i^{(e)}, \qquad i = f+1, ...g \tag{40}$$

which gives Eq. (37).

Examples of the situation where Eq. (37) is useful are shown in Fig. 3. In Fig.3 (a), a semi-permeable membrane separating the colloidal solution from pure solvent is moved with velocity $\dot{L}$. The force $F(t)$ needed to cause such motion is calculated by $\partial R/\partial \dot{L}$. In Fig.3 (b), a gel placed between two plates is squeezed. The force acting at the plate can be calcuted by $\partial R/\partial \dot{h}$. In Fig.3(c), a solution of rod-like polymers is sheared with shear rate $\dot{\gamma}$. The shear stress is calculated by $\partial R/\partial \dot{\gamma}$. Actual calculation for these quantities will be shown later.

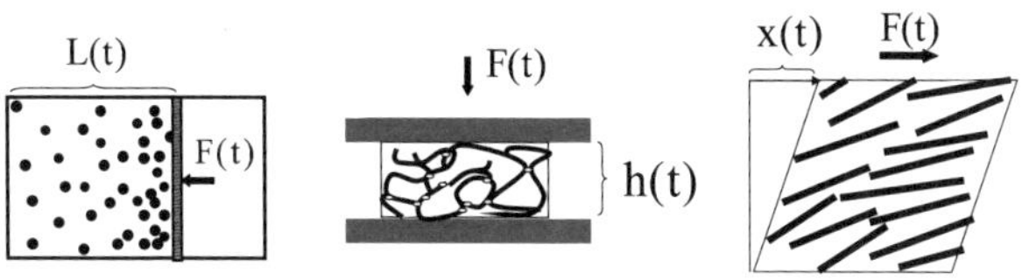

(a) (b) (c)

Fig. 3. Examples of external parameters and their conjugate forces, (a) position $L(t)$ of semi-permeable membrane separating the solution from pure solvent (b) vertical position $h(t)$ of a wall sandwitching a gel, (c) horizontal position $x(t)$ of a wall containing a solution of rod-like molecules. The conjugate forces $F(t)$ are shown in the figure.

4. Brownian Motion

4.1. *Diffusion equation*

As a simple application of Onsager's variational principle, we first discuss the Brownian motion of an isolated particle. The Brownian motion of a particle can be described by the Langevin equation,[3] which represents the stochastic motion of the particle position $x(t)$. The Brownian motion can also be described by the time evolution of the probability distribution function $\psi(x;t)$ which represents the probability of finding the particle at position x. The time evolution equation for $\psi(x;t)$ is deterministic, and can be derived by the variational principle.

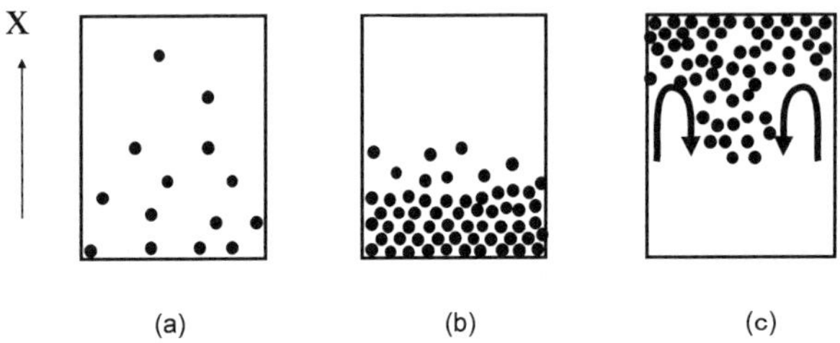

Fig. 4. Examples of the motion of particles in viscous fluids. (a) diffusion in dilute solution, (b) diffusion in concentrated solution (c) sedimentation of particles in concentrated solution.

The essential point of this derivation is to regard $\psi(x)$ as state variables. If we imagine an ensemble of many realization of Brownian motion, we can define the free energy of such ensemble. Alternatively, we can imagine a very dilute dispersion of particles (see Fig. 4 (a)), and regard $\psi(x)$ as a quantity proportional to the number density $n(x)$ of particles i.e., $\psi(x) = n(x)/N_{tot}$ where N_{tot} is the total number of particles in the system. In the following, we shall take the latter view point and derive the time evolution equation for $n(x,t)$.

Now $\dot{n}$ has to satisfy the conservation equation

$$\dot{n} = -\frac{\partial j}{\partial x} \tag{41}$$

where j is the flux of particles. Equation (41) represents a constraint for the time evolution of n: the constraint is that instantaneous transportation of particles from one place to other remote places cannot take place. To account for the constraint, we define the velocity v by

$$v = \frac{j}{n} \tag{42}$$

and determine v by the variational principle.

If the particle at x moves with velocity $v(x)$, the energy dissipation is given by $\zeta v(x)^2$. Thus the energy dissipation function of the system is given by

$$\Phi = \frac{1}{2}\int dx\ \zeta n(x)v(x)^2 \tag{43}$$

On the other hand, if the concentration profile of particles is $n(x)$, the free energy of the system is given by

$$A = \int dx\ [k_B T n(x)\ln n(x) + n(x)U(x)] \tag{44}$$

The timve derivative of A can be calculated by Eqs. (41) and (44);

$$\begin{aligned}\dot{A} &= \int dx\ [k_B T\dot{n}(\ln n + 1) + \dot{n}U(x)] \\ &= -\int dx\ \frac{\partial vn}{\partial x}\ [k_B T(\ln n + 1) + U] \\ &= \int dx\ vn\frac{\partial}{\partial x}(k_B T\ln n + U)\end{aligned} \tag{45}$$

where integral by parts has been used.

From Eq. (43) and (45), the Rayleighian is give by

$$R = \frac{1}{2}\int dx\zeta nv^2 + \int dxnv\frac{\partial}{\partial x}(k_B T\ln n + U) \tag{46}$$

Minimizing this with respect to v, we have

$$v = -\frac{1}{\zeta}\frac{\partial}{\partial x}(k_B T\ln n + U) \tag{47}$$

Equations (41) and (47) give the following time evolution equation for $n(x,t)$

$$\frac{\partial n}{\partial t} = D\frac{\partial}{\partial x}\left(\frac{\partial n}{\partial x} + \frac{n}{k_B T}\frac{\partial U}{\partial x}\right) \tag{48}$$

and the Einstein relation

$$D = \frac{k_B}{\zeta} \tag{49}$$

Equation (48) is the diffusion equation in the potential field $U(x)$.

4.2. *Reciprocal relation in the diffusion equation*

The diffusion equation (48) does not look-like Onsager's kinetic equation Eq. (24), but it is indeed a special form of it. In fact, Eq. (48) can be written as

$$\dot{n}(x) = -\int dx' L(x,x') \frac{\delta A}{\delta n(x')} \tag{50}$$

where the kernel $L(x,x')$ is given by

$$L(x,x') = -\frac{\partial}{\partial x}\left[\frac{n(x)}{\zeta}\frac{\partial}{\partial x'}\delta(x-x')\right], \tag{51}$$

The kernel $L(x,x')$ is symmetric, i.e., for any functions $n_1(x)$ and $n_2(x)$; the following identity holds.

$$\int dx \int dx' \; L(x,x')n_1(x)n_2(x') = \int dx \int dx' \; L(x',x)n_1(x)n_2(x') \tag{52}$$

Thus the the diffusion equation satisfy the reciprocal relation.

4.3. *Forces acting on the semi-permeable membrane*

Let us now discuss the force needed to controll the position of the semi-permeable membrane shown in Fig. 3(a). Since the integral in Eq. (44) is done between 0 and $L(t)$, $\dot{A}$ is calculated as

$$\dot{A} = \dot{L}[k_BTn(L)\ln n(L) + n(L)U(L)] + \int_0^{L(t)} dx \; \dot{n}[k_BT(\ln n + 1) + U(x)] \tag{53}$$

Using Eq. (42) and integral by parts, we have,

$$\begin{aligned}\dot{A} &= \dot{L}[k_BTn(L)\ln n(L) + n(L)U(L)] \\ &\quad -v(L)n(L)[k_BT(\ln n(L) + 1) + U(L)] \\ &\quad + \int_0^{L(t)} dx \; vn(x)\frac{\partial}{\partial x}[k_BT(\ln n + 1) + U(x)] \\ &= -\dot{L}n(L)k_BT + \int_0^{L(t)} dx \; vn(x)\frac{\partial}{\partial x}[k_BT(\ln n + 1) + U(x)]\end{aligned} \tag{54}$$

where we have used the condition that the velocity v satisfies $v(L) = \dot{L}$.

On the other hand, the energy dissipation function is given by

$$\Phi = \frac{\zeta}{2}\int_0^{L(t)} dx \; v^2 n(x) + \frac{1}{2}\xi_m \dot{L}^2 \tag{55}$$

The second term represents the extra energy dissipation caused by the solvent flow through the membrane (ξ_m is the friction constant of the membrane.) The force $F(t)$ acting on the semi-permeable membrane is given by $\partial R/\partial \dot{L}$, and is equal to

$$F(t) = \frac{\partial}{\partial \dot{L}}(\Phi + \dot{A}) = \xi_m \dot{L} - n(L)k_B T \tag{56}$$

If the membrane is fixed (i.e., $\dot{L} = 0$), the force is given by the osmotic pressure of the solution near the membrane. Notice that the potential $U(x)$ does not enter in the expression.

5. Rotational Brownian Motion

5.1. *State variables of a rod-like particle*

To demonstrate the merit of the flexibility in choosing state variables, we now discuss the rotational Brownian motion of a rod-like particle.The orientation of a rod can be specified by two angular coordinates θ and ϕ defined in Fig. 5. Alternatively, the orientation can be specified by a vector $\boldsymbol{u}$ parallel to the rod axis. The x, y, z components of $\boldsymbol{u}$ is given by

$$\begin{aligned} u_x &= u \sin\theta \cos\phi \\ u_y &= u \sin\theta \sin\phi \\ u_z &= u \cos\theta \end{aligned} \tag{57}$$

where $u = |\boldsymbol{u}|$.

The rotational Brownian motion can be described by the time evolution equation for the ditribution function $\psi(\theta, \phi; t)$ or $\psi(\boldsymbol{u}; t)$. We shall first derive these equations.

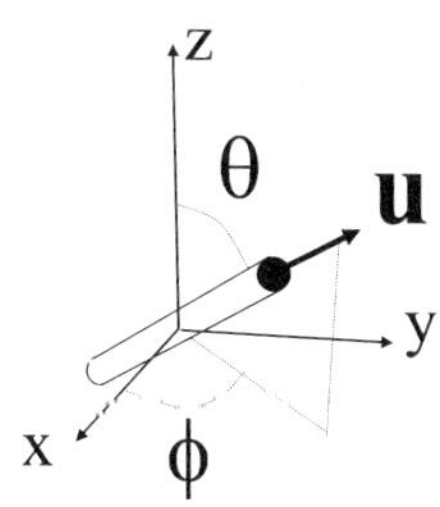

Fig. 5. The state of a rod-like particles is specified by two angular coordinate (θ, ϕ), or by a vector $\boldsymbol{u}$ parallel to the rod-axis.

5.2. *Diffusion equation for $\psi(\theta, \phi)$*

The distribution function $\psi(\theta, \phi)$ is defined as follows. The probability of finding the angular coordinate of rod in the area $dS = \sin\theta d\theta d\phi$ is given by $\psi(\theta, \phi)dS$. The normalization condition for ψ is

$$\int dS\ \psi = \int_0^{\pi} d\theta \int_0^{2\pi} d\phi\ \sin\theta\psi(\theta, \phi) = 1 \tag{58}$$

The conservation equation for such probability is written as

$$\dot{\psi} = -\frac{1}{\sin\theta}\frac{\partial}{\partial\theta}(\sin\theta\dot{\theta}\psi) - \frac{\partial}{\partial\phi}(\dot{\phi}\psi) \tag{59}$$

Now if the rod rotates with angular velocity $\boldsymbol{\omega}$ normal to the axis, the surrounding fluid exerts a frictional torque $-\zeta_r\boldsymbol{\omega}$ on the rod, where ζ_r is the rotational friction constant‡. The energy dissipation for this motion is $\zeta_r\boldsymbol{\omega}^2$ and therefore the energy dissipation function of the system (per unit volume) is given by

$$\Phi = \frac{n\zeta_r}{2}\int dS\ \boldsymbol{\omega}^2\psi \tag{61}$$

where n is the number density of particles in the system.

Since $\boldsymbol{\omega}^2 = \dot{\theta}^2 + \sin^2\theta\dot{\phi}^2$, Equation (61) is written as

$$\Phi = \frac{n\zeta_r}{2}\int dS\ (\dot{\theta}^2 + \sin^2\theta\dot{\phi}^2)\psi \tag{62}$$

On the other hand, the free energy of the system (per unit volume) is given by

$$A = nk_BT\int dS\ \psi\ln\psi \tag{63}$$

(Here, the term of external potential has been dropped for simplicity.) Using Eq. (59), the time derivative of A is calculated as

$$\begin{aligned}\dot{A} &= nk_BT\int dS\ \dot{\psi}(\ln\psi + 1)\\ &= nk_BT\int d\theta d\phi\ \sin\theta\left[-\frac{1}{\sin\theta}\frac{\partial}{\partial\theta}(\sin\theta\dot{\theta}\psi) - \frac{\partial(\dot{\phi}\psi)}{\partial\phi}\right](\ln\psi + 1)\\ &= nk_BT\int d\theta d\phi\ \sin\theta\left(\psi\dot{\theta}\frac{\partial}{\partial\theta} + \psi\dot{\phi}\frac{\partial}{\partial\phi}\right)(\ln\psi + 1)\end{aligned} \tag{64}$$

‡For a rod of length L and diameter b, hydrodynamic calculation gives[3]

$$\zeta_r = \frac{\pi\eta L^3}{3(\ln(L/b) - 0.8)} \tag{60}$$

Minimization of $\Phi + \dot{A}$ with respect to $\dot{\theta}$ and $\dot{\phi}$ gives

$$\zeta_r \dot{\theta} \psi = -k_B T \frac{\partial \psi}{\partial \theta} \tag{65}$$

$$\zeta_r \dot{\phi} \psi = -\frac{k_B T}{\sin^2 \theta} \frac{\partial \psi}{\partial \phi} \tag{66}$$

The time evolution equation for ψ is determined by Eqs. (59), (65) and (66):

$$\frac{\partial \psi}{\partial t} = D_r \left[\frac{1}{\sin\theta} \frac{\partial}{\partial \theta} \left(\sin\theta \frac{\partial \psi}{\partial \theta} \right) + \frac{1}{\sin^2 \theta} \frac{\partial^2 \psi}{\partial \phi^2} \right] \tag{67}$$

where

$$D_r = \frac{k_B T}{\zeta_r} \tag{68}$$

Equation (67) is the rotational diffusion equation for a rod-like particle.[1]

5.3. *Diffusion equation for $\psi(\boldsymbol{u})$*

The rotational diffusion equation can also be written in $\boldsymbol{u}$ space, where $\boldsymbol{u}$ represents a vector parallel to the rod axis. The length $u = |\boldsymbol{u}|$ of the vector does not matter, and can be taken to be unity, but here we assume that u takes some constant value. Let $\psi(\boldsymbol{u})d\boldsymbol{u}$ be the probability of finding the rod in the volume element $d\boldsymbol{u}$ in $\boldsymbol{u}$ space. The conservation equation is written as

$$\dot{\psi} = -\frac{\partial}{\partial \boldsymbol{u}} \cdot (\dot{\boldsymbol{u}} \psi) \tag{69}$$

We shall determine $\dot{\boldsymbol{u}}$ by the variational principle. Since $\boldsymbol{u}^2 = \text{constant}$, $\dot{\boldsymbol{u}}$ has to satisfy the following constraint

$$\dot{\boldsymbol{u}} \cdot \boldsymbol{u} = 0 \tag{70}$$

Since $\dot{\boldsymbol{u}} = \boldsymbol{\omega} \times \boldsymbol{u}$ and $\boldsymbol{\omega} \cdot \boldsymbol{u} = 0$, the energy dissipation function of the system can be written as

$$\Phi = \frac{n}{2} \int d\boldsymbol{u}\ \zeta_r \frac{\dot{\boldsymbol{u}}^2}{\boldsymbol{u}^2} \psi \tag{71}$$

On the other hand, the time derivative of the free energy is calculated as

$$\begin{aligned} \dot{A} &= n k_B T \int d\boldsymbol{u}\ \dot{\psi} (\ln \psi + 1) \\ &= n k_B T \int d\boldsymbol{u}\ \dot{\boldsymbol{u}} \psi \cdot \frac{\partial \ln \psi}{\partial \boldsymbol{u}} \end{aligned} \tag{72}$$

Therefore the Rayleighian of the present system becomes

$$R = \frac{n}{2}\int d\boldsymbol{u}\ \zeta_r \frac{\dot{\boldsymbol{u}}^2}{\boldsymbol{u}^2}\psi + nk_BT\int d\boldsymbol{u}\ \dot{\boldsymbol{u}}\psi \cdot \frac{\partial \ln\psi}{\partial \boldsymbol{u}} - \int d\boldsymbol{u}\ \lambda(\boldsymbol{u})\dot{\boldsymbol{u}}\cdot\boldsymbol{u} \quad (73)$$

The last term in Eq. (73) stands for the constraint (70).

Minimizing R with respect to $\dot{\boldsymbol{u}}$ and determining λ by Eq.(70), we have

$$\dot{\boldsymbol{u}} = -\frac{k_BT}{\zeta_r}(\boldsymbol{u}^2\boldsymbol{I} - \boldsymbol{u}\boldsymbol{u})\cdot\frac{\partial \ln\psi}{\partial \boldsymbol{u}} \quad (74)$$

where $\boldsymbol{I}$ is the unit tensor ($I_{\alpha\beta} = \delta_{\alpha\beta}$). The time evolution equation is now given by

$$\frac{\partial \psi}{\partial t} = D_r \frac{\partial}{\partial \boldsymbol{u}}\cdot\left[(\boldsymbol{u}^2\boldsymbol{I} - \boldsymbol{u}\boldsymbol{u})\cdot\left(\frac{\partial \psi}{\partial \boldsymbol{u}}\right)\right] \quad (75)$$

Using the relation (57), one can confirm that Eq. 75) is equivalent to Eq. (67). It will be shown that Eq. (75) is more convenient than Eq. (67) for practical calculation.

5.4. *Diffusion equation in flow field*

The rotational Browinian motion of particles gives viscoelasticity to the solution. This has been discussed in the molecular theory of rheology.[1–3] Here the main point of the theory is discussed based on the variational principle.

Suppose that a solution of rod-like particles is placed in a velocity field

$$\boldsymbol{v}(\boldsymbol{r}) = \boldsymbol{\kappa}\cdot\boldsymbol{r} \quad (76)$$

or

$$v_\alpha(\boldsymbol{r}) = \kappa_{\alpha\beta} r_\beta \quad (77)$$

where Greek indices α, β stand for the x, y, z components, and Einstein's convention that summation is taken over repeatedly appearing indices has been used. The tensor

$$\kappa_{\alpha\beta} = \frac{\partial v_\alpha}{\partial r_\beta} \quad (78)$$

is called the velocity gradient tensor.

The incompressible condition for the fluid velocity $\boldsymbol{v}(\boldsymbol{r})$ gives the following condition for $\kappa_{\alpha\beta}$

$$\kappa_{\alpha\alpha} = 0 \quad (79)$$

Now, if the solution does not contain particles, the energy dissipation per unit volume of the solution is given by

$$2\Phi_0 = \frac{1}{2}\eta(\kappa_{\alpha\beta} + \kappa_{\beta\alpha})^2 \tag{80}$$

where η is the viscosity of solvent.

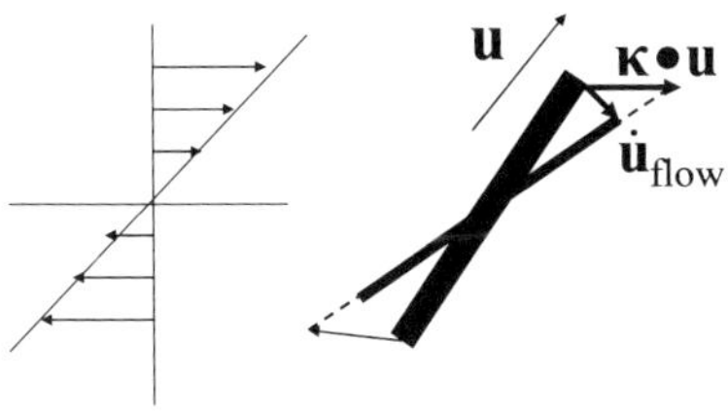

Fig. 6. The change of the vector $\boldsymbol{u}$ when the rod is placed in a velocity gradient $\boldsymbol{\kappa}$. If $\boldsymbol{u}$ follows the flow field completely, it changes with velocity $\boldsymbol{\kappa}\cdot\boldsymbol{u}$, but since the length $|\boldsymbol{u}|$ cannot change, the actual veolcity is given by $\dot{\boldsymbol{u}}_{flow} = (\boldsymbol{I} - \boldsymbol{u}\boldsymbol{u}/\boldsymbol{u}^2)\cdot\boldsymbol{\kappa}\cdot\boldsymbol{u}$.

If the solution contains a rod rotating with speed $\dot{\boldsymbol{u}}$, there is an extra energy dissipation. The energy dissipation can be calculated again by solving the Stokes equation, and is written as a quadratic function of $\dot{\boldsymbol{u}}$ and $\boldsymbol{\kappa}$. For slender rod, hydrodynamic calculation gives the following energy dissipation function[3]

$$\Phi = \Phi_0 + \frac{n}{2}\zeta_r \int d\boldsymbol{u}\psi\left[\frac{(\dot{\boldsymbol{u}} - \dot{\boldsymbol{u}}_{flow})^2}{u^2} + \frac{1}{2}\frac{(\boldsymbol{u}\cdot\boldsymbol{\kappa}\cdot\boldsymbol{u})^2}{u^4}\right] \tag{81}$$

where $\dot{\boldsymbol{u}}_{flow}$ stands for the velocity induced by the macroscopic flow (see Fig. 6)

$$\dot{\boldsymbol{u}}_{flow} = \left(\boldsymbol{I} - \frac{\boldsymbol{u}\boldsymbol{u}}{\boldsymbol{u}^2}\right)\cdot\boldsymbol{\kappa}\cdot\boldsymbol{u} \tag{82}$$

On the other hand, expression for $\dot{A}$ is given by the same form as Eq. (72). Hence the Rayleighian is given by

$$R = \Phi_0 + \frac{n}{2}\zeta_r \int d\boldsymbol{u}\psi\left[\frac{(\dot{\boldsymbol{u}} - \dot{\boldsymbol{u}}_{flow})^2}{u^2} + \frac{1}{2}\frac{(\boldsymbol{u}\cdot\boldsymbol{\kappa}\cdot\boldsymbol{u})^2}{u^4}\right]$$
$$+ k_BTn\int d\boldsymbol{u}\ \dot{\boldsymbol{u}}\psi\frac{\partial\ln\psi}{\partial\boldsymbol{u}} - \int d\boldsymbol{u}\ \lambda(\boldsymbol{u})\dot{\boldsymbol{u}}\cdot\boldsymbol{u} \tag{83}$$

This gives the following equation for $\dot{\boldsymbol{u}}$:

$$\dot{\boldsymbol{u}} = -\frac{k_BT}{\zeta_r}\left(\boldsymbol{u}^2\boldsymbol{I} - \boldsymbol{u}\boldsymbol{u}\right)\cdot\frac{\partial\ln\psi}{\partial\boldsymbol{u}} + \dot{\boldsymbol{u}}_{flow} \tag{84}$$

and the equation for $\dot{\psi}$

$$\frac{\partial \psi}{\partial t} = \frac{\partial}{\partial \boldsymbol{u}} \cdot \left[(\boldsymbol{u}^2 \boldsymbol{I} - \boldsymbol{u}\boldsymbol{u}) \cdot \left(D_r \frac{\partial \psi}{\partial \boldsymbol{u}} - \frac{1}{\boldsymbol{u}^2} \boldsymbol{\kappa} \cdot \boldsymbol{u} \psi \right) \right] \tag{85}$$

Equation (85) determine the orientational distribution of rod-like polymers in flow field.

5.5. *Expression for the stress tensor*

The macroscopic stress tensor $\sigma_{\alpha\beta}$ can be calculated by the formula (37). Since $\sigma_{\alpha\beta}$ is the force conjugate to the velocity gradient tensor $\kappa_{\alpha\beta}$, we have

$$\sigma_{\alpha\beta} = \frac{\partial R}{\partial \kappa_{\alpha\beta}} \tag{86}$$

For the Rayleighian (83), Eq.(86) gives the following expression for the stress tensor

$$\sigma_{\alpha\beta} = n \int d\boldsymbol{u} \psi \left[-\zeta_r \frac{(\dot{u}_\alpha - \dot{u}_{flow,\alpha}) u_\beta}{u^2} + \frac{1}{2} \zeta_r \frac{u_\alpha u_\beta u_\mu u_\nu \kappa_{\mu\nu}}{u^4} \right] + \eta_0 (\kappa_{\alpha\beta} + \kappa_{\beta\alpha}) \tag{87}$$

The first term can be rewritten by the use of Eq.(84):

$$\begin{aligned} -\int d\boldsymbol{u} \psi \zeta_r \frac{(\dot{u}_\alpha - \dot{u}_{flow,\alpha}) u_\beta}{u^2} &= k_B T \int d\boldsymbol{u} \left(\delta_{\alpha\mu} - \frac{u_\alpha u_\mu}{\boldsymbol{u}^2} \right) \frac{\partial \psi}{\partial u_\mu} u_\beta \\ &= -k_B T \int d\boldsymbol{u} \psi \frac{\partial}{\partial u_\mu} \left[u_\beta \left(\delta_{\alpha\mu} - \frac{u_\alpha u_\mu}{\boldsymbol{u}^2} \right) \right] \\ &= k_B T \int d\boldsymbol{u} \psi \left(3 \frac{u_\alpha u_\beta}{\boldsymbol{u}^2} - \delta_{\alpha\beta} \right) \\ &= 3 k_B T \langle \hat{u}_\alpha \hat{u}_\beta - \frac{1}{3} \delta_{\alpha\beta} \rangle \end{aligned} \tag{88}$$

where $\hat{\boldsymbol{u}} = \boldsymbol{u}/|\boldsymbol{u}|$ is the unit vector in the direction of $\boldsymbol{u}$, and $\langle ... \rangle$ stands for the average for $\boldsymbol{u}$ with the distribution function $\psi(\boldsymbol{u})$.

From Eqs.(87) and (88), the stress tensor is given by

$$\sigma_{\alpha\beta} = 3 n k_B T \langle \hat{u}_\alpha \hat{u}_\beta - \frac{1}{3} \delta_{\alpha\beta} \rangle + \frac{n}{2} \zeta_r \langle \hat{u}_\alpha \hat{u}_\beta \hat{u}_\mu \hat{u}_\nu \rangle \kappa_{\mu\nu} + \eta_0 (\kappa_{\alpha\beta} + \kappa_{\beta\alpha}) \tag{89}$$

Equation (89) indicates that the stress tensor of the solution of rod-like particles has two origins. One is the kinetic origin arising from the rotational Brownian motion of particles, and the other is the hydrodynamic origin arising from the extra energy dissipation caused by the rod.

Equations (85) and (89) give the constitutive equation: for given velocity gradient, $\psi(\boldsymbol{u};t)$ can be obtained by solving Eq.(85), and then the stress can be calculated by Eq.(89). The constitutive equation describes the non-linear viscoelasticity of the solution of rod-like polymers.[1,3]

6. Coupling between Diffusion and Flow

6.1. *Diffusion in concentrated solutions*

In Sec. 4, we have discussed the diffusion of particles in dilute solution. Now we discuss the diffusion in concentrated solutions. In dilute solutions, the motion of particles is independent of each other. In concentrated solutions, interaction between particles gives new features to the diffusion.

(a) Due to the interaction, the diffusion constant D now becomes a function of concentration. Accordingly, the diffusion equation becomes a non-linear equation for the particle concentration.
(b) The diffusion of particles generally induces flow of solvent, and can cause macroscopic flow in solutions. If particle concentration is low, this effect can be negligible, but if concentration is high, the effect can be important.

We shall discuss these two issues separately. In this section we shall discuss the effect of interaction on the diffusion constant assuming that there is no macroscopic flow, i.e., we consider the diffusion of particles in a quiescent solutions (see Fig. 4(b)).

To denote the concentration of particles, we use volume fraction $\phi(\boldsymbol{r};t)$, which is related to the number density $n(\boldsymbol{r},t)$ by

$$\phi(\boldsymbol{r},t) = v_c n(\boldsymbol{r},t) \tag{90}$$

where $v_c = 4\pi a^3/3$ is the volume of the particle.

Time evolution equation for ϕ can be obstained in the same way as in the dilute case. Let $\boldsymbol{v}_p$ be the particle velocity, then the conservation equation is written as

$$\dot{\phi} = -\nabla \cdot (\boldsymbol{v}_p \phi) \tag{91}$$

When particles move with velocity $\boldsymbol{v}_p$ in a quiescent solution, the energy dissipation is given by

$$\Phi = \frac{1}{2}\int d\boldsymbol{r}\xi(\phi)\boldsymbol{v}_p^2 \tag{92}$$

where $\xi(\phi)$ is the friction constant (per unit volume) for the collective motion of particles. If the solution is dilute, $\xi(\phi)$ is simply given by $n\zeta = (\phi/v_c)\zeta$. If the solution is concentrated, $\xi(\phi)$ is not expressed by a simple function, but many studies have been done to calculate $\xi(\phi)$ theoretically.[16,17] Also $\xi(\phi)$ can be measured experimentally by the method shown in Fig. 7.

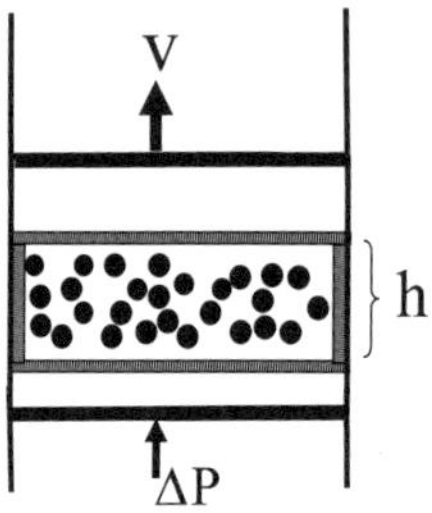

Fig. 7. Experimental set up to measure the friction constant $\xi(\phi)$. The particles are contained in a box made of porous walls fixed to the side wall. The solvent is transported from bottom container to top container by a pressure difference ΔP. If the fluid velocity is v, the friction constant is given by $\xi = \Delta P/(hv)$, where h is the thickness of the box. In practice, the pressure drop in the top and bottom container, and that accorss the porous walls has to be taken into account to obtain $\xi(\phi)$ accurately.

The free energy A of the system can be expressed as a functional of $\phi(\boldsymbol{r})$.

$$A = \int d\boldsymbol{r} f(\phi(\boldsymbol{r})) \tag{93}$$

where $f(\phi)$ stands for the free energy (per unit volume) of the solution of concentration ϕ, and we have assumed that the effect of gravity is negligible. From Eq. (93), $\dot{A}$ can be calculated by using Eq. (91),

$$\dot{A} = \int d\boldsymbol{r} f'(\phi)\dot{\phi} = -\int d\boldsymbol{r} f'(\phi)\nabla \cdot (\boldsymbol{v}_p\phi) = \int d\boldsymbol{r}\boldsymbol{v}_p\phi \cdot \nabla f'(\phi) \tag{94}$$

The minimization of $R = \Phi + \dot{A}$ with respect to $\boldsymbol{v}_p$ gives

$$\boldsymbol{v}_p = -\frac{\phi}{\xi}\nabla f'(\phi) \tag{95}$$

The right hand side can be rewritten by using the osmotic pressure $\Pi(\phi)$ of the solution

$$\Pi(\phi) = \phi f'(\phi) - f(\phi) + f(0) \tag{96}$$

as

$$\boldsymbol{v}_p = -\frac{1}{\xi}\nabla\Pi \tag{97}$$

Equations (91) and (97) give the following diffusion equation

$$\frac{\partial\phi}{\partial t} = \nabla[D(\phi)\nabla\phi] \tag{98}$$

where $D(\phi)$ is given by

$$D(\phi) = \frac{\phi}{\xi}\frac{\partial\Pi}{\partial\phi} \tag{99}$$

This gives a diffusion constant in concentrated solution. In the case of dilute solution, Eq. (99) reduces to the Einstein relation (49).

The force $F(t)$ acting on the semi-permeable membrane can be calculated by the same way as in Sec. 4.3, and it is given by $\xi_m \dot{L} - \Pi(\phi(L))$.

6.2. *Coupling between solute diffusion and solution flow*

In the previous section, it is assumed that the solution is at rest as a whole. This assumption is not always correct. When solute moves relative to solvent, it generally induces a macroscopic flow of solutions. For example, when a colloidal suspension initially placed on top of a pure solvent starts to sediment, the particles do not sediment homogeneously (see Fig. 4(c)). Certain part starts to sediment faster than the other, and form a lane where the particles and solvent move downward, while the solvent in the other part moves upward.

In order to describe such phenomena, it is necessary to take into account of the solution velocity $\boldsymbol{v}(\boldsymbol{r}, t)$ which is defined as the volume averaged velocity

$$\boldsymbol{v} = \phi\boldsymbol{v}_p + (1-\phi)\boldsymbol{v}_s \tag{100}$$

where $\boldsymbol{v}_s$ is the velocity of the solvent. If we assume that the solution is regarded as a Newtonian fluid with effective viscosity $\eta(\phi)$, the energy dissipation function is written as

$$\Phi = \frac{1}{2}\int d\boldsymbol{r}\xi(\phi)(\boldsymbol{v}_p - \boldsymbol{v})^2 + \frac{1}{4}\int d\boldsymbol{r}\eta(\phi)\left(\frac{\partial v_\alpha}{\partial r_\beta} + \frac{\partial v_\beta}{\partial r_\alpha}\right)^2 \tag{101}$$

The first term in Eq.(101) stands for the dissipation due to the relative motion between solute and solution, and the second term represents the dissipation caused by velocity gradient of the solution.

The free energy A of the solution is now given by

$$A = \int d\boldsymbol{r}[f(\phi) - \rho_1 \phi \boldsymbol{g} \cdot \boldsymbol{r}] \tag{102}$$

where the gravitational energy has been added to discuss the problem of sedimentation ($\rho_1 = \rho_p - \rho_s$ is the density difference between the particle and solvent). The time derivative of Eq.(102) is given by

$$\dot{A} = \int d\boldsymbol{r}(f'(\phi) - \rho_1 \boldsymbol{g} \cdot \boldsymbol{r})\dot{\phi}(\boldsymbol{r}) = \int d\boldsymbol{r} \boldsymbol{v}_p \phi \cdot (\nabla f'(\phi) - \rho_1 \boldsymbol{g}) \tag{103}$$

Since the solution is incompressible, $\boldsymbol{v}$ has to satisfy the constraint

$$\nabla \cdot \boldsymbol{v} = 0 \tag{104}$$

Therefore the Rayleighian is given by

$$R = \Phi + \dot{A} - \int d\boldsymbol{r} p(\boldsymbol{r}) \nabla \cdot \boldsymbol{v} \tag{105}$$

where the last term stands for the constraint (104). The conditions $\delta R/\delta \boldsymbol{v}_p = 0$ and $\delta R/\delta \boldsymbol{v} = 0$ give the following equations

$$\xi(\boldsymbol{v}_{\mathrm{p}} - \boldsymbol{v}) = -\phi \boldsymbol{\nabla} f'(\phi) + \rho_1 \phi \boldsymbol{g} \tag{106}$$

$$-\boldsymbol{\nabla}\eta \cdot [\boldsymbol{\nabla v} + (\boldsymbol{\nabla v})^t)] + \xi(\boldsymbol{v} - \boldsymbol{v}_{\mathrm{p}}) = -\boldsymbol{\nabla} p \tag{107}$$

Equations (104)–(107) and Eq. (91) give the following set of equations.

$$\frac{\partial \phi}{\partial t} = -\nabla \cdot (\boldsymbol{v}\phi) + \nabla \left[\frac{\phi}{\xi}(\nabla \Pi - \rho_1 \boldsymbol{g}\phi)\right] \tag{108}$$

$$\boldsymbol{\nabla}\eta \cdot [\boldsymbol{\nabla v} + (\boldsymbol{\nabla v})^t)] = \nabla(p - \Pi) - \rho_1 \boldsymbol{g}\phi \tag{109}$$

and

$$\nabla \cdot \boldsymbol{v} = 0 \tag{110}$$

Again one can show that the above set of equations can be written in the form of Onsager's kinetic equation:

$$\frac{\partial \phi}{\partial t} = -\int d\boldsymbol{r}' L(\boldsymbol{r}, \boldsymbol{r}') \frac{\delta A}{\delta \phi(\boldsymbol{r}')} \tag{111}$$

with symmetric kernel $L(\boldsymbol{r}, \boldsymbol{r}')$.

It should be emphasized that the state variable in the present problem is $\phi(\boldsymbol{r})$, and $\boldsymbol{v}_p$ and $\boldsymbol{v}$ are auxiliary variables introduced to determine $\dot{\phi}(\boldsymbol{r})$. In some literatures, the first term on the right hand side of Eq.(108) is called "reversible part". This terminology is misleading, since there is no reason to discriminate the velocity $\boldsymbol{v}$ from the velocity $\boldsymbol{v}_p$: both velocities are determined by the balance of the dissipative force and the potential force.

6.3. *Phase separation*

A special case of the above coupled equations for diffusion and flow is the equations used to describe the phase separation kinetics in solution. Suppose a homogeneous solution is brought into the state where $\partial\Pi/\partial\phi$ is negative, then the solution starts to separate into two phases, concentrated phase and dilute phase. The kinetics of phase separation has been extensively discussed in statistical mechanics.[7] The equations used there can be obtained by a minor modification of the above equations.

In the final state of phase separation, concentrated region and dilute region coexist with each other. At the boundary between the two regions, solute concentration changes very rapidly over molecular length scale. In such a case, the total free energy cannot be written in the form of Eq.(93). There is an extra energy which corresponds to the interfacial energy between the two regions. The interfacial energy can be accounted for by the following free energy functional

$$A = \int d\boldsymbol{r} \left[f(\phi) + \frac{1}{2}\kappa_s (\nabla\phi)^2 \right] \tag{112}$$

where κ_s is a positive constant. Then $\dot{A}$ is given by

$$\dot{A} = \int d\boldsymbol{r} \frac{\delta A}{\delta\phi(\boldsymbol{r})}\dot{\phi}(\boldsymbol{r}) = \int d\boldsymbol{r}(f'(\phi) - \kappa_s \nabla^2\phi)\dot{\phi}(\boldsymbol{r}) \tag{113}$$

The phase separation kinetics can be discussed by Eqs.(106) and (107) with $f'(\phi)$ being replaced by $f'(\phi) - \kappa_s\nabla^2\phi$. This gives the following set of equations

$$\frac{\partial\phi}{\partial t} = -\nabla\cdot(\boldsymbol{v}\phi) + \nabla\left[\frac{\phi}{\xi}(\nabla\Pi - \rho_1 \boldsymbol{g}\phi - \phi\kappa_s\nabla^2\phi)\right] \tag{114}$$

$$\boldsymbol{\nabla}\eta\cdot[\boldsymbol{\nabla v} + (\boldsymbol{\nabla v})^t)] = \nabla(p - \Pi) - \rho_1\boldsymbol{g}\phi - \phi\kappa_s\nabla^2\phi \tag{115}$$

Outcomes of these equations are discussed in.[7]

7. Gel Dynamics

Gel is a homogeneous mixture of elastic material and fluid.[5] In polymeric gels, gel consists of cross-linked polymer and solvent. The cross-linked polymer forms a three dimensional network and gives an elasticity to the gel.

The deformation of polymer network is coupled with the transport of solvent. For example, when a gel is compressed, solvent bleeds out from

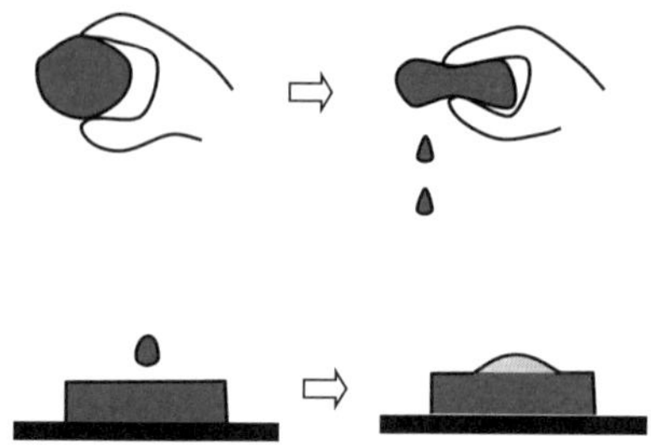

Fig. 8. Examples showing the coupling between solvent motion and deformation of gels. (a) Solvent bleeds out from a squeeed gel. (b) A drop of solvent placed at the top of a gel induces a deformation of the gel.

the gel (see Fig.8 (a)). Conversely, when a drop of solvent permeates into a gel, it deforms the gel (Fig.8 (b)). Equations which describe the coupled phenomena of gel deformation and solvent permeation can be derived from the variational principle.[6,14] The derivation is very similar to that for solutions. The difference is that, in solutions, the free energy is a function of polymer concentration, while in gels, the free energy is a function of the deformation of polymer network.

The state variable in a gel is the displacement field $\boldsymbol{u}(\boldsymbol{r}, t)$ which represents the displacement of the point on the network located at $\boldsymbol{r}$ in a certain reference state. The free energy of the system A is now expressed as a functional of the gradient of the displacement vector, i.e.,

$$A = \int d\boldsymbol{r} f(e_{\alpha\beta}(\boldsymbol{r})) \tag{116}$$

where $e_{\alpha\beta}$ is defined by

$$e_{\alpha\beta} = \frac{\partial u_\alpha}{\partial r_\beta} \tag{117}$$

If the displacement is small, the free energy density f is given by the same form as the elastic energy of deformation

$$f = \frac{1}{2} K e_{\alpha\alpha}^2 + \frac{1}{4} G \left(e_{\alpha\beta} + e_{\beta\alpha} - \frac{2}{3} \delta_{\alpha\beta} e_{\gamma\gamma} \right)^2 \tag{118}$$

where K and G are called the osmotic bulk modulus, and the shear modulus respectively.

The time derivative $\dot{A}$ is calculated as

$$\dot{A} = \int d\boldsymbol{r} \frac{\partial f}{\partial e_{\alpha\beta}} \frac{\partial \dot{u}_\alpha}{\partial r_\beta} = \int d\boldsymbol{r} \sigma_{\alpha\beta}^n \frac{\partial \dot{u}_\alpha}{\partial r_\beta} \tag{119}$$

where

$$\sigma^n_{\alpha\beta} = \frac{\partial f}{\partial e_{\alpha\beta}} = K e_{\gamma\gamma}\delta_{\alpha\beta} + G\left(e_{\alpha\beta} + e_{\beta\alpha} - \frac{2}{3}\delta_{\alpha\beta}e_{\gamma\gamma}\right) \tag{120}$$

is called the osmotic stress tensor. The osmotic stress tensor corresponds to the osmotic pressure in solution, and stands for the thermodynamic force acting on the solute (i.e., the polymer network in the present problem).

On the other hand, the energy dissipation function is given by

$$\Phi = \frac{1}{2}\int d\boldsymbol{r}\xi(\dot{\boldsymbol{u}} - \boldsymbol{v}_s)^2 \tag{121}$$

where $\boldsymbol{v}_s$ stands for the velocity of the solvent. In Eq.(121), it is assumed that the energy dissipation in gels arises dominantly from the relative motion between solvent and polymer network; the energy dissipation associated with the velocity gradient is ignored.

In constructing the Rayleighian, one has to take into account of the fact that gel behaves as an incompressible material; i.e., the volume change of a gel can only occur by taking in (or out) solvent from (or to)surrounding. This constraint is expressed as the incompressible condition $\boldsymbol{\nabla}\cdot\boldsymbol{v} = 0$ for the volume average velocity $\boldsymbol{v} = \phi\dot{\boldsymbol{u}} + (1-\phi)\boldsymbol{v}_s$, or

$$\phi\nabla\cdot\dot{\boldsymbol{u}} + (1-\phi)\nabla\cdot\boldsymbol{v}_s = 0 \tag{122}$$

where the volume fraction of polymer network ϕ is assumed not to change since the deformation is small. The Rayleighian is thus constructed as

$$R = \frac{1}{2}\int d\boldsymbol{r}\left\{\xi(\dot{u}_\alpha - v_{s,\alpha})^2 + \sigma^n_{\alpha\beta}\frac{\partial \dot{u}_\alpha}{\partial r_\beta} - p(\boldsymbol{r})\left[\phi\frac{\partial \dot{u}_\alpha}{\partial r_\alpha} + (1-\phi)\frac{\partial v_{s,\alpha}}{\partial r_\alpha}\right]\right\} \tag{123}$$

The last term comes from the constraint (122). The conditions $\delta R/\delta\dot{u}_\alpha = 0$ and $\delta R/\delta v_{s,\alpha} = 0$ give the following set of equations:

$$\xi(\dot{u}_\alpha - v_{s,\alpha}) - \frac{\partial\sigma^n_{\alpha\beta}}{\partial r_\beta} + \phi\frac{\partial p}{\partial r_\alpha} = 0 \tag{124}$$

$$-\xi(\dot{u}_\alpha - v_{s,\alpha}) + (1-\phi)\frac{\partial p}{\partial r_\alpha} = 0 \tag{125}$$

Equations (124) and (125) give

$$\frac{\partial}{\partial r_\beta}\left(\sigma^n_{\alpha\beta} - p\delta_{\alpha\beta}\right) = 0 \tag{126}$$

and

$$v_{s,\alpha} - \dot{u}_\alpha = -\frac{1-\phi}{\xi}\frac{\partial p}{\partial r_\alpha} \tag{127}$$

Equation (126) stands for the balance of forces acting on each volume element of the gel. Notice that the meachanical stress acting on the volume element is given by $\sigma^n_{\alpha\beta} - p\delta_{\alpha\beta}$. The term $-p\delta_{\alpha\beta}$ represents the contribution from the solvent pressure. In fact, if one calculate the force acting at the wall for the situation shown in Fig.3 (b) using Eq.(37), one can show that the force acting on the wall is given by $(\boldsymbol{\sigma}^n - p\boldsymbol{I}) \cdot \boldsymbol{n}$, where $\boldsymbol{n}$ is a vector normal to the wall.

On the other hand, Eq. (127) represents the permeation of solvent in gel network. Equation (127) indicates that the permeation speed is proportional to the pressure gradient. This is known as Darcy's law for porous materials.

The dynamics of gels (deformation and solvent permeation) can be described by Eqs. (122), (126) and (127). The equations may be written explicitly as partial differential equations for $\boldsymbol{u}$ and p:

$$\left(K + \frac{1}{3}G\right) \boldsymbol{\nabla}(\boldsymbol{\nabla} \cdot \boldsymbol{u}) + G\boldsymbol{\nabla}^2\boldsymbol{u} = \boldsymbol{\nabla} p \tag{128}$$

and

$$\boldsymbol{\nabla}\dot{\boldsymbol{u}} = \kappa \boldsymbol{\nabla}^2 p \tag{129}$$

where

$$\kappa = \frac{(1-\phi)^2}{\xi} \tag{130}$$

Application of these equations is discussed in Ref. 6.

8. Liquid Crystals

Liquid crystal is a fluid which has an intermediate order between liquid and crystal. In a typical class of liquid crystals called nematics, molecular orientation is ordered as in a crystal, while molecular positions are random as in a liquid. The direction of the molecular orientation is represented by a unit vector $\boldsymbol{n}$ called director. The director can change in space and time depending on external forces (of mechanical, electrical or magnetic origin).

The continuum mechanics of nematics was established by Leslie and Ericksen.[4] The Leslie–Ericksen theory describes the time variation of the fluid velocity $\boldsymbol{v}(\boldsymbol{r}.t)$, and the director $\boldsymbol{n}(\boldsymbol{r}, t)$. Here this theory is discussed based on the variational principle.

In the Leslie–Ericksen theory, the state variable is the director field $\boldsymbol{n}(\boldsymbol{r})$ which stands for the average direction of molecules at point $\boldsymbol{r}$. The energy dissipation function Φ is a quadratic function of $\dot{n}_\alpha$ and $\kappa_{\alpha\beta} = \partial v_\alpha / \partial r_\beta$. The explicit form of Φ can be determined by symmetry argument. Using the fact that Φ vanishes for uniform rotation of the system, one can show that Φ is a quadratic function of strain rate

$$\dot{\epsilon}_{\alpha\beta} = \frac{1}{2}(\kappa_{\alpha\beta} + \kappa_{\beta\alpha}) \tag{131}$$

and $\tilde{\dot{n}}_\alpha$ which is defined by

$$\tilde{\dot{n}}_\alpha = \dot{n}_\alpha - \frac{1}{2}(\kappa_{\alpha\beta} - \kappa_{\beta\alpha}) n_\beta \tag{132}$$

$\tilde{\dot{n}}_\alpha$ represents the deviation of $\dot{\boldsymbol{n}}$ from the fluid rotation. If one constructs the energy dissipation function Φ which is scalar and quadratic in $\dot{\epsilon}_{\alpha\beta}$ and $\tilde{\dot{n}}_\alpha$ with coefficients that may depend on $\boldsymbol{n}$, one gets

$$2\Phi = \beta_1(\dot{\epsilon}_{\alpha\beta} n_\alpha n_\beta)^2 + \beta_2 \dot{\epsilon}_{\alpha\beta}^2 + \beta_3(\dot{\epsilon}_{\alpha\beta} n_\beta)^2 + \beta_4 \tilde{\dot{n}}_\alpha^2 + \beta_5 \tilde{\dot{n}}_\alpha \dot{\epsilon}_{\alpha\beta} n_\beta \tag{133}$$

where $\beta_1, ...\beta_5$ are constants.

On the other hand, the free energy of nematics is given by the elastic energy of directors

$$A = \frac{1}{2}K_1(\boldsymbol{\nabla} \cdot \boldsymbol{n})^2 + \frac{1}{2}K_2(\boldsymbol{n} \cdot \boldsymbol{\nabla} \times \boldsymbol{n})^2 + \frac{1}{2}K_3(\boldsymbol{n} \times \boldsymbol{\nabla} \times \boldsymbol{n})^2 \tag{134}$$

If we define

$$g_{\alpha\beta} = \frac{\partial n_\beta}{\partial r_\alpha} \tag{135}$$

the free energy A is written as

$$A = \frac{1}{2}K_1(g_{\alpha\alpha})^2 + \frac{1}{2}K_2(e_{\alpha\beta\gamma} n_\alpha g_{\beta\gamma})^2 + \frac{1}{2}K_3(g_{\alpha\beta} n_\beta - g_{\beta\alpha} n_\beta)^2 \tag{136}$$

The time derivative A is written as

$$\dot{A} = \sigma_{\alpha\beta}^{(e)} \kappa_{\alpha\beta} + h_\alpha \dot{n}_\alpha \tag{137}$$

where $\sigma_{\alpha\beta}^{(e)}$ and h_α are obtained as:[4]

$$\sigma_{\alpha\beta}^{(e)} = -\frac{\partial A}{\partial g_{\alpha\gamma}} g_{\gamma\beta} \tag{138}$$

$$h_\alpha = -\frac{\partial A}{\partial n_\alpha} + \frac{\partial}{\partial r_\beta}\left(\frac{\partial A}{\partial g_{\alpha\beta}}\right) \tag{139}$$

The Raylcighian therefore becomes

$$R = \Phi + \sigma_{\alpha\beta}^{(e)} \kappa_{\alpha\beta} + h_\alpha \dot{n}_\alpha - \lambda n_\alpha \dot{n}_\alpha - p\kappa_{\alpha\alpha} \tag{140}$$

The last two terms arise from the constraints for $\kappa_{\alpha\beta}$ and $\dot{n}_\alpha$.

The time evolution of n_α is determined by $\partial R/\partial \dot{n}_\alpha = 0$, and the stress tensor is given by $\sigma_{\alpha\beta} = \partial R/\partial \kappa_{\alpha\beta}$ Straightforward calculation gives the following expression for the stress tensor $\sigma_{\alpha\beta}$ and the time evolution equation for n_α

$$\begin{aligned}\sigma_{\alpha\beta} &= \alpha_1 \dot{\epsilon}_{\mu\nu} n_\mu n_\nu n_\alpha n_\beta + \alpha_2 n_\alpha \tilde{\dot{n}}_\beta + \alpha_3 \tilde{\dot{n}}_\alpha n_\beta + \alpha_4 \dot{\epsilon}_{\alpha\beta} \\ &\quad + \alpha_5 n_\alpha n_\mu \dot{\epsilon}_{\mu\beta} + \alpha_6 n_\beta n_\mu \dot{\epsilon}_{\alpha\mu} + \sigma^{(e)}_{\alpha\beta} - p\delta_{\alpha\beta}\end{aligned} \tag{141}$$

$$\gamma_1 \tilde{\dot{n}}_\alpha + \gamma_2 \dot{\epsilon}_{\alpha\beta} n_\beta - h_\alpha = \lambda n_\alpha \tag{142}$$

where

$$\begin{aligned}\alpha_1 &= \beta_1 \\ \alpha_2 &= \frac{1}{2}\beta_4 + \frac{1}{4}\beta_5 \\ \alpha_3 &= -\frac{1}{2}\beta_4 + \frac{1}{4}\beta_5 \\ \alpha_4 &= \beta_2 \\ \alpha_5 &= \frac{1}{2}\beta_3 + \frac{1}{4}\beta_5 \\ \alpha_6 &= -\frac{1}{2}\beta_3 - \frac{1}{4}\beta_5 \\ \gamma_1 &= -\beta_4 = \alpha_3 - \alpha_2 \\ \gamma_2 &= -\frac{1}{2}\beta_5 = \alpha_6 - \alpha_5\end{aligned} \tag{143}$$

Equations (141)-(142) are the Leslie–Ericksen's equation.[4] Notice that the Parodi's relation $\alpha_2 + \alpha_3 = \alpha_6 - \alpha_5$ is automatically satisfied in this derivation.

9. Conclusion

In this paper, I have shown that many equations used in soft matter physics can be derived from Onsager's variational principle. The variational principle is equivalent to Onsager's kinetic equation with reciprocal relations, but is has several advantages. The variational principle allows us a great flexibility in choosing state variables, and velocities. Also, the variational principle is quite convenient to deal with systems which have certain constraints, and to calculate the forces needed to change controll parameters.

Various time evolution equations used in soft matter physics, such as the diffusion equation, the flow-diffusion coupled equation, the gel dynamics equation, and the Leslie-Ericksen equations are all derived from this variational principle. Therefore one can say that Onsager's variational principle is giving a unified framework for the soft matter dynamics.

Acknowledgments

This work was supported by JSPS KAKENHI (18540403) and KAKENHI (Grant-in-Aid for Scientific Research) on Priority Area "Soft Matter Physics" from the Ministry of Education, Culture, Sports, Science and Technology of Japan.

Appendix A. Proof of the Hydrodynamic Reciprocal Relation

The friction coefficient ζ_{ij} for particles in viscous fluid can be calculated by the following procedure. Consider that the configuration of the particles changes with the rate $\dot{x} = (\dot{x}_1, \dot{x}_2, ...)$. The velocity of a point $\boldsymbol{r}$ on a surface of a particles is a linear function of $\dot{x}$ and can be written as

$$\boldsymbol{v}(\boldsymbol{r}) = \sum_i \boldsymbol{G}_i(\boldsymbol{r}; x)\dot{x}_i \tag{A.1}$$

The function $\boldsymbol{G}_i(\boldsymbol{r}; x)$ is determined by the shape of the particle. The fluid velocity induced by the particle motion is determined by solving the Stokes equation (1) under the boundary condition (A.1) at the surface of the particles.

The frictional force $\boldsymbol{f}_H(\boldsymbol{r})$ acting on a unit area at the particle surface at $\boldsymbol{r}$ is given by

$$\boldsymbol{f}_H(\boldsymbol{r}; x) = \boldsymbol{\sigma} \cdot \boldsymbol{n} \tag{A.2}$$

where

$$\sigma_{\alpha\beta} = \eta \left(\frac{\partial v_\alpha}{\partial r_\beta} + \frac{\partial v_\beta}{\partial r_\alpha} \right) - p\delta_{\alpha\beta} \tag{A.3}$$

is the $\alpha\beta$ component of the stress tensor of the fluid, and $\boldsymbol{n}$ is a unit vector normal to the surface. The work done to the fluid is therefore calculated by

$$2\Phi = \int dS \sum_i -\boldsymbol{f}_H(\boldsymbol{r}; x) \cdot \boldsymbol{G}_i(\boldsymbol{r}; x)\dot{x}_i = -\sum F_{Hi}\dot{x}_i \tag{A.4}$$

where the integral dS is done for all surfaces, and the force F_{Hi} is defined by

$$F_{Hi} = \int dS \boldsymbol{f}_H(\boldsymbol{r};x) \cdot \boldsymbol{G}_i(\boldsymbol{r};x) \tag{A.5}$$

Equation (A.5) is the definition of the frictional force conjugate to $\dot{x}_i$.

Since the Stokes equation is a linear equation of $\boldsymbol{v}$, and the boundary condition for $\boldsymbol{v}$ is a linear function of $\dot{x}_i$, the frictional force F_{Hi} can also be written as a linear function of $\dot{x}_i$.

$$F_{Hi} = -\sum_j \zeta_{ij}(x)\dot{x}_j \tag{A.6}$$

To prove the reciprocal relation $\zeta_{ij} = \zeta_{ji}$, we consider two situations, one is that the particle state changes with velocity $\dot{x}_i^{(1)}$, and the other with velocity $\dot{x}_i^{(2)}$. Let $F_{Hi}^{(1)}$ $F_{Hi}^{(2)}$ be the frictional forces for each situations. We shall prove the identity

$$\sum_i F_{Hi}^{(1)}\dot{x}_i^{(2)} = \sum_i F_{Hi}^{(2)}\dot{x}_i^{(1)} \tag{A.7}$$

The reciprocal relation follows from this identity.

Let $\boldsymbol{v}^{(a)}(\boldsymbol{r})$ and $\boldsymbol{\sigma}^{(a)}$ be the velocity field and the stress field for the two situations ($(a = 1, 2)$). By Eqs. (A.3) and (A.5), the left hand side of Eq. (A.7) is written as

$$I = -\sum_i F_{Hi}^{(1)}\dot{x}_i^{(2)} = -\sum_i \int dS G_{i\alpha}(\boldsymbol{r};x)\sigma_{\alpha\beta}^{(1)} n_\beta \dot{x}_i^{(2)} \tag{A.8}$$

Since the fluid velocity satisfies the boundary condition (A.1),

$$I = -\int dS v_\alpha^{(2)} \sigma_{\alpha\beta}^{(1)} n_\beta \tag{A.9}$$

Using the Gauss theorem, the surface integral on the right hand side of Eq. (A.9) can be written as

$$I = \int d\boldsymbol{r} \frac{\partial}{\partial r_\beta}(v_\alpha^{(2)}\sigma_{\alpha\beta}^{(1)}) = \int d\boldsymbol{r} \left(\frac{\partial v_\alpha^{(2)}}{\partial r_\beta}\sigma_{\alpha\beta}^{(1)} + v_\alpha^{(2)} \frac{\partial \sigma_{\alpha\beta}^{(1)}}{\partial r_\beta} \right) \tag{A.10}$$

The second term in the intgrand on the right hand side of Eq. (A.10) vanishes by the Stokes equation. The first temr can be written as

$$\begin{aligned} I &= \frac{1}{2}\int d\boldsymbol{r} \left(\frac{\partial v_\alpha^{(2)}}{\partial r_\beta} + \frac{\partial v_\beta^{(2)}}{\partial r_\alpha} \right) \sigma_{\alpha\beta}^{(1)} \\ &= \frac{\eta}{2}\int d\boldsymbol{r} \left(\frac{\partial v_\alpha^{(2)}}{\partial r_\beta} + \frac{\partial v_\beta^{(2)}}{\partial r_\alpha} \right) \left(\frac{\partial v_\alpha^{(1)}}{\partial r_\beta} + \frac{\partial v_\beta^{(1)}}{\partial r_\alpha} \right) \end{aligned} \tag{A.11}$$

The last expression in this equation is invariant for the exchange of the superscript (1), (2). Therefore the hydrodynamic reciprocal relation Eq. (12) has been proven.

If we put $\dot{x}_i^{(1)} = \dot{x}_i^{(2)} = \dot{x}_i$ in Eq. (A.11), we have

$$-\sum_i F_{Hi}\dot{x}_i = \frac{\eta}{2}\int d\boldsymbol{r}\left(\frac{\partial v_\alpha}{\partial r_\beta} + \frac{\partial v_\beta}{\partial r_\alpha}\right)^2 \tag{A.12}$$

The right hand side of Eq. (A.12) is non-negative. Thus Eq. (13) has been proven.

References

1. R. B. Bird, C. F. Curtiss, R.C. Armstrong, and O. Hassager, *The Dynamics of Polymeric Liquids, 2nd ed.* (Wiley–InterScience, 1987).
2. R. G. Larson, *The Structure and Rheology of Complex Fluids* (Oxford University Press, 1999).
3. M. Doi and S. F. Edwards, *The Theory of Polymer Dynamics* (Oxford University Press, 1986).
4. P. G. de Gennes and J. Prost, *The Physics of Liquid Crystals* (Oxford University Press, 1995).
5. K. Dusek and T. Tanaka ed., *Advances in Polymer Science* **109**, *Volume Transitions I* (Springer, 1993).
6. M. Doi, *Gel dynamics, J. Phys. Soc. Jpn.* **78**, 052001 1-19 (2009).
7. A. Onuki, *Phase Transition Dynamics* (Cambridge University Press, 2002).
8. L. Onsager, *Reciprocal relations in irreversible processes I, II, Phys. Rev.* **37** 405-426 (1931); **38**, 2265-2279 (1931).
9. Lord Rayleigh, *Some general theorems relating to vibrations, Proc. Math. Soc. London* **4**, 357-368 (1873).
10. S. R. de Groot and P. Mazur, *Non-equilibrium Thermodynamics* (Dover London, 1984).
11. G. Lebon, D. Jou and J. Casas-Vazquez, *Understanding Non-equilibrium Thermodynamics.* (Springer, 2008).
12. S. Kjelstrup and D. Bedeaux, *Non-equilibrium Thermodynamics of Heterogeneous Systems* (World Scientific, 2008).
13. M. Doi, *Variational principle for the Kirkwood theory for the dynamics of polymer solutions and suspensions, J. Chem. Phys.* **79**, 5080-5087 (1983).
14. M. Doi, *Viscoelastic effect on polymer diffusion*, in *Dynamics and Patterns in Complex Fluids*, edited by A. Onuki and K. Kawasaki (Springer, 1990) 100-112.
15. M. Doi, *Onsager's variational principle in soft matter, J. Phys.: Cond. Matter*, **23**, 284118 (2011).
16. J. Happel and H. Brenner, *Low Reynolds Number Hydrodynamics* (Kluwer, 1963).
17. S. Kim and S. J. Karrila, *Microhydrodynamics* (Butterworth-Heinemann, 1991).

Chapter 2

Rheo-Dielectric Behavior of Soft Matters

Hiroshi Watanabe*, Yumi Matsumiya, Kazushi Horio,
Yuichi Masubuchi and Takashi Uneyama

Institute for Chemical Research, Kyoto University,
Uji, Kyoto 611-0011, Japan

Soft matters are highly susceptible to fast flow without exhibiting macroscopic rupture, and the constituent elements therein can be readily brought into a non-equilibrium state. The dynamics of these elements in this state can be detected with the dielectric method rather easily if the elements have electrical dipoles (and/or net charge). Thus, the rheo-dielectric behavior of the soft matters, detected with a weak electric field superimposed on the flow field, provides us with interesting information for the non-equilibrium dynamics of the constituent elements. This chapter presents a brief summary of the rheo-dielectric behavior of polymers, liquid crystals, and composites to demonstrate the similarities/differences of these materials as well as the usefulness of the rheo-dielectric method.

1. Introduction

Soft matters often exhibit nonlinear responses under fast flow without macroscopic disruption. In a typical experiment to force a soft matter to flow, the material surface is externally displaced and/or subjected to an external force. Under this boundary condition, the flow behavior is determined by the molecules (or constituent elements) in the material. Namely, the interaction between these molecules allows the surface displacement/force to be transmitted into the material body thereby determining the flow profile in the material and the rheological properties of the material of as a whole. Correspondingly, the molecules exhibit the intra- and inter-molecular correlation in their non-equilibrium structures and the dynamics reflecting this interaction under fast flow. Thus, it is of particu-

*hiroshi@scl.kyoto-u.ac.jp

lar interest to compare the non-equilibrium structure and dynamics of the molecules with the rheological properties of the material.

The molecular structures under flow can be detected with several methods. For example, the birefringence optically detects an average orientation of the chemical bonds included in the molecules, the dichroism picks up the orientation of a particular bond, and the static scattering resolves a spatial distribution of the scattering elements (that can be contrasted with isotopes for the case of neutron scattering).[1,2] Thus, for the soft matters under *steady flow*, the rheo-optical and rheo-scattering measurements provide us with the information about the non-equilibrium stationary structure.[1,2] In addition, the optical and scattering signals reflecting the molecular dynamics can be obtained if a small, time-dependent velocity perturbation is added on the steady flow field. For example, the optical/scattering signals oscillating with time are superimposed on the steady signals if a small-amplitude oscillatory velocity field is added, and analysis of the frequency dependence of the oscillatory signals can provide us with the information for the molecular dynamics under steady flow. However, as usual for any measurable signal, those oscillating signals detect just a limited aspect of the dynamics, the *isochronal* anisotropy and/or scattering point distribution that result from the oscillatory flow and relax through the molecular motion under steady flow. (In addition, those oscillatory signals are often masked by noises of the steady signal and not so easily measured.) Thus, some supplemental method is desired to detect a non-isochronal aspect of the dynamics.

The dielectric method is useful for this purpose. The dielectric relaxation basically detects correlation of the microscopic polarization at *two separate times* and thus specifies, by nature, the non-isochronal aspect of the motion of molecules carrying electrical dipoles (and/or net charges).[3,4] In addition, the flow does not distinguish the head and tail of the dipoles (plus and minus charges) thereby giving rise to no steady electrical signal even when the dipole-carrying molecules are flow-oriented. This lack of steady signal allows us to measure rather easily the dielectric signal for a weak electric field even under fast flow.

Thus, the rheo-dielectric measurements have been conducted for a wide variety of soft matters,[5–14] polymers, liquid crystals, and composites, under steady shear flow. The rheo-dielectric data, obtained with the electric field applied in the shear gradient direction, are combined with the rheological data to elucidate interesting dynamics (and structure) of the constituent molecules in those materials. In the remaining part of this chapter, the basics of dielectric relaxation are addressed first. Then, the rheo-dielectric features of respective soft matters are explained and their similarities/differences are discussed. The readers who are familiar with the general framework of dielectric relaxation can skip the next section and directly proceed to Sec. 3.

2. Basics of Dielectric Relaxation

2.1. *Instrumentation*

The rheo-dielectric measurements can be conducted with a dielectric cell mounted on a rheometer. Fig. 1 illustrates a house-made cell utilized by Watanabe et al.[6] The upper part is a plate-type counter electrode fixed on an insulator block (made of machinable ceramics). This block is connected, though a metal attachment, to a stress detector of a laboratory rheometer (ARES, TA Instruments). In the bottom part, the main electrode is fixed on the other insulator block (made of machinable ceramics and Teflon) being mounted in a guard electrode. This assembly, curved as a whole into a truncated cone-shape (of the gap angle $\theta = 2°$)[6] is connected to the rheometer drive through a metal attachment. All electrodes are made of stainless steel. The electrical contact to the counter electrode (upper plate) is made with a thin coaxial wire that does not disturb the stress measurement (for the stress > 10 Pa). The contact to the main electrode (lower cone) is made with a wire through a mercury reservoir. This reservoir allows a noiseless contact to the main electrode even during high-speed rotation of the cone assembly.

A sample is placed between the cone and plate electrodes, the latter being allocated at a given position above the former. (At this position, the cone and plate touch each other if the cone head is not truncated.) In the rheo-dielectric measurements, the cone assembly is rotated at a controlled speed, and AC (or DC) voltage is applied to the electrodes through the wire and reservoir, and the current is measured with a transformer bridge (or a current meter) once the material begins to flow steadily. (The achievement of the steady flow state is monitored through the stress detector of the rheometer). The current density $I(t)$ corresponds to a time derivative of the charge on the electrodes, and the phenomenological framework explained later enables us to determine the rheo-dielectric behavior of the material from the $I(t)$ data. Since the gap angle of the cone is small ($\theta = 2°$), the electric field is in the shear gradient direction with negligibly small tilting,

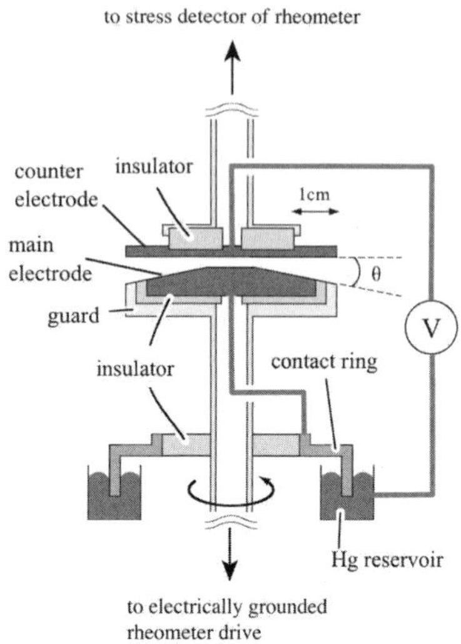

Fig. 1. Schematic illustration of a set-up for rheo-dielectric measurements.

and the rheo-dielectric behavior detects the structure and dynamics in this direction.

2.2. *Phenomenological framework*

Molecules exhibit active thermal motion in a liquid material at temperatures T above the glass transition temperature T_g and/or the melting temperature T_m. The electrical dipoles of the molecules fluctuate accordingly. When a constant electric field E is applied to such a liquid material placed in a capacitor (or a dielectric cell), the capacitor instantaneously stores its bare charge, and the electron cloud and polarized chemical bonds in the molecules quickly respond to the electric field. After this rapid response, the dipoles of the molecules are slowly oriented in the direction of the electric field at a rate determined by the molecular motion. This dipole motion allows the capacitor to store an extra charge thereby resulting in the dynamic dielectric phenomenon. If the material contains ions, the ions are also driven by the electric field. If the ions discharge themselves at an electrode of the capacitor and the same amount of ions are released from the other electrode, the ion motion results in a steady, direct current (DC) conduction. On the other hand, if the ions cannot discharge themselves but are concentrated nearby the electrode attracting the ions, the ion motion leads to a dielectric phenomenon referred to as the electrode polarization.[9,15–17]

The dynamic changes explained above lead to a time (t) evolution of the charge density on the electrode referred to as the electric displacement, $D(t)$. (The current density data $I(t)$ explained earlier are related to $D(t)$ as $I(t) = -\mathrm{d}D(t)/\mathrm{d}t$.) In the time scale of the molecular motion, this evolution can be expressed as:[3,4]

$$D(t) = [P_\infty + \Delta P\{1 - \Phi(t)\}]\,E \qquad (t > 0) \tag{1}$$

Here, P_∞ is the rapidly stored charge density (due to the bare capacitor and the distortion of the electron cloud/chemical bonds) normalized by the field intensity E. ΔP corresponds to the saturated orientation of the dipoles at $t = \infty$. The function $\Phi(t)$ (= 1 and 0 at $t = 0$ and ∞), referred to as the normalized dielectric relaxation function, is most straightforwardly related to the molecular motion, as explained later for Eq. (11): $\Phi(t)$ specifies the decay (or relaxation) of $D(t)$ on removal of the electric field after full saturation of the charge density.

Under a weak electric field, the molecular motion resulting in the time evolution of $D(t)$ is negligibly affected by the electric field (and coincides with the motion under the steady shear flow in the case of the rheo-dielectric measurements). For this case, the linear dielectric response is observed, and $D(t)$ for an electric field $E(t')$ of arbitrary history at $t' < t$ is expressed by a convolution,[3,4]

$$D(t) = P_\infty E(t) + \Delta P \int_{-\infty}^{t} \{1 - \Phi(t - t')\} \frac{\mathrm{d}E(t')}{\mathrm{d}t'} \mathrm{d}t' \tag{2}$$

For a weak sinusoidal electric field, $E(t) = E_0 \sin\omega t$ with ω being the angular frequency, Eq. (2) gives

$$D(t) = \varepsilon_0 E_0 [\varepsilon'(\omega) \sin\omega t - \varepsilon''(\omega) \cos\omega t] \tag{3}$$

where the ε_0 (= 8.85×10^{-12} $\mathrm{C^2 J^{-1} m^{-1}}$) is the absolute permittivity of vacuum. The dynamic dielectric constant $\varepsilon'(\omega)$ and dielectric loss $\varepsilon''(\omega)$ appearing in Eq. (3) are related to the normalized diclectric relaxation function $\Phi(t)$ as[3,4]

$$\varepsilon'(\omega) = \varepsilon'(0) - \Delta\varepsilon \left\{ \omega \int_0^\infty \Phi(t'') \sin\omega t'' \mathrm{d}t'' \right\} \tag{4}$$

$$\varepsilon''(\omega) = \Delta\varepsilon \left\{ \omega \int_0^\infty \Phi(t'') \cos\omega t'' \mathrm{d}t'' \right\} \tag{5}$$

Here, $\Delta\varepsilon$ ($= \Delta P/\varepsilon_0$) is the dielectric relaxation intensity due to the dipole orientation. (The complex dielectric constant ε^* is defined as $\varepsilon^*(\omega) = \varepsilon'(\omega) - i\varepsilon''(\omega)$ with $i = \sqrt{-1}$.)

In general, the molecules have intra- and inter-molecular motional correlation. Correspondingly, $\Phi(t)$ usually has a mode distribution and can be expressed as[4]

$$\Phi(t) = \sum_{q\geq 1} g_q \exp\left(-\frac{t}{\tau_{\varepsilon,q}}\right) \qquad \text{with} \quad \tau_{\varepsilon,1} > \tau_{\varepsilon,2} > \tau_{\varepsilon,3} > \ldots \tag{6}$$

Here, g_q and $\tau_{\varepsilon,q}$ are the normalized intensity ($\sum_q g_q = 1$) and relaxation time of qth dielectric mode. As noted from Eqs. (4)-(6), the decrease of the dynamic dielectric constant from the static (zero-frequency) dielectric constant, $\Delta\varepsilon'(\omega) \equiv \varepsilon'(0) - \varepsilon(\omega)$, and the dielectric loss, $\varepsilon''(\omega)$, are expressed in terms of g_q and $\tau_{\varepsilon,q}$ as

$$\Delta\varepsilon'(\omega) \equiv \varepsilon'(0) - \varepsilon'(\omega) = \omega^2 \Delta\varepsilon \sum_{q\geq 1} g_q \frac{\tau_{\varepsilon,q}^2}{1+\omega^2\tau_{\varepsilon,q}^2} \tag{7}$$

$$\varepsilon''(\omega) = \omega\Delta\varepsilon \sum_{q\geq 1} g_q \frac{\tau_{\varepsilon,q}}{1+\omega^2\tau_{\varepsilon,q}^2} \tag{8}$$

Eqs. (7) and (8) indicate that $\Delta\varepsilon'(\omega)$ and $\varepsilon''(\omega)$ exhibit the power-law type terminal tails at low $\omega \ll 1/\tau_{\varepsilon,1}$ where the relaxation has completed:

$$\Delta\varepsilon'(\omega) = \omega^2 \Delta\varepsilon \sum_{q\geq 1} g_q \tau_{\varepsilon,q}^2 \propto \omega^2, \qquad \varepsilon''(\omega) = \omega\Delta\varepsilon \sum_{q\geq 1} g_q \tau_{\varepsilon,q} \propto \omega \tag{9}$$

Furthermore, an average dielectric relaxation time can be evaluated from these tails as[4]

$$\tau_\varepsilon \equiv \frac{\sum_{q\geq 1} g_q \tau_{\varepsilon,q}^2}{\sum_{q\geq 1} g_q \tau_{\varepsilon,q}} = \left[\frac{\Delta\varepsilon'}{\omega\varepsilon''}\right]_{\omega\to 0} \tag{10}$$

This τ_ε is the average of $\tau_{\varepsilon,q}$ with the weighing factor of $g_q\tau_{\varepsilon,q}$. Since this weighing factor is magnified for slow modes (having large $\tau_{\varepsilon,q}$), the average τ_ε is not identical but close to the longest relaxation time $\tau_{\varepsilon,1}$. Thus, τ_ε is often utilized as the terminal dielectric relaxation time.

The other type of relaxation time evaluated from the angular frequency ω_{peak} for a peak of $\varepsilon''(\omega)$, $\tau_{\text{peak}} = 1/\omega_{\text{peak}}$, is also utilized frequently. This τ_{peak} is regarded as an average relaxation time for a group of the most intensive dielectric modes giving the $\varepsilon''(\omega)$ peak and, in general, τ_{peak} is smaller than τ_ε. Nevertheless, τ_{peak} agrees with τ_ε if the dielectric relaxation function has no mode distribution and is expressed as $\Phi(t) = \exp(-t/\tau_\varepsilon)$. Thus, a τ_ε / τ_{peak} ratio (≥ 1) can be utilized as a measure of the width of the dielectric mode distribution.

If ions are contained in the material to exhibit the direct current (DC) conduction, the expression of $\varepsilon''(\omega)$, Eq. (8), is to be modified to include a DC term, $\varepsilon''_{\text{DC}}(\omega) = \sigma_{\text{DC}}/\omega$ with σ_{DC} being the DC conductivity. No change is necessary for the expression of $\varepsilon'(\omega)$, Eq. (7), for the case of real DC due to ions discharging themselves at the electrode. However, for most of the soft matters, the ions therein cannot discharge. For this case, the electrode polarization (EP)[9,15–17] occurs and a Debye-type single-relaxation term is to be added to Eqs. (7) and (8). Further detail of this EP term is explained later in Sec. 5.

2.3. *Molecular expression of* $\Phi(t)$

Rigorous statistical mechanical analysis[18,19] indicates that the dielectric relaxation function $\Phi(t)$ *at equilibrium* is unequivocally formulated through the Green–Kubo theorem and is given as an auto-correlation function of a microscopic polarization $\boldsymbol{p}$:

$$\Phi(t) = \frac{\langle p_E(t)p_E(0)\rangle_{\text{eq}}}{\langle p_E^2\rangle_{\text{eq}}} \tag{11}$$

In Eq. (11), the subscript E stands for the component of the polarization in the direction of the electric field. The origin of the time ($t = 0$) is taken arbitrarily in the stationary equilibrium state, and $\langle \ldots \rangle_{\text{eq}}$ stands for the ensemble average in this state. The fluctuation of $\boldsymbol{p}$ resulting in a decay of $\Phi(t)$ is activated by the thermal motion of the molecules. Note that the dielectric relaxation at equilibrium is determined by the correlation of the polarization at two separate times, t and 0. This feature makes a contrast between the dielectric and birefringence/dichroism relaxation processes, the latter detecting the *isochronal* anisotropy of the molecular orientation.

The microscopic polarization is given by a sum of molecular dipoles. Thus, if the dipole alignment with respect to the molecular axis is known, Eq. (11) serves as a key relationship for extracting the information for molecular dynamics from the macroscopic dielectric data. This dipole alignment differs from system to system, and the details are explained in later sections for respective systems.

The dielectric intensity $\Delta\varepsilon$ at equilibrium is expressed as[3]

$$\Delta\varepsilon = \frac{4\pi\mu^2}{3k_\mathrm{B}T}\nu F g \tag{12}$$

In Eq. (12), μ and ν denote the magnitude of molecular dipole and the number density of those dipoles, respectively, k_B is the Boltzmann constant, and T is the absolute temperature. F is a factor correcting a difference between the applied and internal electric fields ($F = (\varepsilon_\mathrm{s} + 2)^2/9$ with ε_s = static dielectric constant for non-polar low molecular weight molecules), and g is the Kirkwood–Fröhlich factor[3] that represents the magnitude of the motional correlation of the dipoles (i.e., of the dipole-carrying molecules). Thus, the $\Delta\varepsilon$ data providing this g factor is also useful for examination of the molecular motion.

Now, we have to make a comment on the rheo-dielectric behavior under steady shear flow. In general, the Green–Kubo theorem[18] does not rigorously hold in the non-equilibrium state,[20,21] and the use of Eqs. (11) and (12) cannot be justified in *a priori* sense. Nevertheless, for type-A chains having the dipoles parallel along the chain backbone (later shown in Fig. 5), analysis[22] showed that Eqs. (11) and (12) are valid as a satisfactory approximation under the steady shear flow, as explained in Appendix A. Thus, the rheo-dielectric behavior of those chains can be discussed on the basis of Eqs. (11) and (12).

3. Rheo-Dielectric Behavior of Polymers

3.1. *Glassy relaxation and rubbery relaxation*

Fig. 2 show the angular frequency (ω) dependence of storage and loss moduli, G' and G'', measured for oligomeric styrene (OS-1; weight-average molecular weight $M_\mathrm{w} = 0.95 \times 10^3$, polydispersity index $M_\mathrm{w}/M_\mathrm{n} = 1.13$) and *cis*-polyisoprene (PI-8; $M_\mathrm{w} = 8.2\times10^3$, polydispersity index $M_\mathrm{w}/M_\mathrm{n} = 1.05$) at 42 and 23°C, respectively.[5] (Time-temperature superposition was utilized for those G' and G'' data.) The OS-1 sample (Fig. 2(a)) exhibits the glassy relaxation characterized by the decrease of G' from its high-ω plateau at a level of $\sim 10^9$ Pa. For the OS-1 sample having no polymeric character, this glassy relaxation is immediately followed by the linear viscoelastic flow behavior characterized by the power-law type terminal tails, $G' \propto \omega^2$ and $G'' \propto \omega$. In contrast, the PI-8 sample exhibits the so-called Rouse relaxation between the glassy and terminal flow zones, which is indicative of the polymeric character of this sample. (M_w of PI-8 is not large enough to allow the entanglement plateau to emerge after the Rouse relaxation zone.)

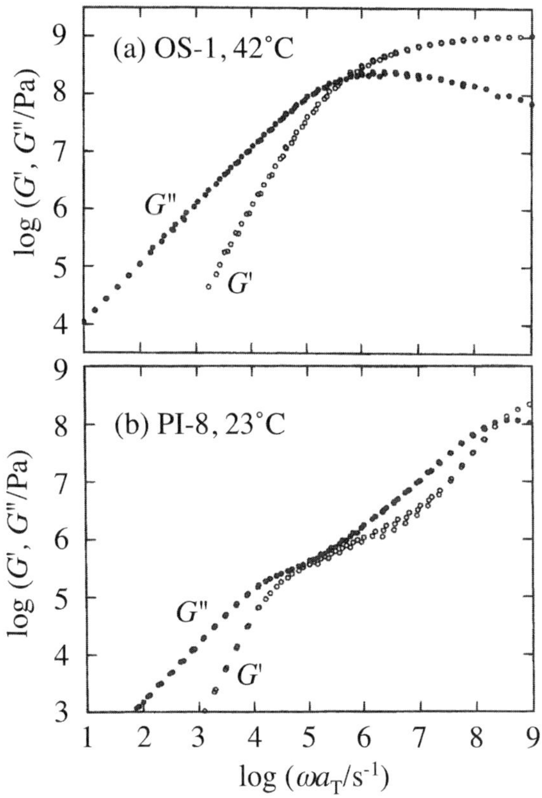

Fig. 2. Linear viscoelastic behavior of (a) OS-1 at 42°C and (b) PI-8 at 23°C. Redrawn, with permission, from Ref. 5.

The terminal viscoelastic relaxation time,[23] τ_G $(= [G'/\omega G'']_{\omega \to 0}$; defined in a way similar to that in Eq. (10)), is evaluated from the G' and G'' data of the OS-1 and PI-8 samples shown in Fig. 2:

$$10^5 \tau_G/s = 1.3\,(\text{OS-1 at } 42°\text{C}), \quad 4.0\,(\text{PI-8 at } 23°\text{C}) \qquad (13)$$

Fig. 3 shows the steady state shear viscosity η of the OS-1 and PI-8 samples measured in a range of shear rate $\dot{\gamma}$ well below the terminal relaxation frequency $\omega_G = 1/\tau_G$ $(\dot{\gamma}/\omega_G < 10^{-2})$. The terminal relaxation of PI-8 is assigned as the rubbery relaxation that corresponds to the relaxation of *axial* orientational anisotropy of the so-called Rouse segments.[23–26] The rubbery relaxation exhibits the nonlinearity only at $\dot{\gamma} > \omega_G$, as usually expected. In contrast, the OS-1 sample exhibits a considerable decrease of η on an increase of the $\dot{\gamma}/\omega_G$ ratio just above 10^{-4} (at $\dot{\gamma} > 10\ s^{-1}$; see Fig. 3). The non-Newtonian behavior seen at such low $\dot{\gamma}$ is characteristic to the glassy relaxation[27,28] that is assigned as the relaxation of the *planar* orientation of the monomeric segments.[23–26] Further details of this non-Newtonian behavior are examined below with the aid of the rheo-dielectric data.

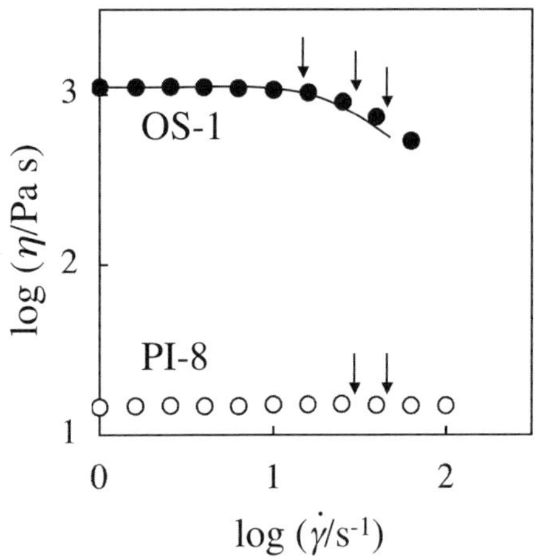

Fig. 3. Steady flow behavior of OS-1 at 42°C and PI-8 at 23°C. Redrawn, with permission, from Ref. 5.

In Fig. 4, the dielectric loss ε'' of OS-1 and PI-8 measured under steady flow are double-logarithmically plotted against the angular frequency ω of the electric field. Those data detect specific mode of motion determined by the alignment of the dipoles in the molecules, as explained below. (Note that the rheo-dielectric data were obtained with the set-up shown in Fig. 1. Thus, those data detect the motion in the shear gradient direction.)

The OS-1 molecules have so-called *type-B* dipoles[3,1] perpendicular to the molecular backbone (cf. Fig. 5), and their rheo-dielectric relaxation reflects the torsional motion around the backbone that allows the orientational memory of these dipoles to decay. The rate of this torsional motion is highly sensitive to the packing of the monomeric units of neighboring molecules, as suggested from rheo-optical data.[23–25] Thus, even a weak distortion of

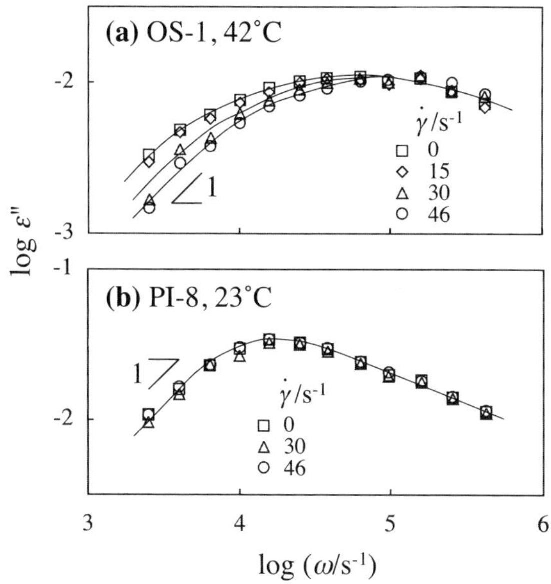

Fig. 4. Rheo-dielectric behavior of OS-1 at 42°C and PI-8 at 23°C. Redrawn, with permission, from Ref. 5.

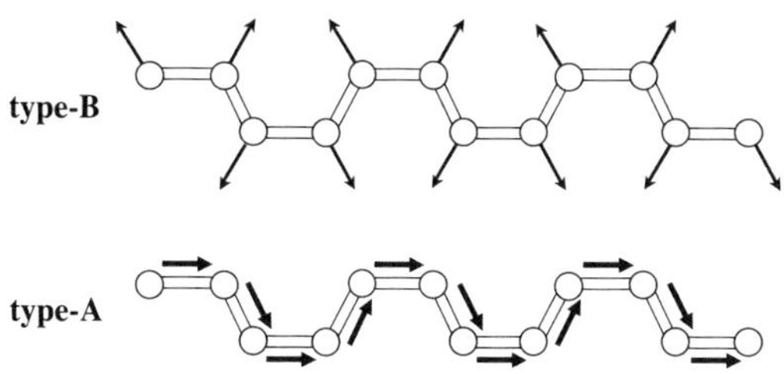

Fig. 5. Schematic illustration of type-B and type-A dipoles. Circles indicate the backbone atoms, and the arrows, the dipoles attached to functional groups of molecules.

the packing due to flow could have accelerated the torsional motion. Fig. 4 demonstrates that the terminal relaxation tail of the rheo-dielectric data of OS-1, characterized by the relationship $\varepsilon'' \propto \omega$ (cf. Eq. (9)), is shifted to high ω with increasing shear rate, which is in harmony with this molecular argument. In fact, the magnitude of this acceleration evaluated from the ε'' data in the terminal tail zone, $\lambda(\dot{\gamma}) = \varepsilon''(\omega;\dot{\gamma})/\varepsilon''(\omega;\dot{\gamma}=0)$ at low ω, is consistent with the decrease of the viscosity; see the solid curve in Fig. 3 that represents a relationship, $\eta(\dot{\gamma}) = \eta_0\lambda(\dot{\gamma})$. Thus, the strong non-

Newtonian behavior of OS-1 is attributable to the flow-induced acceleration of the torsional motion.

For further analysis of this acceleration, the dielectric relaxation function $\Phi(t)$ *under fast shear flow* needs to be quantitatively expressed in terms of the molecular conformation. Unfortunately, this expression has not been formulated yet for the OS-1 molecules having the type-B dipoles. Nevertheless, the rheo-dielectric relaxation of those molecules is unequivocally related to the torsional motion, which allows us to conclude the flow-induced acceleration of this motion.

The PI-8 molecules have so-called *type-A* dipoles[3,4] parallel to the molecular backbone, in addition to the type-B dipoles. The sum of the type-A dipoles is proportional to the end-to-end vector $\boldsymbol{R}$ of the PI chain. Since the motion over the end-to-end distance is much slower than the torsional motion, the rheo-dielectric ε'' data of PI-8 seen in Fig. 4 are related to the former type of motion that allows the decay of the orientational memory of the type-A dipoles. (The relaxation of the type-B dipoles of PI-8 is observed at high ω not covered in Fig. 4.) Clearly, those data of PI-8 including the terminal tail are insensitive to the shear flow at $\dot{\gamma}/\omega_G < 10^{-2}$, indicating that the motion over the end-to-end vector is negligibly affected by such slow flow (as usually expected). This result is consistent with the Newtonian behavior of PI-8 seen in Fig. 3.

Thus, we may experimentally resolve some details of molecular motion under flow through comparison of the rheological and rheo-dielectric data. For PI-8 examined in this section, the flow was too slow to affect the chain motion. In the next section, the rheo-dielectric data are utilized to examine an effect(s) of faster flow on the motion of entangled high-M PI chains.

3.2. *Rheo-dielectric behavior of entangled chain*

3.2.1. *Overview*

Fig. 6(a) shows the steady state viscosity η and the first normal stress coefficient Ψ_1 measured at 30°C for an entangled solution of high-M linear polyisoprene (PI-1190; $M_w = 1.2 \times 10^6$, $M_w/M_n = 1.11$, $w_{PI} = 15$wt%) in an oligomeric butadiene, OB-2 ($M_w = 2.0 \times 10^3$, $M_w/M_n = 2$).[0] OB-2 exhibits no entanglement effect. The entanglement molecular weight for PI-1190 in this solution evaluated from the PI volume fraction therein ($\phi_{PI} = 0.15$) and M_e of bulk PI ($= 5.0 \times 10^3$),[29] $M_{e,sol} = M_e/\phi_{PI}^{1.3} = 5.9 \times 10^4$, is much smaller than M_w of PI-1190. Consequently, PI-1190 chains are highly entangled with each other to exhibit the non-Newtonian behavior well known for entangled linear chains,[29–33] $\eta \propto 1/\dot{\gamma}^{\beta}$ ($\beta \cong 0.8$) and $\Psi_1 \propto 1/\dot{\gamma}^{\beta'}$ ($\beta' \cong 1.5$), at shear rates $\dot{\gamma}$ above the terminal viscoelastic relaxation frequency evaluated from the G' and G'' data, $\omega_G = 1/\tau_G = [\omega G''/G']_{\omega\to 0}$ (shown with the thick arrow in Fig. 6(a)).

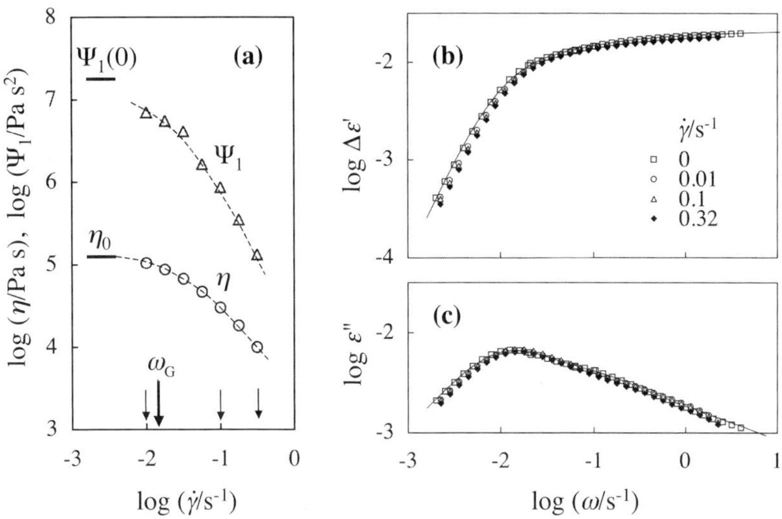

Fig. 6. (a) Non-Newtonian flow behavior, and (b), (c) rheo-dielectric behavior of an entangled solution of linear PI-1190 in oligomeric OB-2 (w_{PI} = 15wt%) at 30°C. In (a), thin arrows show the shear rates where the rheo-dielectric tests were made. Redrawn, with permission, from Ref. 6.

The decrease of the dynamic dielectric constant from the static dielectric constant, $\Delta\varepsilon'(\omega)$ $(= \varepsilon'(0) - \varepsilon'(\omega))$, and the dielectric loss, $\varepsilon''(\omega)$, measured for this solution under steady shear are shown in Figs 6(b) and (c). Clearly, the dielectric behavior of the PI-1190 solution changes only a little on an increase of $\dot{\gamma}$ up to 0.32 s^{-1} ($\dot{\gamma}/\omega_G \cong 30$): The terminal dielectric relaxation time τ_ε (defined by Eq. (10)) and the dielectric intensity $\Delta\varepsilon$ $(= (2/\pi)\int_{-\infty}^{\infty} \varepsilon''(\omega) \mathrm{d}\ln\omega = \Delta\varepsilon'(\infty))$ decrease from respective equilibrium values only by $\cong$ 10% on this increase of $\dot{\gamma}$. (The $\Delta\varepsilon$ data are later shown in Fig. 8.) This shear-insensitivity of the rheo-dielectric behavior, noted also for the other entangled linear PI solution,[6] is in contrast to the significant $\dot{\gamma}$ dependence of η and Ψ_1 seen in Fig. 6(a). This difference of the dielectric and viscoelastic behavior is later discussed in relation to the current molecular model(s).

Fig. 7(a) shows the η and Ψ_1 data measured at 25°C for a solution of high-M 6-arm polyisoprene (PI-(179)$_6$; $M_{\mathrm{w,arm}} = 1.8 \times 10^5$, $M_{\mathrm{w}}/M_{\mathrm{n}} = 1.03$) in OB-2.[6] The PI concentration is $w_{\mathrm{PI}} = 20$wt%. $M_{\mathrm{e,sol}}$ for PI-(179)$_6$ in the solution ($= 4.1 \times 10^4$) is considerably smaller than $M_{\mathrm{w,arm}}$ and thus the PI-(179)$_6$ chains are well entangled with each other. Consequently, the star PI-(179)$_6$ solution shows the non-Newtonian behavior similar to that of the linear PI-1190 solution. Nevertheless, the star solution exhibits moderately $\dot{\gamma}$-dependent rheo-dielectric behavior; cf. Figs. 7(b) and (c): An increase of $\dot{\gamma}$ to 3 s^{-1} ($\dot{\gamma}/\omega_{\mathrm{G}} \cong 40$) results in decreases of dielectric τ_ε and $\Delta\varepsilon$ of PI-(179)$_6$ by $\cong$ 40% and $\cong$ 20%, respectively.[6] (The $\Delta\varepsilon$ data are later shown in Fig. 8.) Thus, the shear effect on the dielectric behavior differs quantitatively for the linear and star PI chains. This difference is discussed later in relation to the current molecular model(s).

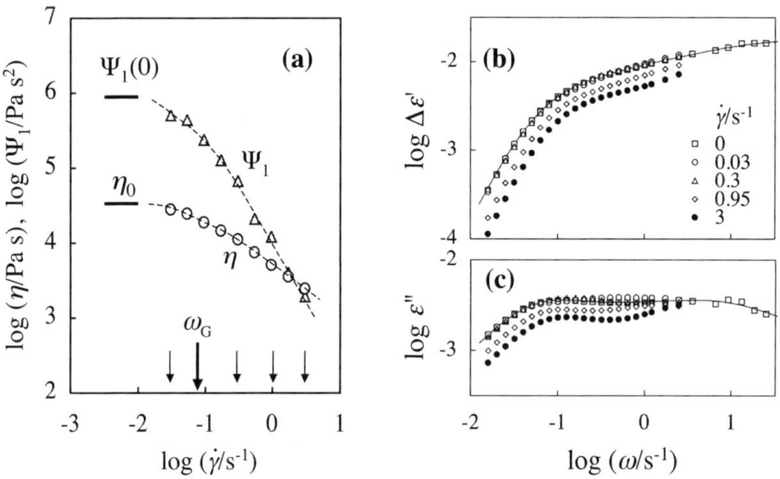

Fig. 7. (a) Non-Newtonian flow behavior, and (b), (c) rheo-dielectric behavior of an entangled solution of star PI-(179)$_6$ in oligomeric OB-2 ($w_{\mathrm{PI}} = 20$wt%) at 25°C. Redrawn, with permission, from Ref. 6.

This discussion requires an expression of the dielectric relaxation function $\Phi(t)$ in terms of the chain conformation. The Green–Kubo type expression of the dielectric relaxation function, Eq. (11), does not rigorously hold under shear flow.[20,21] However, for the relaxation of type-A dipoles of PI chains, an analysis[22] showed that the Green–Kubo type expression is valid as a satisfactory approximation, as explained in Appendix A. Thus, with the aid of a relationship between the polarization p_E of those chains and the component R_y of the end-to-end vector in the direction of the electric field (= y direction), $p_E \propto R_y$, $\Phi(t)$ for the type-A dipole relaxation can be satisfactorily expressed as[22]

$$\Phi(t) = \frac{\langle R_y(t) R_y(0)\rangle_{\mathrm{ss}}}{\langle R_y^2\rangle_{\mathrm{ss}}} \tag{14}$$

where $\langle \ldots \rangle_{\mathrm{ss}}$ represents an ensemble average under steady shear flow. Specifically, $\langle R_y^2\rangle_{\mathrm{ss}}$ is the mean square component of the end-to-end vector in the y direction under steady shear. Correspondingly, the dielectric intensity $\Delta\varepsilon$ is given by (cf. Eq. (12))

$$\Delta\varepsilon = \frac{4\pi\tilde{\mu}}{k_{\mathrm{B}}T}\nu_{\mathrm{chain}}\langle R_y^2\rangle_{\mathrm{ss}} \tag{15}$$

where ν_{chain} is the number density of the chain in the system, and $\tilde{\mu}$ is the magnitude of the type-A dipole per unit length of the chain backbone. (Eq. (12) includes the internal electric field correction factor F and the correlation factor g. However, these factors are close to unity for the global relaxation over the end-to-end distance of the chain[34] and are omitted in Eq. (15).) Eqs. (14) and (15) allow rather quantitative discussion of the rheo-dielectric behavior of PI chains, as explained below.

3.2.2. *Flow-induced equilibration of entanglement segments*

The shear rates $\dot{\gamma}$ examined for the entangled PI chains is higher than the terminal relaxation frequency (cf. ω_{G} in Figs. 6 and 7) but still lower than the Rouse relaxation frequency of the whole chain.[6] Thus, the chains are oriented but not significantly stretched at those $\dot{\gamma}$. For such cases, the entanglement segments are regarded as the basic stress sustaining units,[35] and the stress-optical rule allows us to express the viscosity η and the first and second normal stress coefficients Ψ_1 and Ψ_2 in terms of the components of the unit tangential vector $\tilde{\boldsymbol{u}}$ of those segments as[1,6,35]

$$\eta = \frac{3G_{\mathrm{N}}}{\dot{\gamma}}\langle \tilde{u}_x \tilde{u}_y\rangle_{\mathrm{ss}} \tag{16}$$

$$\Psi_1 = \frac{3G_{\mathrm{N}}}{\dot{\gamma}^2}\left\{\langle \tilde{u}_x^2\rangle_{\mathrm{ss}} - \langle \tilde{u}_y^2\rangle_{\mathrm{ss}}\right\} \tag{17}$$

$$\Psi_2 = \frac{3G_{\mathrm{N}}}{\dot{\gamma}^2}\left\{\langle \tilde{u}_x^2\rangle_{\mathrm{ss}} + 2\langle \tilde{u}_y^2\rangle_{\mathrm{ss}} - 1\right\} \tag{18}$$

Here, G_{N} is the entanglement plateau modulus in the linear viscoelastic regime, $\langle \ldots \rangle_{\mathrm{ss}}$ denotes the ensemble average under steady shear flow, and the x and y directions are chosen as the shear and shear-gradient directions, respectively. Eq. (16) enables us to evaluate the average $\langle \tilde{u}_x \tilde{u}_y\rangle_{\mathrm{ss}}$ from the known G_N value and the η data. Literature data for entangled polymers[32,33] indicate that an empirical relationship $\Psi_2 \cong -0.1\Psi_1$ holds in the non-Newtonian regime at high $\dot{\gamma}$ where the first normal stress difference, $\mathcal{N}_1 = \Psi_1\dot{\gamma}^2$, is close to twice of the shear stress, $2\sigma_s = 2\eta\dot{\gamma}$. In the range of $\dot{\gamma}$ where the rheo-dielectric tests were made (cf. Figs. 6 and 7), $\mathcal{N}_1$ was indeed close to $2\sigma_s$ and thus Ψ_2 appearing in Eq. (18) can be safely replaced by $-0.1\Psi_1$. Then, Eqs. (17) and (18) allow us to evaluate $\langle\tilde{u}_x^2\rangle_{ss}$ and $\langle\tilde{u}_y^2\rangle_{ss}$ from the known G_N value and the Ψ_1 data.

In Fig. 8, the averages $\langle\tilde{u}_x\tilde{u}_y\rangle_{ss}$, $\langle\tilde{u}_x^2\rangle_{ss}$, and $\langle\tilde{u}_y^2\rangle_{ss}$ thus obtained from the viscoelastic data for the linear and star PI (Figs. 6(a) and 7(a)) are plotted against the shear rate normalized by the terminal relaxation frequency ω_G at equilibrium, $\dot{\gamma}/\omega_G$. At equilibrium, the entanglement segments are isotropically oriented to have $\langle\tilde{u}_x\tilde{u}_y\rangle_{ss} = 0$ and $\langle\tilde{u}_x^2\rangle_{ss} = \langle\tilde{u}_y^2\rangle_{ss} = 1/3$. The increases of $\langle\tilde{u}_x\tilde{u}_y\rangle_{ss}$ and $\langle\tilde{u}_x^2\rangle_{ss}$ and the decrease of $\langle\tilde{u}_y^2\rangle_{ss}$ seen for $\dot{\gamma} > \omega_G$ reflect the shear-induced orientation of the entanglement segments. In contrast, the dielectric intensity under steady shear normalized by the intensity at equilibrium, $\Delta\varepsilon(\dot{\gamma})/\Delta\varepsilon(0)$, changes with $\dot{\gamma}$ much less significantly, in particular for the linear PI-1190 chain (Fig. 8(a)). This difference between $\langle\tilde{u}_i\tilde{u}_j\rangle_{ss}$ ($i, j = x$ or y) and $\Delta\varepsilon(\dot{\gamma})/\Delta\varepsilon(0)$ is discussed below in relation to the mutual equilibration of entanglement segments under steady shear.

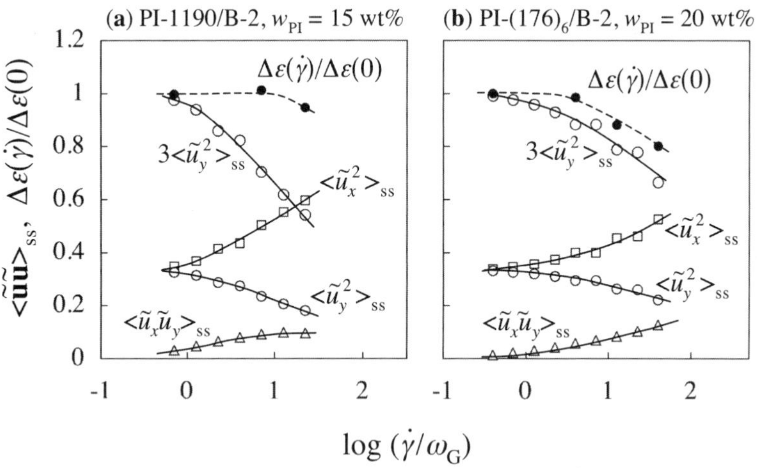

Fig. 8. Second moment averages of unit tangential vector of the entanglement segment, $\langle\tilde{u}_i\tilde{u}_j\rangle_{ss}$, and normalized dielectric intensity, $\Delta\varepsilon(\dot{\gamma})/\Delta\varepsilon(0)$, of PI-1190 and PI-(176)$_6$ chains in the entangled solutions. $\langle\tilde{u}_i\tilde{u}_j\rangle_{ss}$ and $\Delta\varepsilon(\dot{\gamma})/\Delta\varepsilon(0)$ are plotted against the normalized shear rate, $\dot{\gamma}/\omega_G$. Redrawn, with permission, from Ref. 6.

For the oriented but unstretched chain, the end-to-end vector $\boldsymbol{R}$ is expressed in terms of the unit tangential vector of the entanglement segments as $\boldsymbol{R} = a\sum_n \tilde{\boldsymbol{u}}(n)$, where a is the equilibrium size of the entanglement segment, n is the segment index, and the summation $\sum_n$ is taken for all entanglement segments in the chain. With the aid of this expression, Eq. (15) is rewritten as

$$\Delta\varepsilon(\dot{\gamma}) = Ka^2 \sum_{n,n'=1}^{N_{\text{chain}}} \langle \tilde{u}_y(n)\tilde{u}_y(n')\rangle_{\text{ss}} \tag{19}$$

where N_{chain} is the number of entanglement segments per chain, and $K = 4\pi\tilde{\mu}^2\nu_{\text{chain}}/k_BT$. At equilibrium, the entanglement segments have no orientational correlation so that we find $\langle \tilde{u}_y(n)\tilde{u}_y(n')\rangle_{\text{eq}} = \delta_{n,n'}/3$ and $\Delta\varepsilon(0) = Ka^2N_{\text{chain}}/3$. In contrast, under steady shear flow, a sequence of neighboring segments could be internally equilibrated through release and renewal of the topological constraints (entanglements from surrounding chains) due to flow. Then, Eq. (19) can be further rewritten as $\Delta\varepsilon(\dot{\gamma}) = Ka^2N_{\text{chain}}\{\langle\tilde{u}_y^2\rangle_{\text{ss}} + (\beta_{\text{ss}}-1)\langle\tilde{u}_y\tilde{u}'_y\rangle_{\text{ss}}\}$. Here, β_{ss} (≥ 1) is the number of entanglement segments along the chain backbone that are mutually equilibrated under steady shear, and $\langle\tilde{u}_y\tilde{u}'_y\rangle_{\text{ss}}$ is the average of the product $\tilde{u}_y\tilde{u}'_y$ taken for those correlated segments having the y components of the unit tangential vectors, $\tilde{u}_y$ and $\tilde{u}'_y$. Thus, we find

$$\frac{\Delta\varepsilon(\dot{\gamma})}{\Delta\varepsilon(0)} = 3\langle\tilde{u}_y^2\rangle_{\text{ss}} + 3(\beta_{\text{ss}}-1)\langle\tilde{u}_y\tilde{u}'_y\rangle_{\text{ss}} \tag{20}$$

Eq. (20) suggests that the ratio $\Delta\varepsilon(\dot{\gamma})/\Delta\varepsilon(0)$ under steady shear is larger than $3\langle\tilde{u}_y^2\rangle_{\text{ss}}$ by a factor of $3(\beta_{\text{ss}}-1)\langle\tilde{u}_y\tilde{u}'_y\rangle_{\text{ss}}$. In fact, Fig. 8 clearly indicates that $\Delta\varepsilon(\dot{\gamma})/\Delta\varepsilon(0)$ and $3\langle\tilde{u}_y^2\rangle_{\text{ss}}$ are different from each other (in particular for the case of linear PI-1190 chains) and thus the mutual equilibration does occur for entanglement segments under fast shear flow.

The flow-induced release of the constraint activating this mutual equilibration process has been incorporated in the current tube model as a convective constraint release (CCR).[35–38] However, it should be emphasized that the above conclusion of the flow-induced mutual equilibration was deduced just from comparison of the viscoelastic and dielectric data (giving the $\langle\tilde{u}_i\tilde{u}_j\rangle_{\text{ss}}$ and $\Delta\varepsilon(\dot{\gamma})/\Delta\varepsilon(0)$ data in Fig. 8) with the aid of the analysis of the chain conformation, Eq. (20), i.e., without relying on the tube model. This fact is a good example showing the usefulness of the rheo-dielectric method.

3.2.3. *Mutual equilibration number of entanglement segments*

The number β_{ss} of the entanglement segments mutually equilibrated under fast shear flow (cf. Eq. (20)) can be estimated from the $\Delta\varepsilon(\dot{\gamma})/\Delta\varepsilon(0)$ and $3\langle\tilde{u}_y^2\rangle_{ss}$ data if the $\langle\tilde{u}_y\tilde{u}'_y\rangle_{ss}$ value is known. Intuitively, we expect $\langle\tilde{u}_y\tilde{u}'_y\rangle_{ss} = \langle\tilde{u}_y^2\rangle_{ss}$ if successive β_{ss} entanglement segments are mutually equilibrated to have the highly correlated orientational anisotropy thereby behaving as a coarse-grained stress-sustaining unit. In fact, analysis of the orientational distribution of $\tilde{\boldsymbol{u}}(n)$ $(1 \leq n \leq \beta)$ and a unit tangential vector of a sequence of these $\tilde{\boldsymbol{u}}(n)$, $\tilde{\boldsymbol{r}} = \sum_{n=1}^{\beta}\tilde{\boldsymbol{u}}(n)/|\sum_{n=1}^{\beta}\tilde{\boldsymbol{u}}(n)|$, on the basis of the freely-jointed rod model[39] showed that $\langle\tilde{u}_y\tilde{u}'_y\rangle_{ss}$ is very close to $\langle\tilde{u}_y^2\rangle_{ss}$.[6] Thus, β_{ss} can be estimated from the $\Delta\varepsilon(\dot{\gamma})/\Delta\varepsilon(0)$ and $3\langle\tilde{u}_y^2\rangle_{ss}$ data as $\Delta\varepsilon(\dot{\gamma})/\{3\Delta\varepsilon(0)\langle\tilde{u}_y^2\rangle_{ss}\}$; cf. Eq. (20).

Fig. 9 shows β_{ss} thus estimated for the linear PI-1190 and star PI-$(179)_6$ solutions of the PI content w_{PI} as indicated. β_{ss} is plotted against the normalized shear rate, $\dot{\gamma}/\omega_G$. For a given value of $\dot{\gamma}/\omega_G$, β_{ss} is insensitive to w_{PI}, as naturally expected from a fact the nonlinear behavior of either linear or star chains is essentially determined by competition between the flow and chain motion specified by the $\dot{\gamma}/\omega_G$ ratio. More importantly, we note that β_{ss} is considerably larger for the linear chain than for the star chains.

The data in Fig. 8 also suggest that β_{ss} is larger for the linear chain *even if* the $\langle\tilde{u}_y\tilde{u}'_y\rangle_{ss}$ value does not exactly coincide with the $\langle\tilde{u}_y^2\rangle_{ss}$ value. Namely, an equilibration factor $\Delta\varepsilon(\dot{\gamma})/\Delta\varepsilon(0)-3\langle\tilde{u}_y^2\rangle_{ss}$ is larger for the linear chain than for the star chain, as clearly noted in Fig. 8. Since the linear and star chains would have a common value of $\langle\tilde{u}_y\tilde{u}'_y\rangle_{ss}$ for the mutually equilibrated segments (as suggested from the similarity of their η and Ψ_1 data), β_{ss} evaluated from Eq. (20) is larger for the linear chains just because of this difference of the equilibration factor.

The molecular origin of the above difference of β_{ss} for the star and linear chains has not been fully elucidated yet. However, one possible molecular scenario has been proposed within the context of the tube model for entangled chains.[6] In this model, the topological constraint between uncrossable chains is represented by an impenetrable tube, and the chain is constrained in the tube thereby being forced to move along the tube axis.[30,31,35,38,40] The tube for a given chain is not fixed in space but moves according to the motion of the tube-forming (surrounding) chains. This tube motion results in a local constraint release (CR) for the chain in the tube.[30,31,35,38,40] Accumulation of such local CR processes allows succes-

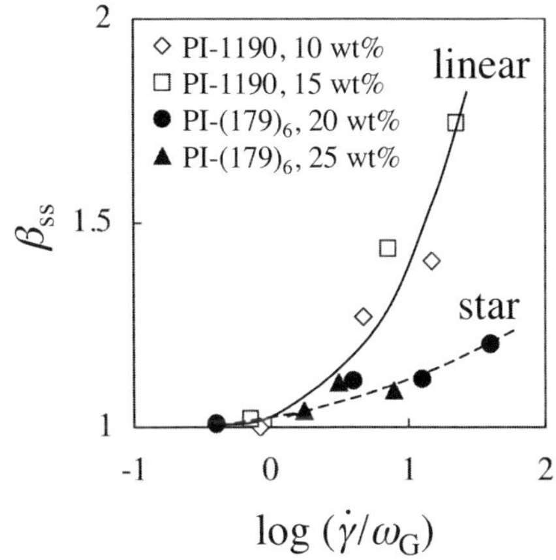

Fig. 9. Flow-induced equilibration number β_{ss} of entangled linear PI-1190 and star PI-$(179)_6$ chains in the solutions with the concentrations as indicated. β_{ss} is plotted against the normalized shear rate, $\dot{\gamma}/\omega_G$. Redrawn, with permission, from Ref. 6.

sive entanglement segments to mutually equilibrate to behave as a coarse-grained stress-sustaining unit. This coarse-graining corresponds to dynamic tube dilation (DTD)[26,35,38,41] considered in the tube model. In the absence of flow, the equilibrium motion of the tube-forming chain activates the *thermal* CR/DTD process. Under flow, the chain contraction following the transient stretch due to the flow activates the convective CR (CCR)[35–38] and the corresponding DTD.

Since the mutual equilibration under the shear flow is achieved through both of the thermal and convective constraint release (TCR and CCR) mechanisms and these mechanisms would contribute to the net equilibration number β_{net} in an essentially independent way, β_{net} under the flow can be reasonably approximated in terms of the contributions β_{TCR} and β_{CCR} from these mechanisms as[6]

$$\beta_{net} \cong \beta_{TCR}\beta_{CCR} \tag{21}$$

At equilibrium, β_{TCR} entanglement segments are mutually equilibrated through the TCR mechanism, and β_{ss} shown in Fig. 9 is evaluated on the basis of the dielectric $\Delta\varepsilon(0)$ data at equilibrium; cf. Eq. (20). Thus, the experimentally estimated β_{ss} corresponds to β_{CCR} appearing in Eq. (21). Under the fast shear flow, the topological difference between the linear and star chains would not affect too much the flow behavior, as suggested from the similarity of the viscosity and first normal stress coefficient data for the linear and star chains. Thus, $\beta_{\rm net}$ reflecting the flow behavior would be similar for those chains. In contrast, at equilibrium, the mutual equilibration of the entanglement segments due to the TCR mechanism is more significant for the star chain than for the linear chain (because the star chain exhibits a much broader relaxation mode distribution compared to the linear chain), as revealed from extensive viscoelastic/dielectric studies for linear and star PI.[26,42–46] These arguments suggests that $\beta_{\rm ss}$ ($= \beta_{\rm CCR} = \beta_{\rm net}/\beta_{\rm TCR}$) is larger for the linear chain than for the star chain, as observed in Fig. 9.

Of course, the above argument is qualitative because it is based on an approximation of the product form in Eq. (21) (corresponding to perfectly independent TCR and CCR contributions to $\beta_{\rm net}$), and a more quantitative argument of $\beta_{\rm net}$ supported by detailed analysis of the equilibration dynamics is highly desired. Nevertheless, the above argument appears to serve as a good starting point for this detailed analysis.

3.2.4. *Lack of flow-induced dielectric acceleration for linear chain*

Within the context of the tube model, the non-Newtonian behavior of entangled polymers has been attributed to the CCR mechanism that accelerates the chain motion under fast flow (and also affects the viscoelastic mode distribution under flow). In fact, some viscoelastic data are consistent with this CCR mechanism. As an example, Fig. 10 shows the storage modulus $G'_{/\!/}$ of an isotactic polypropylene sample ($M_{\rm w} = 3.8 \times 10^5$, $M_{\rm w}/M_{\rm n} = 6.8$) at 200°C (in melt state) measured against a small-amplitude oscillatory shear stress superimposed parallel to the steady shear stress $\sigma_{\rm ss}$.[47] Clearly, the terminal tail of $G'_{/\!/}$ ($\propto \omega^2$ with ω = angular frequency of the oscillatory stress) shifts to high ω on an increase of $\sigma_{\rm ss}$, *i.e.*, on an increase of the rate $\dot{\gamma}$ of the background steady shear. Although the framework of linear viscoelasticity does not rigorously apply to the $G'_{/\!/}$ data associated with the nonlinear thinning behavior, this shift of the terminal tail is consistent with the acceleration of viscoelastic relaxation due to the CCR mechanism. (In fact, the $G'_{/\!/}$ data are close to the prediction of a CCR-type constitutive equation shown with the solid curves.[47])

Nevertheless, no corresponding acceleration is noted for the rheo-dielectric data for the entangled linear PI-1190 chain; see Fig. 6. The current framework of the molecular models (including that of the tube model) does not appear to describe the viscoelastic and rheo-dielectric data consistently. As an example, in Fig. 11, the data of the rheo-dielectric relaxation time and intensity, τ_ε and $\Delta\varepsilon$, and the viscosity, η, obtained for

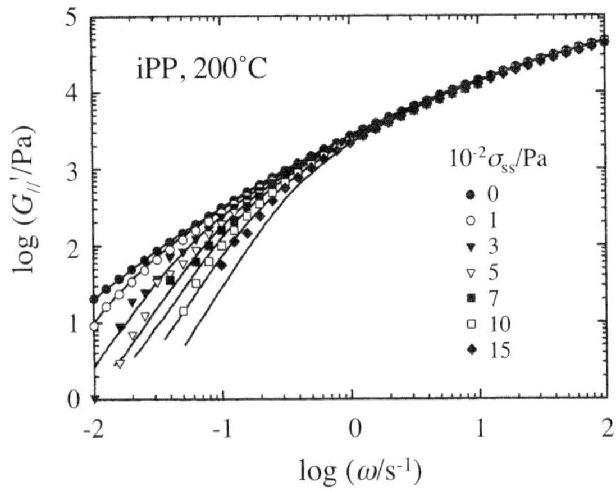

Fig. 10. Storage modulus of iPP at 200°C measured against small oscillatory stress parallel superimposed on steady stress σ_{SS}. The modulus is plotted against the angular frequency of oscillatory stress. Redrawn, with permission, from Ref. 47.

the PI-1190/B-2 solution (w_{PI} = 15wt%; cf. Fig. 6) are compared with the prediction of NAPLES simulation.[22] For convenience of the comparison, these quantities are normalized by their zero-shear (equilibrium) values and plotted against the shear rate normalized by the dielectric relaxation frequency $1/\tau_\varepsilon^\circ$ at equilibrium, $\dot{\gamma}\tau_\varepsilon^\circ$. The NAPLES simulation considers a network of Gaussian chains temporarily connected through slip-links (representing topological constraints) and allows the chain to move according to a balance of the elastic force due to the thermal tension along the chain, the frictional force, the thermal Brownian force, and the thermodynamic force resulting from a bias of the chemical potential of the monomers (along the chain as well as in space).[48–51] The slip-link is removed when a chain end passes through it, and the link is created when a dangling end of a chain hooks the other chain. Thus, all mechanisms of the chain motion (including CCR) considered in the current tube model are incorporated in the NAPLES simulation. In addition, the simulation also considers a hidden topological constraint for a given chain due to surrounding but not linked chains: This hidden constraint becomes effective once the links existing for the given chain is released.[51,52]

As seen in Fig. 11, the NAPLES simulation based on such detailed molecular consideration describes the η and $\Delta\varepsilon$ data of the entangled linear PI-1190 chains very well. (In addition, the simulation reproduces the storage and loss moduli in the linear viscoelastic regime.[48–51]) Nevertheless, the lack of flow-induced acceleration of the dielectric relaxation ($\dot{\gamma}$-insensitivity of τ_ε) cannot be reproduced by the simulation. This result strongly suggests that some important mechanism is still missing in the simulation (as well as in the current tube model). One possibility is related to the creation of entanglements (slip-links) immediately following the removal of entanglements. The created entanglements could be distributed isotropically in space to give a negligible effect on the viscoelastic stress (detecting the orientational anisotropy) but significantly constrain the chain motion thereby giving $\dot{\gamma}$-insensitive τ_ε. However, no further detail of this isotropic entanglement creation has been clarified. A theoretical study is strongly desired for the consistent description of the nonlinear viscoelastic and rheo-dielectric behavior of entangled chains.

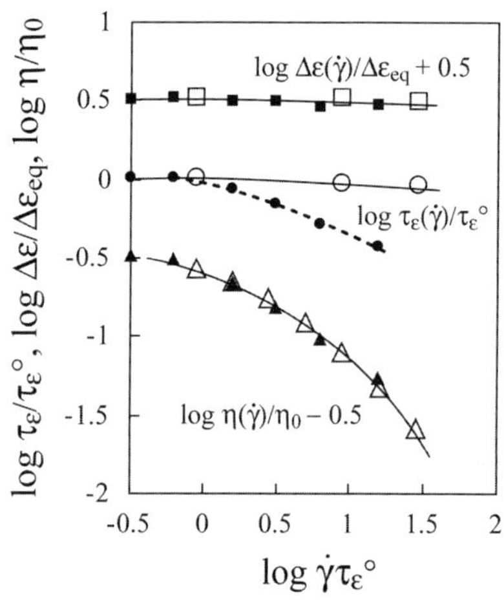

Fig. 11. Comparison of data (unfilled symbols) of viscosity and dielectric relaxation time/intensity of entangled linear PI-1190/B-2 solution (w_{PI} = 15wt%) with results of NAPLES simulation (filled symbols). Redrawn, with permission, from Ref. 22.

4. Rheo-Dielectric Behavior of Liquid Crystalline Materials

In general, the liquid crystalline (LC) texture is highly susceptible to an external fields.[53–62] The flow effects on the texture and molecular alignment have been extensively investigated for LC materials, in particular for alkyl cyanobiphenyl (nCB)[55–60] having a chemical structure shown in Fig. 12. The nCB molecule behaves as a rigid molecule except for the alkyl (C_nH_{2n+1}) group, and a large dipole of the cyano group ($C \equiv N$) is always directed parallel to the axis vector $\boldsymbol{b}$ of the biphenyl group. Thus, the motion of nCB molecules activates pronounced dielectric relaxation. Since this motion is anisotropic with respect to the director (or the local symmetry axis) of the LC texture, the rheo-dielectric behavior of LC materials provides us with the information for the LC texture and molecular motion under flow. The rheo-dielectric data of nCB giving this information[7,8] are summarized below.

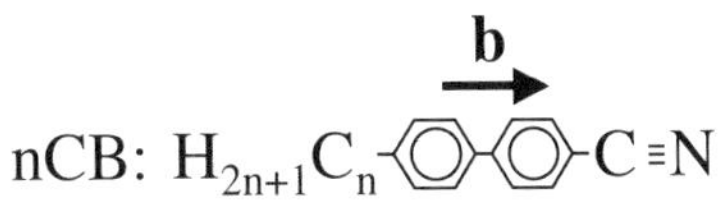

Fig. 12. Chemical structure of alkyl cyanobiphenyl (nCB) molecule having a large dipole parallel to the axis vector $\boldsymbol{b}$ of the biphenyl group.

4.1. *Rheo-dielectric behavior of nematic 7CB*

4-4′-heptyl cyanobiphenyl (7CB; n = 7 in Fig. 12) has the nematic-to-isotropic and crystalline-to-nematic (melting) transition temperatures, $T_{\mathrm{ni}} = 42°\mathrm{C}$ and $T_{\mathrm{m}} = 28.5$–$30°\mathrm{C}$.[55,63] Fig. 13 shows the rheo-dielectric data for the 7CB system quiescently cooled from 50°C ($> T_{\mathrm{ni}}$) to 30°C (just above T_{m}) and annealed at 30°C;[7] see unfilled symbols. These data detect fluctuation of the molecular axis vector $\boldsymbol{b}$ shown in Fig. 12.

In general, the orientational order of the 7CB molecules in the nematic texture is limited only within a mesoscopic length scale ($\sim \mu$m). However, these molecules have a rather large magnetic susceptibility to form a monodomain having a well-ordered nematic director under a magnetic field.[55–57] In Fig. 13, the filled symbols and the curves, respectively, indicate the complex dielectric constant $\varepsilon^*(\omega)$ $(= \varepsilon'(\omega) - i\varepsilon''(\omega))$ of the monodomain 7CB in the quiescent state reported by Davies et al.[57] and Lippens et al:[56] $\varepsilon^*_{/\!/}(\omega)$ is the data of the monodomain measured in the direction parallel to the director (i.e., the electric field is parallel to the director), and $\varepsilon^*_{\perp}(\omega)$, the data measured in the direction perpendicular to the director. These $\varepsilon^*_{/\!/}$ and $\varepsilon^*_{\perp}$ data are quite different from each other, confirming that the motion/fluctuation of the 7CB molecular motion is highly anisotropic with respect to the director. (The fluctuation is slower for the component of molecular axis vector $\boldsymbol{b}$ parallel to the director than for the component perpendicular to the director, as can be noted from the difference of the relaxation frequencies for the $\varepsilon^*_{/\!/}$ and $\varepsilon^*_\perp$ data.)

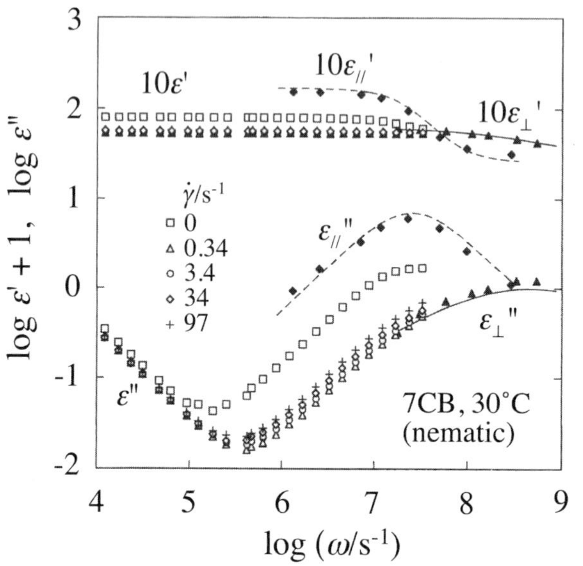

Fig. 13. Rheo-dielectric behavior of nematic 7CB at 30°C (unfilled symbols). The filled symbols and curves indicate the dielectric response of nematic monodomain of 7CB measured with the electric field being parallel ($\varepsilon^*_{/\!/}$) and/or perpendicular ($\varepsilon^*_\perp$) to the nematic director.[56,57] Redrawn, with permission, from Ref. 7.

In Fig. 13, we note that the quiescently ordered 7CB system exhibits the ε^* data (unfilled squares) between the $\varepsilon^*_{/\!/}$ and $\varepsilon^*_\perp$ data of the monodomains. This result reflects the nematic texture having no long-range order. This texture can be dielectrically regarded as a polydomain composed of randomly oriented grains. In fact, the ε^* data were found to be rather close to the complex dielectric constant expected for randomly stacked parallel and perpendicular domains having the volume fractions of 1/3 and 2/3,[7] $\varepsilon^*_{\mathrm{pd}}(\omega) = \{1/3\,\varepsilon^*_{/\!/}(\omega) + 2/3\,\varepsilon^*_\perp(\omega)\}^{-1}$.

More importantly, the ε^* data become significantly smaller than those in the quiescent state under the steady shear: Under the shear at a rate of $\dot{\gamma} = 0.34\ \mathrm{s}^{-1}$ (unfilled triangles), the ε^* data are very close to $\varepsilon^*_\perp$ of the monodomain having the director perpendicular to the electric field. This rheo-dielectric behavior indicates that the 7CB molecules are shear-aligned in the flow plane perpendicular to the shear gradient direction (perpendicular to the electric field direction), as also suggested from rheo-scattering measurements.[61,62] (This alignment hardly affected the viscosity.) We also note that the ε^* data begin to moderately deviate from $\varepsilon^*_\perp$ on an increase of $\dot{\gamma}$ from 0.34 s^{-1} to 97 s^{-1}; see unfilled circles, diamonds, and plus symbols. Thus, the optimum shear-alignment of the 7CB molecules is achieved at a rather low shear rate, which is suggested also from the rheo-scattering measurements.[61,62]

The above rheo-dielectric changes occurred reversibly, and the data in the quiescently annealed state (unfilled squares in Fig. 13) were recovered within 40 min rest after cessation of shear.[7] It should be emphasized that the rheo-dielectric changes occurred at the shear rates *much lower* than the relaxation frequency of the 7CB molecules, $\omega_r \cong 10^7$ s^{-1} (estimated from the terminal tail of the $\varepsilon^*_\perp$ data). Such slow flow does not affect the molecular motion itself, as can be noted from the close coincidence of the ε^* data at $\dot{\gamma} = 0.34$ s^{-1} and the $\varepsilon^*_\perp$ data, the latter being measured in the quiescent state. However, the slow flow still orients the nematic director thereby changing the direction of dielectric detection of the molecular motion. This type of rheo-dielectric change can be noted for a variety of texture fluids.

4.2. *Rheo-dielectric behavior of smectic 8CB*

4-4′-octyl cyanobiphenyl (8CB; n = 8 in Fig. 12), exhibits the nematic-to-isotropic, smectic-to-nematic, and crystalline-to-smectic (melting) transitions at $T_{ni} = 40°C$, $T_{sn} = 32.5°C$, and $T_m = 21°C$.[63] For the smectic 8CB system quiescently cooled from 45°C ($> T_{ni}$) to 25°C (between T_m and T_{sn}) and annealed at 25°C, the rheo-dielectric behavior is demonstrated in Fig. 14.[8] The corresponding steady state viscosity data are shown in Fig. Fig. 15.[8] (The rheo-dielectric behavior of 8CB in the nematic state at T between T_{ni} and T_{sn} was similar to that of 7CB examined in the previous section.[8])

The quiescently ordered smectic 8CB system behaves as a plastic solid because of many deflects between the smectic grains, as reported by Colby et al.[64] Correspondingly, the linear viscoelastic complex moduli $G^*(\omega)$ $(= G'(\omega) + iG''(\omega))$ exhibited unrelaxed low-ω tails due to these defects:[8,58,64] The motion of grains/grain boundaries required for the relaxation is very slow thereby providing these tails. The grains are enlarged (grain boundaries are removed) to enhance the relaxation of the system on

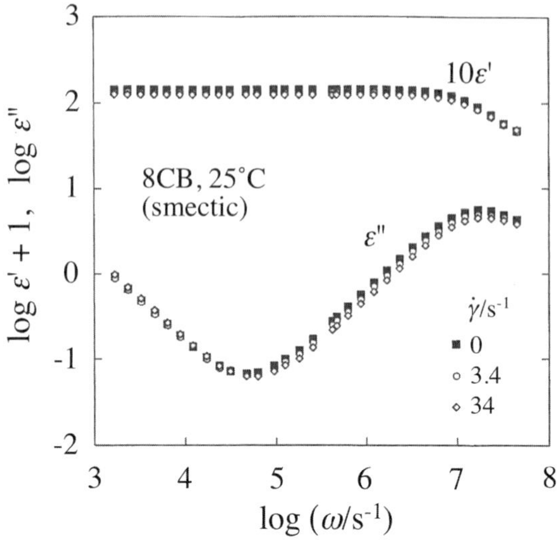

Fig. 14. Rheo-dielectric behavior of smectic 8CB at 25°C. Redrawn, with permission, from Ref. 8.

application of large-amplitude oscillatory shear at appropriate frequencies and amplitudes, as found by Larson et al.[58]

The significant non-Newtonian behavior of the smectic 8CB system seen in Fig. 15 (never observed in the nematic state) was associated with enhancement of the unrelaxed tails of $G^*(\omega)$ that prevailed after cessation of the steady flow.[8] Thus, the non-Newtonian behavior is attributable to massive disruption of the smectic grains into smaller fragments. The steady state viscosity for the lowest $\dot{\gamma}$ examined in Fig. 15 was obtained after a gradual decay of the shear stress for $\sim$ 10 min,[8] which suggests that the massive disruption took long time.

The dielectric relaxation in the quiescent state (filled squares in Fig. 14) is attributable to the relaxation of $\varepsilon^*_{/\!/}(\omega)$ of the smectic layers (or grains) having the director parallel to the direction of the electric field.[8] This relaxation moderately reduces its intensity (by $\simeq$ 30%) under the steady shear at $\dot{\gamma} = 3.4$–$34\ \mathrm{s}^{-1}$. This result corresponds to a moderately enhanced orientation of the grains in the direction perpendicular to the shear gradient direction. However, it should be emphasize that this shear orientation is associated with massive disruption of the nematic grains, as noted from the rheological behavior. For this reason, the rheo-dielectric changes of the

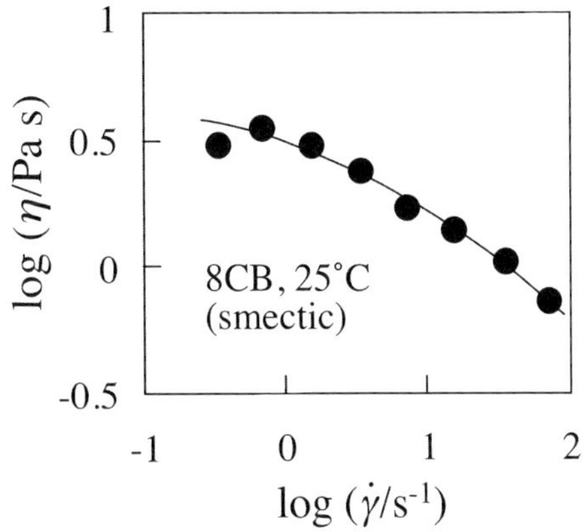

Fig. 15. Steady state viscosity of smectic 8CB at 25°C. Redrawn, with permission, from Ref. 8.

smectic 8CB is far less significant compared to the changes seen for the nematic 8CB and 7CB.[7,8]

Finally, we note an additional feature related to the ion motion. Both smectic 7CB (Fig. 13) and nematic 8CB (Fig. 14) exhibit the direct current (DC) conduction due to ionic impurities characterized by a relationship $\varepsilon'' \propto \omega^{-1}$ at low ω ($< 10^5$ s^{-1}). This DC conduction is insensitive to the shear flow, suggesting that the ions are not strongly interacting with (bound to) the 7CB and 8CB molecules and thus the ion motion is not significantly affected by the shear-orientation of these molecules. A different type of conduction is noted for ions strongly bound to the molecules, as explained below for a polyelectrolyte-type composite.

5. Rheo-Dielectric Behavior of Salt/PEO Composite Systems

Electrical properties of polyelectrolyes composed of polar polymers and dissolved salts have been extensively investigated because of its practical importance as well as the scientific interest for the dynamics of ions (dissociated salt) under strong interaction with the polymers.[65–74] In particular, extensively studies have been made for composites of lithium perchlorate salt ($LiClO_4$) and polyethylene oxide (PEO). The Li^+ cation dissociated from the salt is bound to the ether oxygen of PEO:[71,72,75] Typically, three-to-six oxygen atoms form a cage binding each Li^+.[72] The conductivity due to the motion of Li^+ reflects this Li^+-O-Li^+ bonds and is affected by the shear flow that can transiently disrupt the bonds. The corresponding rheo-dielectric behavior of the $LiClO_4$/PEO system[9] is summarized below.

5.1. *Overview of rheo-dielectric behavior of $LiClO_4$/PEO System*

Fig. 16 shows the rheo-dielectric data at 25°C obtained for a composite of $LiClO_4$ (with the concentration of 2.3×10^3 mol m^{-3}) and PEO-1 ($M_w = 1.0 \times 10^3$, $M_w/M_n = 2$) under steady shear flow.[9] The corresponding data of the viscosity growth function $\eta^+(t)$ ($= \sigma^+(t)/\dot{\gamma}$ with $\sigma^+(t)$ being the stress after start-up of the shear flow) are shown in Fig. 17.[9] The rheo-dielectric data shown in Fig. 16 were obtained at long times after $\eta^+(t)$ reached the steady state. The base polymer, PEO-1, was crystalline at 25°C but the composite was in the liquid state because of the melting point depression due to $LiClO_4$.[9]

In Fig. 16, we first note that the $LiClO_4$/PEO-1 system in the quiescent equilibrium state (cf. plus symbols) exhibits the dielectric loss $\varepsilon''(\omega)$ satisfying a relationship $\varepsilon'' \propto \omega^{-1}$ at $\omega > 10^4$ s^{-1}. This relationship is characteristic to the DC conduction (due to the dissociated salt). However, at lower ω, the ω dependence of ε'' weakens a little. More importantly, the system exhibits a strong increase of the dynamic dielectric constant $\varepsilon'(\omega)$ on a decrease of ω below 10^5 s^{-1}. This increase of ε', never observed for the real DC conduction, is indicative of a huge electrostatic energy stored in the system at low ω. Thus, the dielectric behavior of the $LiClO_4$/PEO-1 system in the quiescent equilibrium state is attributed to the electrode polarization (EP)[3,9,15,16] due to ions (Li^+ and ClO_4^- in the current system) that cannot discharge at the electrodes. At low ω, those ions have a sufficient time to follow the oscillatory electric field and move from one electrode to the other, thereby being concentrated near the electrode to store the huge electrostatic energy. For this EP phenomenon, ε'' exhibits a peak and then decreases while ε' becomes constant as ω approaches zero. (This low-ω asymptote is not observed in the range of ω covered in Fig. 16). Note that the apparent DC conduction, characterized by the relationship $\varepsilon'' \propto \omega^{-1}$ at $\omega > 10^4$ s^{-1}, corresponds to the high-ω asymptote of the EP phenomenon.

Fig. 16 shows that the ε' and ε'' data increase significantly under steady shear. This rheo-dielectric change corresponds to the non-Newtonian behavior seen in Fig. 17. Namely, in the quiescent state, Li^+ is bound to the

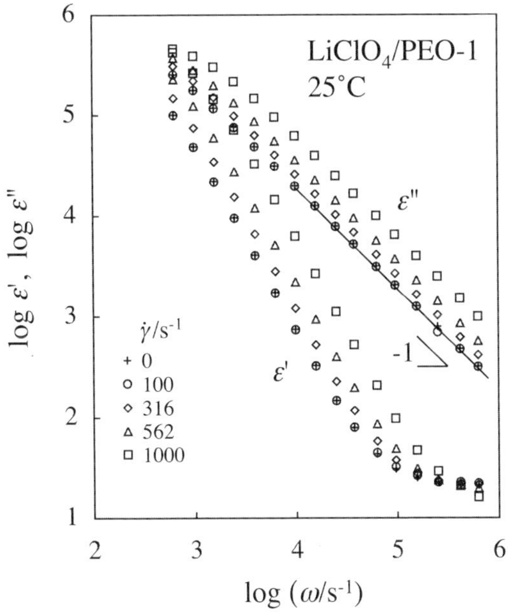

Fig. 16. Rheo-dielectric behavior of the $LiClO_4$/PEO-1 system at 25°C. Redrawn, with permission, from Ref. 9.

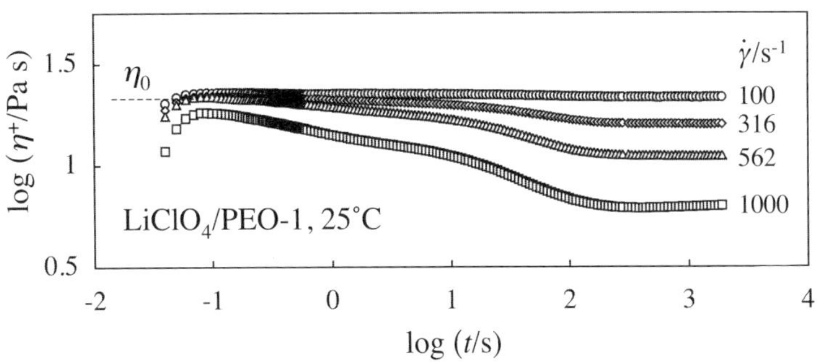

Fig. 17. Viscosity growth behavior of the $LiClO_4$/PEO-1 system at 25°C. Redrawn, with permission, from Ref. 9

ether oxygen of PEO thereby forming transient physical crosslinks (bridges) between PEO chains, as schematically illustrated in Fig. 18(a). (Each Li^+ is bound in a cage composed of three-to-six oxygen atoms of the PEO chains,[72] but this coordination number is not important for the argument in this section.) These transient crosslinks are thermally disrupted (and reformed) to allow Li^+ to hop in space, which results in the conduction due to Li^+. It is well established that Li^+ and the PEO segments (crosslinked by Li^+) move cooperatively and the Li^+ conductivity is well correlated with the segment mobility.[65,66,68,71,73,74] Under the shear, the PEO chains transiently crosslinked by Li^+ are oriented/stretched to mechanically disrupt the Li^+-O-Li^+ bonding, as schematically shown in Fig. 18(b). (This mechanical disruption takes long time, as judged from the very gradual decrease of η^+ seen in Fig. 17.) Then, the hopping motion of Li^+ is accelerated to increase $\varepsilon'(\omega)$ and $\varepsilon''(\omega)$, as observed in Fig. 16.

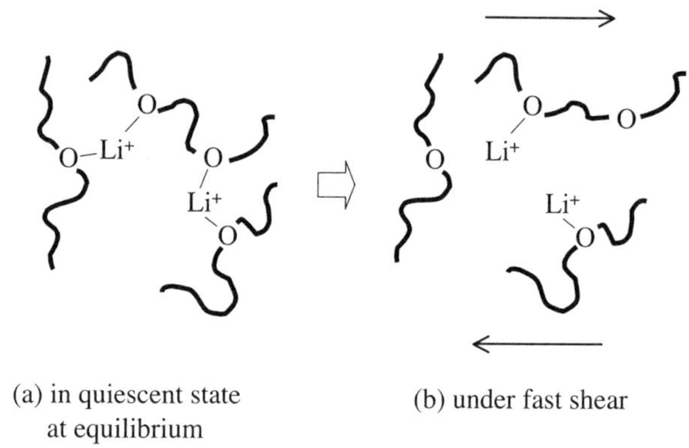

Fig. 18. Schematic illustration of the mechanism of rheo-dielectric changes observed for the $LiClO_4$/PEO-1 system.

In this mechanism of rheo-dielectric changes, the flow does not directly affect the Li^+ motion but it orients/stretches the PEO chains crosslinked by Li^+ thereby forcing the chains to disrupt the Li^+-O-Li^+ crosslink and accelerate the Li^+ motion as well as the motion of the PEO segments constrained by the crosslinks. Thus, we expect that the acceleration of the Li^+ motion is well correlated with the non-Newtonian behavior of the $LiClO_4$/PEO-1 system. This expectation is examined below with the aid of the Macdonald theory[15] for the electrode polarization.

5.2. *Flow-induced enhancement of Li^+ mobility*

For the electrode polarization (EP) under oscillatory electric field, Macdonald considered dragging of the ions due to the electric field and the competing thermal diffusion of the ions to formulate an expression of the complex admittance.[15] His results, obtained under the conditions of (i) no discharge of ions at the electrode (blocking condition) and (ii) lack of electric field-induced dissociation of salt, are summarized in Appendix B together with the corresponding expression of the complex dielectric constant $\varepsilon^*_{\rm EP}$ for EP. For the polyelectrolyte system containing monovalent ions (dissociated from $LiClO_4$, for example), this expression can be satisfactorily cast in the Debye-type single-relaxation form:[9,16,17]

$$\varepsilon'_{\rm EP}(\omega) = \varepsilon_{\rm m} + \Delta\varepsilon_{\rm EP}\frac{1}{1+\omega^2\tau_{\rm EP}^2} \tag{22}$$

$$\varepsilon''_{\rm EP}(\omega) = \Delta\varepsilon_{\rm EP}\frac{\omega\tau_{\rm EP}}{1+\omega^2\tau_{\rm EP}^2} \tag{23}$$

Here, $\varepsilon_{\rm m}$ is the dielectric constant of the medium (assumed to be independent of ω), and $\tau_{\rm EP}$ and $\Delta\varepsilon_{\rm EP}$ denote the rheo-dielectric relaxation time and relaxation intensity, respectively. $\tau_{\rm EP}$ depends on the mobility μ averaged for the anion and cation and the number density ν of the cation (identical to that of anions) as well as on the gap L between the electrodes, and $\Delta\varepsilon_{\rm EP}$, on ν and L:[9,17]

$$\tau_{\rm EP} \propto L\mu^{-1}\nu^{-1/2}, \qquad \Delta\varepsilon_{\rm EP} \propto L\nu^{1/2} \tag{24}$$

The L dependence of $\tau_{\rm EP}$ emerges because the time required for the ions to move from one electrode to the other changes with L, and the L dependence of $\Delta\varepsilon_{\rm EP}$ reflects an increase of electrostatic energy of the cation and anion separated at larger L. Thus, $\varepsilon^*_{\rm EP}$ is not a pure material function but changes with the experimental set up, L.

From Eqs. (22) and (23), we expect that the rheo-dielectric data can be superposed on the data in the quiescent equilibrium state *if* the shear affects only the ion mobility μ and does not change the cation number density ν. We can test this expectation for the data shown in Fig. 16 by shifting the data along the ω axis by an adequately chosen factor λ^{-1} (< 1). The results, *i.e.*, the plots of the $\varepsilon^*(\omega)$ data against $\omega\lambda^{-1}$, are shown in Fig. 19. Clearly, the rheo-dielectric data for $\dot\gamma/{\rm s}^{-1} = 100$–$1000$ (unfilled symbols) are excellently superposed on the data in the quiescent equilibrium state (plus symbol), which confirms that the ion mobility μ in the $LiClO_4$/PEO-1 system is increased while the cation number density ν is not affected by the shear flow at those $\dot\gamma$ and that the shift factor λ^{-1} coincides with the $\mu(0)/\mu(\dot\gamma)$ ratio. Similar results were found for the other $LiClO_4$/PEO systems containing high-M PEO.[9]

In Fig. 20, the shift factor λ^{-1} utilized in Fig. 19 is plotted against the normalized shear rate $\dot\gamma\tau^\circ_{\rm G}$, with $\tau^\circ_{\rm G}$ being the linear viscoelastic terminal

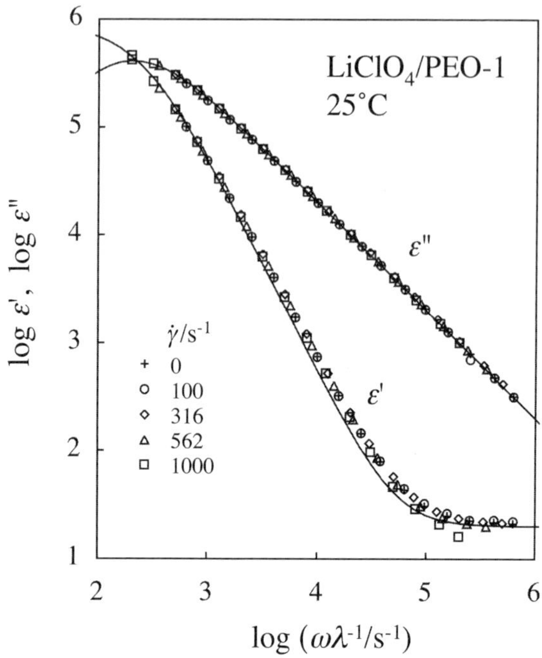

Fig. 19. Comparison of the dielectric data of the $LiClO_4$/PEO-1 system in the quiescent state and under steady shear, the latter being plotted against a normalized frequency $\omega\lambda^{-1}$. Solid curves indicate results of fitting with the Macdonald theory. Redrawn, with permission, from Ref. 9.

relaxation time of the $LiClO_4$/PEO system in the absence of the steady shear flow.[9] For comparison, the normalized viscosity of the system, η/η_0, evaluated from the η^+ data at long t (cf. Fig. 17) is also shown. The λ^{-1} factor (filled circles) excellently agrees with the η/η_0 ratio (unfilled squares). This result confirms the cooperativity of the ion motion and the PEO segmental motion. Namely, the segmental motion is accelerated by the factor of λ (> 1) on the mechanical disruption of the transient Li^+-O-Li^+ crosslink under flow and thus the Li^+ mobility (reflecting the Li^+ hopping coupled with the segmental motion) is increased by the same factor.

Finally, we attempt to fit the dielectric data shown in Fig. 19 with the Macdonald theory for the electrode polarization. Good fitting was obtained in the range of ω examined, as shown with the solid curves in Fig. 19. (At lower ω, the fitting should become worse because of a surface capacitor layer on the electrodes not considered in the theory.[9]) The fitting parameters for

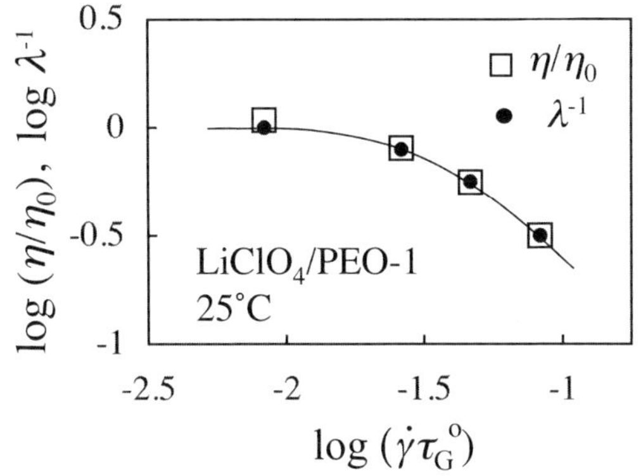

Fig. 20. Comparison of the normalized viscosity η/η_0 and the acceleration factor λ^{-1} for the EP relaxation of the $LiClO_4$/PEO-1 system at 25°C. Redrawn, with permission, from Ref. 9.

those curves were:[9,76] $\nu = 5.5 \times 10^{23}$ m^{-3}, $\mu = 1.0 \times 10^{-8}$ m^{2}V^{-1}s^{-1}, and $\varepsilon_m = 20$. Since the fitting was achieved under an assumption that the cation (Li^+) and anion (ClO_4^-) have the same mobility (to avoid uncertainty that the same quality of fitting was achieved for various sets of the cation and anion mobilities),[9] these parameter values should not be regarded as very quantitative estimates of ν and μ in the actual $LiClO_4$/PEO system. Nevertheless, the above μ value is in a typical range of the Li^+ mobility expected for the $LiClO_4$/PEO systems.[68,71] In addition, the ν value is much smaller than the nominal ν value ($= 1.4 \times 10^{27}$ m^{-3}) expected for a case of full dissociation of $LiClO_4$ (having the concentration of 2.3×10^3 mol m^{-3}). Thus, the fitting suggests that just a small fraction of $LiClO_4$ worked effectively as a source of the conducting ions, which is in harmony with the results of spectroscopic analysis.[71,74]

6. Rheo-Dielectric Behavior of Carbon Black Suspensions

Rheological behavior of suspension of aggregating particles sensitively reflects the shape and connectivity of the aggregates determined by the interparticle interaction.[77,78] The shape can be estimated from scattering experiments if the aggregates are dilute, but this is not the case in usual suspensions. Thus, a method sensitively detecting the shape and connectivity of the aggregates is desired.

If the particles are conductive, the rheo-dielectric method serves as such a sensitive method. As an example of the rheological behavior of suspensions of such particles, Fig. 21(a) shows the G' and G'' data measured at 30°C for a 20wt% suspension of carbon black particles with non-polar surface (CB25A; average diameter = 25 nm, heterogeneity index = 1.3) in a polar varnish (Varnish-p).[10] Since the non-polar CB25A particles have a poor affinity toward the polar medium, the particles form aggregates to exhibit the unrelaxed tails of G' and G'' at low ω. The steady state viscosity η of this suspension decreases significantly with increasing shear rate $\dot{\gamma}$, as shown in Fig. 21(b). This non-Newtonian thinning behavior is attributable to flow-induced disruption of the aggregates.

The above disruption behavior is reflected in the rheo-dielectric data of the suspension shown in Fig. 22.[10] In the quiescent state ($\dot{\gamma} = 0$), both ε' and ε'' increase significantly on a decrease of the angular frequency ω of the electric field. The increase of ε'' may appear to be similar to that for the DC electronic conduction through the conductive paths of the aggregated CB25A particles. However, the increase of ε', somewhat similar to that seen for the electrode polarization (Fig. 19), indicates that the suspension does not exhibit the simplest, electronic DC conduction possibly because the aggregates are not dense enough to allow the electron to freely hop within the aggregates and also because the aggregates are not long enough to firmly form conductive bridges between the electrodes.

Thus, the suspension in the quiescent state appears to contain anisotropic and loosely packed aggregates of finite size, and the increase of ε' at low ω is attributable to electronic polarization in respective aggregates that is hindered at high ω due to an electron hopping barrier between the loosely aggregating CB particles. (Note that the polarization P and the corresponding ε' at low ω depend on the shape of the aggregates: P and ε' are larger for anisotropic aggregates oriented in the direction of the electric field than for spherical aggregates.) The slow viscoelastic relaxation seen in Fig. 21 appears to correspond to the motion of such anisotropic/loose aggregates.

Under shear flow at $\dot{\gamma} = 3\ \mathrm{s}^{-1}$, ε' and ε'' exhibit orders-of-magnitude decreases from their values in the quiescent state; see Fig. 22. This rheo-dielectric change is attributable to massive, shear-induced disruption of the loose/anisotropic aggregates into less anisotropic fragments that weakens the polarization in respective aggregates/fragments. At a higher shear rate, $\dot{\gamma} = 30\ \mathrm{s}^{-1}$, the aggregates appear to be further disrupted so that ε' of the suspension almost coincides with $\varepsilon'_{\mathrm{m}}$ of the medium (Varnish-p) shown with the dashed curve in Fig. 22. In addition, ε'' of the suspension under the flow at $\dot{\gamma} = 30\ \mathrm{s}^{-1}$ is moderately smaller than $\varepsilon''_{\mathrm{m}}$ of the medium possibly

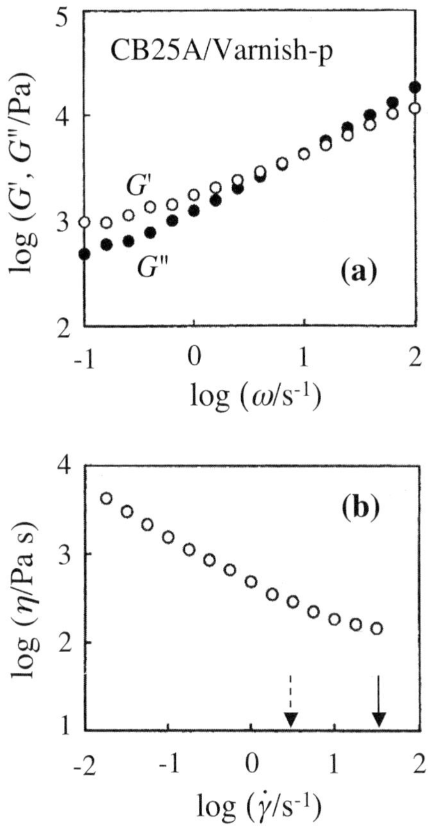

Fig. 21. (a) Linear viscoelastic behavior, and (b) non-Newtonian flow behavior of 20wt% CB25A/Varnish-p suspension at 30°C. Redrawn, with permission, from Ref. 10.

because the ionic impurities giving $\varepsilon''_{\mathrm{m}}$ of the medium are adsorbed on the CB particles and immobilized. This adsorption weakens the electrode polarization due to the ionic impurities of the medium in the suspension, which could lead to the ε' value of the suspension slightly smaller than $\varepsilon'_{\mathrm{m}}$ of the pure medium, as noted in Fig. 22. These rheo-dielectrically deduced structural changes are in harmony with the η data. Namely, η at $\dot{\gamma} = 30$ s^{-1} (cf. solid arrow in Fig. 21(b)) is just moderately smaller than η at $\dot{\gamma} = 3$ s^{-1} (cf. dashed arrow) possibly because the anisotropic/loose aggregates have been already disrupted massively at $\dot{\gamma} = 3$ s^{-1}.

The above example demonstrates the usefulness of the rheo-dielectric method for structural analysis for the suspensions of conductive particles.

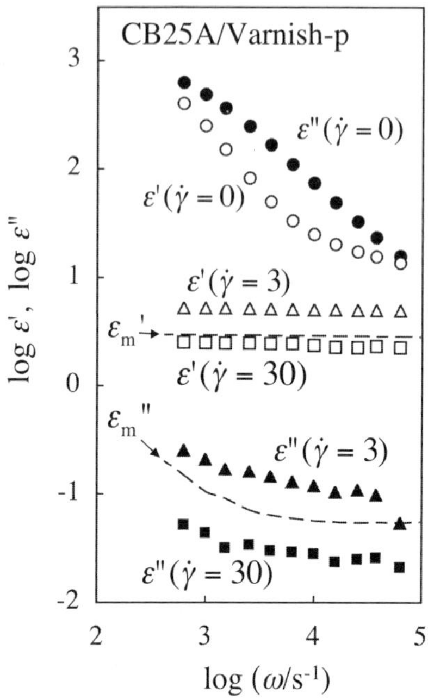

Fig. 22. Rheo-dielectric behavior of 20wt% CB25A/Varnish-p suspension at 30°C. Redrawn, with permission, from Ref. 10.

This method can be applied also for composites containing conductive carbon fibers and/or carbon nanotubes to estimate the number of the firm contacts between the fibers and/or nanotubes.[79–82]

7. Concluding Remarks

This chapter has demonstrated the usefulness of rheo-dielectric method for investigation of the molecular dynamics and structure in soft matters under flow. For the soft matters having no higher-order structure (such as OS and PI examined in Sec. 3), this method allows us to directly examine the flow effect on the molecular motion. In contrast, for those having the higher-order structures, rather slow flow does not straightforwardly change the molecular motion but still affects the structures. This structural change indirectly affects the dynamics in respective systems, for example, the dielectrically detected direction of the LC molecular motion (cf. Sec. 4), the ion release from the PEO chains (cf. Sec. 5), and the electron hopping in the CB aggregates (cf. Sec. 6). Thus, the rheo-dielectric method is applicable to a wide variety of soft matters given that the basic dynamics (dipole motion, ion motion, electron hopping, *etc.*) is specified.

Acknowledgment

The supports from Grant-in-Aid for Scientific Research on Priority Area "Soft Matter Physics" from MEXT (grant #18068009) and Grant-in-Aid (KAKENHI) for Young Scientists B (grant #22740273) were helpful for summarizing the work presented in this chapter.

Appendix A. Rheo-Dielectric Telaxation Function of Type-A Chain

A.1. General

We consider a system of type-A chains (*e.g.*, PI). The dipole moment ($\tilde{\mu}$ per unit length of the chain backbone) is weak to exhibit negligibly small dipole-dipole interaction. In the (rheo)-dielectric measurement, a weak, time dependent electric field $E(t)$ is applied as a perturbation in the y direction and the macroscopic polarization $P(t)$ in the same direction is measured. The end-to-end dipole of the jth chain in this direction is expressed as $\tilde{\mu}R_{y,j}$, with $R_{y,j}$ being the y component of the end-to-end vector of this chain. $P(t)$ is given by a the sum of these dipoles averaged over a conformational distribution function $f(\boldsymbol{p}, \boldsymbol{q}, t)$ in the presence of the electric field, where f is defined in the phase space specified by generalized momentum $\boldsymbol{p}$ and coordinate $\boldsymbol{q}$. This $f(\boldsymbol{p}, \boldsymbol{q}, t)$ is given by a sum of the distribution function in the absence of the electric field (either at equilibrium or under steady shear), $f^{\circ}(\boldsymbol{p}, \boldsymbol{q})$, and a perturbation due to the electric field, $\Delta f(\boldsymbol{p}, \boldsymbol{q}, t)$. Since the average of the dipoles over $f^{\circ}(\boldsymbol{p}, \boldsymbol{q})$ is zero (no spontaneous polarization in the absence of the electric field), $P(t)$ is expressed in terms of $\Delta f(\boldsymbol{p}, \boldsymbol{q}, t)$ as[19]

$$P(t) = \int \mathrm{d}\boldsymbol{p}\, \mathrm{d}\boldsymbol{q} \left\{ \sum_j \tilde{\mu} R_{y,j} \right\} \Delta f(\boldsymbol{p}, \boldsymbol{q}, t) \tag{A.1}$$

In the absence of flow at equilibrium, the time evolution equation of $f(\boldsymbol{p}, \boldsymbol{q}, t)$ is given by $\partial f/\partial t = [\mathcal{H}, f]$ with $\mathcal{H}(\boldsymbol{p}, \boldsymbol{q}, t) = \mathcal{H}^{\circ}(\boldsymbol{p}, \boldsymbol{q}) - \{\sum_j \tilde{\mu}R_{y,j}\}E(t)$. Here, $\mathcal{H}^{\circ}$ is the unperturbed Hamiltonian at equilibrium, the $-\{\sum_j \tilde{\mu}R_{y,j}\}E(t)$ term represents the electrostatic energy due to the interaction of the dipoles with the electric field, and $[A, B] \equiv$

$(\partial A/\partial \boldsymbol{q})\cdot(\partial B/\partial \boldsymbol{p})-(\partial A/\partial \boldsymbol{p})\cdot(\partial B/\partial \boldsymbol{q})$ is the Poisson bracket. Considering the t-independence of $f^{\mathrm{eq}}(\boldsymbol{p},\boldsymbol{q})$ $(= f^{\circ}(\boldsymbol{p},\boldsymbol{q})$ defined in the absence of flow), we find

$$\frac{\partial \Delta f}{\partial t} = [\mathcal{H}^{\circ}, \Delta f] + \left[\left\{\sum_j \tilde{\mu} R_{y,j}\right\}, \mathcal{H}^{\circ}\right] \frac{f^{\mathrm{eq}}(\boldsymbol{p},\boldsymbol{q})}{k_{\mathrm{B}}T} E(t) \tag{A.2}$$

Eq. (A.2) is solved to give

$$\Delta f(\boldsymbol{p},\boldsymbol{q},t) = \int_{-\infty}^{t} \mathrm{d}s\, E(s) e^{(t-s)i\mathcal{L}} \left\{ -i\mathcal{L} \sum_j \tilde{\mu} R_{y,j} \right\} \frac{f^{\mathrm{eq}}(\boldsymbol{p},\boldsymbol{q})}{k_{\mathrm{B}}T} \tag{A.3}$$

where $i\mathcal{L}$ is the Liouville operator $(i\mathcal{L}A = [\mathcal{H}^{\circ}, A])$. Substituting Eq. (A.3) into Eq. (A.1) and replacing the Liouville operator by the time derivative, we can express $P(t)$ as

$$\begin{aligned} P(t) = \frac{\tilde{\mu}^2}{k_{\mathrm{B}}T} \int_{-\infty}^{t} \mathrm{d}s\, E(s) \times \\ \int \mathrm{d}\boldsymbol{p}\,\mathrm{d}\boldsymbol{q} \left\{ \sum_j R_{y,j}(t) \right\} \frac{\mathrm{d}\left\{\sum_j R_{y,j}(s)\right\}}{\mathrm{d}s} f^{\mathrm{eq}}(\boldsymbol{p},\boldsymbol{q}) \end{aligned} \tag{A.4}$$

Eq. (A.4) is known as the Green–Kubo expression.[19,20] We note that the Green–Kubo expression is also obtained for a system whose dynamics is described by the Langevin equation (or the Fokker-Planck equation).[83]

For a given configuration of the ith chain having $R_{y,i}$, the configurations of jth chain having $R_{y,j}$ and $-R_{y,j}$ emerge with the same probability to give $\langle R_{y,j}(t) R_{y,i}(s)\rangle = 0$ for $i \neq j$. Thus, Eq. (A.4) is rewritten as

$$P(t) = \frac{\tilde{\mu}^2 \nu_{\mathrm{chain}}}{k_{\mathrm{B}}T} \int_{-\infty}^{t} \mathrm{d}s\, E(s) \frac{\partial}{\partial s} \langle R_y(t) R_y(s)\rangle_{\mathrm{eq}} \tag{A.5}$$

For a step-wise electric field E imposed at a time $s = 0$, Eq. (A.5) gives

$$P(t) = \Delta\varepsilon\{1 - \Phi(t)\} \tag{A.6}$$

with

$$\Delta\varepsilon = \frac{4\pi\tilde{\mu}^2}{k_{\mathrm{B}}T} \nu_{\mathrm{chain}} \langle R_y^2\rangle_{\mathrm{eq}} \tag{A.7}$$

and

$$\Phi(t) = \frac{\langle R_y(t) R_y(0)\rangle_{\mathrm{eq}}}{\langle R_y^2\rangle_{\mathrm{eq}}} = \frac{\langle \boldsymbol{R}(t)\cdot\boldsymbol{R}(0)\rangle_{\mathrm{eq}}}{\langle \boldsymbol{R}^2\rangle_{\mathrm{eq}}} \tag{A.8}$$

(For the dielectric intensity $\Delta\varepsilon$ shown in Eq. (A.7), we have introduced a factor of 4π to represent $\Delta\varepsilon$ in the MKSA units.[3,22]) The expression of the dielectric relaxation function $\Phi(t)$ given by Eq. (A.8) is known as the Cole expression[19] (a special case of Green–Kubo expression).

A.2. Analysis under steady shear

Under steady shear flow, Eq. (A.2) with f^{eq} replaced by the stationary distribution function f° does not rigorously hold[20,21] (because of the advection), and thus the Cole expression is not rigorously valid. However, an analysis of the Langevin equation for the chain motion indicates that the Cole expression (Eq. (14)) is still valid as a good approximation under the steady shear,[22] as explained below.

The Langevin equation for the global motion of a linear entangled chain can be written as

$$\zeta\left\{\frac{\partial \boldsymbol{r}(n,t)}{\partial t} - \boldsymbol{K}\cdot\boldsymbol{r}(n,t)\right\} = \boldsymbol{\Lambda}(n;\dot{\gamma})\cdot\boldsymbol{r}(n,t) + \boldsymbol{F}^{\text{B}}(n,t) + \boldsymbol{F}^{\text{E}}(n,t) \quad \text{(A.9)}$$

Here, $\boldsymbol{r}(n,t)$ (with $0 \le n \le N$) is the position of nth segment of the type-A chain at time t, ζ is the segmental friction coefficient, $\boldsymbol{K}$ is the shear rate tensor (having the non-zero value of $\dot{\gamma}$ only for the xy component), $\boldsymbol{F}^{\text{B}}(n,t)$ is the Brownian force for nth segment at time t characterized by $\langle \boldsymbol{F}^{\text{B}}(n,t)\boldsymbol{F}^{\text{B}}(n,t)\rangle = 2\boldsymbol{I}\zeta k_{\text{B}}T\delta(n-n')\delta(t-t')$, and $\boldsymbol{F}^{\text{E}}(n,t)$ is the electric force acting on this segment due to the constant electric field E (at time $t>0$). From the alignment of the type-A dipole along the chain backbone, $\boldsymbol{F}^{\text{E}}(n,t)$ is modeled as[22]

$$\boldsymbol{F}^{\text{E}}(n,t) = \begin{bmatrix} 0 \\ \tilde{\mu}E\{\delta(n-N)-\delta(n)\} \\ 0 \end{bmatrix} \quad \text{(A.10)}$$

The 3-D tensorial operator with respect to n, $\boldsymbol{\Lambda}(n;\dot{\gamma})$ appearing in Eq. (A.9), represents the interaction of the chain segments and may explicitly depend on the shear rate $\dot{\gamma}$. This $\boldsymbol{\Lambda}$, giving the potential forces for a given chain conformation $\boldsymbol{r}(n,t)$ to govern the chain motion, is determined by an effective Hamiltonian of the chain under steady shear.

As noted from the dielectric behavior of a special class of linear PI having inverted type-A dipoles,[84–86] the actual chain motion can be decomposed into the motional eigenmodes. Thus, $\boldsymbol{\Lambda}$ for actual chains can be safely assumed to be a linear operator associated with the eigenmodes. In fact, the dielectric mode distribution of linear PI hardly changes under steady shear (cf. Fig. 6), suggesting that the eigenmodes are well defined and $\boldsymbol{\Lambda}$ is linear under the conditions examined experimentally. Then, Eq. (A.9) can be analyzed straightforwardly. Namely, expanding $\boldsymbol{r}(n,t)$ with respect to the eigenfunctions associating to $\boldsymbol{\Lambda}$, we can formally solve Eq. (A.9) to calculate the unnormalized dielectric relaxation function $\Phi_{\rm ss}^{\rm un}(t,\dot{\gamma})$ and the end-to-end correlation function $\Theta_{\rm ss}(t,\dot{\gamma})$ defined by[22]

$$\Phi_{\rm ss}^{\rm un}(t;\dot{\gamma}) = \frac{\tilde{\mu}\nu_{\rm chain}}{E}\left\{\langle R_y(\infty)\rangle_{\rm ss}^{E} - \langle R_y(t)\rangle_{\rm ss}^{E}\right\} \tag{A.11}$$

$$\Theta_{\rm ss}(t;\dot{\gamma}) \equiv \langle R_y(t)R_y(0)\rangle_{\rm ss} \tag{A.12}$$

Here, the subscript "ss" stands for the quantities under steady shear, and the superscript "E", those under the electric field. $\Phi_{\rm ss}^{\rm un}(t;\dot{\gamma})$ directly reflects the polarization due to the chain orientation driven by the electric field. Thus, the Cole expression (Eq. (14)) is deduced under steady shear if $\Phi_{\rm ss}^{\rm un}(t,\dot{\gamma})$ is proportional to $\Theta_{\rm ss}(t;\dot{\gamma})$.

The eigenfunction expansion gives[22]

$$\Phi_{\rm ss}^{\rm un}(t;\dot{\gamma}) = \frac{2\tilde{\mu}^2\nu_{\rm chain}}{N\zeta}\sum_{\beta\geq 1} g_\beta^{[\Phi]}\exp(-\lambda_\beta t) \tag{A.13}$$

with

$$g_\beta^{[\Phi]} = \frac{1}{\lambda_\beta}\left\{\sum_{p\geq 1}(\cos p\pi - 1)Q_{3p-1,\beta}\right\}\left\{\sum_{p'\geq 1}(\cos p'\pi - 1)Q^{-1}_{\beta,3p'-1}\right\} \tag{A.14}$$

and

$$\Theta_{\rm ss}(t;\dot{\gamma}) = \frac{2k_BT}{N\zeta}\sum_{\beta\geq 1} g_\beta^{[\Theta]}\exp(-\lambda_\beta t) \tag{A.15}$$

with

$$\begin{aligned} g_\beta^{[\Theta]} = \sum_{\beta'\geq 1}&\frac{2}{\lambda_\beta+\lambda_{\beta'}}\left\{\sum_{\alpha\geq 1}Q^{-1}_{\beta,\alpha}Q^{-1}_{\beta',\alpha}\right\}\times \\ &\left\{\sum_{p\geq 1}(\cos p\pi - 1)Q_{3p-1,\beta}\right\}\left\{\sum_{p'\geq 1}(\cos p'\pi - 1)Q^{-1}_{\beta',3p'-1}\right\} \end{aligned} \tag{A.16}$$

The parameters λ_β, $Q_{i,j}$, and $Q^{-1}_{i,j}$ are related to a matrix $\boldsymbol{A}$ with the components being defined by

$$\begin{aligned} &A_{3p-2,3q-2} = a^{xx}_{p,q}, \quad A_{3p-2,3q-1} = a^{xy}_{p,q} + \zeta\dot{\gamma}, \quad A_{3p-2,3q} = a^{xz}_{p,q}, \\ &A_{3p-1,3q-2} = a^{yx}_{p,q}, \quad A_{3p-1,3q-1} = a^{yy}_{p,q}, \quad A_{3p-1,3q} = a^{yz}_{p,q}, \\ &A_{3p,3q-2} = a^{zx}_{p,q}, \quad A_{3p,3q-1} = a^{zy}_{p,q}, \quad A_{3p,3q} = a^{zz}_{p,q} \\ &(p,q = 1,2,\ldots) \end{aligned} \tag{A.17}$$

Here, $a_{i,j}^{\alpha\beta}$ $(i, j = 1, 2, \ldots; \alpha, \beta = x, y, z)$ are the expansion coefficients of the Rouse eigenfunction[a] defined with respect to the $\alpha\beta$ component of the operator, $\Lambda_{\alpha\beta}$, as[22]

$$\Lambda_{\alpha\beta}(n; \dot{\gamma}) \cos\left(\frac{j\pi n}{N}\right) = \sum_{i \geq 1} a_{i,j}^{\alpha\beta} \cos\left(\frac{i\pi n}{N}\right) \qquad (i, j = 1, 2, \ldots) \quad \text{(A.18)}$$

The parameter λ_β is the βth eigenvalue of a matrix $\zeta^{-1}\boldsymbol{A}$, $Q_{i,j}$ is the i, j component of a matrix $\boldsymbol{Q}$ that diagonalizes $\boldsymbol{A}$, and $Q_{i,j}^{-1}$ is the i, j component of the inverse matrix $\boldsymbol{Q}^{-1}$; namely, $\{\boldsymbol{Q}^{-1} \cdot \boldsymbol{A} \cdot \boldsymbol{Q}\}_{\alpha\beta} = -\zeta\lambda_\beta\delta_{\alpha\beta}$.

As can be noted from Eqs. (A.13) and (A.15), the dielectric relaxation function $\Phi_{\rm ss}^{\rm un}(t; \dot{\gamma})$ and the end-to-end correlation function $\Theta_{\rm ss}(t; \dot{\gamma})$ under steady shear have the same relaxation time, $1/\lambda_\beta$. Thus, as long as $\boldsymbol{r}(n, t)$ is expanded with respect to the eigenfunctions, we can unequivocally conclude that the shear-insensitivity of the terminal dielectric relaxation time is equivalent to the shear-insensitivity of the end-to-end vector fluctuation time. Furthermore, an analysis considering the shear symmetry indicated that the intensity factors for $\Phi_{\rm ss}^{\rm un}(t; \dot{\gamma})$ and $\Theta_{\rm ss}(t; \dot{\gamma})$, $g_\beta^{[\Phi]}$ and $g_\beta^{[\Theta]}$, are proportional to each other at least up to the first order perturbation expansion.[22] Thus, to a good approximation, the normalized dielectric relaxation function of type-A chains under steady shear, $\Phi_{\rm ss}(t; \dot{\gamma}) = \Phi_{\rm ss}^{\rm un}(t; \dot{\gamma})/\Phi_{\rm ss}^{\rm un}(0; \dot{\gamma})$, coincides with the auto-correlation function of the end-to-end vector component, $\langle R_y(t)R_y(0)\rangle_{\rm ss}/\langle R_y^2\rangle_{\rm ss} = \Theta_{\rm ss}(t; \dot{\gamma})/\Theta_{\rm ss}(0, \dot{\gamma})$, indicating the validity of Eqs. (14) and (15) under steady shear. (In fact, NAPLES simulation showed that $\Phi_{\rm ss}(t; \dot{\gamma})$ and $\langle R_y(t)R_y(0)\rangle_{\rm ss}/\langle R_y^2\rangle_{\rm ss}$ are indistinguishable in the high-$\dot{\gamma}$ regime for the rheo-dielectric tests.[22]) An analysis for a dumbbell having an anisotropic segmental friction also suggested the above coincidence.[87]

As mentioned above, the Green–Kubo relation does not hold in general at non-equilibrium steady states. The violation of the Green–Kubo relation is considered to be related to the entropy production or the advection velocity[88–91] characterizing non-equilibrium systems. Although the Green–Kubo relation approximately holds for the rheo-dielectric response, this does not mean that the system is at (or near) equilibrium. The magnitude of violation is determined by the inner product of the advection velocity and the gradient of the perturbation potential.[90,91] For the case of the rheo-dielectric response, the contribution from this inner product is known

[a]The Rouse eigenfunctions are not the eigenfunctions associating $\boldsymbol{\Lambda}(n, \dot{\gamma})$ but were simply utilized as the base functions for calculation of $\Phi_{\rm ss}^{\rm un}(t; \dot{\gamma})$ and $\Theta_{\rm ss}(t; \dot{\gamma})$.

to be small and practically negligible.[22,87] Therefore, the rheo-dielectric response is quite insensitive to the entropy production, even if the system is at non-equilibrium.

A.3. Analysis under LAOS

The analysis was also carried out for the rheo-dielectric responses under large-amplitude shear oscillation (LAOS).[22] It turned out that the responses under LAOS are not so simple even though the Green–Kubo expression is valid as a good approximation under LAOS. Namely, under LAOS, the distribution function $f^{\mathrm{eq}}(\boldsymbol{p}, \boldsymbol{q})$ appearing in Eq. (A.4) is to be replaced by a distribution function $f(\boldsymbol{p}, \boldsymbol{q}, t)$ oscillating at the angular frequency Ω of LAOS. For this case, the oscillation of the electric field (introduced for detecting the rheo-dielectric signal) is coupled with the LAOS oscillation, which results in a deviation of the rheo-dielectric ε' and ε'' data from a signal reflecting the true statistical average under LAOS.[22]

As an example, Fig. A.1 compares the square average of the y component of the end-to-end vector $\langle R_y^2 \rangle_{\mathrm{LAOS}}$ directly obtained from NAPLES simulation in the absence of the electric field (circles) with $\langle R_y^2 \rangle_{\mathrm{LAOS}}^{\mathrm{base}}$ evaluated, with the aid of Eq. (15), from the dielectric intensity $\Delta\varepsilon$ obtained in the same simulation under a small electric field. ($\langle R_y^2 \rangle_{\mathrm{LAOS}}^{\mathrm{base}}$ corresponds to the average of R_y^2 over a non-oscillating base part of $f(\boldsymbol{p}, \boldsymbol{q}, t)$ under LAOS.) $\langle R_y^2 \rangle_{\mathrm{LAOS}}^{\mathrm{base}}$ becomes considerably smaller than the real average, $\langle R_y^2 \rangle_{\mathrm{LAOS}}$, with increasing LAOS amplitude γ_0, as seen in Fig. A.1. This result demonstrates a significant effect of coupling between the oscillation of the electric field and LAOS on the rheo-dielectric ε' and ε'' data, which is quite different from the situation under steady shear (where $\Delta\varepsilon$ gives the correct $\langle R_y^2 \rangle_{\mathrm{ss}}$.) This result indicates that of the rheo-dielectric data[13,14] under LAOS need to be interpreted very carefully.

Appendix B. Macdonald Theory for Electrode Polarization

Macdonald formulated a time-evolution equation for the concentration profile of ions in a medium containing a dissociated monovalent salt under a weak oscillatory electric field.[15] The profile forms standing wave that oscillates at the angular frequency ω of the applied field.

For the $LiClO_4$/PEO systems placed between the electrodes made of stainless steel, we can safely assume the blocking condition and neglect the recombination of the cation (Li^+) and anion (ClO_4^-). For this case, the complex admittance y^* due to the electrode polarization per unit area of

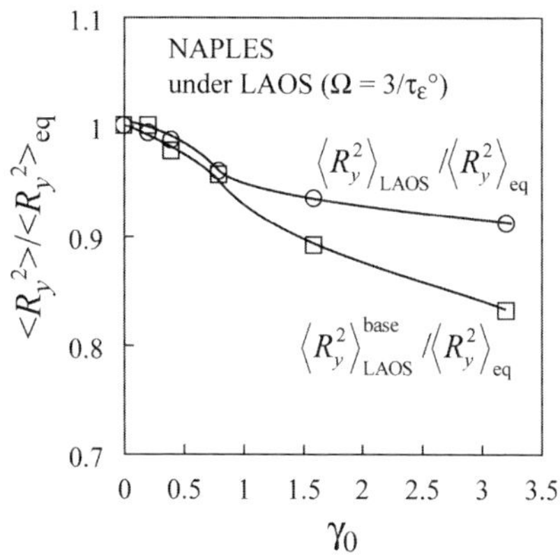

Fig. A.1. Comparison of $\langle R_y^2\rangle_{\rm LAOS}$ (real average) and $\langle R_y^2\rangle_{\rm LAOS}^{\rm base}$ obtained from NAPLES simulation for entangled PI (9.6 entanglements per chain) under LAOS. The LAOS condition, γ_0 ($\leq$ 3.2) and Ω ($= 3/\tau_\varepsilon^\circ$), is similar to the experimental condition reported in literature.[13] Redrawn, with permission, from Ref. 22.

the electrode, deduced from the Macdonald theory, can be written in a form,[9]

$$y^* = \frac{e\nu(\mu_+ + \mu_-)}{L} + \frac{e\nu}{L}\frac{W_3}{W_1 - W_2} \tag{B.1}$$

with

$$W_1 = \frac{(\gamma_+ - 1)(\gamma_- + 1)\vartheta_-(\vartheta_+ - \tanh\vartheta_+)}{\xi_+} - \gamma_+(\gamma_- + 1)\vartheta_+\vartheta_- \tag{B.2}$$

$$W_2 = \frac{(\gamma_+ + 1)(\gamma_- - 1)\vartheta_+(\vartheta_- - \tanh\vartheta_-)}{\xi_-} - (\gamma_+ + 1)\gamma_-\vartheta_+\vartheta_- \tag{B.3}$$

$$\begin{aligned} W_3 = {} & (\gamma_- + 1)(\gamma_+\mu_- - \mu_+)\vartheta_- \tanh\vartheta_+ \\ & - (\gamma_+ + 1)(\gamma_-\mu_- - \mu_+)\vartheta_+ \tanh\vartheta_- \end{aligned} \tag{B.4}$$

Here, ν is the cation number density (identical to the anion number density), e is the absolute value of the electron charge ($e = 1.602\times 10^{-19}$C), μ_+ and μ_- are the mobilities of the cation and anion, and L is the gap between the electrodes. The parameters $\xi_\pm$, $\gamma_\pm$, and $\vartheta_\pm$ appearing in Eqs. (B.2)-

(B.4) are given by[9]

$$\xi_{\pm} = 1 + i\frac{(\mu_+ + \mu_-)Q\omega}{2\mu_+\mu_-} \pm \left[1 - \frac{(\mu_+ + \mu_-)^2 Q^2\omega^2}{4\mu_+^2\mu_-^2}\right]^{1/2} \tag{B.5}$$

$$\gamma_{\pm} = 1 + i\frac{Q\omega}{\mu_+} - \xi_{\pm} \tag{B.6}$$

$$\vartheta_{\pm} = \frac{L\xi_{\pm}^{1/2}}{\sqrt{2}}\left[\frac{e^2\nu}{2\varepsilon_0\varepsilon_{\mathrm{m}}k_{\mathrm{B}}T}\right]^{1/2} \tag{B.7}$$

with $i = \sqrt{-1}$, $\varepsilon_0 = 8.85 \times 10^{-12}$ $\mathrm{C^2J^{-1}m^{-1}}$ (absolute permittivity of vacuum), ε_{m} = dielectric constant of the medium PEO (assumed to be independent of ω), and $Q = \varepsilon_0\varepsilon_{\mathrm{m}}/e\nu$.

The dynamic dielectric constant $\varepsilon'(\omega)$ and dielectric loss $\varepsilon''(\omega)$ are related to the imaginary and real parts of y^* given by Eqs. (B.1)-(B.7):

$$\varepsilon'(\omega) = \varepsilon_{\mathrm{m}} + \frac{L}{\omega\varepsilon_0}\,\mathrm{Im}\{y^*(\omega)\} \tag{B.8}$$

$$\varepsilon''(\omega) = \frac{L}{\omega\varepsilon_0}\,\mathrm{Re}\{y^*(\omega)\} \tag{B.9}$$

Fig. B.1(a) shows $\varepsilon'(\omega)$ and $\varepsilon''(\omega)$ calculated from Eqs. (B.1)-(B.9) with typical values of $\varepsilon_{\mathrm{m}} = 10$, $\mu = 1.0 \times 10^{-9}\mathrm{m^2V^{-1}s^{-1}}$ $(= \mu_+ = \mu_-)$, $\nu = 1.0 \times 10^{24}\mathrm{m^{-3}}$, and various values of the electrode gap L.[17] At high ω, ε' is independent of ω and coincides with ε_{m} while ε'' is proportional to ω^{-1}. These features correspond to the (apparent) DC conduction, and ε' and ε'' in this DC region are independent of L. However, at low ω, ε' and ε'' exhibit characteristic features of the electrode polarization (EP): Namely, ε' significantly increases and then approaches the zero-frequency plateau whereas ε'' exhibits a peak with decreasing ω. It should be noted that ε' and ε'' are dependent on L in this EP region.

In Fig. B.1(b), the ε' and ε'' shown in Fig. B.1(a) are normalized by an intensity factor I being proportional to L and plotted against a normalized frequency $\omega\tau_{\mathrm{EP}}$, where τ_{EP} is the EP relaxation time being proportional to L. In this normalized plots, the ε' and ε'' for various L values are collapsed into a master curve in the EP region, and this master curve is well described by the Debye-type single relaxation function shown with the solid curve. Thus, ε' and ε'' deduced from the Macdonald theory can be satisfactorily

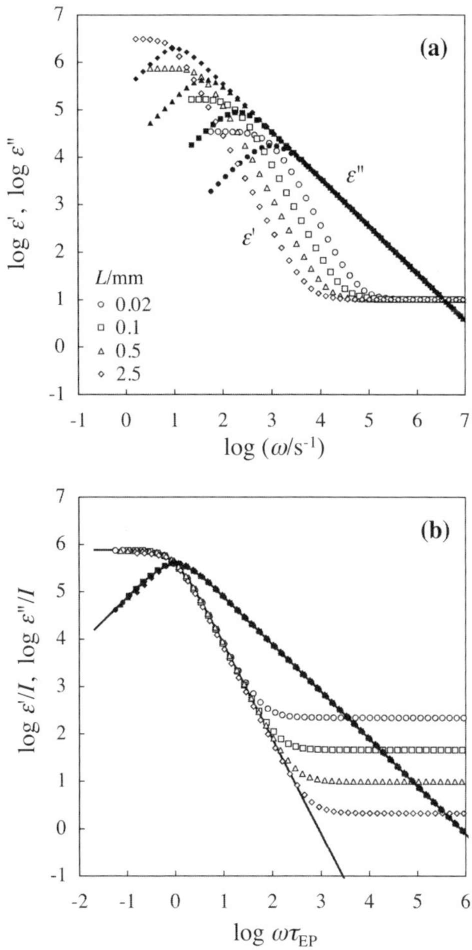

Fig. B.1. Dynamic dielectric constant and dielectric loss calculated with the Macdonald theory.[15] Redrawn, with permission, from Ref. 17.

cast in the Debye form,[9,16,17]

$$\varepsilon'_{\mathrm{EP}}(\omega) = \varepsilon_{\mathrm{m}} + \Delta\varepsilon_{\mathrm{EP}} \frac{1}{1+\omega^2\tau_{\mathrm{EP}}^2} \tag{B.10}$$

$$\varepsilon''_{\mathrm{EP}}(\omega) = \Delta\varepsilon_{\mathrm{EP}} \frac{\omega\tau_{\mathrm{EP}}}{1+\omega^2\tau_{\mathrm{EP}}^2} \tag{B.11}$$

The EP relaxation time and intensity, τ_{EP} and $\Delta\varepsilon_{\mathrm{EP}}$ appearing in

Eqs. (B.10) and (B.11), depend on μ, ν, and L as[9,17]

$$\tau_{\rm EP} \propto L\mu^{-1}\nu^{-1/2}, \qquad \Delta\varepsilon_{\rm EP} \propto L\nu^{1/2} \tag{B.12}$$

Note that the high-ω DC conduction behavior being independent of L is also described by Eq. (B.11); $\varepsilon''_{\rm EP} = \Delta\varepsilon_{\rm EP}/\tau_{\rm EP}\omega \propto \mu\nu/\omega$ at $\omega \gg 1/\tau_{\rm EP}$.

References

1. G. G. Fuller, *Optical Rheometry* (Oxford, New York, 1995).
2. R. J. Roe, *Methods of X-Ray and Neutron Scattering in Polymer Science* (Oxford, New York, 2000).
3. F. Kremer and A. Scönhals, *Broadband Dielectric Spectroscopy* (Springer, Berlin, 2003).
4. H. Watanabe, Dielectric Relaxation of Type-A Polymers in Melts and Solutions, *Macromol. Rapid Commun.* **22**, 127 (2001).
5. Y. Matsumiya, H. Watanabe, T. Inoue, and K. Osaki, Rheo-Dielectric Behavior of Oligostyrene and Polyisoprene, *Macromolecules* **21**, 7973 (1998).
6. H. Watanabe, S. Ishida, and Y. Matsumiya, Rheodielectric Behavior of Entangled *cis*-Polyisoprene under Fast Shear, *Macromolecules* **35**, 8802 (2002).
7. H. Watanabe, T. Sato, M. Hirose, K. Osaki, and M. L. Yao, Rheo-Dielectric Behavior of Low Molecular Weight Liquid Crystals. 1. Behavior of Nematic 5CB and 7CB, *Rheol. Acta* **37**, 519 (1998).
8. H. Watanabe, T. Sato, M. Hirose, K. Osaki, and M. L. Yao, Rheo-Dielectric Behavior of Low Molecular Weight Liquid Crystal 2. Behavior of 8CB in Nematic and Smectic States, *Rheol. Acta* **38**, 100 (1999).
9. Y. Matsumiya, N. P. Balsara, J. B. Kerr, T. Inoue, and H. Watanabe, In-Situ Dielectric Characterization of Poly(ethylene oxide) Melts Containing Lithium Perchlorate under Steady Shear Flow, *Macromolecules* **37**, 544 (2004).
10. H. Watanabe, Y. Matsumiya, M. Kakiuchi, and Y. Aoki, Rheo-dielectric Behavior of Carbon Black Suspensions, *J. Soc. Rheol. Japan (Nihon Reoroji Gakkaishi)* **29**, 77 (2001).
11. H. Watanabe, T. Sato, Y. Matsumiya, T. Inoue, and K. Osaki, Rheo-Dielectrics: Its Applicability, *J. Soc. Rheol. Japan (Nihon Reoroji Gakkaishi)* **27**, 121 (1999).
12. H. Watanabe, Y. Matsumiya, and T. Inoue, Rheo-Dielectrics in Oligomeric and Polymeric Fluids: A Review of Recent Findings, *J. Phys.: Condens. Matter* **15**, S909 (2003).
13. S. Höfl, F. Kremer, H. W. Spiess, M. Wilhelm, and S. Kahle, Effect of Large Amplitude Oscillatory Shear (LAOS) on the Dielectric Response of 1,4-*cis*-polyisoprene, *Polymer* **47**, 7282 (2006).
14. S. Capaccioli, D. Prevosto, A. Best, A. Hanewald, and T. Pakula, Applications of the Rheo-dielectric Technique, *J. Non-Cryst. Solids* **353**, 4267 (2007).
15. J. R. Macdonald, Theory of ac Space-charge Polarization Effects in Photoconductors, Semiconductors, and Electrolytes, *Phys. Rev.* **92**, 4 (1953).

16. D. Fragiadakis, S. Dou, R. H. Colby, and J. Runt, Molecular Mobility, Ion Mobility, and Mobile Ion Concentration in Poly(ethylene oxide)-based Polyurethane Ionomers, *Macromolecules* **41**, 5723 (2008).
17. A. Lee, Y. Matsumiya, H. Watanabe, K. H. Ahn, and S. J. Lee, Multiple Representation of Linear Dielectric Response, *J. Soc. Rheol. Japan (Nihon Reoroji Gakkaishi)* **38**, 149 (2010).
18. R. Kubo, Statistical-Mechanical Theory of Irreversible Processes. I. General Theory and Simple Applications to Magnetic and Conduction Problems, *J. Phys. Soc. Japan* **12**, 570 (1957).
19. R. Cole, Correlation Function Theory of Dielectric Relaxation, *J. Chem. Phys.* **42**, 637 (1965).
20. D. Evans and P. Morris, *Statistical Mechanics of Nonequilibrium Liquids* (Academic Press, London, 1990).
21. D. J. Evans, A. Baranyai and S. Sarman, Response Theory of Symmetry Restricted Interactions, *Mol. Phys.* **76**, 661 (1992).
22. T. Uneyama, Y. Masubuchi, K. Horio, Y. Matsumiya, H. Watanabe, J. A. Pathak, and C. M. Roland, A Theoretical Analysis of Rheodielectric Response of Type-A Polymer Chains, *J. Polym. Sci. Part B: Polym. Phys.* **47**, 1039 (2009).
23. T. Inoue, H. Okamoto, and K. Osaki, Birefringence of Amorphous Polymers. I. Dynamic Measurement on Polystyrene, *Macromolecules* **24**, 5670 (1991).
24. T. Inoue, H. Matsui, and K. Osaki, Molecular Origin of Viscoelasticity and Chain Orientation of Glassy Polymers, *Rheol. Acta* **36**, 239 (1997).
25. T. Inoue, T. Uematsu, and K. Osaki, The Significance of the Rouse Segment: Its Concentration Dependence, *Macromolecules* **35**, 820 (2002).
26. H. Watanabe, Slow Dynamics in Homopolymer Liquids, *Polym. J.* **41**, 929 (2009).
27. T. Inoue, H. Okamoto, and K. Osaki, Large Deformation of Polycarbonate near the Glass-transition Temperature, *Macromolecules* **25**, 7069 (1992).
28. T. Inoue, D. Ryu, and K. Osaki, A Rheo-optical Study on Polystyrene under Large Tensile Deformation around the Glass Transition Temperature, *Macromolecules* **31**, 6977 (1998).
29. J. D. Ferry, *Viscoelastic Properties of Polymers*, 3rd. Ed. (Wiley, New York, 1980).
30. M. Rubinstein and R. Colby, *Polymer Physics* (Oxford, New York, 2003).
31. W. W. Graessley, *Polymeric Liquids and Networks: Dynamics and Rheology* (Garland Science, New York, 2008).
32. J. J. Magda, C. S. Lee, S. J. Muller, and R. G. Larson, Rheology, Flow Instabilities, and Shear-Induced Diffusion in Polystyrene Solutions, *Macromolecules* **26**, 1696 (1993).
33. E. F. Brown, W. R. Burghardt, H. Kahvand, and D. C. Venerus, Comparison of Optical and Mechanical Measurements of Second Normal Stress Difference Relaxation Following Step Strain, *Rheol. Acta* **34**, 221 (1995).
34. K. Adachi and T. Kotaka, Dielectric Normal Mode Relaxation, *Prog. Polym. Sci.* **18**, 585 (1993).
35. H. Watanabe, Viscoelasticity and Dynamics of Entangled Polymers, *Prog.*

Polym. Sci. **24**, 1253 (1999).
36. G. Ianniruberto and G. Marrucci, On Compatibility of the Cox-Merz Rule with the Model of Doi and Edwards, *J. Non-Newtonian Fluid Mech.* **65**, 241 (1996).
37. D. W. Mead, R. G. Larson, and M. Doi, A Molecular Theory for Fast Flows of Entangled Polymers, *Macromolecules* **31**, 7895 (1998).
38. T. C. B. McLeish, Tube Theory of Entangled Polymer Dynamics, *Adv. Phys.* **51**, 1379 (2002).
39. H. Yamakawa, *Modern Theory of Polymer Solutions* (Harper & Row, New York, 1971).
40. M. Doi and S. F. Edwards, *The Theory of Polymer Dynamics* (Clarendon, Oxford, 1986).
41. G. Marrucci, Relaxation by Reptation and Tube Enlargement: A Model for Polydisperse Polymers, *J. Polym. Sci. Polym. Phys. Ed.* **23**, 159 (1985).
42. H. Watanabe, Y. Matsumiya, and T. Inoue, Dielectric and Viscoelastic Relaxation of Highly Entangled Star Polyisoprene: Quantitative Test of Tube Dilation Model, *Macromolecules* **35**, 2339 (2002).
43. H. Watanabe, S. Ishida, Y. Matsumiya, and T. Inoue, Viscoelastic and Dielectric Behavior of Entangled Blends of Linear Polyisoprenes Having Widely Separated Molecular Weights: Test of Tube Dilation Picture, *Macromolecules* **37**, 1937 (2004).
44. H. Watanabe, S. Ishida, Y. Matsumiya, and T. Inoue, Test of Full and Partial Tube Dilation Pictures in Entangled Blends of Linear Polyisoprenes, *Macromolecules* **37**, 6619 (2004).
45. H. Watanabe, T. Sawada, and Y. Matsumiya, Constraint Release in Star/Star Blends and Partial Tube Dilation in Monodisperse Star Systems, *Macromolecules* **39**, 2553 (2006).
46. X. Qiao, T. Sawada, Y. Matsumiya, and H. Watanabe, Constraint Release in Moderately Entangled Monodisperse Star Polyisoprene Systems, *Macromolecules* **39**, 7333 (2006).
47. E. Somma, O. Valentino, G. Titomanlio, and G. Ianniruberto, Parallel Superposition in Entangled Polydisperse Polymer Melts: Experiments and Theory, *J. Rheol.* **51**, 987 (2007).
48. Y. Masubuchi, J. Takimoto, K. Koyama, G. Ianniruberto, F. Greco, and G. Marrucci, Brownian Simulations of A Network of Reptating Primitive Chains, *J. Chem. Phys.* **115**, 4387 (2001).
49. Y. Masubuchi, G. Ianniruberto, F. Greco, and G. Marrucci, Entanglement Molecular Weight and Frequency Response of Sliplink Networks, *J. Chem. Phys.* **119**, 6925 (2003).
50. Y. Masubuchi, G. Ianniruberto, F. Greco, and G. Marrucci, Quantitative Comparison of Primitive Chain Network Simulations with Literature Data of Linear Viscoelasticity for Polymer Melts, *J. Non-Newtonian Fluid Mech.* **149**, 87 (2008).
51. Y. Masubuchi, H. Watanabe, G. Ianniruberto, F. Greco and G. Marrucci, Primitive Chain Network Simulations on Dielectric Relaxation of Linear Polymers under Shear Flow, *J. Soc. Rheol. Japan (Nihon Reoroji Gakkaishi)*

32, 197 (2004).
52. G. Ianniruberto and G. Marrucci, Convective Orientational Renewal in Entangled Polymers, *J. Non-Newtonian Fluid Mech.* **95**, 363 (2000).
53. P. E. Cladis and S. Torza, Stability of Nematic Liquid-crystals in Couette Flow, *Phys. Rev. Lett.* **35**, 1283 (1975).
54. W. H. de Jeu and J. van der Veen, Instabilities in Electric Fields of a Nematic Liquid Crystal with Large Negative Dielectric Anisotropy, *Phys. Lett.* **44A**, 277 (1973).
55. W. H. de Jeu, *Physical Properties of Liquid Crystalline Materials* (Gordon and Breach, New York, 1980).
56. M. Davies, R. Moutran, A. H. Price, M. S. Beevers, and G. Williams, Dielectric and Optical Studies of a Nematogen (4,4-n-heptyl-cyanobiphenyl), *J. Chem. Soc. Faraday II* **72**, 1447 (1976).
57. D. Lippens, J. P. Parneix, and A. Chapoton, Study of 4 Heptyl 4′ Cynanobiphenyl Using Analysis of Its Dielectric Properties, *J. Phys. (Paris)* **38**, 1465 (1977).
58. R. G. Larson, K. I. Winey, S. S. Patel, H. Watanabe, and R. Bruinsma, The Rheology of Layered Liquids: Lamellar Block Copolymers and Smectic Liquid Crystals, *Rheol. Acta* **32**, 245 (1993).
59. P. T. Mather, D. S. Pearson, and R. G. Larson, Flow Patterns and Disclination-Density Measurements in Sheared Nematic Liquid Crystals I: Flow-aligning 5CB, *Liquid Crystals* **20**, 527 (1996).
60. P. T. Mather, D. S. Pearson, and R. G. Larson, Flow Patterns and Disclination-Density Measurements in Sheared Nematic Liquid Crystals II: Tumbling 8CB, *Liquid Crystals* **20**, 539 (1996).
61. C. R. Safinya, E. B. Sirota, R. Plano, and R. F. Bruinsma, A Non-equilibrium Synchrotron X-ray Study of A Liquid-crystal Phase-transition under Shear Flow, *J. Phys.: Condensed Matter* **2**, SA365 (1990).
62. C. R. Safinya, E. B. Sirota, and R. J. Plano, Nematic to Smectic-A Phase-transition under Shear Flow: A Nonequilibrium Synchrotron X-ray Study, *Phys. Rev. Lett.* **66**, 1986 (1991).
63. K. Okano ed., *Encyclopedia of Liquid Crystals (Ekisho-Jiten)*, (Baifu-kan, Tokyo, 1989).
64. R. H. Colby, C. K. Ober, J. R. Gillmor, R. W. Connelly, T. Duong, G. Galli, and M. Laus, Smectic Rheology, *Rheol. Acta* **36**, 498 (1997).
65. M. Armand, Polymer Solid Electrolytes — An Overview, *Solid State Ionics* **9**, 745 (1983).
66. J. M. G. Cowie and K. Sadaghianizadeh, Effect of Side Chain Length and Crosslinking on the AC Conductivity of Oligo(ethyleneoxide) Comb branch Polymer Salt Mixtures, *Solid State Ionics* **42**, 243 (1990).
67. C. D. Robitaille and D. Fauteux, Phase Diagrams and Conductivity Characterization of Some PEO-LiX Electrolytes, *J. Electrochem. Soc.* **133**, 315 (1986).
68. M. Watanabe and N. Ogata, Ionic Conductivity of Polymer Electrolytes and Future Applications, *Br. Polym. J.* **20**, 181 (1988).
69. M. Watanabe, T. Endo, A. Nishimoto, K. Miura, and M. Yanagida, High

Ionic Conductivity and Electrode Interface Properties of Polymer Electrolytes Based on High Molecular Weight Branched Polyether, *J. Power Sources* **81**, 786 (1999).
70. M. Kakihana, S. Schantz, L. M. Torell, and J. R. Stevens, Dissociated Ions and Ion-Ion Interactions in Poly(ethylene oxide) Based $NaCF_3SO_3$ Complexes, *Solid State Ionics* **40/41**, 641 (1990).
71. K. Mitani and K. Adachi, Ionic Conduction in LiSCN/dimethylformamide poly(propylene oxide) System I. Dissociation Equilibria of LiSCN, *J. Polym. Sci. Part B: Polym. Phys.* **33**, 937 (1995).
72. O. Borodin and G. D. Smith, Molecular Dynamics Simulations of Poly(ethylene oxide)/LiI Melts. 1. Structural and Conformational Properties, *Macromolecules* **31**, 8396 (1998).
73. O. Borodin, R. Doughlas, G. Smith, E. M. Eyring, and S. Petrucci, Microwave Dielectric Relaxation, Electrical Conductance, and Ultrasonic Relaxation of $LiPF_6$ in Poly(ethylene oxide) Dimethyl Ether-500, *J. Phys. Chem. B* **106**, 2140 (2002).
74. S. Petrucci and E. M. Eyring, Microwave Dielectric Relaxation, Electrical Conductance and Ultrasonic Relaxation of $LiClO_4$ in Polyethylene Oxide Dimethyl Ether-500 (PEO-500), *Phys. Chem. Chem. Phys.* **4**, 6043 (2002).
75. N. J. R. v. E. Hommes, M. Bühl, and P. v. R. J. Schleyer, and Y. D. Wu, A Comparative Theoretical Study of the Allyl Alkali Metals, *J. Organomet. Chem.* **409**, 307 (1991).
76. Y. Matsumiya, N. P. Balsara, J. B. Kerr, T. Inoue, and H. Watanabe, In Situ Dielectric Characterization of Poly(ethylene oxide) Melts Containing Lithium Perchlorate under Steady Shear Flow, *Macromolecules* **37**, 7064 (2004).
77. W. B. Russel, D. A. Savile, and W. R. Showalter, *Colloidal Dispersions* (Cambridge, London, 1989).
78. R. G. Larson, *The Structure and Rheology of Complex Fluids* (Oxford University Press, New York, 1999).
79. W. A. Deheer, W. S. Bacsa, A. Chatelain, T. Gerfin, R. Humphrey-Baker, L. Forrol, and D. Ugarte, Aligned Carbon Nanotube Films: Production and Optical and Electronic Properties, *Science* **268**, 845 (1995).
80. P. Potschke, M. Abdel-Goad, I. Alig, S. Dudkin, and D. Lellinger, Rheological and Dielectrical Characterization of Melt Mixed Polycarbonate-Multiwalled Carbon Nanotube Composites, *Polymer* **45**, 8863 (2004).
81. A. Nogales, G. Broza, Z. Roslaniec, K. Schulte, I. Sics, B. S. Hsiao, A. Sanz, M. C. Garcia-Gutierrez, D. R. Rueda, C. Domingo, and T. A. Ezquerra, Low Percolation Threshold in Nanocomposites Based on Oxidized Single Wall Carbon Nanotubes and Poly(butylene terephthalate), *Macromolecules* **37**, 7669 (2004).
82. Z. M. Dang, L. Wang, Y. Yin, Q. Zhang, and Q. Q. Lei, Giant Dielectric Permittivities in Functionalized Carbon-Nanotube/Electroactive-Polymer Nanocomposites, *Adv. Mater.* **19**, 852 (2007).
83. H. Risken, *The Fokker-Planck Equation*, 2nd edn. (Springer, Berlin, 1989).
84. H. Watanabe, O. Urakawa, and T. Kotaka, Slow Dielectric Relaxation of Entangled Linear *cis*-Polyisoprenes with Asymmetrically Inverted Dipoles.

1. Bulk Systems, *Macromolecules* **26**, 5073 (1993).

85. H. Watanabe, O. Urakawa, and T. Kotaka, Slow Dielectric Relaxation of Entangled Linear *cis*-Polyisoprenes with Asymmetrically Inverted Dipoles. 2. Behavior in a Short Matrix, *Macromolecules* **27**, 3525 (1994).
86. H. Watanabe, Y. Matsumiya, K. Osaki, and M. L. Yao, Dynamics of Dipole-Inverted *cis*-Polyisoprene Chains in a Matrix of Long, Entangling Chains. 2. Effects of Constraint Release on the Coherence of the Subchain Motion, *Macromolecules* **31**, 7538 (1998).
87. T. Uneyama, K. Horio, and H. Watanabe, Anisotropic Mobility Model for Polymers under Shear and its Linear Response Functions, to appear in *Phys. Rev. E*, arXiv:1104.3744 (2011).
88. T. Harada and S. Sasa, Equality Connecting Energy Dissipation with a Violation of the Fluctuation-Response Relation, *Phys. Rev. Lett.* **95**, 130602 (2005).
89. T. Speck and U. Seifert, Restoring a Fluctuation-Dissipation Theorem in a Nonequilibrium Steady State, *Europhys. Lett.* **74**, 391 (2006).
90. R. Chetrite, G. Falkovich, and K. Gawędzki, Fluctuation Relations in Simple Examples of Non-Equilibrium Steady States, *J. Stat. Mech.* **2008**, P08005 (2008).
91. M. Baiesi, C. Maes, and B. Wynants, Fluctuations and Response of Nonequilibrium States, *Phys. Rev. Lett.* **103**, 010602 (2009).

Chapter 3

Morphology and Rheology of Immiscible Polymer Blends in Electric and Shear Flow Fields

Hiroshi Orihara*

Division of Applied Physics, Hokkaido University, Sapporo, Hokkaido 060-8628, Japan

Three-dimensional images are presented of immiscible polymer blends under external fields such as electric field and shear flow, obtained using a new system combining a rheometer and a confocal scanning laser microscope. This system, which enables structural observation and shear stress measurement to be performed simultaneously, is explained in detail. Immiscible polymer blends exhibit interesting rheological behavior related to morphological changes. In this study, the relationship between morphology and rheology is investigated, and the interfacial tension and the Maxwell stress are shown to play important roles. Four topics are presented, which were investigated by using the new system. First, under an electric field without shear flow, we study the droplet coalescence process caused by the induced dipole-dipole interaction. Second, we show results for the storage shear modulus of the columnar structure formed by droplet coalescence and elongation along the applied electric field. Next, the relationship is investigated between the structural evolution and shear stress during the transient process, when a blend is subjected to a step electric field under shear flow. Lastly, we discuss scaling properties within the above transient process.

1. Introduction

Immiscible fluid blends (emulsions) exhibit many interesting rheological characteristics, which are closely related to changes in morphology. When subjected to a flow, blends undergo changes through complex processes, such as deformation, breakup and coalescence. Needless to say, in immiscible blends, the interface between two phases plays a crucial role in determining the rheological properties, and this interface has excess energy, giving

*orihara@eng.hokudai.ac.jp

rise to surface tension. Recent theoretical and experimental advances in understanding the relationships are reviewed by Tucker and Moldenaers.[1] The excess stress, due to the surface tension, can be expressed in a simple form:[2–4]

$$\sigma_{\alpha\beta}^{(\mathrm{if})} = -\Gamma q_{\alpha\beta}, \tag{1}$$

with

$$q_{\alpha\beta} = \frac{1}{V}\int_S \left(n_\alpha n_\beta - \frac{1}{3}\delta_{\alpha\beta}\right) dS, \tag{2}$$

where Γ is the interfacial tension and $\mathbf{n}$ is the unit vector normal to the interface, V is the system volume and the integral is done over the whole interface S in the system. $q_{\alpha\beta}$ is a traceless and symmetric tensor, called the interface tensor, and represents the anisotropy of the blend containing the interface. Recently, a detailed experimental evaluation of Eq. (1) was made by Takahashi *et al.*,[5,6] in which the transient shear stress, after applying a large step strain, was measured using a rheometer. They also observed the shape change of a droplet under the same magnitude of step strain, using a sliding-plate apparatus specially designed to conduct this observation. The droplet shape observed in the process was classified into three types: ellipsoid, rod-like shape and dumbbell,[7] and the corresponding interface tensors were calculated. Good agreement was obtained between the measured stress and the stress calculated from the interface tensor using Eq. (1).

In general, the stress tensor for a mixture of two Newtonian fluids, where inertial force is negligible, can be written as[2,8]

$$\sigma_{\alpha\beta} = -p\delta_{\alpha\beta} + 2\eta_{\mathrm{m}} D_{\alpha\beta} - \frac{(\eta_{\mathrm{m}} - \eta_{\mathrm{d}})}{V}\int_S (u_\alpha n_\beta + u_\beta n_\alpha)\, dS + \sigma_{\alpha\beta}^{(\mathrm{if})}, \tag{3}$$

where p is the pressure, $D_{\alpha\beta}$ the rate-of-deformation, $\mathbf{u}$ the velocity at the interface, and η_{m} and η_{d} are the viscosities of matrix and droplet, respectively. Apart from $\sigma_{\alpha\beta}^{(\mathrm{if})}$, which is given in Eq. (1), there are three terms on the right-hand side. The first term represents the isotropic pressure, and the second term represents the viscous stress of the matrix. The third term, called the interface velocity term, comes from the viscous flows within the matrix and the dispersed phases, due to local flow caused by the displacement of the interface. The interface velocity term has been studied both theoretically and experimentally.[9–13] Note that, for a blend of two fluids with equal viscosity, this term vanishes and the total stress can be obtained by determining the interface tensor.

In 1991, Doi and Ohta[14] developed a semi-phenomenological theory for a blend of two Newtonian fluids with equal viscosity and a volume fraction close to $\phi = 0.5$. To do this, they used the interface tensor, $q_{\alpha\beta}$, and the interfacial area per unit volume, $Q = V^{-1} \int_s dS$, as state variables and derived evolution equations for these variables. One can then get the time dependence of the total stress by solving the equations and substituting the solutions into Eq. (3). In the Doi-Ohta model there is no intrinsic length scale, implying the absence of an intrinsic timescale. This is a remarkable feature of the model and predicts interesting scaling properties. Details of them will be described in Sec. 6.

As mentioned above, flows bring about various morphological changes in immiscible blends. However, in addition to flows, an electric field may also be effective in bringing about morphological changes. In fact, under application of an electric field, when two blended fluids have mismatched electrical properties such as dielectric constant and conductivity remarkable changes have been reported.[15–19] At the same time, the apparent viscosity increases. This phenomenon is called the electrorheological (ER) effect. ER fluids have been extensively investigated because of their potential applicability in active control devices such as dampers, shock absorbers, clutches, and brakes. Besides immiscible blends having the ER effect, there are mainly two types of ER fluids: the particle-dispersed and homogeneous types. The former, for example, the suspension of silica particles in a mineral oil, is well known.[20,21] In the latter category, polar liquids,[22] nonpolar liquids,[23] low molecular liquid crystals[24] and lyotropic polymeric liquid crystals[25] have been reported. Immiscible blend ER fluids may be classified into a third type. This was first found by Inoue *et al.* for a blend of a liquid crystalline polymer (LCP) and a dimethylsiloxane (DMS).[15] However, in the initial stages of their study, it was proposed that the ER effect within the blend was due to a mechanism whereby the liquid crystalline domains formed a network, connected by flexible polymer chains which anchor neighboring domains using one or more liquid crystalline groups within the chain. Under an electric field, the network prevents the liquid crystalline domains from disordering, and increases the viscosity of the system. Later on, it was clarified that the ER effect appears in the isotropic phase and, furthermore, microscopic observations found a macroscopic domain structure consisting of droplets which change size and shape under the electric field. The results suggest that this macroscopic structure consisting of droplets may play an important role in the ER effect.

Detailed studies, using blends containing different kinds of LCP and

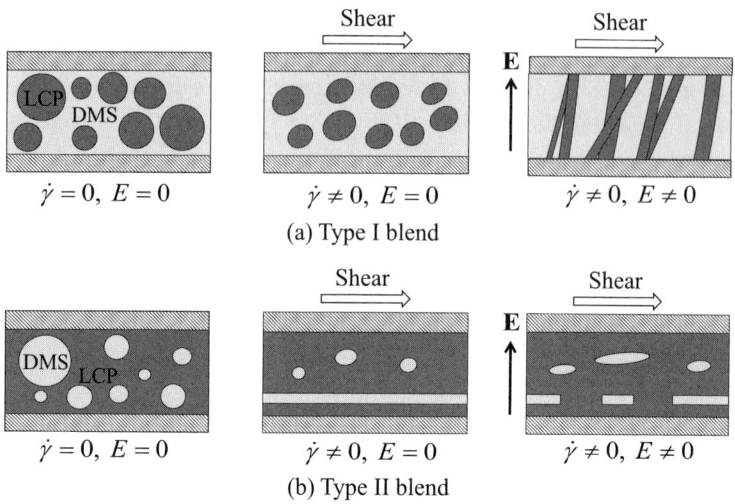

Fig. 1. Illustrations of (a) Type I blend and (b) Type II blend.

DMS, revealed two different types of immiscible blend ER fluids.[26] Here, we assume that the viscosity of LCP is larger than that of DMS. For both types there are small droplets dispersed in the matrix; however in one type (Type I, Fig. 1(a)) the droplets consist of LCP, while in the other (Type II, Fig. 1(b)) they consist of DMS. In Type I, the apparent viscosity is low, as the droplets can flow easily, sliding over one another in the matrix of low viscosity DMS. When subjected to an electric field, the droplets of the high viscosity LCP become elongated along the field, due to the Maxwell stress exerted on the interface between the LCP and DMS, and join to form bridges between the electrodes. This results in an apparent increase of viscosity.[16,27,28] Conversely, for a Type II blend it was found that, under a shear flow and in the absence of an electric field, thin layers of DMS are formed between the plates. These layers are possibly formed by the shear flow, which makes the DMS droplets stretch and coalesce such that the layers grow. The area of the thin layers increases as the shear rate is raised. These DMS layers reduce the apparent viscosity. When subjected to an electric field, the layers become unstable and degrade, resulting in the ER effect.[17,29]

Direct observations of any structural change are obviously important in the study of immiscible blends. For simple and sparse structures, normal optical microscopy can be used effectively.[7,28,30] In contrast, for complex structures such as a network structure, confocal scanning laser microscopy (CSLM) becomes a powerful tool and characteristic quantities related to interface shape such as the mean and Gaussian curvatures have been successfully obtained.[31–33] Recent development of three-dimensional (3D) imaging using CSLM even enables us to perform real-time three-dimensional (3D) observations.

Recently, we have constructed a new system consisting of a rheometer and a CSLM which, under shear flows and electric fields, enables us to observe 3D structures and measure shear stress at the same time. In the next section this system is explained in detail, after which, four topics are presented where the results are obtained by using the new system. First, we study the droplet coalescence process under an electric field without shear flow. Second, we present results of the storage shear modulus of a columnar structure formed by the droplet coalescence. Next, the relationship between the structure evolution and the shear stress during the transient process, when a blend is subjected to a step electric field under shear flow, is investigated. Finally, we discuss scaling properties within the above transient process.

2. Experimental System for Observing Three-Dimensional Structures

We have combined a rheometer (MCR301, Anton Paar) and a CSLM (Fig. 2). The rheometer was specially designed for observations. We placed the CSLM under a rotating plate of the rheometer. Our CSLM consists of an inverted optical microscope (IX71, Olympus) and a confocal scanner unit with a Nipkow spinning disk (CSU22, Yokogawa). Images are captured with a video camera with an image intensifier (Kinor, Seika). In this system the maximum frame rate is 1000 fps (frames per second) and the frame size is 512 $\times$ 512 pixels. A bottom metal stage of the rheometer was constructed just below the upper rotating plate of the rheometer. Since we used a water-immersed objective lens with a short working distance (40x, NA 1.2, Olympus) to obtain high-resolution images, we employed a 100 μm-thick glass plate with a transparent electrode ITO (indium tin oxide) as a bottom plate, which was supported by the metal stage with a small hole (see Fig. 2(a)). The bottom glass plate was temperature-controlled within $\pm 0.5°$C. The sample was sandwiched between the bottom glass plate and the rotating plate of the rheometer. The shear rate was defined as the one at the edge of the rotating disk and the shear stress there was calculated

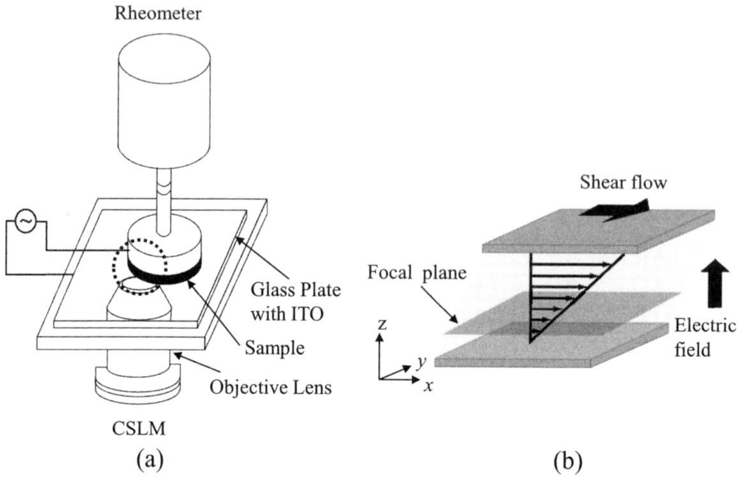

Fig. 2. (a) Schematic illustration of system combining a rheometer and CSLM, and (b) details of the sample space.

from the measured torque by assuming that the fluid is Newtonian because a parallel-plate rheometer was used in the experiment.

We constructed 3D images, by using a piezo-actuator (P-721.10, PI) attached to the objective lens for vertical displacement, where the x, y and z axes are taken along the flow, vorticity and velocity gradient (the electric field) directions, respectively, as shown in Fig. 2 (b). We applied a triangular voltage to the piezo-actuator so that the focal plane could move at a constant speed. For example, if we apply a triangular voltage of 5 Hz and set the frame rate at 500 fps, 50 2D images can be captured per half period, from which one 3D image is constructed. Therefore, we can obtain 5 3D images per second. However, it should be noted that during the capture of one 3D image, a shear rate of, to say, 1 s^{-1} brings about a total strain of 10 %. This is not negligible, and correction should be made as described in what follows. It is assumed that the focal plane is scanned between $z = z_1$ and z_2, where $z_2 > z_1$, and the glass surface is located at $z = 0$, as shown in Fig. 3. First, let us consider the upward scan from z_1 to z_2 (Fig. 3(a)). It is assumed that the focal plane is at $z = z_1$ at $t = 0$. At time t, the focal plane is at $z = z_1 + v_s t$, where v_s is the scan speed or the vertical speed of the focal plane. As the flow velocity at z is $\dot{\gamma} z$, the horizontal displacement is given as $\Delta x = \dot{\gamma} z t = \dot{\gamma}/v_s \cdot z(z - z_1)$. Thus, the 2D image at z should be moved back by Δx. Note that this correction

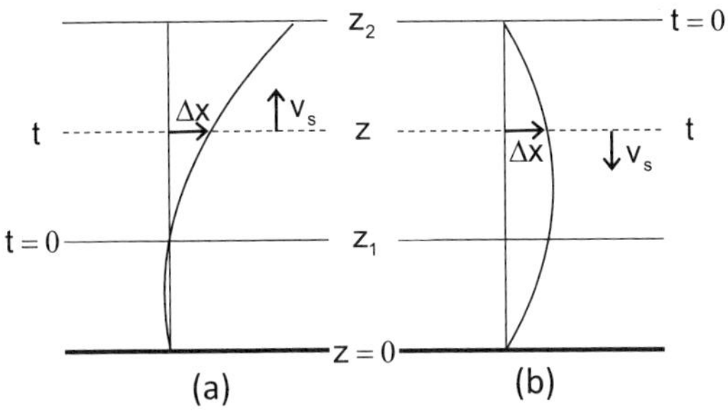

Fig. 3. (a) Upward scan and (b) downward scan, where the focal plane is located at z and v_s is the scan speed.

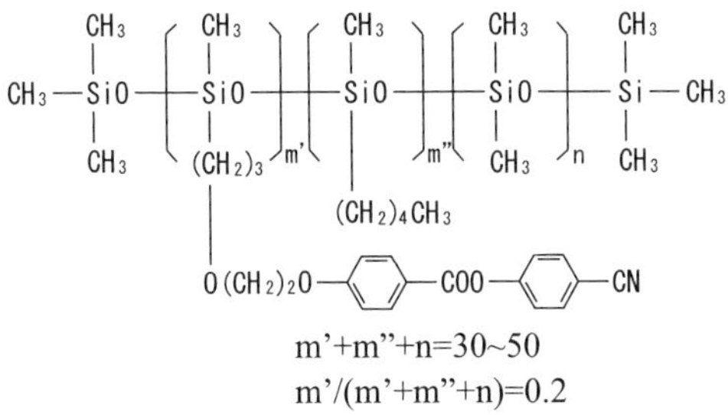

Fig. 4. Chemical structure of LCP.

is most effective when all of the points in the blend move in accordance with the shear strain. For the downward scan (Fig. 3(b)), Δx is given as $\dot{\gamma}/v_s \cdot z(z_2 - z)$. It is clear that the correction for the downward scan is smaller than that for the upward one. As a result, the 3D images are constructed from the 2D images obtained by scanning downward.

We employed liquid crystalline polymers (LCP)[15] (Fig. 4) as one component of blends. All the experiments were performed at around 25°C and LCP is in the isotropic phase at the temperature, indicating that the ER effect should not be related to the orietational change of mesogens like the liquid crystalline ER fluids. However, it should be noted that mesogens of LCP play a role in enhancing the dielectric constant and conductivity. In experiments, two LCPs with different mesogen fractions were used. As the other component polyisobutylene (PIB) was employed, which was a mixture of two PIBs with different molecular weights, because it has low dielectric constant and conductivity in comparison to LCP, and the refractive indices of the fluids are closely matched, which is necessary for clear observations of structures with a CSLM. We used different blends of LCP and PIB, the details of which will be described in each section. For CSLM observations a small amount of fluorescent dye, IANBD amide (Molecular Probes), was added to LCP before mixing with PIB. It was confirmed that the dye did not dissolve in PIB. The wavelength of excitation was 488 nm.

3. Droplet Coalescence Process Under Electric Fields

In this section, we present the results of observing droplet coalescence processes when immiscible polymer blends are subjected to electric fields and investigate the mechanism of droplet growth.[34] Under an electric filed an electric dipole moment is induced in each droplet if there is mismatch of electric properties such as dielectric constant and conductivity between the two phases, and then due to the dipole-dipole interaction the droplets attract one another to coalesce. One proposed application of this phenomenon is, for example, to remove water droplets from oils. Several studies have observed experimentally the coalescence process of a pair of droplets subjected to an electric field in order to elucidate the coalescence mechanism.[35–37] However, the coalescence process of multiple droplets has not yet been studied, which is furthermore one of the most important elementary processes in the ER effect of the immiscible blend ER fluid.

In colloidal systems, on the other hand, the formation process of the columnar structure (chain or bridge) under electric fields, where clusters consisting of particles grow larger with time via aggregation, has been well studied because of the application for the particle-dispersed ER fluid.[21,38–40] The aggregation kinetics of particles was studied by Vorob'eva *et al.*[41,42] for a dispersion of melamine-formaldeyde resin particles. They observed with a microscope the growth of aggregates after an electric field was applied, and found the following relationship between average cluster size s_{av}, concentration ϕ and time t:

$$s_{\mathrm{av}} \propto \phi^{0.72} t^{0.50}. \tag{4}$$

This relationship was theoretically derived on the basis of the "hierarchical model"[43] by Soo and Doi.[44] For a system in d-dimensions, they showed

$$s_{\mathrm{av}} \propto \phi^{3/(d+3)} t^{d/(d+3)}, \tag{5}$$

which agrees reasonably well with the experimetal result when $d = 3$.

The droplet coalescence process may be similar to the aggregation process of the colloidal particles, which implies that this model may be applicable to the immiscible polymer blends studied here. In this section, we first present sequential 3D images captured during the droplet coalescence process after applying a step electric field. Then we derive the time dependence of the average droplet size from the images, and from the results obtained by changing the strength of the electric field applied, a scaling property is derived. Lastly, we show that the hierarchical model taking into account the elongation of droplets under an electric field can successfully represent the time evolution of the droplet coalescence process as observed experimentally.

3.1. *3D observation of coalescence process*

We used an immiscible polymer blend of a liquid crystalline polymer (LCP) and polyisobutylene (PIB), as described in Sec. 2. All the experiments were performed at 25 °C at which LCP was in the isotropic phase. The viscosity, relative dielectric constant, conductivity and refractive index of LCP at 25 °C are 65 Pa s, 15, $4\times10^{-9}\ \Omega^{-1}\ m^{-1}$, and 1.486, respectively; those of PIB are 7.8 Pa.s, 2.6, $5\times10^{-11}\,\Omega^{-1}\ m^{-1}$, and 1.485, respectively. For CSLM observations a small amount of fluorescent dye, IANBD amide, was added to LCP before mixing with PIB. Mixtures were made with volume ratios of LCP:PIB = 1:6, 1:10, and 1:14. In the absence of an electric field, LCP was dispersed in PIB as droplets in these blends.

The blends were sandwiched between the bottom glass plate with an ITO-coated electrode and the rotating metal plate of the rheometer, as shown in Fig. 2. The diameter of the rotating plate was 35mm, and the gap between the plates was 200 μm. Observations were made through the bottom glass plate with the CSLM. To avoid ionic migration we applied ac electric fields of 512 Hz. We applied a triangular voltage of 0.3 Hz to the piezo-actuator so that the objective lens or the focal plane of the microscope oscillated vertically between 32 μm and 145 μm above the bottom glass surface. The frame rate of the CSLM was 60 fps so that 100 images were captured per half period, from which one 3D image was constructed. Furthermore, from these 3D images a clear 3D animation was made.

In the absence of an electric field LCP was dispersed in PIB as droplets in this polymer blend. Before each measurement a pre-shear was applied for 20 min at 200 s^{-1} to attain a steady state. Figures 5(a) and (b) show the sequential 3D images obtained at 2 kV_{amp}/mm and 4 kV_{amp}/mm, respectively, for ϕ=0.14 (LCP:PIB = 1:6), where the subscript "amp" indicates the amplitude of the ac voltages. At $t = 0$ before applying the electric field without shear, the droplet radius was about 2 μm. Immediately after ap-

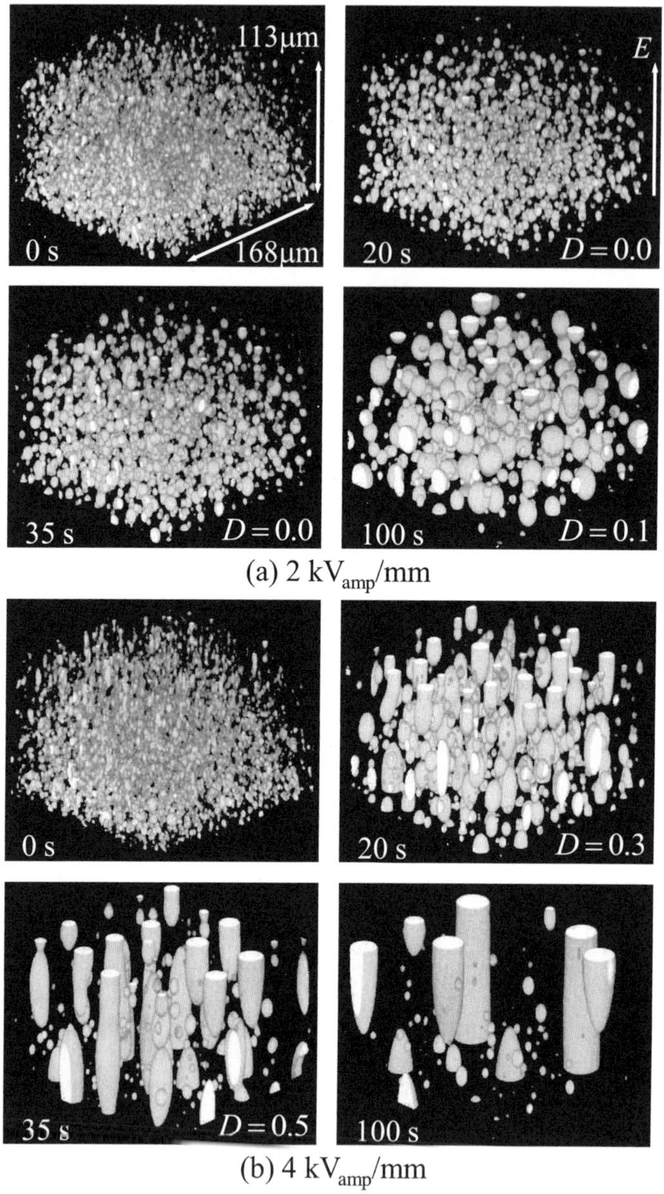

Fig. 5. Morphological changes after applying ac electric fields of (a) 2 kV_{amp}/mm and (b) 4 kV_{amp}/mm with a frequency of 512Hz. The electric fields were applied at $t=0$. The deformation defined as $D = (c - a)/(c + a)$ was calculated for each image. In (a) $D = 0.0$ at 0, 20, and 35 s, $D = 0.1$ at 100 s, and in (b) $D = 0.0$ at 0 s, $D = 0.32$ at 20 s, $D = 0.5$ at 35 s, D is impossible to calculate at 100 s because a columnar structure is formed. Reproduced with permission from Ref. 34. Copyright (2010) by the American Physical Society.

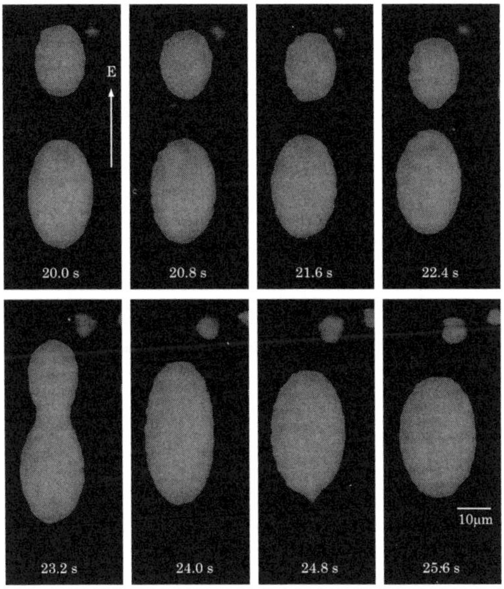

Fig. 6. Sequential images taken during the coalescence process of a pair of droplets in an electric field of 3 kV_{amp}/mm. Reproduced with permission from Ref. 34. Copyright (2010) by the American Physical Society.

plying the electric field, the droplets began to coalesce due to the induced dipole-dipole interactions, resulting in coarsening of the structure. When this blend was subjected to higher electric fields, not only the coalescence of droplets, but also the formation of columns took place (Fig. 5(b)). The elastic property of the columnar structure will be investigated in the next section. The coalescence rate increased with both the amplitude of the ac electric field and the density of droplets. This is because the dipole-dipole interaction is proportional to the square of the applied electric field and is inversely proportional to the fourth power of the distance between two droplets, as will be described later.

In these experiments we have succeeded in observing the coalescence process of two droplets, as shown in Fig. 6. A pair of droplets elongated along the electric field approach each other to reduce the inter-droplet distance and then coalesce into a large droplet. Immediately after coalescence the droplet is strongly deformed, but then gradually relaxes to an equilibrium shape.

Under electric fields droplets are elongated and become prolate spheroids. The lengths of the major and minor axes were estimated from the normalized spatial correlation function, defined as

$$C(\mathbf{r}) = \frac{\langle S(\mathbf{r}) S(0) \rangle - \langle S(0) \rangle^2}{\left\langle S(0)^2 \right\rangle - \langle S(0) \rangle^2}, \tag{6}$$

where $S(\mathbf{r}) = 1$ in the bright region (LCP) and 0 in the dark region (PIB), and $\langle \cdots \rangle$ means the spatial average. Clearly, $C(\mathbf{r})$ is anisotropic for spheroids. We can theoretically estimate the spatial correlation function of a spheroid-dispersed system by using the theory developed from the Avrami model[45] by Sekimoto.[46] This theory allows for the overlap of spheroids, but clearly in our system there is no overlap. In the theory, the spheroids (in the original theory the shape is arbitrary) are randomly distributed so that some of them must overlap each other. The discrepancy between the model and our system may be resolved by only considering the case when the spheroids are sparsely dispersed. Therefore, we derive the correlation function at small volume fractions of spheroids. We assume that the spheroids are of the same aspect ratio and size. According to the theory, the volume fraction of the spheroids ϕ, is given as

$$\phi = 1 - \exp(-nV_0), \tag{7}$$

where n is the number density of spheroids, and V_0 is the volume of a spheroid. Note that this equation is valid for all volume fractions. The spatial correlation function is expressed as

$$\langle S(\mathbf{r}) S(\mathbf{0}) \rangle = 1 - 2(1-\phi) + (1-\phi)^2 \exp[nV_{\text{over}}(\mathbf{r})], \tag{8}$$

where $V_{\text{over}}(\mathbf{r})$ is the overlapped volume of two spheroids whose centers are at a distance $\mathbf{r}$ from each other. Therefore, the normalized correlation function defined in Eq. (6), $C(\mathbf{r})$, becomes

$$C(\mathbf{r}) = \frac{\exp[nV_{\text{over}}(\mathbf{r})] - 1}{\exp(nV_0) - 1} \tag{9}$$

When the volume fraction is small, that is, $nV_{\text{over}}(\mathbf{r}) \leq nV_0 << 1$, this equation can be approximated as

$$C(\mathbf{r}) = \frac{V_{\text{over}}(\mathbf{r})}{V_0} \tag{10}$$

This equation is independent of the number density, and therefore the two correlation lengths parallel and perpendicular to the applied electric field can be related to the length of the major axis, c, and that of the minor axis, a, of the spheroid as follows. In 3D space, $C(\mathbf{r})$ takes the maximum 1 at the origin and monotonically decreases. The correlation lengths ξ_c and

ξ_a are defined as the distances along the major and minor axes at which the correlation function equals 1/2. This definition yields $c \cong 1.44\xi_c$ and $a \cong 1.44\xi_a$. In the above deviation, we have assumed that spheroids with the same aspect ratio and size are randomly and sparsely dispersed. In our experimental results, however, the droplet shape and size have some degree of distribution, so the obtained values of c and a should be regarded as averages.

Figures 7(a) and (b) show the time dependencies of c and a, respectively, at several amplitudes of the ac electric field. In Fig. 7(b), we could not obtain the length of the major axis at extended times because of the formation of columns. In terms of the deformation, defined as $D = (c - a)(c + a)$, it was found that D increased with time and the column formation started at approximately $D = 0.4$ almost irrespective of the applied electric field. Data for $D > 0.4$ in Fig. 7(b) were not plotted, while in Fig. 7(a) all the data were plotted, and the data for $D \leq 0.4$ and $D > 0.4$ are distinguished by open and closed symbols, respectively.

The growth rates of the lengths of both the major and minor axes increased when increasing the electric field. As can be seen in Fig. 7(a), the length of the minor axis, a, increases exponentially with time in the early stage ($D \leq 0.4$), but in the late stage ($D > 0.4$) the growth rate slows and the size is almost constant. This indicates that once the columnar structure is formed there is hardly any coalescence of these columns. The initial slopes in Fig. 7(b) are steeper than those in Fig. 7(a) for the same electric field. This difference may be due to elongation. The droplet deformation D when subjected to an electric field is proportional to the radius without the electric field,[47] hence as droplets grow larger with time the growth rate of the major axis c is accelerated while that of minor axis a is reduced. The exponential growth is discussed in detail in the next subsection.

3.2. *Scaling property and hierarchical model*

First, let us consider a simple model without elongation, that is, all the droplets maintain a spherical shape during the coalescence process. From dimensional analysis the radius r is given as

$$r = r_0 f\left(\frac{\varepsilon_0 E_{\mathrm{rms}}^2 t}{\eta}\right), \tag{11}$$

where r_0 is the initial droplet radius, η the viscosity of PIB, ε_0 the dielectric constant of vacuum, E_{rms} the root-mean-square of the ac electric field, and $f(x)$ is a scaling function. In Fig. 8 we replotted the data of Fig. 7 as

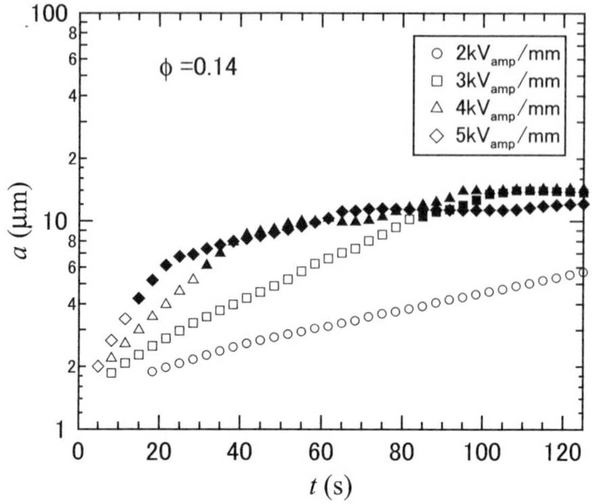

(a)

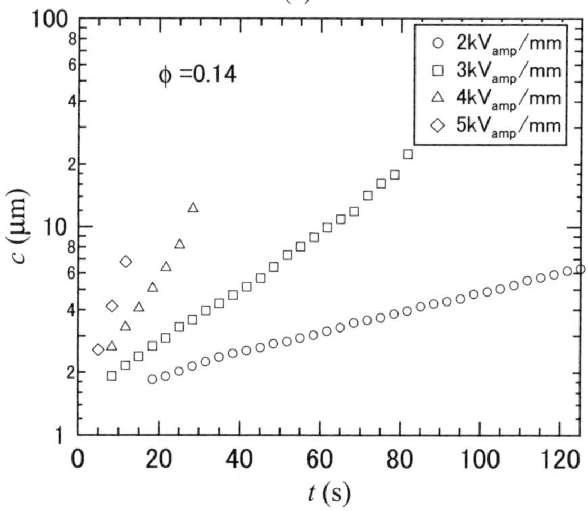

(b)

Fig. 7. Time dependencies of (a) a and (b) c at several amplitudes of ac electric field for $\phi = 0.14$ (LCP:PIB = 1:6). Open and closed symbols in (a) are for $D \leq 0.4$ and $D > 0.4$, respectively. Reproduced with permission from Ref. 34. Copyright (2010) by the American Physical Society.

a function of $\varepsilon_0 E_{\mathrm{rms}}^2 t/\eta$. Note that the initial radius r_0 at each value of electric field is almost the same in our experiments due to the applied pre-shear. The growth of a seems to obey the scaling relation as shown in Fig. 8(a), while in Fig. 8(b) it appears that of c does not. This is natural

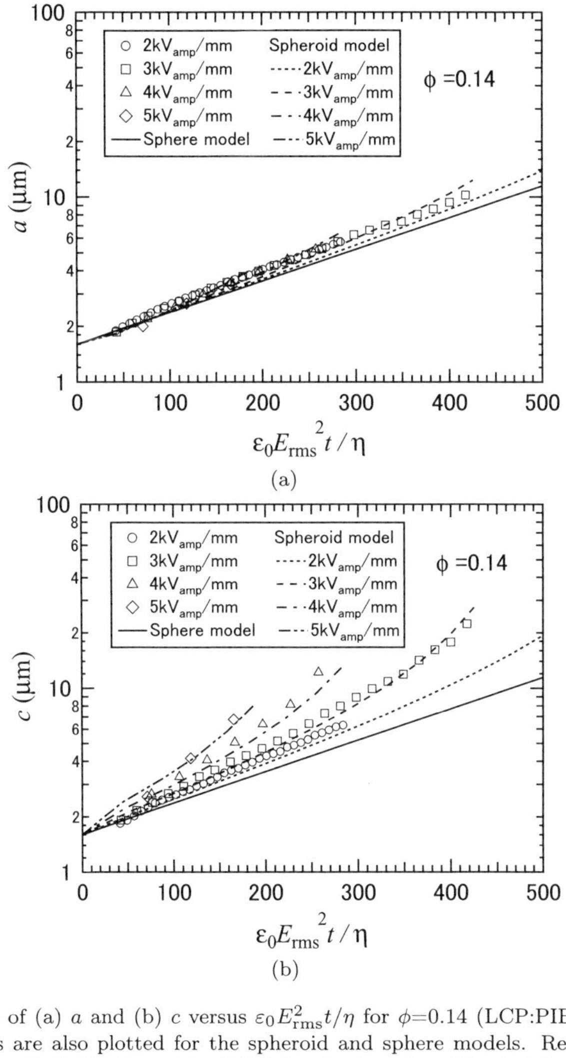

Fig. 8. Plots of (a) a and (b) c versus $\varepsilon_0 E_{\rm rms}^2 t/\eta$ for ϕ=0.14 (LCP:PIB = 1:6). Theoretical values are also plotted for the spheroid and sphere models. Reproduced with permission from Ref. 34. Copyright (2010) by the American Physical Society.

time is not similar to the others, resulting in a breakdown of the scaling relation. Although the above simple model is not perfectly applicable to the present system, we can calculate the growth rate from the model to examine the volume fraction dependence, and then compare these results to that of an improved model, which we propose, that takes into account this deformation.

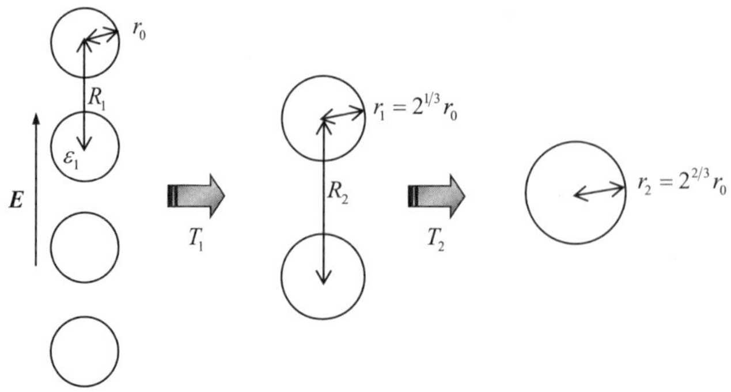

Fig. 9. Hierarchical model representing the sequential coalescence of droplets.

In the calculation, we used a "hierarchical model".[43] In this model, as shown in Fig. 9, it is assumed that at $t = 0$ spherical droplets of the same radius are uniformly dispersed, and that after time T_1 pairs of droplets coalesce to form larger spherical droplets with double the volume, and then these doublets coalesce to form droplets with quadruple the initial volume after a further time T_2, and so on. In order to determine the time T_n required for droplets of radius $r_{n-1} = 2^{(n-1)/3} r_0$ to coalesce and form droplets of radius $r_n = 2^{n/3} r_0$, we consider the time evolution of the average distance between nearest neighbor droplets, R.

The electrostatic interaction energy necessary to estimate the electrostatic attraction force between a pair of droplets is expressed by using the point dipole approximation:

$$u(r, \theta) = -\frac{1}{4\pi\varepsilon_0\varepsilon_2} \frac{d^2}{R^3} \left(3\cos^2\theta - 1\right), \tag{12a}$$

where θ is the angle between the electric field and a line between the two droplets, ε_2 is the relative dielectric constant of the matrix, and d is the induced dipole moment:

$$d = 4\pi\varepsilon_0\varepsilon_2\beta E r^3, \tag{12b}$$

with

$$\beta(\varepsilon_1, \varepsilon_2) = \frac{\varepsilon_1 - \varepsilon_2}{\varepsilon_1 + 2\varepsilon_2}, \tag{12c}$$

where ε_1 is the relative dielectric constant of the droplets. In general, the effective polarizability $\beta(\varepsilon_1, \varepsilon_2)$ depends on the conductivities σ_1 (droplet) and σ_2 (matrix) and the frequency f of the applied electric field; so ε_j

$(j = 1, 2)$ in Eq. (12a) should be replaced by $\tilde{\varepsilon}_j = \varepsilon_j + \sigma_j / (i2\pi\varepsilon_0 f)$. To avoid conduction we used ac electric fields of 512 Hz in our experiments, at which $|\beta(\tilde{\varepsilon}_1, \tilde{\varepsilon}_2)| / |\beta(\varepsilon_1, \varepsilon_2)| = 1.00004$.

For simplicity, putting $\theta = 0$ in Eq. (12a), the electrostatic attraction force can be estimated as follows:

$$F_{\mathrm{el}} \approx 24\pi \frac{\varepsilon_0 \varepsilon_2 \beta^2 E_{\mathrm{rms}}^2 r^6}{R^4}, \tag{13}$$

where we have replaced E in Eq. (12b) by E_{rms} because the droplets cannot follow the oscillating electric field. Since the friction coefficient of a droplet of radius r is $6\pi\eta r$, the time evolution of R is given by

$$6\pi\eta r \frac{dR}{dt} = -48\pi\varepsilon_0\varepsilon_2\beta^2 E_{\mathrm{rms}}^2 r^2 \left(\frac{r}{R}\right)^4. \tag{14}$$

Now if we assume that after the nth step, that is, at time $t_n = \sum_{i=1}^{n} T_i$, the droplets are homogeneously dispersed, then the average distance between droplets, R_n, is expressed as $R_n = h\phi^{-1/3} r_n$, where ϕ and r_n are the volume fraction of droplets and the radius of droplets after the nth step, respectively, and h is a constant depending on the spatial distribution of the droplets. The solution of Eq. (14) is

$$t - t_{n-1} = -\frac{\eta}{40\varepsilon_0\varepsilon_2\beta^2 E_{\mathrm{rms}}^2 r_{n-1}^5} \left(R(t)^5 - R_{n-1}^5\right), \tag{15}$$

where we have used an initial condition of $R = R_{n-1}$ at $t = t_{n-1}$. From this equation $T_n = t_n - t_{n-1}$ is easily obtained by putting $R(t_n) = 0$:

$$T_n = \frac{\eta}{40\varepsilon_0\varepsilon_2\beta^2 E_{\mathrm{rms}}^2} \left(\frac{R_{n-1}}{r_{n-1}}\right)^5 = C \frac{\eta}{\varepsilon_0\varepsilon_2\beta^2 E_{\mathrm{rms}}^2} \phi^{-5/3}, \tag{16}$$

where $C = h^5/40$. Note that T_n is independent of the step for the sphere model so that we define the characteristic time as $\tau = T_n$. Thus, $t(= t_n) = n\tau$ and $r(= r_n) = 2^{n/3} r_0$ yield

$$r = 2^{t/3\tau} r_0 = r_0 \exp\left(p \frac{\varepsilon_0 E_{\mathrm{rms}}^2 t}{\eta}\right), \tag{17a}$$

with

$$p = \frac{\ln 2}{3C} \varepsilon_2 \beta^2 \phi^{5/3}, \tag{17b}$$

indicating that the radius increases exponentially with time, as seen in Fig. 8(a), and the growth rate depends on the volume fraction as $\phi^{5/3}$. It should be noted that this result is quite different from Eq. 5 for the

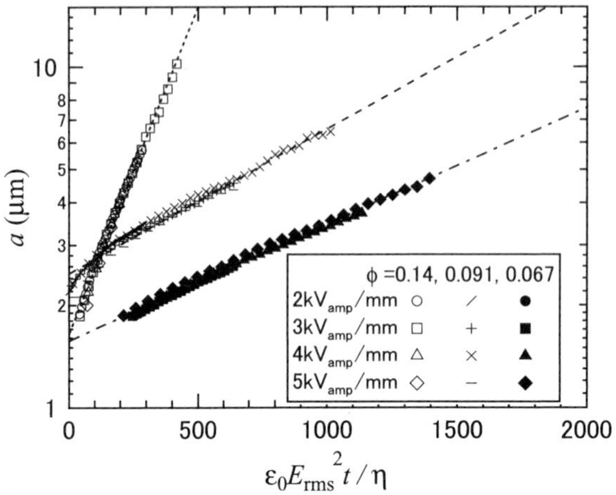

Fig. 10. Plots of a vs $\varepsilon_0 E_{\mathrm{rms}}^2 t/\eta$ under electric fields $E_0 = 2, 3, 4, 5$ $\mathrm{kV_{amp}/mm}$ and at volume fractions $\phi = 0.14$ (LCP:PIB = 1:6), 0.091 (1:10), and 0.067 (1:14). Reproduced with permission from Ref. 34. Copyright (2010) by the American Physical Society.

particle-dispersed system forming elongated clusters. We measured the volume fraction dependence of p to confirm whether Eq. (17b) holds, despite the aforementioned limitations of this simplified model. Figure 10 shows a semi-log plot of the length of the minor axis a versus the scaled time $\varepsilon_0 E_{\mathrm{rms}}^2 t/\eta$ under various electric fields E_0=2, 3, 4, 5 $\mathrm{kV_{amp}/mm}$ and for volume fractions ϕ= 0.14 (LCP:PIB = 1:6), 0.091 (1:10), and 0.067 (1:14). The data at each volume fraction fall almost on a straight line irrespective of the electric field, and the slope increases with the volume fraction, as predicted from the above theory. Figure 11 shows the volume fraction dependence of the slope p obtained from Fig. 10, where the measurements were repeated twice at the same volume fraction. The slope p seems to obey a power law with an exponent of 5/3. By using the least squares method the coefficient C in Eq. (17b) was determined as $C \approx 2.2$, which is used for the following analyses.

Calculated values from Eqs. (17a) and (17b) with $C = 2.2$ and ϕ=0.14 are plotted as solid lines in Figs. 8(a) and (b), where the two lines are the same because we assumed that droplets are spherical. The discrepancy between the experiment and the theory in Fig. 8(b) may arise from this assumption. In the discussion that follows, we will take into account the droplet deformation under electric fields.

According to Torza et al., the time-averaged deformation of a droplet

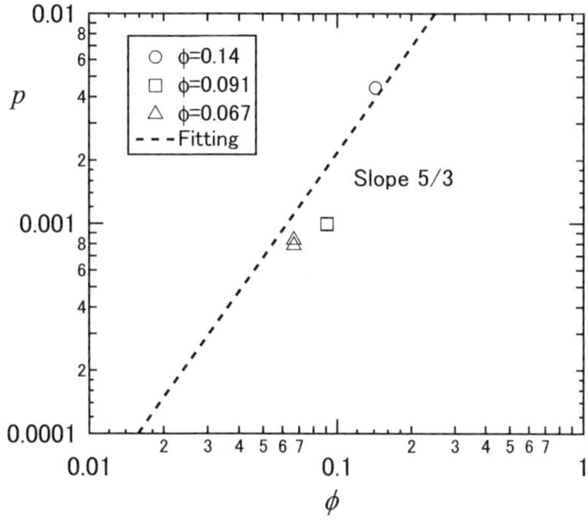

Fig. 11. Dependence of the slope p on the volume fraction ϕ in Eq. (17b). Reproduced with permission from Ref. 34. Copyright (2010) by the American Physical Society.

under an ac electric field, $D = (c - a)\,/\,(c + a)$, is given theoretically as[47]

$$D = \frac{9\varepsilon_0\varepsilon_2}{16\Gamma}\Phi E_{\mathrm{rms}}^2 r, \tag{18}$$

where Φ is a function of the dielectric constants, the conductivities, the viscosities and the frequency of the ac electric field, Γ is the interfacial tension, and r the radius of the droplet without an electric field. Note that the deformation D oscillates in an ac electric field, but it is negligible in our experiments. Equation (18) means that the deformation is in proportion to the square of the electric field and the radius of droplet. From the experimental values of c and a we calculated $D = (c - a)\,/\,(c + a)$ and $r = \left(a^2c\right)^{1/3}$ at various electric fields and volume fractions, and plotted D vs $E_{\mathrm{rms}}^2 r$ in Fig. 12. D is almost proportional to $E_{\mathrm{rms}}^2 r$, though the data is scattered to some degree. The proportionality constant was determined as $(5.3 \pm 0.1) \times 10^{-9}\ \mathrm{mV}_{\mathrm{rms}}^{-2}$ by the least-squares method. We will use this constant for the following numerical calculations to obtain the deformation at a given electric field.

For a spheroid the effective polarizability β in Eq. (12c) should be modified:[48]

$$\beta = \frac{1}{3}\frac{\varepsilon_1 - \varepsilon_2}{\varepsilon_2 + (\varepsilon_1 - \varepsilon_2)n^{(z)}}, \tag{19a}$$

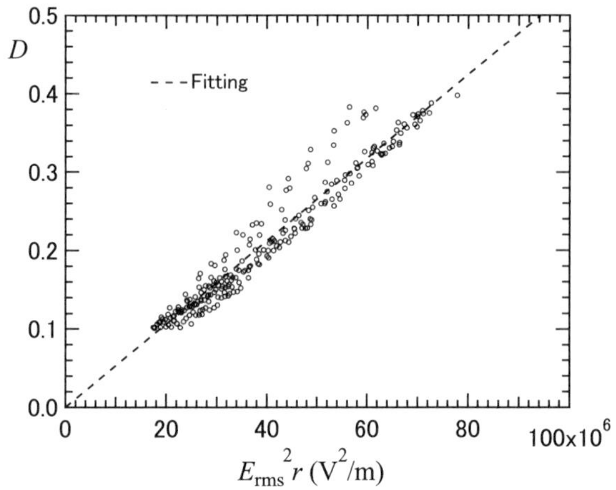

Fig. 12. Plot of D vs $E_{\mathrm{rms}}^2 r$ under electric fields E_0= 3, 4, 5 kV$_{amp}$/mm and at volume fractions ϕ=0.14, 0.091, 0.067. Reproduced with permission from Ref. 34. Copyright (2010) by the American Physical Society.

with

$$n^{(z)} = \frac{1-e^2}{2e^3}(\ln\frac{1+e}{1-e} - 2e), \tag{19b}$$

where e is the eccentricity, defined as

$$e = \sqrt{1-\frac{a^2}{c^2}} = \sqrt{1-\left(\frac{1-D}{1+D}\right)^2}. \tag{19c}$$

Since β is a function of D, which increases with time, T_n in Eq. (16) depends on the step. From the above equations, it is easily understood that β is an increasing function of D so that the growth rate should be accelerated with time, resulting in a derivation from the exponential growth for the simple sphere model. In the spheroid model we have to add T_i. At $t_n = \sum_{i=1}^{n} T_i$, the lengths of the major and minor axes are obtained from $D = (c-a)\,/\,(c+a)$, $r = \left(a^2 c\right)^{1/3}$ and $r(=r_n) = 2^{n/3} r_0$ as

$$a = 2^{n/3} r_0 \left(\frac{1-D}{1+D}\right)^{1/3}, \tag{20a}$$

$$c = 2^{n/3} r_0 \left(\frac{1-D}{1+D}\right)^{2/3}, \tag{20b}$$

where D can be calculated using the proportionality constant between D and $E_{\mathrm{rms}}^2 r$, $(5.3 \pm 0.1) \times 10^{-9}$ mV$_{\mathrm{rms}}^{-2}$. In the calculation of T_i, for simplicity

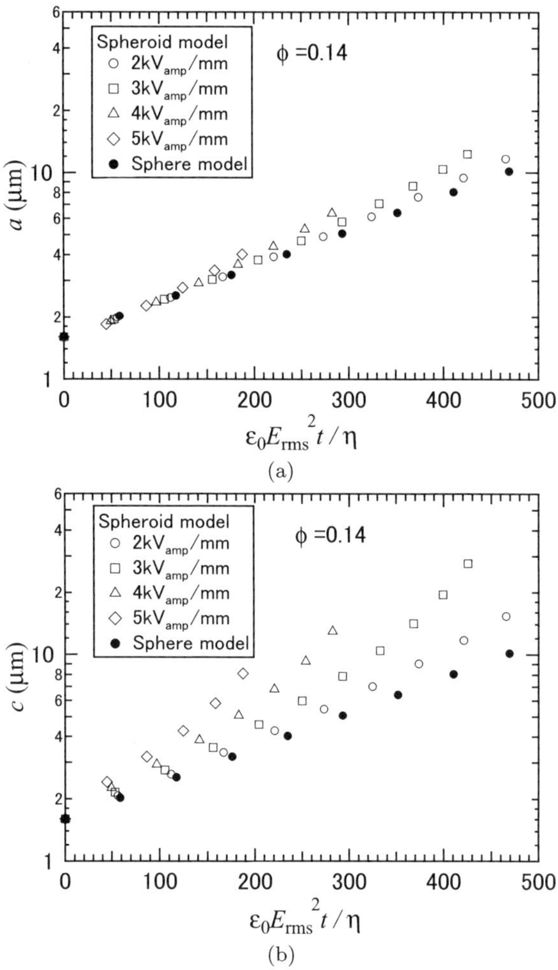

Fig. 13. Time dependencies of (a) a and (b) c at several amplitudes of ac electric fields calculated from the spheroid model for ϕ=0.14 (LCP:PIB = 1:6). Results for the sphere model are also plotted. Reproduced with permission from Ref. 34. Copyright (2010) by the American Physical Society.

we use Eq. (16) for the sphere model, though the friction coefficient should be modified in Eq. (14).

Figures 13(a) and (b) show the time dependencies of a and c, respectively, calculated at several amplitudes of the ac electric field for $\phi =$ 0.14. Results for the sphere model are also plotted. As for the length of the minor axis a, the deviation from the sphere model is small, whereas the length of the major axis c, deviates from the sphere model significantly as the electric field is increased. It is now clear why a almost obeys exponential growth but c does not: the increase of the deformation D accelerates the growth rate of droplet volume while it reduces the growth rate of a, resulting in a cancellation. In contrast, the growth rate of c is further accelerated. We plotted these theoretical values as broken lines in Fig. 8 to compare to the experimental results, and a good agreement is clearly seen, validating the spheroid model.

4. Shear Modulus of Columnar Structure Formed in an Immiscible Polymer Blend Under Electric Fields

In the previous section we have shown that a columnar structure can be formed when a strong electric field is applied to an immiscible polymer blend, as shown in Fig. 5. This structure is stable enough so far as the electric field is maintained, and furthermore it has elasticity against shear deformation. The restoring force comes from the surface tension and the Maxwell stress, as will be clarified later. They also play important roles in the ER effect when an immiscible polymer blend is subjected to both electric and shear flow fields. However, under shear flow it is not easy to quantitatively investigate them in the complicated structures such as a network structure. Therefore, as a first step to do so, in this section we turn our attention to a quiescent columnar structure formed along the electric field without shear flow, before proceeding to the network structure studied in the next section. A quantitative investigation of them may be performed through oscillation measurements of the columnar structure and theoretical considerations are not so difficult.

Columnar structures have been constructed in different ways so far.[49–53] One interesting way is to use the instability of a flat liquid/liquid interface subjected to an electric field perpendicular to the interface; fluctuations with a characteristic wavelength grow to lead the formation of well-defined columnar structures.[51–53] Meanwhile, the simplest way is to apply an electric field to a droplet-dispersed phase. The droplets are elongated along the electric field to form bridges between two electrodes, where it should be noted that certain conditions are necessary for the elongation of droplets in electrical properties.[47] Ha and Yang made dynamic frequency sweep measurements by imposing a small-amplitude oscillatory shear to a columnar structure thus obtained under electric fields.[19] However, the columnar structure was not perfect according to a photo presented in their paper;

columns were not straight and broken, and moreover many small droplets were dispersed. They have shown that the storage modulus at low frequencies increases as the applied electric field is increased, but there was no quantitative analysis.

We adopted the latter method and constructed perfect columnar structures suitable for performing oscillation measurements. The structures were observed with our system and three-dimensional images were constructed to confirm the quality of the columns and furthermore to measure the radius and the density of columns, which are necessary for theoretical analyses. In this section we present the dynamic modulus of an ideal columnar structure under electric fields and compare the results with a theory developed by taking into account the Maxwell stress and interfacial tension.[54]

4.1. *3D observation of columnar structure and shear modulus*

We used a blend consisting of LCP and PIB with different physical quantities from those used in the previous section. The viscosities, relative dielectric constants and conductivities of LCP and PIB are, respectively, 47.0 Pa s, 30.5 Pa s, 15, 2.8, $8.1\times10^{-9}\Omega^{-1}\mathrm{m}^{-1}$, and $3.3\times10^{-11}\Omega^{-1}\mathrm{m}^{-1}$ at 25°. All the experiments were done at 25°, where the LCP was in the isotropic phase. LCP and PIB were mixed in the weight ratio of LCP:PIB=1:6. In this blend, LCP was dispersed in PIB as droplets in the absence of an electric field. The average droplet size is inversely proportional to the shear rate. In our experiment, therefore, first we applied a preshear to the blend to make small droplets, stopped the shear flow and then applied an electric field. When subjected to the electric field, as described in the previous section, many columns were formed. The formation of columns after applying the field is shown in Fig. 14, where a preshear of 200 s^{-1} was applied for 1 min, and then an ac electric field with an amplitude 5 kV/mm and a frequency 2 Hz was applied. Immediately after applying the electric field, thin columns are formed by and elongation of droplets (Fig. 14(a)). They gradually grow and finally remain static (Fig. 14(b)).

Unlike the droplet-dispersed structure, in the present columnar structure the bridges of LCP were formed between the top and bottom plates so that the storage modulus was expected to increase at low frequencies. We measured the dynamic modulus of the columnar structure. In order to keep the columnar structure, successive application of the electric field was necessary. At each oscillatory measurement, the same procedure mentioned

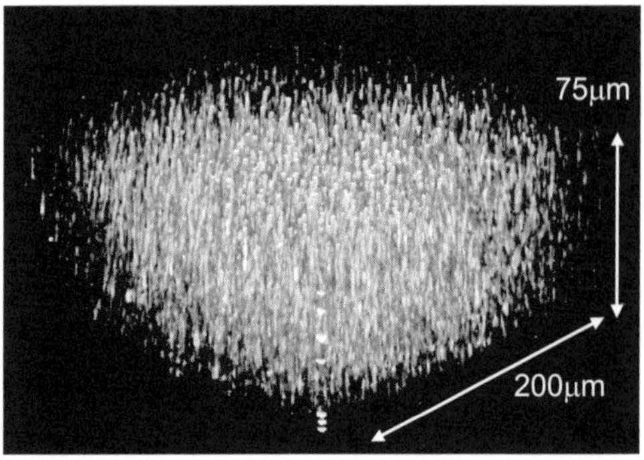

(a)

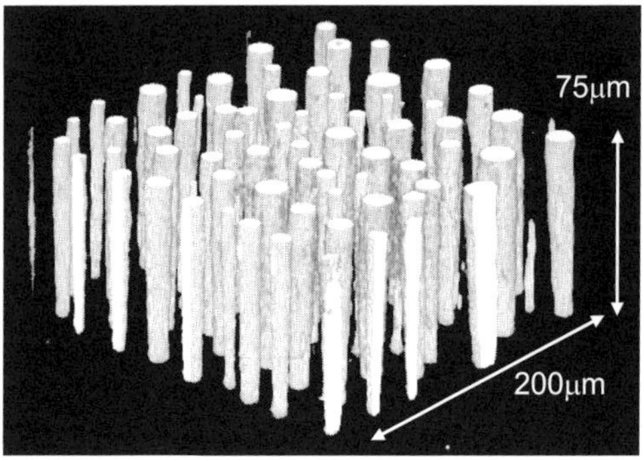

(b)

Fig. 14. Formation of columns after applying an ac electric field with an amplitude of 5kV/mm and a frequency of 2Hz. 3D images were taken (a) 1 sec later and (b) 100 sec later. Reproduced with permission from Ref. 54. Copyright (2009) by the American Physical Society.

above was repeated for constructing the columnar structure and then we waited 15 min before the measurement so that the coalescence of columns almost ceased. A small oscillating strain with an amplitude of 5% was applied so that the columnar structure could be maintained. It took about 3 min for each oscillation measurement, during which the structure was observed and it was confirmed that the structure did not change.

Figure 15 shows the frequency dispersions of the complex modulus measured at several amplitudes of ac electric field, E_0, with a frequency of 2Hz, including the modulus in the droplet-dispersed structure without electric field, which was prepared by applying a shear flow of 200 s^{-1} for 1 min. In the droplet-dispersed state, the storage modulus is small at low angular frequencies, while at high frequencies it increases due to the elastic defor-

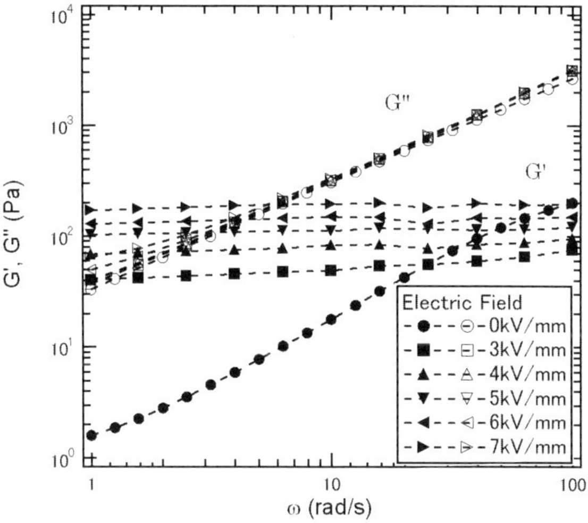

Fig. 15. Frequency dispersions of the storage modulus (G'(closed)) and the loss modulus (G''(open)) measured at several amplitudes of ac electric field with a frequency of 2Hz. The oscillation amplitude is 5 %. Reproduced with permission from Ref. 54. Copyright (2009) by the American Physical Society.

mation of droplets.[55] On the other hand, in the columnar structure the storage modulus takes a large value at low frequencies and increases with increasing the electric field. On the other hand, the loss modula are almost the same irrespective of the electric field, indicating that it may be governed only by the viscosities of LCP and PIB, independent of the structure.

The above elasticity appearing in the columnar structure may originate in the interfacial tension between LCP and PIB, and the electric stress at the interface. The interfacial energy is proportional to the surface area of columns. Therefore, when the columns tilt, the surface area, i.e., the surface energy increases, resulting in a restoring force. The electric stress comes from the mismatch in electric properties such as the dielectric constant and conductivity between LCP and PIB, as has been mentioned before. In our blend, LCP is larger than PIB both in dielectric constant and conductivity. The electric stress is expressed in terms of the Maxwell stress tensor, $T_{ij} = \varepsilon E_i E_j - \varepsilon E^2 \delta_{ij}/2$, where ε is the dielectric constant. This indicates that the storage modulus related to the electric stress should be proportional to the square of the applied electric field. This is illustrated in Fig. 16, where G' at $\omega = 10$ s^{-1} is plotted against E_0^2 at 2 and 512 Hz. There seems to be a linear relation between G' and E_0^2. The value of G' extrapolated at

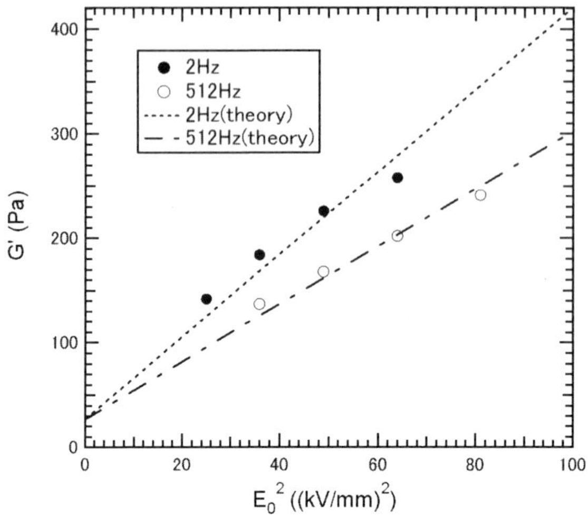

Fig. 16. Plot of the storage modulus at ω =10s^{-1} vs the square of the amplitude of ac electric field. Reproduced with permission from Ref. 54. Copyright (2009) by the American Physical Society.

$E_0 = 0$ is the contribution of the interfacial tension. The slope decreases as the frequency of ac electric field is increased. This frequency dependence is caused by the conductivity, as will be shown.

4.2. *Theoretical derivation of static shear modulus*

Let us theoretically derive the static elastic modulus, $G'(0)$. The contribution of the interfacial tension, Γ, to it can be calculated as follows. When a column with a radius a and a height d is tilted by a shear strain γ, as shown in Fig. 17 (a), the interfacial area becomes $S = 2\pi ad\left(1+\gamma^2/4\right)$ for a small strain. Therefore, the force acting on the upper plate is $F_t = -\partial\left(S\Gamma\right)/\partial\left(d\gamma\right) = -\pi a\Gamma\gamma$, and so

$$G'_t(0) = \pi na\Gamma, \tag{21}$$

where n is the number of columns per unit area. It is natural that the same result can be obtained also from Eq. 1. The calculation of the electrostatic contribution is more tedious. We should solve the Laplace equation for the electrostatic potential ϕ,

$$\Delta\phi = 0, \tag{22}$$

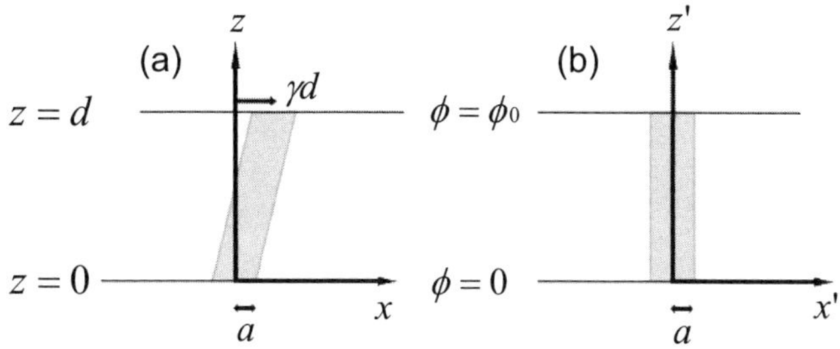

Fig. 17. (a) Schematic illustration of a slant column. (b) A new coordinate system of $x' - y' - z'$, in which the slant column is transformed to the vertical one.

under the boundary condition at the interface,

$$[\phi] = 0, \tag{23a}$$

$$[\tilde{\varepsilon}\nabla\phi\cdot\mathbf{n}] = 0, \tag{23b}$$

with

$$\tilde{\varepsilon} = \varepsilon + \frac{\sigma}{i\omega_e} \tag{24}$$

where $\mathbf{n}$ is the outward unit vector normal to the interface, and ε, σ and ω_e are the dielectric constant, the conductivity, the angular frequency of the applied ac electric field (not to be confused with that of oscillatory shear strain), respectively, and $[A]$ indicates the subtraction of the value of A inside from that outside the interface. In addition, we put $\phi\,(z = 0) = 0$ and $\phi(z = d) = \phi_0$. Here, we introduce a new coordinate system of x'-y'-z', in which the slant column is transformed back to the vertical one, as shown in Fig. 17(b). The coordinates of x-y-z and x'-y'-z' are related as

$$x = x' + \gamma z', \tag{25a}$$

$$y = y', \tag{25b}$$

$$z = z'. \tag{25c}$$

In the x'-y'-z' coordinate system, Eqs. (22), (23a) and (23b) become

$$\Delta'\phi - 2\gamma\frac{\partial^2\phi}{\partial x'\partial z'} + \gamma^2\frac{\partial^2\phi}{\partial x'^2} = 0, \tag{26}$$

$$[\phi] = 0, \tag{27a}$$

$$\left[\tilde{\varepsilon}\left\{\nabla'\phi\cdot\mathbf{n}' - \gamma\left(\frac{\partial\phi}{\partial z'}n'_x + \frac{\partial\phi}{\partial x'}n'_z\right) + \gamma^2\frac{\partial\phi}{\partial x'}n'_x\right\}\right] = 0. \tag{27b}$$

The above equations can be solved by the perturbation method. We expand ϕ with respect to γ:

$$\phi = \phi^{(0)} + \gamma\phi^{(1)} + \gamma^2\phi^{(2)} + \cdots . \tag{28}$$

After some calculations, we obtain the solution in the cylindrical coordinate,

$$x' = \rho\cos\theta \ , y' = \rho\sin\theta, \tag{29}$$

as

$$\phi_1^{(0)} = \phi_2^{(0)} = \phi_0/d \cdot z', \tag{30}$$

$$\phi_1^{(1)} = \sum_{m=1}^{\infty} c_{1m} I_1\left(m\pi\rho/d\right)\cos\theta\sin m\pi z'/d, \tag{31a}$$

$$\phi_2^{(1)} = \sum_{m=1}^{\infty} c_{2m} K_1\left(m\pi\rho/d\right)\cos\theta\sin m\pi z'/d, \tag{31b}$$

with

$$c_{1m} = \frac{2[1-(-1)^m]\phi_0\left(\tilde{\varepsilon}_1 - \tilde{\varepsilon}_2\right)/\left[\left(m\pi\right)^2 I_1\left(m\pi a/d\right)\right]}{\tilde{\varepsilon}_1 I_1'\left(m\pi a/d\right)/I_1\left(m\pi a/d\right) - \tilde{\varepsilon}_2 K_1'\left(m\pi a/d\right)/K_1\left(m\pi a/d\right)}, \tag{32a}$$

$$c_{2m} = \frac{2[1-(-1)^m]\phi_0\left(\tilde{\varepsilon}_1 - \tilde{\varepsilon}_2\right)/\left[\left(m\pi\right)^2 K_1\left(m\pi a/d\right)\right]}{\tilde{\varepsilon}_1 I_1'\left(m\pi a/d\right)/I_1\left(m\pi a/d\right) - \tilde{\varepsilon}_2 K_1'\left(m\pi a/d\right)/K_1\left(m\pi a/d\right)}, \tag{32b}$$

where subscripts '1' and '2' refer to the inside and outside of the column, respectively, and $I_1(x)$ and $K_1(x)$ are, respectively, the modified Bessel functions of the first and second kinds, and $I_1'(x)$ and $K_1'(x)$ are their derivatives.

The force acting on the upper plate can be derived from the balance between two torques acting on the column; one is from the plates and the other is the Maxwell stress exerted on the interface, expressed in terms of the Maxwell stress tensor, $T_{ij} = \varepsilon\left(E_iE_j - E^2/2\cdot\delta_{ij}\right)$, as

$$F_e = \frac{1}{d}\int\sum_j\left[T_{1j}n_j\right]zdS, \tag{33}$$

where the integration is taken over the surface of the slant column. The right-hand side of the above equation is rewritten up to the first order of γ as

$$F_e = -\frac{1}{2}\gamma\left(\varepsilon_1 - \varepsilon_2\right)\phi_0\mathrm{Re}\left[\int\phi_1^{(1)}\cos\theta dS'\left(1 + \exp\left(i2\omega_e t\right)\right)\right], \tag{34}$$

where the integration is taken over the surface of the vertical column. It should be noted that the force consists of dc and $2\omega_e$ components. In our experiments, however, the contribution may average out to zero in our oscillation measurements. Thus, we have obtained the elastic modulus owing to the Maxwell stress. The total static elastic modulus is, therefore,

$$G'(0) = \pi na\Gamma + 2nad(\varepsilon_1 - \varepsilon_2)E_0^2 \times \mathrm{Re}\left[\sum_{m=1}^{\infty} \frac{[1-(-1)^m](\tilde{\varepsilon}_1 - \tilde{\varepsilon}_2)/m^3\pi^2}{\tilde{\varepsilon}_1 I_1'(m\pi a/d)/I_1(m\pi a/d) - \tilde{\varepsilon}_2 K_1'(m\pi a/d)/K_1(m\pi a/d)}\right], \tag{35}$$

where $E_0 = \phi_0/d$ is the amplitude of the applied ac electric field. Here, we have omitted the oscillating term coming from the $2\omega_e$ term in Eq. (34). It is readily understood that the electrostatic contribution disappears for the column and the matrix being the same in dielectric constant, that is, $\varepsilon_1 = \varepsilon_2$, and it depends on the frequency of the ac electric field through the conductivity.

From Eq. (35) we can numerically calculate the static modulus at a given ac electric field. We used the following values; $\Gamma = 3 \times 10^{-3}$ Nm^{-1}, $a = 8.6 \times 10^{-6}$ m, $d = 2 \times 10^{-4}$ m, $n = 3.3 \times 10^8$ m^{-2}, $\varepsilon_1 = 15\varepsilon_0$ Fm^{-1} (ε_0: dielectric constant of vacuum), $\varepsilon_2 = 2.8\varepsilon_0$ Fm^{-1}, σ_1 =8.1×10^{-9} Ω^{-1}m^{-1}, σ_2 =3.3×10^{-11} Ω^{-1}m^{-1}. In Fig. 16, the calculated results are shown in a dotted line (2 Hz) and a dashed line (512 Hz), which are in good agreement with the experimental data. Next, we investigate the frequency dependence in detail. Figure 18 shows the dependence of the storage modulus at $\omega = 10$ s^{-1} on the frequency of ac electric field with a constant amplitude of 8 kV/mm, where dots and a line denote, respectively, the experimental data and the calculated result. It is seen that the second term in the right-hand side of Eq. (35) is a decreasing function of the frequency of ac electric field. At the high-frequency limit the conductivity contribution disappears; $\tilde{\varepsilon_i} \to \varepsilon_i$. The conductivity effect was investigated also in ER fluids.[19,21] The experimental result in Fig. 18 shows such behavior, but the theoretical curve shifts to lower frequency compared with the experimental one. The origin of the discrepancy is not clear at present. It might be ascribed to the fact that the radius of the real column is dependent on the height; it becomes large near the plates.

The interfacial tension and the Maxwell stress have been both experimentally and theoretically clarified to play important roles for the elastic behavior in a columnar structure stabilized by an electric field. In the next

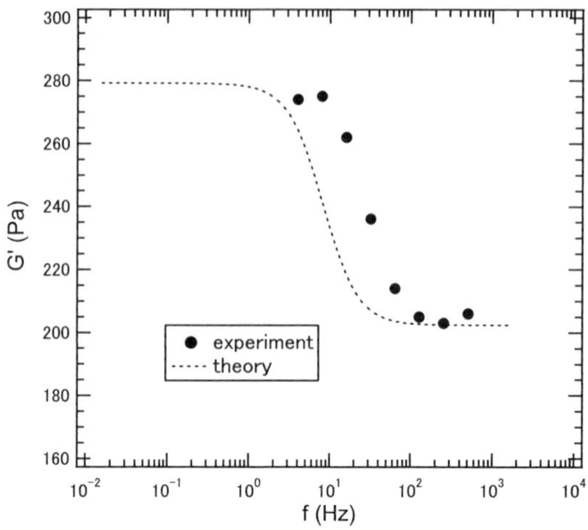

Fig. 18. Dependence of the storage modulus at $\omega = 10\ s^{-1}$ on the frequency of ac electric field with an amplitude of 8 kV/mm. Reproduced with permission from Ref. 54. Copyright (2009) by the American Physical Society.

section, they will be shown to do so also for the ER effect under electric and shear flow fields.

5. Transient Response of an Immiscible Polymer Blend to a Step Electric Field Under Shear Flow

In this section, we present our results about the relationship between morphology and rheology of an immiscible polymer blend when it is subjected to a step electric field under steady shear flow.[56] As described in Sec. 1, the stress tensor under flow can be written in Eq. (3). In a columnar structure, the surface tension and the Maxwell stress contribute to the elastic modulus. They also contribute to the stress under flow. Therefore, the Maxwell stress should be added to Eq. (3), though the theoretical expression is not yet known, which will be derived here. In the presence of shear flow, the most important contribution is the viscous stress, which exists in all fluids. The third term in the right-hand side of Eq. (3) depends on the interface velocity, and therefore it is difficult to study the contribution, which has been investigated only for droplets.[9–13] In order to avoid this difficulty we used an immiscible blend with equal viscosity, in which the surface velocity term should vanish, as is easily seen from the third term with $\eta_{\rm m} = \eta_{\rm d}$. Thus, we have three contributions in this blend; the viscous

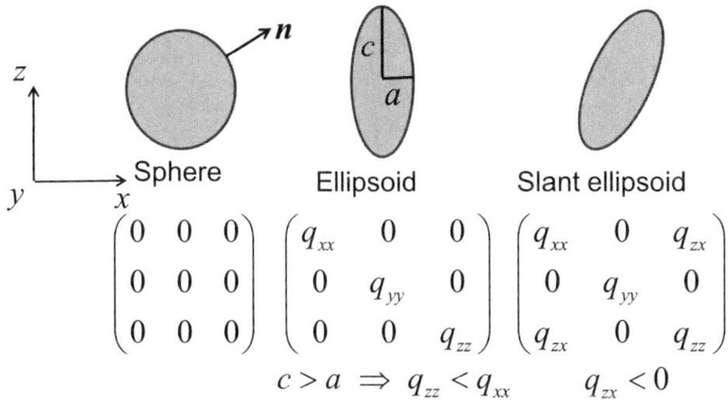

Fig. 19. Illustration of interface tensor for sphere, ellipsoid and slant ellipsoid.

stress, the interfacial stress and the Maxwell stress. The first and second ones may be expressed as $\eta(= \eta_{\mathrm{m}} = \eta_{\mathrm{d}})\gamma$ and $-\Gamma q_{zx}$, respectively, for our experimental geometry, in which the flow velocity and the velocity gradient (the electric field) are along the x and z axes, respectively. In the present study, the three contributions are experimentally separated in the transient process after a step electric field is applied under a steady shear flow. At the same time, three-dimensional (3D) structures are observed during the process with our system. The interface tensor is calculated, and we show that not only the interfacial stress, but also the Maxwell stress, can be well described by the interface tensor defined in Eq. (2).

Before presenting our results, let us briefly explain the interface tensor, which represents the anisotropy of a structure consisting of interfaces. This tensor is symmetrical and traceless. As illustrated in Fig. 19, all the elements are zero in the case of a sphere, whereas the diagonal elements are nonzero for a general ellipsoid with semi-axes along the reference frame axes. If the lengths of the semi-axes are a, b and c (parallel to the x, y, and z axes, respectively), note that $q_{xx} < q_{yy} < q_{zz}$ if $a > b > c$; in other words, the diagonal element along the elongated direction becomes small. Furthermore, when the ellipsoid tilts toward the x direction, the off-diagonal elements, q_{xz}=q_{zx}, become nonzero.

5.1. *3D observation of transient process and stress measurement*

We used a blend consisting of LCP and PIB with different physical properties from those used in the previous sections. All experiments were conducted at 28°C, where the viscosities of LCP and PIB are the same,

33.5 Pa s. The relative dielectric constant and conductivity of LCP are 16.0 and 4×10^{-9} Ω^{-1}m^{-1}, and those of PIB are 2.7 and 3.3×10^{-11} Ω^{-1}m^{-1}, respectively. The polymers were blended at a volume ratio of LCP:PIB=1:8. For this blend, in the absence of an electric field, the LCP was dispersed in PIB as droplets.

3D observations have already been made in the previous sections. This time, however, we apply a constant shear flow. This makes the observation more difficult because the flow conveys structures and the capture speed of CSLM is limited. The diameter of the rotating plate was 35 mm, and the gap between the plates was 200 μm. The shear rate was measured at the edge of the rotating disk, and the shear stress was calculated from the measured torque under the assumption of a Newtonian fluid because a parallel-plate rheometer was used in the experiment. Although immiscible polymer blends under an electric field are not generally Newtonian,[57] the above assumption is not considered to seriously affect our conclusions because the contribution to the stress around the edge is dominant and we only roughly estimate the shear stress.

3D images were constructed as described in Sec. 2 by using a piezo-actuator attached to the objective lens for vertical displacement, with the x, y and z axes taken as the flow, vorticity and velocity gradient (the electric field) directions, respectively. A triangular voltage of 5 Hz was applied to the piezo-actuator so that the objective lens or the focal plane oscillated vertically between 11 μm and 67 μm above the bottom glass plate. We used a water-immersion objective lens ($\times$40). The gap between the lens and the bottom glass plate was filled with water such that the actual height of 3D images was about 56 μm. To construct each 3D image, 50 images were captured per half period (0.1 s) at a frame rate of 500 fps. It should be noted that during the capture of one 3D image in this experiment, a shear rate of 1 s^{-1} brings about a total strain of 10 %. This is not negligible, and correction should be made as described in Sec. 2. Before each measurement, we appled a preshear of 1 s^{-1} for 3 h to ensure that the blend was in a steady state under the constant shear. We then applied a step ac electric field to the blend under steady shear flow. An ac electric field of 1000 Hz was used to avoid the conductivity effect.

Figure 20 shows the transient shear stress, when the blend was subjected to an ac electric field with amplitude of 6 kV/mm at a constant shear rate of 1 s^{-1}. The shear stress gradually increases after the electric field is applied at $t = 0$. It passes through a local maximum and minimum and then gradually increases until reaching a steady value. The transient 3D

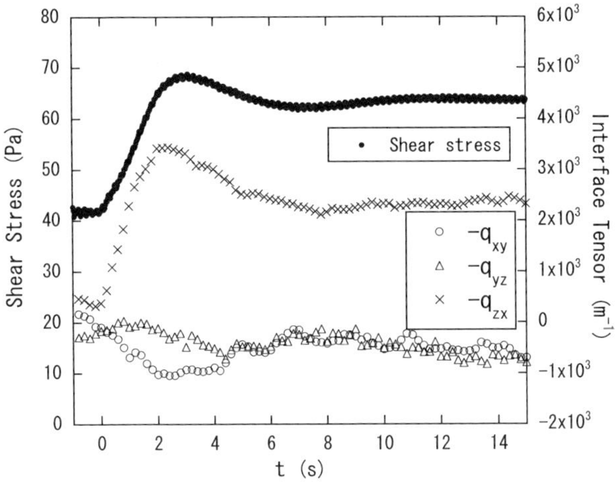

Fig. 20. Transient shear stress under step electric field. Off-diagonal elements of the interface tensor are also shown. Reproduced with permission from Ref. 56. Copyright (2011) by the American Physical Society.

images captured together with the stress measurement are shown in Fig. 21. The bright parts are LCP, into which a small amount of fluorescent dye was added. In the absence of an electric field, the droplets of LCP are dispersed within PIB matrix (0 s). Immediately after application of the step electric field, all the droplets are elongated along the electric field and some coalesce with one another to form columns (1 s). The same behavior has been observed without shear flow, and the elongation and coalescence are attributed to the Maxwell stress exerted on the interface and the induced dipole-dipole interaction between droplets, respectively,[34] as described in Sec. 3. Then, the columns gradually tilt toward the flow direction (2-3 s), causing the shear stress to increase. Further tilting may decrease the shear stress (4 s). In the meantime, the columns become linked with each other to form a network structure (4-8 s).

Now, let us quantitatively investigate the structural change in terms of the interface tensor defined in Eq. (2). Figure 22 shows the time dependences of the diagonal elements corresponding to the shear stress in Fig. 20. In the absence of an electric field, all the elements are almost zero because droplets are nearly spherical. Immediately after the step electric field is applied, q_{zz} decreases while both q_{xx} and q_{yy} increase, indicating that the droplets are elongated along the electric field (the z axis). Next, q_{xx} decreases and q_{zz} increases. The time dependence of the off-diagonal

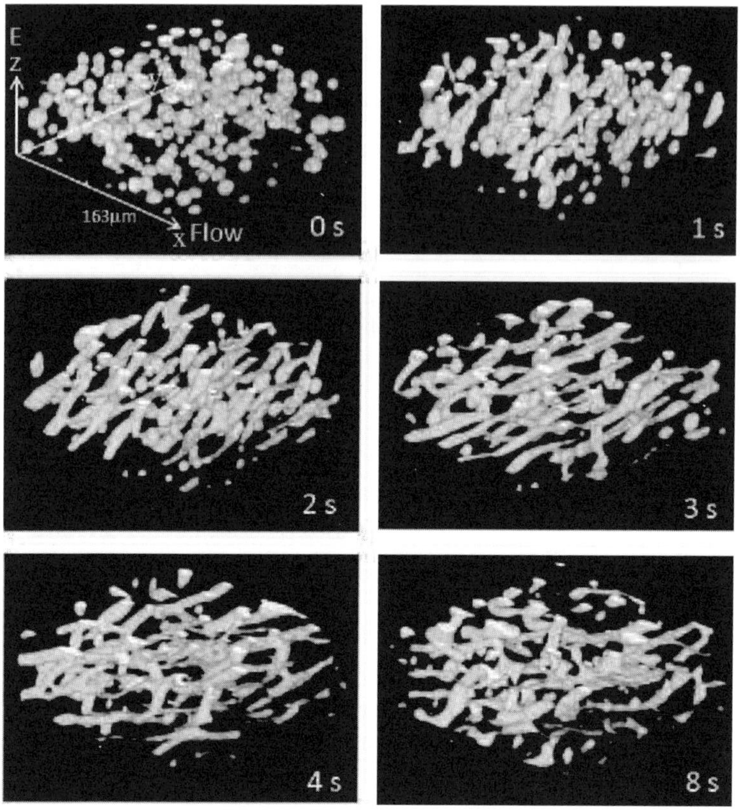

Fig. 21. 3D images in transient process, where the x, y and z axes are taken along the flow, vorticity and velocity gradient (the electric field) directions, respectively. Reproduced with permission from Ref. 56. Copyright (2011) by the American Physical Society.

elements $-q_{\alpha\beta}$ is plotted in Fig. 20. Both $-q_{xy}$ and $-q_{yz}$ are almost zero, as expected from the symmetry. On the other hand, $-q_{zx}$ increases immediately after the electric field is applied, passes through a maximum and minimum, and then gradually increases until reaching a steady value. This behavior is quite similar to that of the shear stress. In fact, as mentioned above, the interfacial stress is expressed as $-\Gamma q_{\alpha\beta}$.[2–4] However, it should be noted that the total shear stress consists of contributions other than the interfacial tension. We will attempt to experimentally separate the contributions later, but next let us investigate the structural anisotropy in more detail.

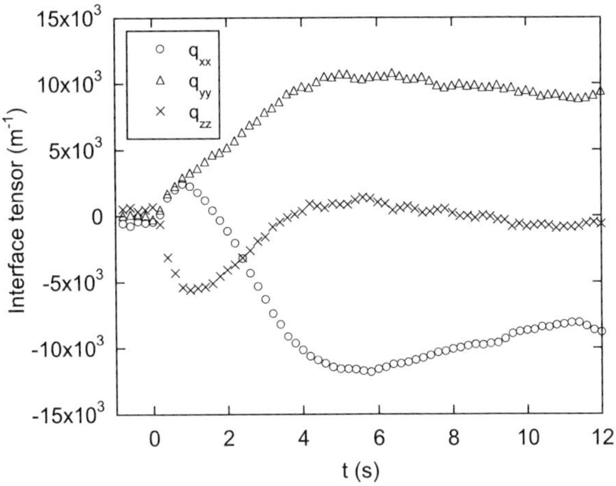

Fig. 22. Time dependence of diagonal elements of the interface tensor. Reproduced with permission from Ref. 56. Copyright (2011) by the American Physical Society.

To visualize the interface tensor, we use an ellipsoid, which is assumed to be present in a volume V_e. The interface tensor for the ellipsoid of the principal axes parallel to the reference frame axes can be expressed as

$$q_{\alpha\alpha} = Q_{\alpha\alpha} - \frac{1}{3}Q, \quad (\alpha = x, y, z) \tag{36}$$

where

$$Q = Q_{xx} + Q_{yy} + Q_{zz}. \tag{37}$$

Here, Q is the surface area density, and Q_{xx}, Q_{yy} and Q_{zz} are

$$\begin{aligned} Q_{xx} &\equiv \frac{1}{V_e}\int_S n_x^2 dS \\ &= \frac{1}{V_e}\int_0^{\pi} d\theta \int_0^{2\pi} d\phi \frac{abc\,(\sin\theta\cos\phi/a)^2}{\sqrt{(\sin\theta\cos\phi/a)^2 + (\sin\theta\sin\phi/a)^2 + (\cos\theta/c)^2}}, \end{aligned} \tag{38a}$$

$$\begin{aligned} Q_{yy} &\equiv \frac{1}{V_e}\int_S n_y^2 dS \\ &= \frac{1}{V_e}\int_0^{\pi} d\theta \int_0^{2\pi} d\phi \frac{abc\,(\sin\theta\sin\phi/a)^2}{\sqrt{(\sin\theta\cos\phi/a)^2 + (\sin\theta\sin\phi/a)^2 + (\cos\theta/c)^2}}, \end{aligned} \tag{38b}$$

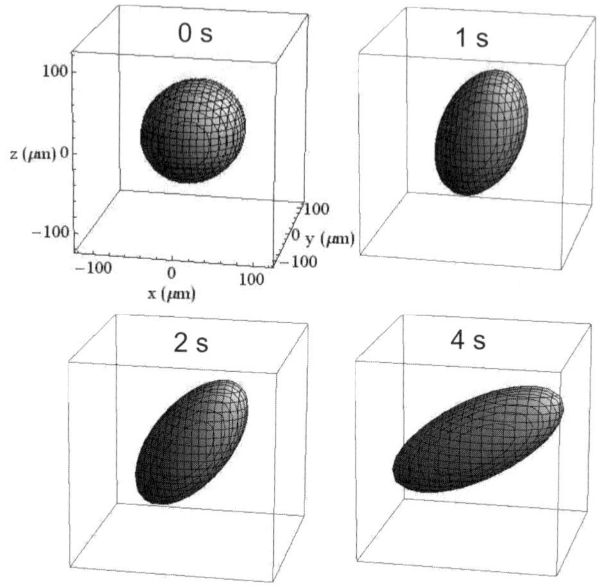

Fig. 23. Time evolution of the ellipsoid corresponding to that of the surface tensor in Figs. 20 and 22. Reproduced with permission from Ref. 56. Copyright (2011) by the American Physical Society.

$$\begin{aligned} Q_{zz} &\equiv \frac{1}{V_e}\int_S n_z^2 dS \\ &= \frac{1}{V_e}\int_0^{\pi} d\theta \int_0^{2\pi} d\phi \frac{abc\,(\cos\theta/c)^2}{\sqrt{(\sin\theta\cos\phi/a)^2 + (\sin\theta\sin\phi/a)^2 + (\cos\theta/c)^2}}, \end{aligned} \tag{38c}$$

and the off-diagonal elements $q_{xy} = q_{yz} = q_{zx} = 0$. For a tilting ellipsoid the interface tensor can be obtained by rotating the above diagonal tensor. Inversely, the ellipsoid corresponding to a given interface tensor is uniquely determined. Thus, the interface tensor can be expressed by an ellipsoid. For $V_e = 100^3$ μm^3, the time evolution of the ellipsoid corresponding to Figs. 20 and 22 is shown in Fig. 23. In Fig. 23, the features of the structural change described above can be seen clearly: the elongation of droplets and the gradual tilt of the columns. It should be noted that although the structural anisotropy can be described by an ellipsoid, the real structure is not ellipsoidal except in the initial time stage. More detail can be seen in Fig. 24, where the lengths of the semi-axes and the tilt angle are plotted as

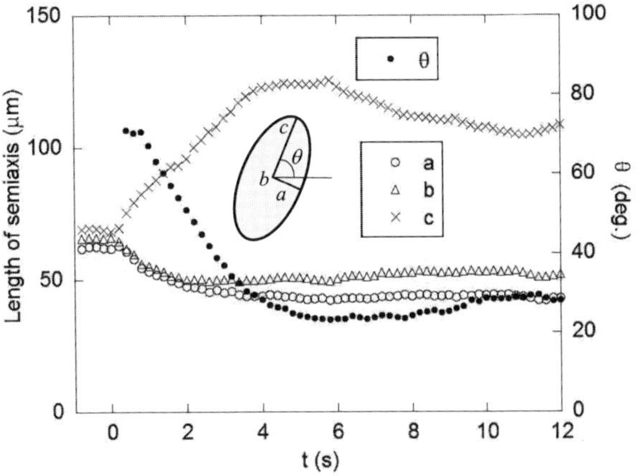

Fig. 24. Time dependence of a, b, c and θ. Reproduced with permission from Ref. 56. Copyright (2011) by the American Physical Society.

a function of time. As expected in the absence of an electric field, all the lengths are almost the same. Immediately after application of the electric field, the c axis increases while the a and b axes initially decrease at the same rate, indicating that the structure is axially symmetric (prolate spheroid). At around 1 s, the degeneracy of a and b is lifted, giving rise to a general ellipsoid. In the meantime, the tilt angle θ gradually decreases. It should be noted that if the lengths of the semi-axes are constant, q_{zx} should take a maximum at $\theta = 45°$, however, in our result θ reaches 45° at 2.4 s (Fig. 24), whereas q_{zx} peaks at about 3.1 s (Fig. 20). This is because c is still increasing around 2.4 s, as is seen from Fig. 24.

Now we turn our attention to the shear stress. For a blend of two Newtonian fluids with equal viscosity, the shear stress can be separated into three parts:

$$\sigma = \sigma_{\text{v}} + \sigma_{\text{if}} + \sigma_{\text{e}}, \tag{39}$$

where σ_{v}, σ_{if} and σ_{e} are the viscous stress, the interfacial stress and the electric stress (the Maxwell stress), respectively. The first two of these stresses can be expressed by the equations $\sigma_{\text{v}} = \eta\dot{\gamma}$ and $\sigma_{\text{if}} = -\Gamma q_{zx}$, where η is the viscosity, $\dot{\gamma}$ is the shear rate, and Γ is the interfacial tension. The expression for σ_{e} will be derived later. The experimental separation is performed as follows. As can be seen from $\sigma_{\text{v}} = \eta\dot{\gamma}$, the viscous stress vanishes when the shear flow is stopped. Similarly, σ_{e} should vanish when the electric field is removed, because σ_{e} must come from the Maxwell stress tensor, $T_{ij} = \varepsilon E_i E_j - \varepsilon E^2 \delta_{ij}/2$, where ε is the dielectric constant. Hence, we can devise a procedure for the separation of the three stresses.

During a measurement, after applying the step electric field under a steady shear flow, both the electric field and the shear flow are intermittently turned off and on within a short interval (duration: 0.24 s). The result is shown in Fig. 25(a), where the experimental conditions are the same as those shown in Fig. 20 except for the intermittent removal of the electric field and the shear flow. The details at around 6.8 s are shown in Fig. 25(b), where the time scale is expanded. First, the electric field is removed and the sudden decrease in shear stress corresponds to σ_{e}. Next, 0.06 s later, the shear flow is stopped to obtain σ_{v}. The remaining shear stress is assumed to correspond to σ_{if}. Lastly, both the shear flow and the electric field are applied again at an interval of 0.06 s. The outline of the stress response in Fig. 25(a) is comparable to that in Fig. 20 except for the sudden changes, implying that the intermittent removal of the electric field and shear flow should not affect the structural evolution. From the changes shown in Fig. 25, we can separately obtain the three stresses in Eq. (39). Figure 26 shows the time dependence of the stresses, together with $-q_{zx}$. The viscous stress σ_{v} is almost constant, as expected, giving a value of around 35 Pa, which is in good agreement with the value 33.5 Pa calculated from $\eta = 33.5$ Pa s and $\dot{\gamma} = 1\ \mathrm{s}^{-1}$. On the other hand, the interfacial stress σ_{if} depends on time and its behavior is similar to that of $-q_{zx}$. Interestingly, the electric stress σ_{e} also shows a similar time dependence as $-q_{zx}$. As shown in Fig. 27, both σ_{if} and σ_{e} appear to be proportional to $-q_{zx}$. As for σ_{if}, this result indicates the validity of $\sigma_{\mathrm{if}} = -\Gamma q_{zx}$ and the slope of the straight line in Fig. 27 gives interfacial tension Γ of approximately $3.6 \times 10^{-3}\ \mathrm{Nm}^{-1}$, which is comparable with the value of 2.5×10^{-3} Nm^{-1} determined by the deformed drop retraction method.[58]

The proportional relation between σ_{e} and $-q_{zx}$ can be explained by supposing that there are ellipsoidal droplets dispersed in electric and shear flow fields having the same size and shape, and that the hydrodynamic interactions between them are negligible. In a steady state the balance between the hydrodynamic and electric torques on each ellipsoid yields

$$\sigma_{\mathrm{e}} = -K_y^{(e)} \phi / v, \tag{40}$$

where $K_y^{(\mathrm{e})}$ is the y component of the electric torque exerted on each ellipsoid, and ϕ is the volume fraction of droplets and v is the volume of a droplet.[39] Strictly speaking, the system under consideration here is not in a steady state, but Eq. (40) approximately holds if the structure changes

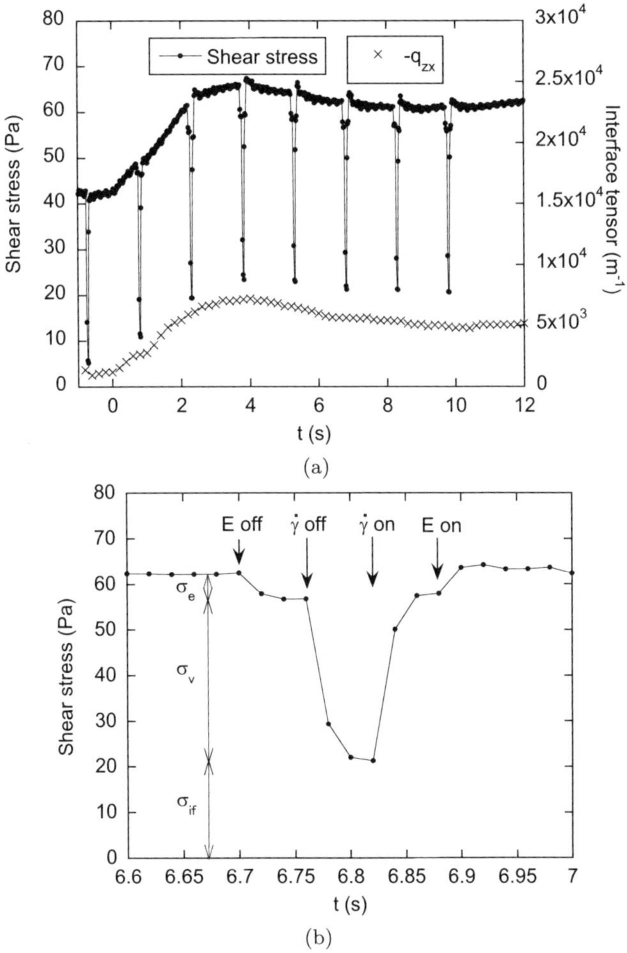

Fig. 25. (a) Time dependence of the shear stress when the electric field and the shear flow are removed intermittently and (b) a part expanded between 6.6 s and 7 s. Reproduced with permission from Ref. 56. Copyright (2011) by the American Physical Society.

slowly. The electric torque $K_y^{(e)}$ is written as

$$K_y^{(e)} = p_x E, \tag{41}$$

where p_x is the x component of the induced dipole moment and E the electric field applied along the z direction. The dipole moment p_x is easily

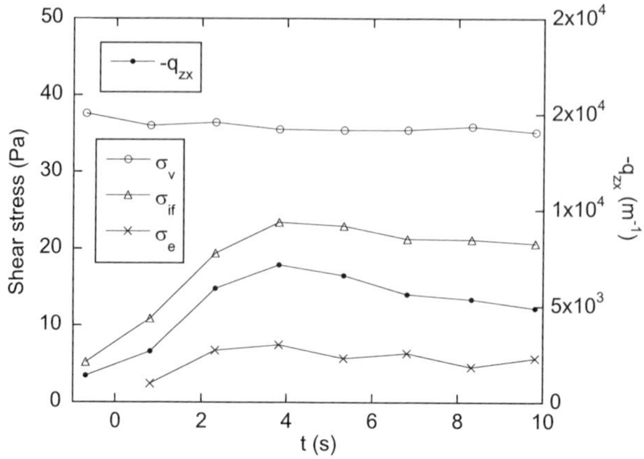

Fig. 26. Time dependence of viscous stress σ_{v}, interfacial stress σ_{if} and electric stress σ_{e}, obtained from data shown in Fig. 25. Reproduced with permission from Ref. 56. Copyright (2011) by the American Physical Society.

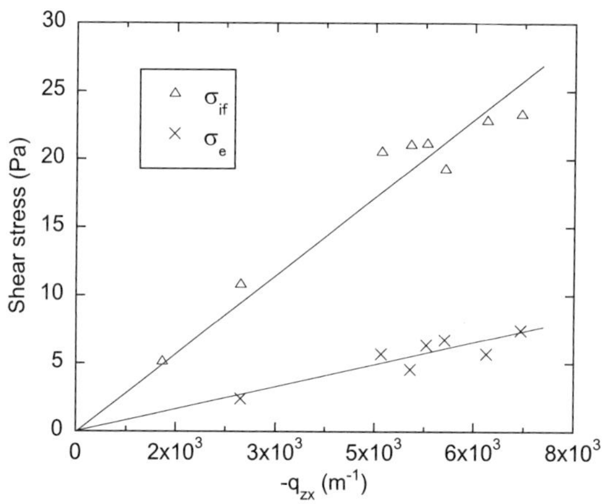

Fig. 27. Dependence of σ_{if} and σ_{e} on $-q_{zx}$. Reproduced with permission from Ref. 56. Copyright (2011) by the American Physical Society.

calculated as follows[48]

$$p_x = \frac{v\varepsilon_2\left(\varepsilon_1-\varepsilon_2\right)^2\left(n^{(a)}-n^{(c)}\right)\sin\theta\cos\theta E}{\left\{\varepsilon_2+\left(\varepsilon_1-\varepsilon_2\right)n^{(a)}\right\}\left\{\varepsilon_2+\left(\varepsilon_1-\varepsilon_2\right)n^{(c)}\right\}}. \tag{42}$$

Here ε_1 and ε_2 are the dielectric constants of LCP and PIB, respectively, and $n^{(a)}$ and $n^{(c)}$ are the depolarization factors defined as

$$n^{(a)} = \frac{abc}{2} \int_0^\infty \frac{ds}{(s+a^2)\sqrt{(s+a^2)(s+b^2)(s+c^2)}}, \tag{43a}$$

$$n^{(c)} = \frac{abc}{2} \int_0^\infty \frac{ds}{(s+c^2)\sqrt{(s+a^2)(s+b^2)(s+c^2)}}. \tag{43b}$$

$n^{(b)}$ is similarly defined. The depolarization factors can be related to the surface tensor in the principal axis frame as follows. We have similar equations for $n^{(\alpha)}$ and $Q_{\alpha\alpha}$: $n^{(a)}+n^{(b)}+n^{(c)} = 1$ and $Q_{aa}/Q+Q_{bb}/Q+Q_{cc}/Q = 1$ (Note that $Q_{\alpha\alpha}/Q = \langle n_\alpha^2 \rangle$, where $\langle \cdots \rangle$ denotes the average). Furthermore, for a sphere $n^{(a)} = n^{(b)} = n^{(c)} = 1/3$ and $Q_{aa}/Q = Q_{bb}/Q = Q_{cc}/Q = 1/3$, and for a cylinder $n^{(a)} = n^{(b)} = 1/2$, $n^{(c)} = 0$ (the unique axis is taken along the c axis) and $Q_{aa}/Q = Q_{bb}/Q = 1/2$, $Q_{cc}/Q = 0$, implying that the depolarization factors may be approximated by $n^{(\alpha)} \simeq Q_{\alpha\alpha}/Q = \langle n_\alpha^2 \rangle$, which has been numerically proven. By using this approximation and assuming that the ellipsoid is nearly spherical, the electric stress is given by the following:

$$\sigma_{\mathrm{e}} = \frac{9\varepsilon_2 (\varepsilon_1 - \varepsilon_2)^2}{(\varepsilon_1 + 2\varepsilon_2)^2} E^2 \phi \left(-q_{zx}/Q\right), \tag{44}$$

where the relation $(Q_{aa} - Q_{cc})/Q \sin\theta \cos\theta = -q_{zx}/Q \; (= -\langle n_z n_x \rangle)$ is used.

The electric stress is plotted as a function of $-q_{zx}/Q$ in Fig. 28. The slope is about 64 Pa, comparable with 156 Pa obtained from Eq. (44). This equation can also apply to a columnar structure in Sec. 4. For it, $\phi = \pi a^2 n$ and $-q_{zx}/Q = \gamma/2$ yields

$$\sigma_{\mathrm{e}}^{(\mathrm{col})} = \frac{9\varepsilon_2 (\varepsilon_1 - \varepsilon_2)^2}{2(\varepsilon_1 + 2\varepsilon_2)^2} E^2 \pi a^2 n \gamma, \tag{45}$$

where a and n are the radius and the number of columns per unit area, respectively, and γ is the shear strain. Using the values of ε_1, ε_2, a and n obtained in Sec. 4, we get $\sigma_{\mathrm{e}}^{(\mathrm{col})}/\gamma = 96$ Pa s when $E = 8/\sqrt{2}$ kV/mm. This should correspond to G' in Fig. 18 at high frequencies, where the contribution of the interfacial tension of about 30 Pa s (see Fig. 16) should be subtracted. The contribution of the electric stress thus experimentally obtained is about 175 Pa s, which reasonably agrees with the theoretical value of 96 Pa s.

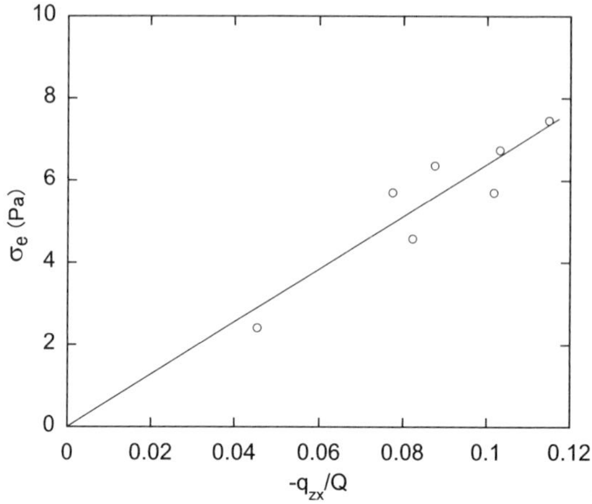

Fig. 28. Plot of σ_e vs $-q_{zx}/Q$ $(= -\langle n_z n_x \rangle)$. Reproduced with permission from Ref. 56. Copyright (2011) by the American Physical Society.

With $Q = S/V$, where S is the total interface area, Eq. (44) can be rewritten as

$$\sigma_{\mathrm{e}} = -\Gamma_{\mathrm{e}} q_{zx} \tag{46}$$

with

$$\Gamma_{\mathrm{e}} = \frac{9\varepsilon_2 (\varepsilon_1 - \varepsilon_2)^2}{(\varepsilon_1 + 2\varepsilon_2)^2} E^2 V_{\mathrm{d}}/S, \tag{47}$$

where V_{d} is the total volume of droplets. Equation (46) corresponds to $\sigma_{\mathrm{if}} = -\Gamma q_{zx}$ for the interfacial stress. Therefore, Γ_{e} in Eq. (47) can be regarded as the electric interfacial tension. However, it should be noted that Γ_{e}, unlike Γ, is not always constant because S can change. As shown in Fig. 29, $Q = S/V$ is almost independent of time, indicating that Γ_{e} can be regarded as constant in our system. For this reason, σ_{e} is proportional to q_{zx} in Fig. 27. At the end of this section, we should emphasize that Eq. (46) is theoretically derived for an ellipsoid-dispersed structure, but may equally be valid for a network structure, as has been shown experimentally.

6. Scaling Properties When Subjected to a Step Electric Field under shear flow

In the previous section we have presented the simultaneous 3D observation and shear stress measurement of an immiscible polymer blend when it is

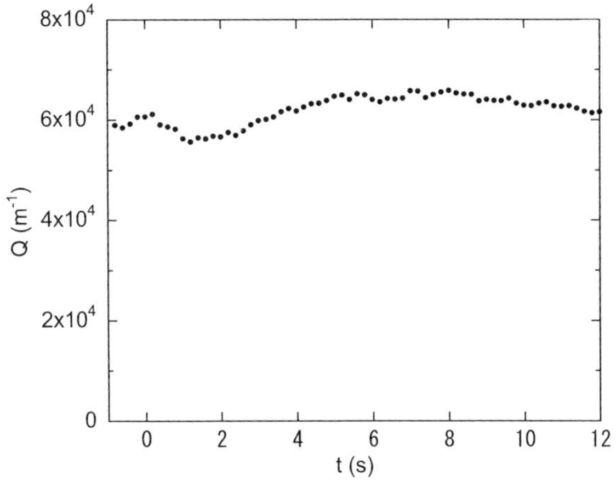

Fig. 29. Time dependence of the interface area density Q. Reproduced with permission from Ref. 56. Copyright (2011) by the American Physical Society.

subjected to a step electric field. In this section, we demonstrate scaling properties in the transient process when changing the applied shear rate and electric field.[59] In immiscible fluids, an interesting scaling property was first predicted by Doi and Ohta.[14] They developed a semiphenomenological theory and derived it. Let $\boldsymbol{\sigma}\,(t, [\boldsymbol{\kappa}(t)])$ be the stress tensor for a given time-dependent velocity gradient tensor $\boldsymbol{\kappa}(t)$, where $[\boldsymbol{\kappa}(t)]$ indicates that $\boldsymbol{\sigma}$ is a functional of $\boldsymbol{\kappa}(t)$. They derived the following scaling relation for a blend of Newtonian fluids:

$$\boldsymbol{\sigma}\,(t, [c\boldsymbol{\kappa}(ct)]) = c\boldsymbol{\sigma}\,(ct, [\boldsymbol{\kappa}(t)])\,, \tag{48}$$

meaning that the stress at time t under the velocity gradient $c\boldsymbol{\kappa}(ct)$ is c times larger than the stress at time ct under the velocity gradient $\boldsymbol{\kappa}(t)$, where c is an arbitrary number. This was experimentally examined using a shear rate increasing step wisely by Takahashi *et al.*[60] Hereafter, we derive scaling properties both in shear flow and electric fields from dimensional analysis, and furthermore prove the validity by actually measuring the transient shear stress to a step electric field under steady flow. In addition, the structural change is observed in the transient process, which also has a scaling property.

6.1. *Derivation of scaling properties in electric and shear flow fields*

Here, we assume that the two fluids are Newtonian and perfect dielectrics without frequency dispersion. The effect of conductivity will be discussed later. When the electric field is a function of time, $E(t)$, the shear stress at time t under a constant shear flow may be written as

$$\sigma = \sigma\left(t, \eta_1, \eta_2, \varepsilon_1, \varepsilon_2, \dot{\gamma}, [E(t)]\right), \tag{49}$$

where η_1 and η_2 are, respectively, the viscosities of the dispersed phase and the matrix phase, ε_1 and ε_2 are, respectively, the dielectric constants of them, and $\dot{\gamma}$ is the shear rate. We discretize the time domain with an interval Δt and apply dimensional analysis:

$$\begin{aligned}
&\sigma\left(t, \eta_1, \eta_2, \varepsilon_1, \varepsilon_2, \dot{\gamma}, [E(t)]\right) = \sigma\left(t, \eta_1, \eta_2, \varepsilon_1, \varepsilon_2, \dot{\gamma}, E(-\infty), \cdots, E(0),\right.\\
&\qquad\qquad \left. E(\Delta t), \cdots, E(t), \Delta t\right)\\
&= \eta_1\dot{\gamma} f\left(\dot{\gamma}t, \eta_1/\eta_2, \varepsilon_1/\varepsilon_2, \varepsilon_1 E^2(-\infty)/\eta_1\dot{\gamma}, \cdots, \varepsilon_1 E^2(0)/\eta_1\dot{\gamma},\right.\\
&\quad \left.\varepsilon_1 E^2(\Delta t)/\eta_1\dot{\gamma}, \cdots, \varepsilon_1 E^2(t)/\eta_1\dot{\gamma}, \dot{\gamma}\Delta t\right)\\
&= 1/c\cdot\eta_1 c\dot{\gamma} f\left(c\dot{\gamma}\cdot t/c, \eta_1/\eta_2, \varepsilon_1/\varepsilon_2, \varepsilon_1\left(\sqrt{c}E(-\infty)\right)^2/\eta_1 c\dot{\gamma}, \cdots,\right.\\
&\quad \left.\varepsilon_1\left(\sqrt{c}E(t)\right)^2/\eta_1 c\dot{\gamma}, c\dot{\gamma}\Delta t/c\right)\\
&= 1/c\cdot\sigma\left(t/c, \eta_1, \eta_2, \varepsilon_1, \varepsilon_2, c\dot{\gamma}, \sqrt{c}E(-\infty), \cdots, \sqrt{c}E(0), \sqrt{c}E(\Delta t), \cdots,\right.\\
&\quad \left.\sqrt{c}E(t), \Delta t/c\right)\\
&= 1/c\cdot\sigma\left(t/c, \eta_1, \eta_2, \varepsilon_1, \varepsilon_2, c\dot{\gamma}, \left[\sqrt{c}E(ct)\right]\right),
\end{aligned} \tag{50}$$

where c is an arbitrary number and $f(\cdots)$ is a scaling function. It should be noted that σ is also a function of the time interval Δt and it was reduced by c in the above 4th equation. Replacing t by ct in the above equation we get

$$\begin{aligned}
&\sigma\left(t, \eta_1, \eta_2, \varepsilon_1, \varepsilon_2, c\dot{\gamma}, \left[\sqrt{c}E(ct)\right]\right)\\
&\quad = c\sigma\left(ct, \eta_1, \eta_2, \varepsilon_1, \varepsilon_2, \dot{\gamma}, [E(t)]\right).
\end{aligned} \tag{51}$$

This is a scaling relation under a varying electric field and a constant shear rate corresponding to Eq. (48) obtained by Doi and Ohta. Similarly, we can derive an extended scaling relation under a varying electric field and a varying shear rate including the Doi-Ohta scaling relation:

$$\begin{aligned}
&\sigma\left(t, \eta_1, \eta_2, \varepsilon_1, \varepsilon_2, [c\dot{\gamma}(ct)], \left[\sqrt{c}E(ct)\right]\right)\\
&\quad = c\sigma\left(ct, \eta_1, \eta_2, \varepsilon_1, \varepsilon_2, [\dot{\gamma}(t)], [E(t)]\right).
\end{aligned} \tag{52}$$

In order to experimentally prove the above scaling relation, we adopted a step electric field at a constant shear rate; $E(t) = 0$ for $t < 0$ and $E(t) = E$ for $t \geq 0$. For this simple case, the scaling relation becomes

$$\begin{aligned} &\sigma\left(t, \eta_1, \eta_2, \varepsilon_1, \varepsilon_2, \dot{\gamma}, E\right) \\ &\quad = \eta_1 \dot{\gamma} f\left(\dot{\gamma} t, \eta_1/\eta_2, \varepsilon_1/\varepsilon_2, \eta_1 \dot{\gamma}/\varepsilon_1 E^2\right), \end{aligned} \tag{53}$$

where we have introduced a dimensionless number, $\eta_1 \dot{\gamma}/\varepsilon_1 E^2$, similar to the Mason number, M_n, used for the particle-dispersed ER fluid:[61]

$$M_n = \frac{\eta_2 \dot{\gamma}}{2\varepsilon_1 \beta^2 E^2}, \tag{54}$$

where $\beta = (\varepsilon_1 - \varepsilon_2)/(\varepsilon_1 + 2\varepsilon_2)$ and the subscripts "1" and "2" refer to the quantities of the dispersed phase and the matrix phase, respectively. The Mason number has the clear physical meaning that it is a ratio of the viscous force driven by the imposed shear flow to the electric polarization interaction force, but in our case better dimensionless numbers with proper physical meanings may be constructed from $\eta_1 \dot{\gamma}/\varepsilon_1 E^2$, η_1/η_2 and $\varepsilon_1/\varepsilon_2$. For example, Pan and McKinley added a viscous drag contribution to the Mason number:[18]

$$M_n' = \frac{3\eta_1 + 2\eta_2}{3\eta_1 + 3\eta_2} M_n. \tag{55}$$

However, for constant η_1, η_2, ε_1 and ε_2, the switching properties depend only on $\dot{\gamma}/E^2$. Thus, Eq. (53) predicts that the plot of the viscosity $\eta = \sigma/\dot{\gamma}$ against the strain $\gamma = \dot{\gamma} t$ becomes a universal curve independent of $\dot{\gamma}$ and E, as long as $\dot{\gamma}/E^2$ is constant. It should be noted that Eq. (53) can be directly derived by usual dimensional analysis.

6.2. *Experimental tests*

As fluids, we used a liquid crystalline polymer (LCP) and a methyl phenyl silicone oil (MPS).[62] In the previous sections PIB was used, while in the present one MPS with low viscosity was adopted because a large stress change could be obtained. All the experiments were done at 22.5°C, where the LCP was in the isotropic phase. The viscosities, relative dielectric constants and conductivities of the LCP and MPS are, respectively, 26.8 Pa s, 0.04 Pa s, 15.7, 2.8, $2.0 \times 10^{-8}\ \Omega^{-1}\text{m}^{-1}$, less than $10^{-10}\ \Omega^{-1}\text{m}^{-1}$ at 22.5°C. From these we prepared a blend of LCP:MPS=1:4 in weight. The experimental procedure was almost the same with that in the previous section. Here, we should mention that in the parallel-plate rheometer the shear rate depends on the position, while in the scaling relation Eq. (53) it was assumed to be constant. However, it can be shown that the scaling relation is still valid in our case as well. For the parallel-plate rheometer, we usually define the shear rate as the one at the edge of the rotating disk and the shear stress there is calculated from the measured torque by assuming that the fluid is Newtonian. It is easily shown that the shear stress thus defined, σ_p, obeys the following scaling relation similar to Eq. (53):

$$\sigma_p\left(t, \eta_1, \eta_2, \varepsilon_1, \varepsilon_2, \dot{\gamma}_p, E\right)$$
$$= \eta_1 \dot{\gamma}_p F\left(\dot{\gamma}_p t, \eta_1/\eta_2, \varepsilon_1/\varepsilon_2, \eta_1 \dot{\gamma}_p/\varepsilon_1 E^2\right) \quad (56)$$

with

$$F\left(x, \eta_1/\eta_2, \varepsilon_1/\varepsilon_2, y\right) \equiv \frac{1}{4}\int_0^1 p^3 f\left(xp, \eta_1/\eta_2, \varepsilon_1/\varepsilon_2, yp\right) dp, \quad (57)$$

where $\dot{\gamma}_p$ is the shear rate at the edge. For simplicity, we will omit the subscript p hereafter.

We performed measurements at several shear rates and electric fields; (a) $(\dot{\gamma}, E)$ = (30 1/s, 0.33 kV/mm), (60 1/s, 0.46 kV/mm), (90 1/s, 0.57 kV/mm) for $\dot{\gamma}/E^2 \simeq 9.2 \times 10^2$ $(\text{mm/kV})^2$/s and (b) $(\dot{\gamma}, E)$ = (30 1/s, 0.57 kV/mm), (60 1/s, 0.8 kV/mm), (90 1/s, 0.99 kV/mm) for $\dot{\gamma}/E^2 \simeq 2.8 \times 10^2$ $(\text{mm/kV})^2$/s. Before each measurement, a preshear was applied to attain a steady state for 300 min at 30 1/s, and for 150 min both at 60 and 90 1/s. Figure 30 shows the transient shear stresses for (a) $\dot{\gamma}/E^2 \simeq 9.2 \times 10^2$ $(\text{mm/kV})^2$/s and (b) $\dot{\gamma}/E^2 \simeq 2.8 \times 10^2$ $(\text{mm/kV})^2$/s, where step electric fields were applied at $t = 0$. It is apparent that there are two steps in the increase of shear stress: a fast rise immediately after applying a step electric field and then a slow rise including an overshoot, called the first and second modes, respectively.[57] Note that in the blend of equal viscosity the first mode was not observed, as shown in Fig. 20. For the same $\dot{\gamma}/E^2$, both the modes increase in shear stress as the shear rate (the electric field) is raised.

Figure 31 shows the plots of the viscosity η vs the strain γ for (a) $\dot{\gamma}/E^2 \simeq 9.2 \times 10^2$ $(\text{mm/kV})^2$/s and (b) $\dot{\gamma}/E^2 \simeq 2.8 \times 10^2$ $(\text{mm/kV})^2$/s. In Fig. 31(a) all the three curves at different shear rates almost coincide, clearly indicating the validity of the scaling relation. Also before applying the field, the viscosities are the same. This is a result of the Doi-Ohta scaling relation, i.e., the viscosity is independent of the shear rate. While in Fig. 31(b) the curves become separate at large strains, though the coincidence is good at small strains. This may be explained as follows. When an electric field is

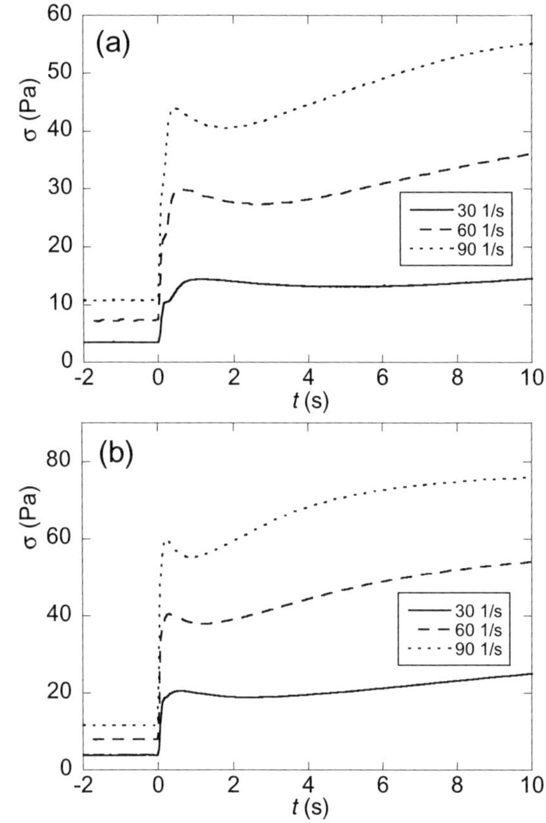

Fig. 30. Transient shear stresses for (a) $\dot{\gamma}/E^2 \simeq 9.2 \times 10^2$ $(\mathrm{mm/kV})^2$ /s and (b) $\dot{\gamma}/E^2 \simeq 2.8 \times 10^2$ $(\mathrm{mm/kV})^2$ /s, where step electric fields were applied at $t = 0$. Reproduced with permission from Ref. 59. Copyright (2008) by the Physical Society of Japan.

applied, LCP droplets near the glass surface adhere to it, and the amount of LCP on the glass surface may increase gradually with time, leading to the breakdown of the scaling relation. Furthermore, the adhesion rate may become large as $\dot{\gamma}/E^2$ is decreased, and therefore the breakdown may become remarkable for small $\dot{\gamma}/E^2$. From the comparison between Figs. 31(a) and (b) it is seen that the ratio of the first mode to the second one increases as $\dot{\gamma}/E^2$ is reduced.

In immiscible blends we have complex interfaces between two fluids, which are crucial to determine the rheological properties. Curiously, however, there appears no interfacial tension Γ in Eq. (49). This is because

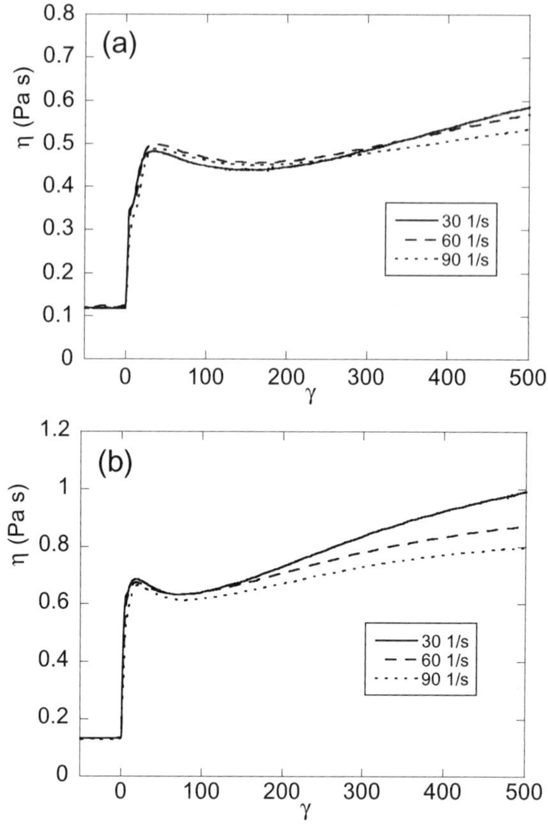

Fig. 31. Plots of the viscosity η vs the strain γ for (a) $\dot{\gamma}/E^2 \simeq 9.2 \times 10^2$ $(\mathrm{mm/kV})^2/\mathrm{s}$ and (b) $\dot{\gamma}/E^2 \simeq 2.8 \times 10^2$ $(\mathrm{mm/kV})^2/\mathrm{s}$. Reproduced with permission from Ref. 59. Copyright (2008) by the Physical Society of Japan.

one can construct neither any quantity with units of stress nor any dimensionless quantity including Γ. However, one might still think that Eq. (1) has Γ so that the stress should depend on the interfacial tension. This is resolved as follows. The interface tensor $q_{\alpha\beta}$ in Eq. (1) has a unit of inverse length. We can construct a quantity with units of inverse length including Γ, for example, $\eta_1\dot{\gamma}/\Gamma$. Thus, the scaling relation for q_{zx} is given as

$$\begin{aligned} q_{zx}&\left(t, \eta_1, \eta_2, \varepsilon_1, \varepsilon_2, \Gamma, \dot{\gamma}, E\right) \\ &= \eta_1\dot{\gamma}/\Gamma \cdot g\left(\dot{\gamma}t, \eta_1/\eta_2, \varepsilon_1/\varepsilon_2, \eta_1\dot{\gamma}/\varepsilon_1 E^2\right), \end{aligned} \tag{58}$$

where $g(\cdots)$ is a scaling function. This equation multiplied by Γ is independent of Γ.

We have a similar scaling relation for a characteristic length:

$$\begin{aligned}\xi &(t, \eta_1, \eta_2, \varepsilon_1, \varepsilon_2, \dot{\gamma}, E) \\ &= \Gamma/\eta_1\dot{\gamma} \cdot h\left(\dot{\gamma}t, \eta_1/\eta_2, \varepsilon_1/\varepsilon_2, \eta_1\dot{\gamma}/\varepsilon_1 E^2\right),\end{aligned} \tag{59}$$

where $h(\cdots)$ is a scaling function. This scaling relation indicates that the structures at different times become statistically the same when they are magnified by a factor proportional to $\dot{\gamma}$ and compared at the same strain γ, as long as $\dot{\gamma}/E^2$ remains constant. We examined it experimentally by observing the structure. The CSLM observation was made 3 mm far from the edge of the rotating disk and 45 μm high from the glass surface, and the frame rate was 250 frames/s. Note that no 3D image could not be obtained because the shear rates were so high in the present experiment. Figure 32 shows the transient images captured at the same time with the stress measurements in Fig. 30(a). The images are arrayed at each shear strain for 30 and 60 1/s, and the flow and vorticity directions are, respectively, along the horizontal and vertical axes and so the velocity gradient and the electric field are both perpendicular to the image plane. Before applying a step electric field ($t < 0$), spherical droplets are seen and the size depends on the shear rate; it seems to be inversely proportional to the shear rate, as is expected from Eq. (59). Also, after applying the step electric field, the images seem to satisfy the above-mentioned scaling property, though it was difficult to make quantitative analysis using the limited number of images. It is seen from Fig. 31(a) that the first mode changes into the second one at around $\gamma = 5$. We could not observe any deformation of droplets for $\gamma < 5$, indicating that the droplets may only translate and/or they may be elongated along the electric field in the first mode. While in the second mode the droplets are deformed, elongated also along the flow direction and coalesce.

Lastly, we should examine the effect of conductivity. In the above derivation of the scaling relations, we assumed that the two fluids were perfect dielectrics. In actual, however, the LCP is weakly conductive (2.0×10^{-8} $\Omega^{-1}\text{m}^{-1}$), though the MPS can be regarded as perfect dielectrics. In such a case, β in Eq. (54) becomes $\beta_c = (\sigma_1 - \sigma_2)/(\sigma_1 + 2\sigma_2)$ instead of $\beta_d = (\varepsilon_1 - \varepsilon_2)/(\varepsilon_1 + 2\varepsilon_2)$ for a DC electric field.[18] In our blend, $\beta_c = 1$ is larger than $\beta_d = 0.61$. In general, the Maxwell-Wagner model for an AC electric field of frequency ω gives

$$\beta = \frac{(\sigma_1 + i\omega\varepsilon_1) - (\sigma_2 + i\omega\varepsilon_2)}{(\sigma_1 + i\omega\varepsilon_1) + 2(\sigma_2 + i\omega\varepsilon_2)}. \tag{60}$$

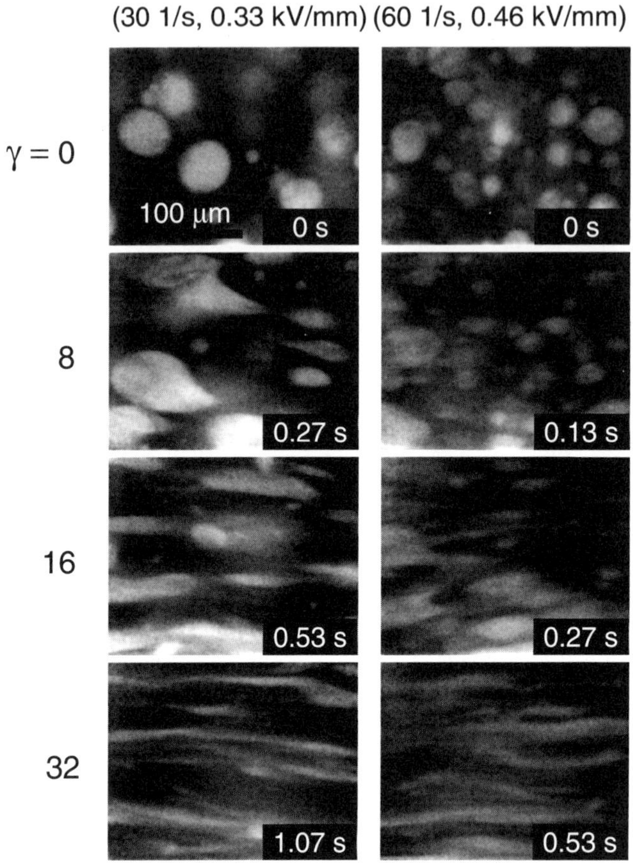

Fig. 32. Transient images captured at the same time with the stress measurements in Fig. 30(a). The images are arrayed at each shear strain for 30 and 60 1/s, and the electric field is perpendicular to the image plane. Reproduced with permission from Ref. 59. Copyright (2008) by the Physical Society of Japan.

sphere as $\tau = (\varepsilon_1 + 2\varepsilon_2) / (\sigma_1 + 2\sigma_2)$. For a step electric field, β changes from β_d to β_c at around $t = \tau$. The presence of this characteristic time adds another dimensionless quantity $\dot{\gamma}\tau$ to the right-hand side of Eq. (53), implying the breakdown of the above scaling relations. However, they are still valid for the following reason. In our blend, τ is about 0.01 s. While the rise time of the first mode is about 0.1 s in our experiments. Therefore, the charge relaxation occurs faster than the structural change. It is expected that this situation may hold for complex structures different from the isolated sphere.

7. Summary

A new system has been introduced which has succeeded in clearly observing the three-dimensional structure of immiscible polymer blends. The relationships between this structure and rheological properties have been discussed. Without shear flow, droplet coalescence was observed under low electric fields, and the process was well described in terms of the hierarchical model. Exponential and non-exponential growths were derived for spherical and ellipsoidal droplets, respectively. Shear modulus of a columnar structure appearing at high electric fields was measured and the emergence of elasticity was found. The dependence of the elasticity on electric field and frequency was well theoretically explained. Under shear flow subjected to a step electric field, on the other hand, a structural change was observed from a droplet-dispersed phase to a network one. Three stress contributions were successfully separated. It was shown that not only the interfacial stress but also the Maxwell stress can be described in terms of the interface tensor, and that scaling properties hold in the transient process. Thus, we have successfully determined the relationship between the structure (the interface tensor) and the shear stress in the transient process of immiscible polymer blends. However, the mechanism of network formation under shear flow and electric fields remains unclear. This should be studied based upon evolution equations such as the Doi-Ohta model.[14] One important point, we think, is that in the structural change from a droplet-dispersed phase to a network one the topology does change; in the droplet-dispersed phase only the matrix phase (PIB) is continuous, whereas in the network one both LCP and PIB are continuous. This indicates that we need in phenomenological equations other state variables such as the Gaussian curvature, which specify the topological properties.

The author would like to thank Yang Ho Na, Tomoyuki Nagaya, Seiji Ujiie, Kohei Aida, Yuki Nishimoto and Tanabu Kana for their collaboration and Takao Ohta for fruitful discussions. This work was supported by a Grant-in-Aid for Scientific Research [Priority Area Soft Matter Physics (Contract No. 463)].

References

1. C. L. Tucker and P. Moldenaers, Microstructural evolution in polymer blends, *Annu. Rev. Fluid Mech.* **34**, 177–210, (2002).
2. G. K. Batchelor, Stress system in a suspension of force-free particles, *J. Fluid Mech.* **41**, 545–570, (1970).

3. A. Onuki, Viscosity enhancement by domains in phase-separating fluids near the critical-point - proposal of critical rheology, *Phys. Rev. A.* **35**, 5149–5155, (1987).
4. A. Onuki, Domain growth and rheology in phase-separating binary-mixtures with viscosity difference, *Europhys. Lett.* **28**, 175–179, (1994).
5. M. Takahashi, P. H. P. Macaubas, K. Okamoto, H. Jinnai, and Y. Nishikawa, Stress prediction for polymer blends with various shapes of droplet phase, *Polymer.* **48**, 2371–2379, (2007).
6. M. Takahashi and K. Okamoto, Retraction of rod-like and spheroidal droplets and stress relaxation after step shear strain in polymer blends, *Nihon Reoroji Gakkaishi.* **35**, 199–205, (2007).
7. H. Yamane, M. Takahashi, R. Hayashi, K. Okamoto, H. Kashihara, and T. Masuda, Observation of deformation and recovery of poly(isobutylene) droplet in a poly(isobutylene)/poly(dimethyl siloxane) blend after application of step shear strain, *J. Rheol.* **42**, 567–580, (1998).
8. J. Mellema and M. W. M. Willemse, Effective viscosity of dispersions approached by a statistical continuum method, *Physica A.* **122**, 286–312, (1983).
9. E. D. Wetzel and C. L. Tucker, Droplet deformation in dispersions with unequal viscosities and zero interfacial tension, *J. Fluid Mech.* **426**, 199–228, (2001).
10. W. Yu and M. Bousmina, Ellipsoidal model for droplet deformation in emulsions, *J. Rheol.* **47**, 1011–1039, (2003).
11. W. Yu, M. Bousmina, C. X. Zhou, and C. L. Tucker, Theory for drop deformation in viscoelastic systems, *J. Rheol.* **48**, 417–438, (2004).
12. K. Okamoto and M. Takahashi, Effects of interface velocity on the stress tensor in immiscible polymer blends: Retraction of spheroidal droplets and stress relaxation, *Nihon Reoroji Gakkaishi.* **36**, 43–49, (2008).
13. K. Okamoto and M. Takahashi, Shear and normal stress relaxation in polymer blends: Stress predictions from interface velocity and laplace pressure terms, *Rheol. Acta.* **47**, 797–805, (2008).
14. M. Doi and T. Ohta, Dynamics and rheology of complex interfaces .1, *Journal of Chemical Physics.* **95**, 1242–1248, (1991).
15. A. Inoue and S. Maniwa, Electrorheological effect of liquid-crystalline polymers, *J. Appl. Polym. Sci.* **55**, 113–118, (1995).
16. K. Tajiri, K. Ohta, T. Nagaya, H. Orihara, Y. Ishibashi, M. Doi, and A. Inoue, Electrorheological effect in immiscible polymer blends, *J. Rheol.* **41**, 335–341, (1997).
17. H. Kimura, K. Aikawa, Y. Masubuchi, J. Takimoto, K. Koyama, and K. Minagawa, Phase structure change and er effect in liquid crystalline polymer/dimethylsiloxane blends, *Rheol. Acta.* **37**, 54–60, (1998).
18. X. D. Pan and G. H. McKinley, Characteristics of electrorheological responses in an emulsion system, *J. Colloid Interface Sci.* **195**, 101–113, (1997).
19. J. W. Ha and S. M. Yang, Rheological responses of oil-in-oil emulsions in an electric field, *J. Rheol.* **44**, 235–256, (2000).

20. F. E. Filisko and L. H. Radzilowski, An intrinsic mechanism for the activity of aluminosilicate based electrorheological materials, *J. Rheol.* **34**, 539–552, (1990).
21. C. F. Zukoski, Material properties and the electrorheological response, *Annu. Rev. Mater. Sci.* **23**, 45–78, (1993).
22. T. Honda and T. Sasada, The mechanism of electroviscosity. i. electrohydrodynamic effect on polar liquids, *Jpn. J. Appl. Phys.* **16**, 1783–1977, (1977).
23. T. Honda and T. Sasada, Electroviscous effect in non-polar liquids, *Jpn. J. Appl. Phys.* **18**, 1031–1037, (1979).
24. K. Skarp, S. T. Lagerwall, and B. Stebler, Measurements of hydrodynamic parameters for nematic 5cb, *Mol. Cryst. Liquid Cryst.* **60**, 215–236, (1980).
25. I. K. Yang and A. D. Shine, Electrorheology of a nematic poly(normal-hexyl isocyanate) solution, *J. Rheol.* **36**, 1079–1104, (1992).
26. H. Orihara, M. Doi, and Y. Ishibashi, Two types of mechanisms of electrorheological effect in polymer blends, *in Proc. of the 6th Int. Conf. on ER Fluids, MR Suspensions and Their Applications, Yonezawa,1997, edited by M. Nakano and K. Koyama (World Scientific, Singapore, 1998).* pp. 367–373, (1998).
27. H. Kimura, K. Aikawa, Y. Masubuchi, J. Takimoto, K. Koyama, and T. Uemura, Electrically induced phase inversion in polyurethane / dimethylsiloxane blend, *in Proc. of the 6th Int. Conf. on ER Fluids, MR Suspensions and Their Applications, Yonezawa, 1997, edited by M. Nakano and K. Koyama (World Scientific, Singapore, 1998).* pp. 429–435, (1998).
28. H. Kimura, K. Aikawa, Y. Masubuchi, J. Takimoto, K. Koyama, and T. Uemura, 'positive' and 'negative' electro-rheological effect of liquid blends, *J. Non-Newton. Fluid Mech.* **76**, 199–211, (1998).
29. H. Kito, K. Tajiri, H. Orihara, Y. Ishibashi, M. Doi, and A. Inoue, A novel mechanism of electrorheological effect in polymer blends, *in Proc. of the 6th Int. Conf. on ER Fluids, MR Suspensions and Their Application, Yonezawa, 1997, edited by M. Nakano and K. Koyama (World Scientific, Singapore, 1998).* pp. 401–407, (1998).
30. I. Delaby, B. Ernst, Y. Germain, and R. Muller, Droplet deformation in polymer blends during uniaxial elongational flow - influence of viscosity ratio for large capillary numbers, *J. Rheol.* **38**, 1705–1720, (1994).
31. H. Verhoogt, J. Vandam, A. P. Deboer, A. Draaijer, and P. M. Houpt, Confocal laser scanning microscopy — a new method for determination of the morphology of polymer blends, *Polymer.* **34**, 1325–1329, (1993).
32. H. Jinnai, Y. Nishikawa, T. Koga, and T. Hashimoto, Direct observation of 3-dimensional bicontinuous structure developed via spinodal decomposition, *Macromolecules.* **28**, 4782–4784, (1995).
33. H. Jinnai, T. Koga, Y. Nishikawa, T. Hashimoto, and S. T. Hyde, Curvature determination of spinodal interface in a condensed matter system, *Phys. Rev. Lett.* **78**, 2248–2251, (1997).
34. K. Aida, Y. H. Na, T. Nagaya, and H. Orihara, Droplet coalescence process under electric fields in an immiscible polymer blend, *Phys. Rev. E.* **82**, 031805, (2010).

35. J. S. Eow, M. Ghadiri, A. O. Sharif, and T. J. Williams, Electrostatic enhancement of coalescence of water droplets in oil: a review of the current understanding, *Chem. Eng. J.* **84**, 173–192, (2001).
36. J. S. Eow, M. Ghadiri, and A. Sharif, Experimental studies of deformation and break-up of aqueous drops in high electric fields, *Colloids and Surfaces a-Physicochemical and Engineering Aspects.* **225**, 193–210, (2003).
37. M. Chiesa, J. A. Melheim, A. Pedersen, S. Ingebrigtsen, and G. Berg, Forces acting on water droplets falling in oil under the influence of an electric field: numerical predictions versus experimental observations, *Eur. J. Mech. B-Fluids.* **24**, 717–732, (2005).
38. D. J. Klingenberg, F. Vanswol, and C. F. Zukoski, Dynamic simulation of electrorheological suspensions, *Journal of Chemical Physics.* **91**, 7888–7895, (1989).
39. T. C. Halsey, J. E. Martin, and D. Adolf, Rheology of electrorheological fluids, *Phys. Rev. Lett.* **68**, 1519–1522, (1992).
40. J. E. Martin, J. Odinek, and T. C. Halsey, Evolution of structure in a quiescent electrorheological fluid, *Phys. Rev. Lett.* **69**, 1524–1527, (1992).
41. T. A. Vorob'eva, I. N. Vlodavets, and P. I. Zubov, The size distribution of oriented aggregates formed in suspension with the application of an alternating electric field, *Colloid Journal of the USSR.* **31**, 533–537, (1969).
42. T. A. Vorob'eva and I. N. Vlodavets, Kinetics of the change in the average size of oriented aggregates upon application of an alternating electric field to disperse systems, *Colloid Journal of the USSR.* **36**, 1646, (1974).
43. R. Botet, R. Jullien, and M. Kolb, Hierarchical model for irreversible kinetic cluster formation, *J. Phys. A-Math. Gen.* **17**, L75–L79, (1984).
44. H. See and M. Doi, Aggregation kinetics in electrorheological fluids, *J. Phys. Soc. Jpn.* **60**, 2778–2782, (1991).
45. M. Avrami, Kinetics of phase change. iii: Granulation, phase change an microstructures, *J. Chem. Phys.* **9**, 177–184, (1941).
46. K. Sekimoto, Evolution of the domain-structure during the nucleation-and-growth process with nonconserved order parameter, *Physica A.* **135**, 328–346, (1986).
47. S. Torza, R. G. Cox, and S. G. Mason, Electrohydrodynamic deformation and burst of liquid drops, *Philos. Trans. R. Soc. London, Ser. A.* **269**, 269–319, (1971).
48. L. D. Landau, E. M. Lifshitz, and L. P. Pitaevskii, *Electrodynamics of Continuous Media, 2nd ed.* (Pergamon, Oxford, 1984).
49. M. Trau, S. Sankaran, D. A. Saville, and I. A. Aksay, Electric-field-induced pattern formation in colloidal dispersions, *Nature.* **374**, 437–439, (1995).
50. M. Trau, S. Sankaran, D. A. Saville, and I. A. Aksay, Pattern formation in nonaqueous colloidal dispersions via electrohydrodynamic flow, *Langmuir.* **11**, 4665–4672, (1995).
51. S. Y. Chou, L. Zhuang, and L. J. Guo, Lithographically induced self-construction of polymer microstructures for resistless patterning, *Appl. Phys. Lett.* **75**, 1004–1006, (1999).

52. E. Schaffer, T. Thurn-Albrecht, T. P. Russell, and U. Steiner, Electrically induced structure formation and pattern transfer, *Nature.* **403**, 874–877, (2000).
53. Z. Q. Lin, T. Kerle, T. P. Russell, E. Schaffer, and U. Steiner, Structure formation at the interface of liquid liquid bilayer in electric field, *Macromolecules.* **35**, 3971–3976, (2002).
54. K. Aida, Y. H. Na, T. Nagaya, and H. Orihara, Storage shear modulus of columnar structure formed in an immiscible polymer blend under electric fields, *Phys. Rev. E.* **80**, 041807, (2009).
55. D. Graebling, R. Muller, and J. F. Palierne, Linear viscoelastic behavior of some incompatible polymer blends in the melt - interpretation of data with a model of emulsion of viscoelastic liquids, *Macromolecules.* **26**, 320–329, (1993).
56. H. Orihara, Y. Nishimoto, K. Aida, and Y. H. Na, Three-dimensional observation of an immiscible polymer blend subjected to a step electric field under shear flow, *Phys. Rev. E.* **83**, 026302, (2011).
57. H. Orihara, N. Miwa, M. Doi, and A. Inoue, Response properties of immiscible polymer blend electrorheological fluids, *J. Rheol.* **45**, 773–781, (2001).
58. A. Luciani, M. F. Champagne, and L. A. Utracki, Interfacial tension coefficient from the retraction of ellipsoidal drops, *J. Polym. Sci. Pt. B-Polym. Phys.* **35**, 1393–1403, (1997).
59. H. Orihara, K. Tanabu, and Y. H. Na, Scaling properties of immiscible fluid blends in electric and shear flow fields, *J. Phys. Soc. Jpn.* **77**, 093802, (2008).
60. Y. Takahashi, N. Kurashima, I. Noda, and M. Doi, Experimental tests of the scaling relation for textured materials in mixtures of 2 immiscible fluids, *J. Rheol.* **38**, 699–712, (1994).
61. M. Parthasarathy and D. J. Klingenberg, Large amplitude oscillatory shear of er suspensions, *J. Non-Newton. Fluid Mech.* **81**, 83–104, (1999).
62. H. Orihara, T. Shibuya, T. Nagaya, and S. Ujiie, Topological transition of an immiscible polymer blend in electric and shear flow fields, *J. Phys. Soc. Jpn.* **75**, 063802, (2006).

Chapter 4

Dynamical Aspects of Two-Dimensional Soft Matter

F. Sagués*, J. Claret and J. Ignés-Mullol

Departament de Química Física, Facultat de Química, Universitat de Barcelona, Martí i Franquès 1 08028 Barcelona, Catalunya, Spain

Two-dimensional soft matter permits to unveil new physical phenomena which may trigger the attention of an interdisciplinary community. This is the ultimate goal of this chapter that reviews our recent work on a wide variety of dynamical behaviors of forced Langmuir monolayers, from pattern forming scenarios to rheological properties of confined flows. Such diverse aspects are here covered in most cases using as a model system a photosensitive monolayer prepared from an azobenzene surfactant. The approach is mostly experimental although, when appropriate, a convenient level of modelization is provided to better interpret the observed results.

1. Introduction

Soft matter is an increasingly important discipline both in relation to fundamental physico-chemical problems as well as regarding its many daily life practical applications. Under this category we identify a wide variety of labile condensed systems self-assembled from its individual molecular components. Among other properties systems under this name share three important features: they are *complex* (sometimes the term complex fluids is equivalently employed to refer to them), they are *flexible* in the sense given by P.G. de Gennes, "a very mild chemical action induces a drastic change in its mechanical properties"[1] and finally they are *addressable*, meaning that they display large and tunable responses when facing external forces (electrical, magnetical, optical, etc). A particularly studied group of soft materials is represented by ultrathin films (in the paradigmatic case one

*f.sagues@ub.edu

molecule thick) organized either at the interface between two fluids or anchored on solid supports.[2,3] Particularly interesting for us is the example provided by (insoluble) surfactant-like compounds deposited at the water/air interface, the so-called Langmuir monolayers.[4] Although known for at least more than one century in laboratory tailored experiments, they still offer new challenges to researchers both from the point of view of their equilibrium configurations, but singularly for their non-equilibrium properties, our main interest on what follows. From this latter point of view, Langmuir monolayers are ideal systems for two-dimensional physics encompassing aspects such diverse as those reviewed here: from distinctive patterns of self-organization to peculiar rheological properties or newly discovered effects related to the appearance and dynamics of topological defects. Underlying all these developments, classical observation procedures, mainly X-ray, fluorescence and reflection-based optical microscopies, offer a contrasted set of techniques to reveal the intimate organization of such assembled films from the nano to the micrometer scale.[5] From a more applied point of view and only briefly tackled on what follows, Langmuir–Blodgett monolayers, resulting from the transfer of Langmuir monolayers onto solid supports are presently engineered towards the preparation of magnetic or optoelectronic devices and chemical and biochemical sensors.[6] Finally and although not discussed at all in this contribution, we should not forget the implications of Langmuir monolayers as simple models for lipidic membranes or other cellular components.[7]

As anticipated, the choice of the material contained in this chapter goes beyond the classical, equilibrium-based, description of Langmuir monolayers to concentrate on dynamic phenomena that have been and are currently being investigated in our laboratory. For the most part we will refer to a particular example of an azobenzene-based photosensitive monolayer. In Sec. 2 we review different modes of pattern formation under polarized light. In Sec. 3, we report on different scenarios of defect recombination in relation to a phase-separation and coarsening process. In Sec. 4 we discuss a striking discovery of chiral selection at the level of the monolayer induced by a vortical stirring applied to the aqueous subphase. In Sec. 5, we turn finally into a more rheological perspective to propose an unprecedented strategy mimicking microfluidic developments in two-dimensional flows.

In the remaining part of this introduction we will briefly refer to the main characteristics of the azobenzene based monolayers object of our study and their most important features relative to their preparation, observation and characterization. In general, the state of a Langmuir monolayer is characterized by the temperature, the molecular area and the surface pressure, the latter defined as the decrease of the surface tension of the air-water interface relative to its value for the clean free surface. Monolayer preparation follows standard protocols. First a solution of the amphiphile in an organic volatile solvent is prepared and further poured on the aqueous subphase contained in a thermostated hydrophobic trough (typically made of Teflon). This simple equipment is completed in its standard configuration with a movable barrier in order to gain control on the total area available to the monolayer. In this way the molecular area may be changed in a controlled way and the subsequent variation of the surface pressure is monitored with a Wilhelmy balance. The monolayer can be observed with different techniques depending on the required length scale. Here we will mainly resort on the use of Brewster angle microscopy (BAM), designed to elucidate the structure at a mesoscopic (submillimeter) level.[8,9] Unless otherwise stated the analyzer of the Brewster angle microscope is set at 60° with respect to the polarizer. A detailed discussion on the different observation techniques can be found elsewhere.[10] BAM images reproduce the long range orientational order of the monolayers. In their condensed state monolayers behave optically as uniaxial anisotropic media. The elicited birefringence gives rise to typical textures whose quantitative interpretation is far from trivial in most practical situations. In our case the textures show a high degree of symmetry, as it will become clear in what follows, a feature that renders their interpretation quantitative. As a matter of fact, we will consistently assume that the tilt angle of the amphiphilic molecules with respect to the monolayer normal is uniform and thus BAM images can be directly mapped to the distribution of the azimuthal angle.

In most of what follows the chosen monolayer is based on 4-[4-[(4-octylphenyl)azo]-phenoxy]butanoic acid, denoted in short as 8Az3COOH, kept at low surface pressures. The characteristics of the monolayer strongly depend on the presence of the azobenzene group (a double benzene ring connected through a -N=N- (azo) bridge) in the aliphatic chain of the amphiphilic molecule. This chemical functionality endows the compound with a well-known photoisomerization capability upon appropriate irradiation.[11] Radiation in the UV range drifts the isomerization equilibrium into the *cis* form and contrariwise with blue light (or simply by thermal relaxation). Monolayers prepared entirely from the *cis* (kinked) isomer do not display any signature of structured mesophase. Conversely, when spreading a solution from the *trans* (stretched) isomer, BAM images reproduce the formation of extended domains of different morphologies and smooth reflectivity gradients, depending on the preparation and operation conditions.[12] It is worth stressing at this point the observation of long range orientational order at higher area per molecule values of the *trans* amphiphile as compared to relatively similar fatty-acid compounds. This feature is attributed to labile collective interactions between individual components, in particular the formation of H-aggregates by stacking of the benzene rings.[13–16]

When a monolayer is spread from a solution containing both isomers, or from the pure *cis* isomer left to evolve under room light, a phase separation process takes place. This is explained in terms of the different dipolar moment of the two species (higher in the *cis* conformation). The result is a two-dimensional emulsion-like state composed of circular *trans* domains embedded within an isotropic matrix of *cis* isomer. BAM analysis reflects a perfectly well-defined molecular assembling inside the domains that depend on the domain size, apart from the basic thermodynamic state variables of the monolayer, i.e. the temperature and surface pressure.[13,17] The most important and common feature of all these arrangements is the assembling of the molecules around a central defect (topological singularity). At low temperatures, molecules are organized in a bending mode (azimuths perpendicular to the radial direction) winding clockwise and counterclockwise with equal probability. At higher temperatures ($T > 30°C$), this configuration still appears for small domains but for larger ones it changes substantially. In the inner regions of the domains molecules continue to bend around the central core defect, but the molecular azimuths tend to align along the radial direction when approaching the domain boundary. In this case the distortion has a dominant splay contribution. Observations of both splay-in (molecules pointing inwards) and splay-out (molecules pointing outwards) appear depending on the thermodynamic state of the monolayer.[13] This state of the 8Az3COOH Langmuir monolayer is not thermodinamically stable and slowly evolves to the formation of irregular large domains displaying the reflectivity gradients typical of a pure *trans* Langmuir monolayer. This process occurs by means of two different mechanisms: the slow thermal conversion of *cis* to *trans* isomers, and the fusion of bend domains, which will permit to analyze the dynamics of topological defects in Sec. 3.

In Fig. 1, we present BAM observations of an extended region of the monolayer and a focused image of a singled-out domain. Notice the encircled area showing the fusion of two bend domains during the evolution of the monolayer. Finally in Fig. 2 we compose different panels corresponding to the reconstructed textures of isolated domains comprising the whole set of bend and splay configurations observed in our azo-benzene monolayers.

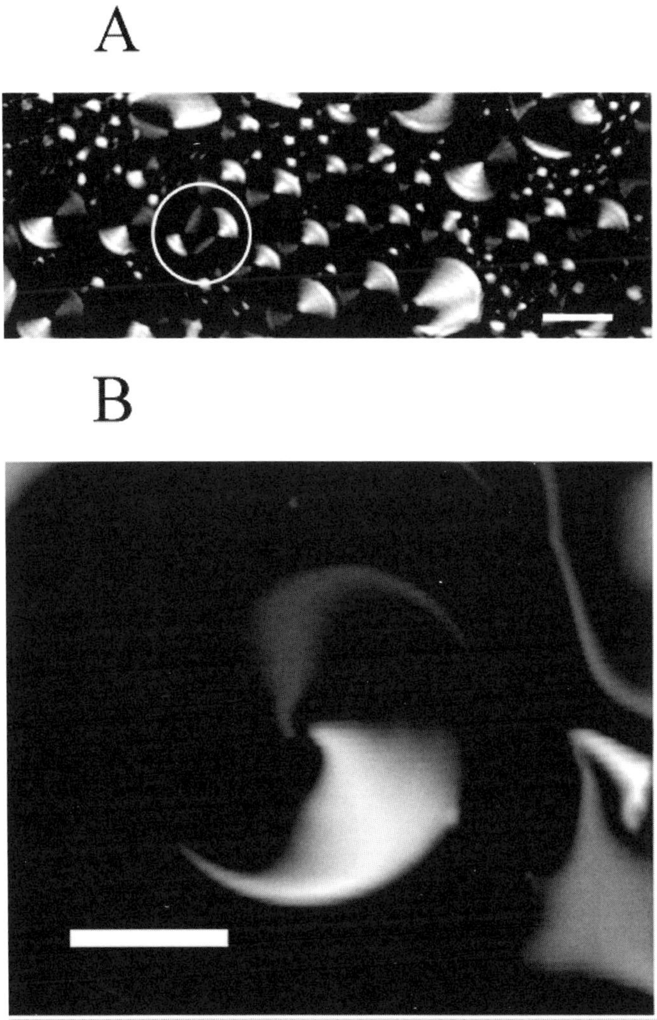

Fig. 1. (A) Bend domains in a mixed *cis-trans* 8Az3COOH monolayer 2 min after spreading at 34°C and 10mN m^{-1}. The encircled area shows the fusion of two bend domains. The line represents 100 μm. (B) CW splay-out domain of the trans-isomer embedded in the isotropic phase of the cis-isomer at 35°C and 2mN m^{-1}. The line represents 100 μm.

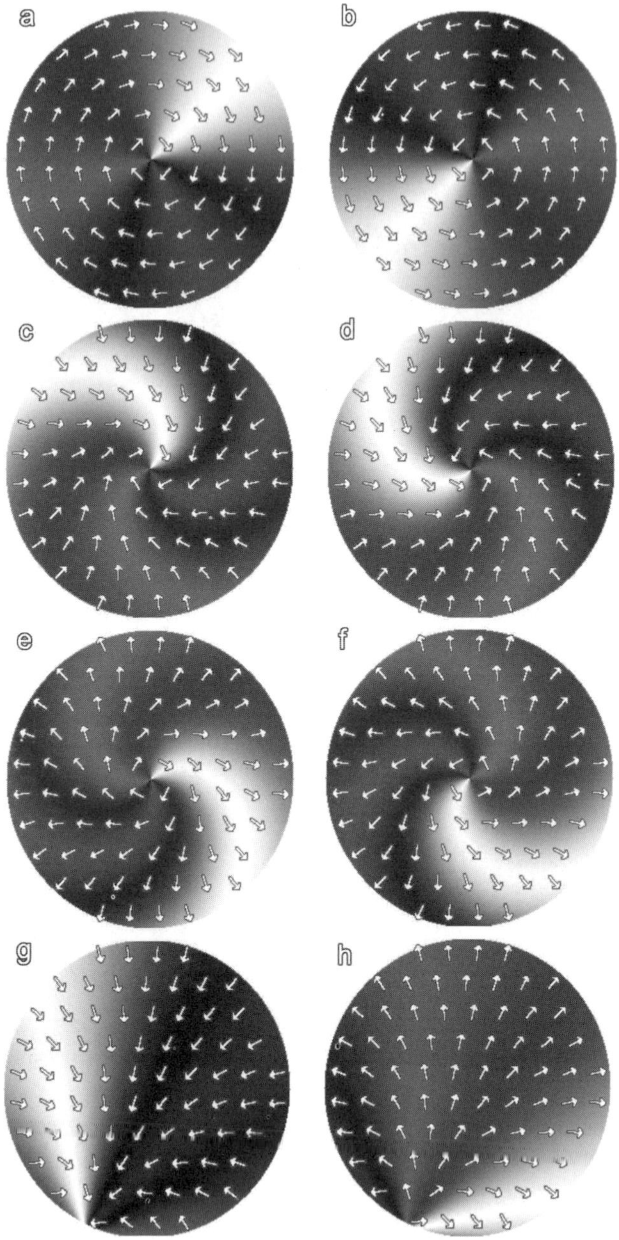

Fig. 2. Sketches of confined azimuth fields with corresponding generated BAM images. (a)-(f): six basic motifs for the observed centrosymmetric configurations with a core texture either clockwise (left column) or counterclockwise bend (right column). The bend texture is either preserved in the whole droplet [(a), (b)] or it evolves linearly towards splay (inwards or outwards) at the droplet boundary. Sketch (e) corresponds to the texture in 1B. Sketches (g) and (h) illustrate two examples of boojum textures.

2. Photoalignment and Rotation

In this section we report on experiments that were designed to prove the emergence of cooperative phenomena in Langmuir monolayers when subjected to external stimuli and thus kept under non-equilibrium conditions. Following previous model-based studies,[18] our declared goal was to analyze how stimulated at a molecular level these self-assembled systems transfer information spanning disparate length scales up to the possibility to elicit patterns at meso or macrosocopic ranges.[19] In a sense we were motivated to parallel the common scenarios discovered in self-organizing systems where instabilities from unstructured states are known to give rise to patterns endowed with characteristic time and length scales.[20]

Photoalignment experiments in azobenzene Langmuir monolayers had been performed originally in a system composed nearly exclusively of *trans* molecules.[21] In this paper, the effect of irradiation was studied by illuminating striped, Smectic-C-like, textures which develop spontaneously at long times. Although a photoalignment effect had been reported following these initial observations, its analysis was hindered by the spatial gradients in molecular tilt inherent to the stripe patterns. In our case, we preferred to study dynamical modes of self-organization following from controlled illumination of mixed *cis/trans* monolayers.[22,23] In this case, as mentioned in the introduction, the system organizes as a two-dimensional emulsion composed of circular *trans* domains embedded in an isotropic *cis* matrix.[12] This choice permits to study the photoalignment process by monitoring the instantaneous orientation of the molecular field inside the domains during irradiation of the monolayer. Extending previous theoretical approaches referring to simpler dynamical modes of such monolayers,[17] we proposed a phenomenological approach that unveils the nature of the coupling between irradiation and long-range orientational order in this system.

2.1. *Patterns of photoalignment: Experiments*

In the experiments reviewed here, monolayers were prepared by spreading a room light photostationary chloroform solution of 8Az3-COOH on a water subphase. Experiments were performed in a custom-built Teflon trough thermostated at 35.0±0.2°C, equipped with two barriers for symmetric compression, while surface pressure was kept below 2 mN m^{-1}. Based on the assumption of a uniaxial dielectric layer and a uniform molecular tilt with respect to the monolayer normal, BAM imaging allows the extraction of the distribution of the azimuth angle from the spatial gradients of the monolayer reflectivity.[24] Monolayers were irradiated with power densities in the range of 0.1-5 mW cm^{-2} by means of a 50-W halogen lamp, filtered through a bandpass filter (480±5 nm,) and linearly polarized with an absorption filter. Under these conditions, isomerization is practically absent within the timescale of our experiments, and thus we can assume that monolayer composition remains unmodified.

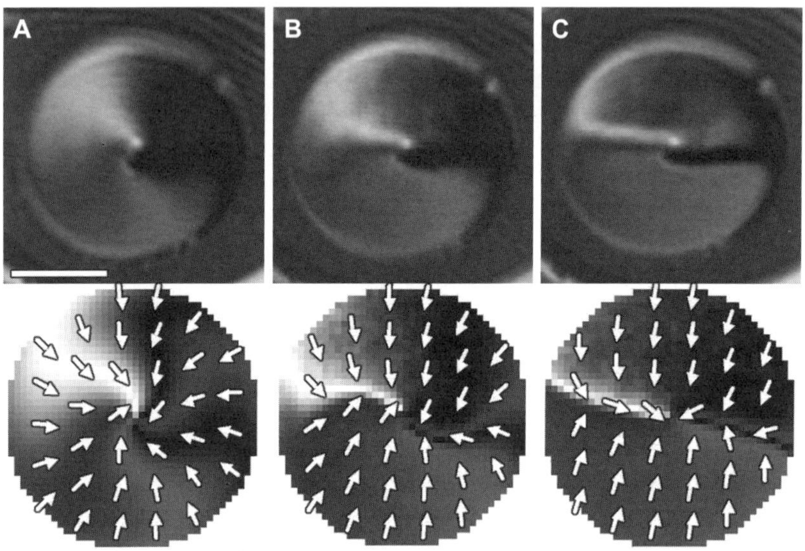

Fig. 3. Evolution towards the photoaligned state of splay-in droplets. Light is switched on in A with a power density 0.4 mW cm^{-2}. Elapsed times are 2.0 s in B, and 4.0 s in C. The line segment in A is 100 μm long. The bottom row contains the results of the numerical experiment. Light polarization plane is horizontal.

In these experiments, we took as initial condition monolayers composed of domains with inner textures resulting from a continuous splay-like distortion. Irradiation of labile *trans* droplets displaying a (metastable) splay-in configuration (molecules pointing towards a central defect) supposes the simplest scenario. In this case, linear polarized light induces a collective reorientational dynamics, at the end of which most molecules align perpendicular to the polarization plane. If the intensity of light irradiation exceeds a threshold, or the rotational mobility is hindered (under conditions of low temperature or high surface pressure conditions), irradiation results in a reversible fragmentation of the droplet texture. Due to the initial axisymmetric droplet textures, a photoalignment perpendicular to the polarization direction leads to a symmetry change of the molecular orientational field, so that the droplet gets divided into two semicircles (see Fig. 3C) separated by a "wall" roughly parallel to the polarization direction. Inside each semicircle, molecules point, on average, towards the wall. The pathway leading to the final configuration is in this case rather simple (Fig. 3). The splay distortion loses its axial symmetry and concentrates in the equatorial region, where molecules are roughly parallel to the $\vec{E}$ field. The reorientation process starts in the domain poles, where molecules are already properly oriented, and propagates to the rest of the droplet in a few seconds.

The reorientation process from splay-out droplets, although leading to the same final configuration, is much more complex and in fact can even follow different pathways as illustrated in Fig. 4 (pathway I) and Fig. 5 (pathway II). This shouldn't come as a surprise since the initial splay-out texture is indeed very different from the final photoaligned state. In particular, the overall sign of the splay distortion must change sign from positive to negative. Independently of the reorientational pathway, all splay-out droplets are first divided along the diameter perpendicular to the $\vec{E}$ field. On either side, the 2d director orients, on average, parallel to the polarization (see Figs. 4B and 5B). In pathway I, we observe the simultaneous propagation of two distortion lines from east and west towards the central diameter. Molecular orientation in the external part of the moving walls is perpendicular to the $\vec{E}$ field and is consistent with the chirality of the central bend defect. Eventually, the three walls merge, resulting in a configuration with molecules on either side perpendicular to the polarization direction (Fig. 4D) The final step involves a quarter-turn rotation of the dividing diameter, opposed to the droplet chirality (Fig. 4E), leading to the final photoaligned configuration (Fig. 4F).

In pathway II, the initial centrally dividing line is preserved for a more extended period of time, and no secondary vertical walls appear (Fig. 5B). The next step leads to a droplet fragmented in four regions, each one containing molecules roughly perpendicular to the polarization direction (Fig. 5C). Next, two of the sectors (in opposite quadrants) recede until they vanish, while the other two expand (Fig. 5D). The sectors that prevail are those with molecules oriented according to the initial chirality, which is preserved. The result is again a configuration in which the vertical diameter separates two regions with molecules perpendicular to the $\vec{E}$ field, which

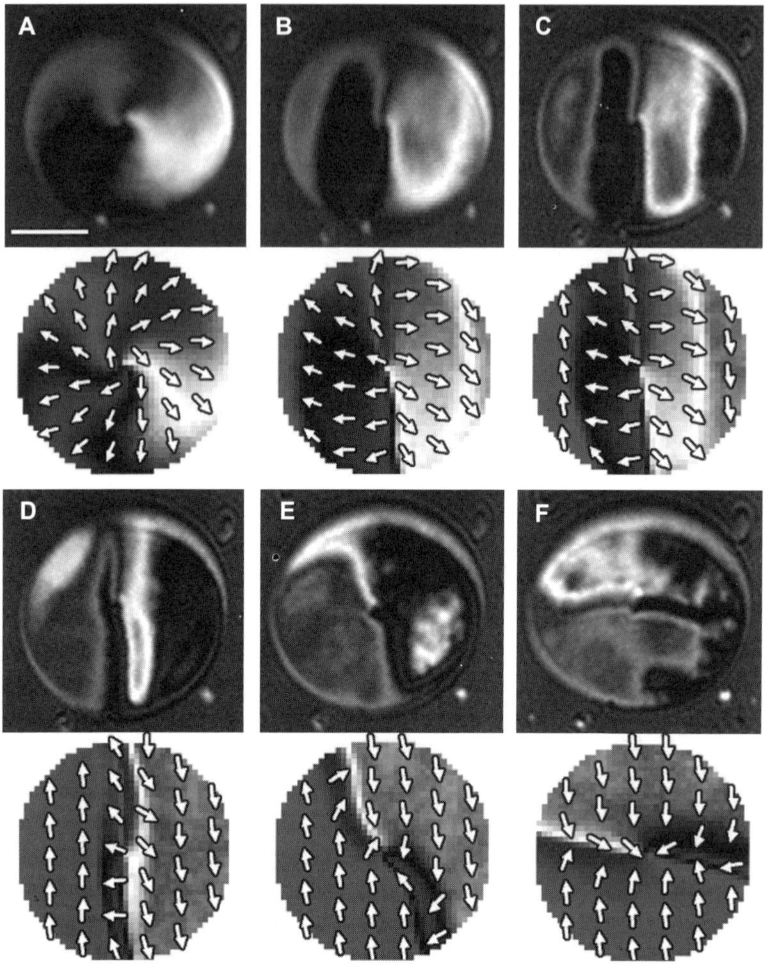

Fig. 4. Evolution of a splay-out domain towards the photostationary state following pathway I. Light is switched on in A with a power density 0.8 mW cm^{-2}. Elapsed times are 2.0 s in B, 3.0 in C, 4.5 s in D, 7.5 s in E, and 12.5 s in F. The line segment in A is 100 μm long. The bottom row contains the results of the numerical experiments. The polarization plane is horizontal.

later turns 90° continuously, opposed again to the initial droplet chirality (Fig. 5E), leading to the final photoaligned configuration (Fig. 5F). Notice that the main differences between pathways I and II are in the intermediate regimes.

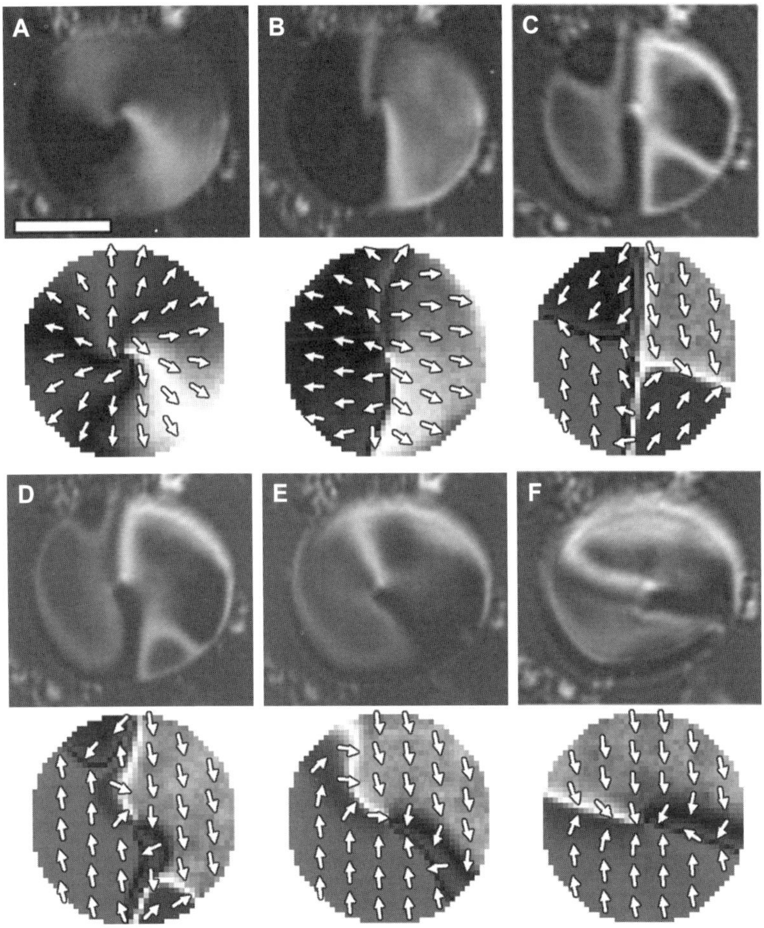

Fig. 5. Evolution of a splay-out domain towards the photostationary state following pathway II. Light is switched on in A with a power density 0.4 mW cm^{-2}. Elapsed times are 6.0 s in B, 13.0 s in C, 15.4 s in D, 16.5 s in E, and 18.4 s in F. The line segment in A is 100 μm long. The bottom row contains the results of the numerical experiment. The polarization plane is horizontal.

In spite of accumulated observations we were not able to unambiguously relate a given initial splay-out texture to a given pathway upon photoalignment. We tend to think that inhomogeneities in the molecular orientation at the domains boundary lead to choose one of the two pathways, thus making the outcome of a given experiment rather unpredictable. Moreover, it is not unusual to encounter droplets exhibiting mixed pathways.

We should also comment on the collective relaxation inside the domains when light is turned off. In this situation the configuration first evolves towards a splay-in axisymmetric texture, in all the cases presented above. In fact this is easily interpreted since the latter is the closest distortion in configuration space to the photoaligned state. However, since the splay-in symmetry is metastable in front of the splay-out texture under the experimental conditions here considered, the relaxed splay-in configuration finally evolves to change the sign of its inner distortion.[13]

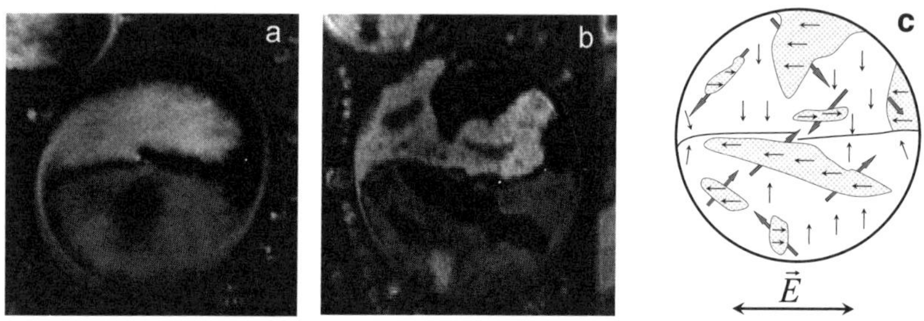

Fig. 6. (a) BAM image of a photoaligned domain after 37 s of irradiation of a droplet with polarized light. (b) BAM image showing traveling waves propagating inside the photoaligned domain. The time elapsed between both pictures is 156 s. The width of the pictures is 600 μm. The illumination power is 0.8 mW cm^{-2}. (c) Scheme representing the azimuthal orientation of trans-8Az3COOH molecules inside the waves (thin arrows) and their advancing directions (wide arrows) of the waves of (b).

Under prolonged illumination a more complex situation was observed corresponding to spots propagating within the hemispherical areas previously aligned.[25] They nucleate as localized reflectivity distortions and move invariably towards the dividing equatorial line (see Fig. 6a). Simultaneous and apparently uncoherent repeated nucleation from different locations was a common outcome of our experiments (see Fig. 6b). Some remarkable features were however unambiguously established from our BAM observations. (i) Spots propagate as localized azimuthal distortions with molecules parallel to the polarization direction (see Fig. 6c). (ii) The direction of propagation is always fixed inwards at roughly 45ž with respect to polarization (see Fig. 6c). (iii) When traveling spots encounter the dividing line the latter embeds them when the azimuth inside the spots matches that of the landing side of the equatorial line. (iv) When light is turned off, spots stop and their reflectivity changes to that of the underlying texture prior to global relaxation of the whole droplet. (v) Finally when spots with oppositely oriented molecules meet, they do not mutually annihilate but the smaller approaching piece either slows down to give way to the larger one or disappears if sufficiently small. These results have permitted to unveil the characteristics of the propagation of the orientational waves found in extended domains of *trans* 8Az3COOH Langmuir monolayers.[26]

2.2. *Patterns of photoalignment: Model and numerical results*

Photon absorption depends on the orientation of the transition dipole moment of the *trans*-azobenzene molecule, $\overrightarrow{\mu}$, relative to the external field $\overrightarrow{E}$. Assuming the simplest law $(\overrightarrow{E} \cdot \overrightarrow{\mu})^2$, photoexcitation would proceed until most of the transition moments are perpendicular (and thus inert) to the **E** field. According to the experiments mentioned above we can safely assume that the 2d projections of the optical axis and the transition moments coincide. Although different physicochemical phenomena at a molecular level might participate in the photoalignment process here described, we concentrate on the effect of irradiation in the (face-to-face parallel) H-aggregation known to occur in azobenzene-based monolayers.[13,17,27,28] The latter phenomenon shifts the absorption bands of the cromophores to higher energies resulting in a decrease of the absorbance for aggregates.

After these considerations a continuum model can be proposed to describe the patterns of photoalignment observed within the domains of the excited monolayer. This model is based on two spatio-temporal varying order parameters: one stands for the composition in terms of the fraction of aggregated molecules $c(x, y; t)$ (not to be confused with the *cis/trans* fraction), while the other is the usual projection of the 2d director **a**. The latter is better represented in terms of the azimuthal angle $\varphi(x, y; t)$. Considering first-order kinetics for the light-induced aggregation process and the polarization direction along the x-axis

$$\dot{c} = -\left(k_1 \cos^2 \varphi + k_2\right) c + k_2. \tag{1}$$

The second part of the model is based on a free-energy functional that accounts for the distortion of the orientational field

$$\begin{aligned}\mathcal{F} = &\int_{bulk} \left[\frac{K_s}{2}\left(\vec{\nabla}_{xy}\cdot\vec{a}\right)^2 + \frac{K_b}{2}\left|\vec{\nabla}_{xy}\times\vec{a}\right|^2 + \lambda(1-c)\left(\vec{\nabla}_{xy}\cdot\vec{a}\right)\right] dx dy\\ &- \int_{boun} \beta\,(\vec{a}\cdot\vec{n}_b)\, dl,\end{aligned} \tag{2}$$

where the first two terms stand for the usual splay and bend elastic contributions, the third one incorporates the coupling of the azimuthal field with the c-variable, and the last contribution accounts for the line tension of the domain boundary ($\vec{n}_b$ denotes the outward normal unit vector). Notice that the linear divergence term in the free energy functional has a spatially-dependent coefficient and can not be trivially transformed into a boundary contribution as it would be otherwise.

The dynamics of the azimuth is assumed as usual to be purely dissipative

$$\frac{\partial \varphi}{\partial t} = -\Gamma \frac{\delta \mathcal{F}}{\delta \varphi} + \zeta \tag{3}$$

where Γ represents a kinetic coefficient introducing a typical relaxation time scale for the molecular reorientation and ζ denotes a white noise that accounts for thermal fluctuations. As usual, we assume it obeys a fluctuation-dissipation relation

$$\langle \zeta(t)\zeta(t') \rangle = 2\Gamma k_B T \delta(t - t'). \tag{4}$$

The terms that come from the functional derivative of the elastic and boundary terms can be found elsewhere.[17] The remaining contribution reads in polar coordinates (R denotes the domain radius)

$$\begin{aligned}\frac{\partial \varphi}{\partial t} &= -\Gamma \left[-\lambda \frac{\partial c}{\partial x} \sin \varphi \right. \\ &\left. + \lambda \frac{\partial c}{\partial y} \cos \varphi - \lambda(1-c)\sin(\varphi - \phi)\delta(r - R) \right],\end{aligned} \tag{5}$$

Numerical results corresponding to the photoalignment process, reproducing the different pathways mentioned above, are shown in the corresponding panels in Figs. 3-5. Since the initial textures display a splayed configuration we choose $K_b = 0.8 < K_s < |\beta|R = 4$ in units of $K_s = R = \Gamma = 1$. On what respects to the aggregation reaction we assume it kinetically unbalanced with $k_2 = 10k_1$. We start with a fixed uniform concentration $c(x, y) = 1$ and the numerical accuracy of the simulation is guaranteed by choosing discretization steps $\Delta x = 0.05$ and $\Delta t = 10^{-6}$. Temperature is chosen to be $k_B T = 2.5 \cdot 10^{-3}$ as corresponds to the experimental relation $K_s/k_b T \approx 400$. The remaining parameters are chosen after exploring the parameter space. It turns out that the best choice is given by $\lambda = 200$ and $k_1 = 50$. The model with $k_1 = 0$ also describes how the original splay texture is recovered after relaxation following switch off of the illumination. In Fig. 7 we present numerical simulations corresponding to prolonged illumination and spot propagation.

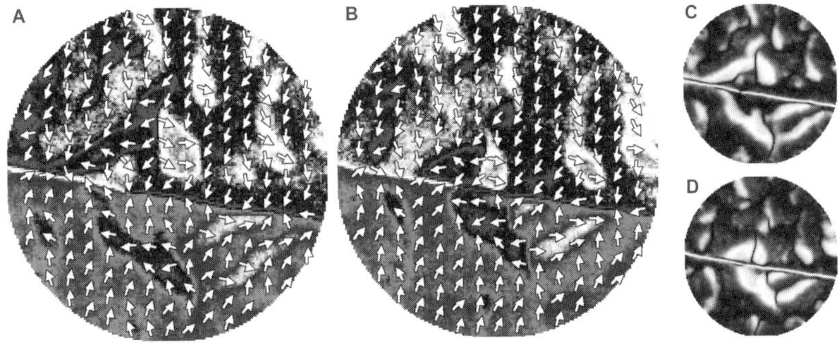

Fig. 7. Numerical simulations of model equations 3-5. Two successive snapshots for a photoaligned droplet showing the propagation of orientational (A and B) and compositional (C and D) pulses. In A and B arrows represent the in-plane molecular orientation and the gray color emulates the BAM observations. In C and D, brighter regions stand for larger fractions of aggregated species ($c < 1$). The set of parameters is K_b=0.8, k_BT=2.5·10^{-3}, k_1=50, k_2=200, λ =600, and β =8 in the unit system where $K_s = R$ (droplet radius)=Λ=1.

2.3. *Collective precession under rotating polarized illumination*

Another variant of the experiments reviewed above consists in photoaligning the domains and later rotating continuously the polarizer at a constant angular speed. We will briefly refer to experiments conducted with splay-in domains.[29]

For low enough angular speeds (typically below $10^\circ s^{-1}$ for the droplet sizes reported here), the dividing wall of the photoaligned texture rotates continuously with the molecular field. Conversely, rotation of the polarizer at high angular speeds, typically above $20^\circ s^{-1}$, destroys the photoalignment: the dividing wall widens up, and the contrast fades away, resulting in a texture similar to the nonirradiated one. As we explain on what follows, both regimes are actually the same, with the perceived differences being attributed to a progressive decrease in the amplitude of the periodic oscillation of the azimuth field as the angular frequency increases. Continued exposure of the system to the exciting light (the same droplet may be irradiated for several minutes) does not significantly alter the material properties. Eventually, however, this process may render the studied droplet useless for our purpose. Particularly for small droplets (diameters smaller that 150 μm) or for low rotational speeds, the central defect is un-

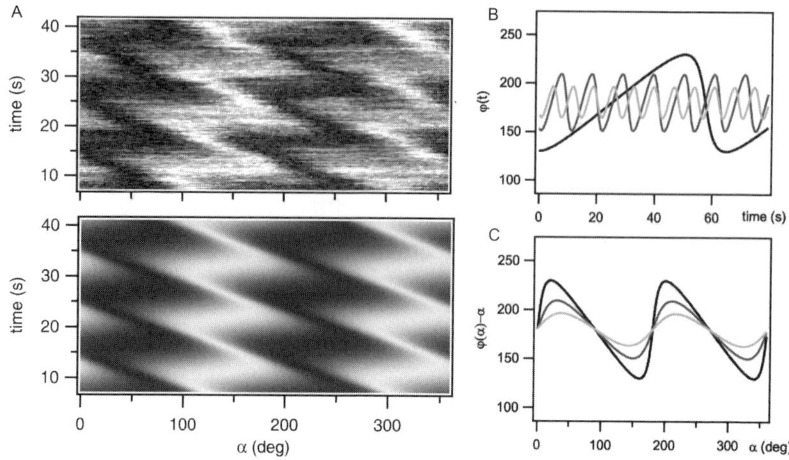

Fig. 8. (A) Top: gray level along a corona centered with the droplet defect, half the droplet radius, and 10 pixels thick is measured as a function of time. The BAM analyzer is set to 0°. The polarizer rotates at 16.9° s^{-1} clockwise. Bottom: space-time data fitted to the model described in the text. (B) Time change and (C) angular change of the azimuth field on a centered circumference, as determined from a fit of the space-time data. The three series correspond to angular speeds of the polarizer of 2.78 (dark gray), 16.9 (medium gray), and 25.8° s^{-1} (light gray). In each graph, the horizontal axis is shifted for ease of comparison so that the three sets of data have the same initial value.

stable and drifts from the droplet center. For large droplets, where the central position of the defect is stable, repeated on/off cycles result in the progressive development of a rigid structure at the droplet center.

To understand the apparently different dynamical behavior observed in the low- and in the high-speed regimes, we seek a model that approximately quantifies the evolution of the azimuth field. We propose the following expression to describe the azimuth field along a circumference of radius r, centered on the central defect

$$\varphi(\alpha, \alpha_1) = \alpha_1 + \frac{3\pi}{2} + \sum_{n=-N}^{N} \tan^{-1}[c(\alpha - \alpha_1 + n\pi)] \tag{6}$$

where α_1 is the angle at which the dividing wall intersects the imaginary circumference. The parameter c, with inverse angular units, quantifies the sharpness of the wall and the angular position of the wall at a distance r from the defect is taken as $\alpha_1 = \alpha_0(r) + \omega t$. To overcome observation difficulties inherent to BAM analysis, we have resorted to constructing a space-time plot of the reflectivity along a circumference of constant radius (typically a corona centered at half the droplet radius and a radial thickness of 10 pixels). Typical results are shown in Fig. 8. Once these BAM images are deconvoluted,[12] the model above permits to extract not only the angular variation of the azimuth field along a reference circumference but also its temporal evolution at a fixed location (Fig. 8). At low velocities, the azimuth field performs local anharmonic oscillations. For a given position inside the droplet, the azimuth precession velocity is small most of the time and follows the sense of rotation of the polarizer. When the dividing wall approaches the current location, the precession velocity inverts and quickly brings the azimuth orientation to its minimum value. Most important and contrary to what one might have expected, the azimuth field does not perform a full rotation. Rather, its orientation spans a range close to 100° during these oscillations for small rotational speeds. As the latter increase, the oscillations become more harmonic and display rapidly decreasing amplitudes. A detailed analysis of such a dynamics can be found in the original reference.[29]

3. Defect Dynamics in Two-Dimensional Soft Matter

3.1. *Formation of defects during domain coalescence*

Topological defects are structures that locally break the symmetry in systems with long-range order. In Condensed Matter Physics, defects are relevant in dynamic phenomena such as the plastic behavior of metals or phase transitions in two-dimensional systems.[30] In Soft Condensed Matter Physics, defects play an essential role in dynamic phenomena in systems such as liquid crystals or biological structures, where frustrated phases and confined geometries are often encountered.[31] In this context, Langmuir monolayers are interesting model systems to study defect topology and dynamics[32] due to the ubiquity of phases with long range orientational order.[5] In monolayers of simple amphiphiles, such as long-chain fatty acids or alcohols, orientational order is accompanied by short-range positional order (hexatic phases), which often results in structures with frozen or slow dynamics.

Quasi-two-dimensional systems with interesting defect dynamics can also be obtained by preparing free-standing thin Smectic-C liquid crystal films. In these systems, layered molecules are assembled with a uniform tilt with respect to the layer normal, and only spatial variations of the azimuth field can be observed as textures in polarizing microscopy images.

As a result, these textures can be characterized by a two dimensional azimuth field. A similar ordering can be obtained in Langmuir monolayers of a family of fatty acids with an azobenzene moiety in the alkyl chain. The labile mesophases formed by the *trans* isomer are topologically equivalent to Smectic-C layers, except that in monolayers one can distinguish between head and tail, so the azimuth field is a true vector field ($\mathbf{n}$ and $-\mathbf{n}$ are no longer equivalent). This will be relevant in the type (topological charge) of defects that can be found in the different systems.

Defect formation and dynamics in Langmuir monolayers of 8Az3COOH can be studied with increased precision in the two-dimensional emulsion formed spontaneously upon spreading a mixed *cis-trans* monolayer of 8Az3COOH. For a wide range of experimental conditions, circular *trans*-rich domains feature a well defined inner texture, and molecular ordering can be described by a two-dimensional azimuth field that includes inner point defects of total charge +1 arising from constant-angle anchoring conditions at the domain boundary (see Sec. 1). Spontaneous ripening of the two-dimensional emulsion formed spontaneously upon spreading a mixed *cis-trans* monolayer of 8Az3COOH leads to a process of domain coalescence driven by a reduction of the interfacial energy that arises from the finite line tension between the mesophase *trans*-rich domains and the isotropic *cis*-rich continuous phase.

Merging of domains results in a certain topological mismatch between the azimuth fields in the two droplets. Actually, this acts as a steric hindrance to merging in certain cases. For instance, while merging of bend domains with opposed orientational chirality occurs readily upon contact (Fig. 9.A), domains with the same chirality seldom merge. In the latter case, the azimuth field in the contact point would be antiparallel, thus creating a barrier of elastic energy against coalescence. Merging of domains leads to two dynamical processes evolving in parallel. In the one hand, the new domain quickly evolves to a circular shape driven by line tension. On the other hand, the merging azimuth fields must adapt to the boundary conditions of the combined domains. The former process typically proceeds much faster, so often one can analyze the later stages in the evolution of the inner azimuth field by assuming a circular boundary.

The outcome of the merging process is a combined domain featuring an azimuth field with the same topological charge as each of the parent domains, *i. e.*, $s = +1$. Since topological charge must be conserved inside a closed domain, one or more defects must be created during the merging process, with a total charge $s = -1$. This way, the merging of two

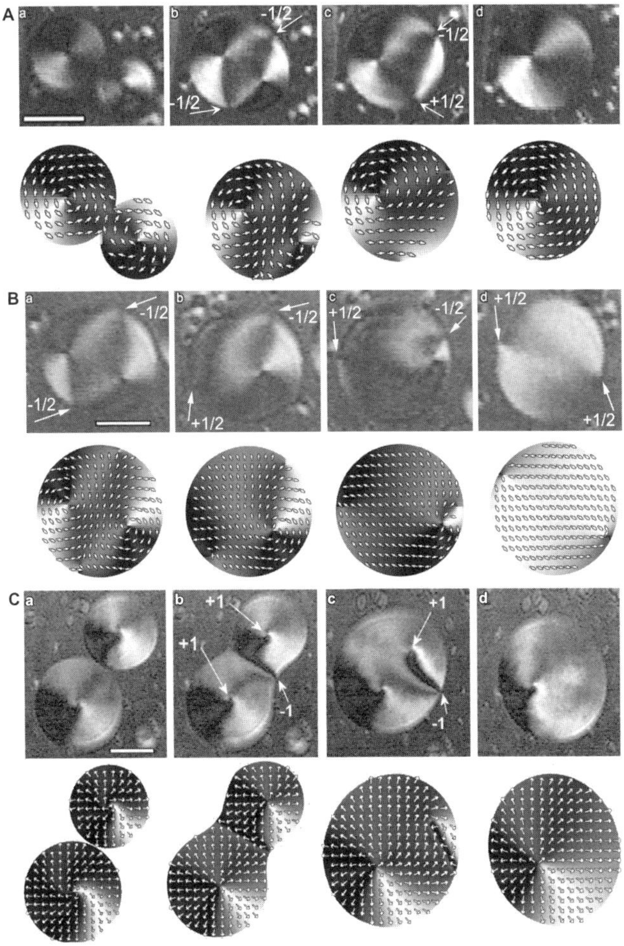

Fig. 9. Three different scenarios of defect creation and annihilation during the coalescence of two mesophase domains in monolayers of 8Az3COOH. A sketch of the azimuth field is shown below each experimental BAM image. Experimental conditions are $T = 32.0°\text{C}$, $\pi = 0.3$ mN m^{-1} (A), $T = 32.7°\text{C}$, $\pi = 4.5$ mN m^{-1} (B), $T = 35.0°\text{C}$, $\pi = 3.0$ mN m^{-1} (C) Elapsed times from the leftmost to the rightmost frame are 125 s (A), 300 s (B), and 26 s (C).

domains each with a charge $s = +1$ results in a single domain also with topological charge $s = +1$. The process of defect formation upon domain coalescence, and the subsequent annihilation dynamics offers a variety of different scenarios, some of them amenable to quantitative analysis.

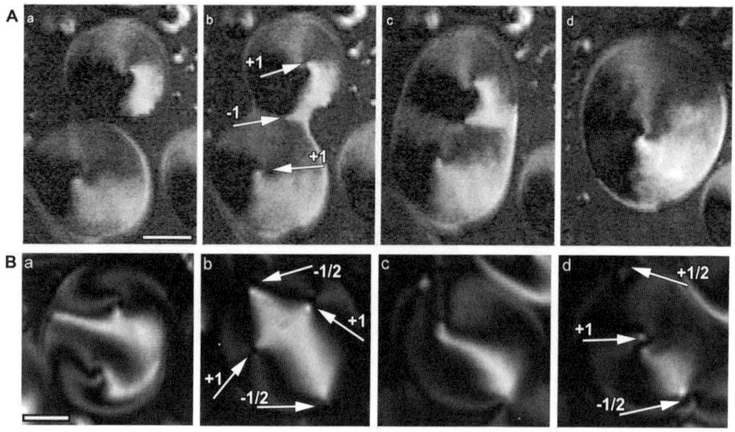

Fig. 10. Alternative scenarios for the coalescence of splay-out domains. Experimental conditions are $T = 34.8°\text{C}$, $\pi = 1.2$ mN m^{-1} (A), $T = 35.0°\text{C}$, $\pi = 4.0$ mN m^{-1} (B). Elapsed times from the leftmost to the rightmost frame are 22 s (A), 118 s (B).

Three of these scenarios are illustrated in Fig. 9. Cases A and B correspond to the two possible outcomes of the merging of two bend domains with opposite chirality. Upon coalescence, two defects are created at the contact points of the intersecting boundaries. Preservation of total charge lets us consistently assign a charge $s = -1/2$ to each defect. These semi-integer defects are necessarily confined to the domain boundary, since they are prohibited inside a two-dimensional vector field. The subsequent dynamics proceeds in similar ways in cases A and B, with one of the original inner $s = +1$ defect drifting towards the boundary until it merges with one of the newly created $s = -1/2$ defects. The result is a $s = 1/2$ defect that, again, will be confined to the domain boundary. The subsequent stages is what differentiates scenarios A and B. In case A, the second $s = +1$ defect occupies the domain center while the $s = +1/2$ and $s = -1/2$ boundary defects attract and eventually annihilate. In case B, the second $s = +1$ defect moves towards the remaining $s = -1/2$ defect until finally combining with it. This results in a second $s = +1/2$ boundary defect. Repulsion between the identical boundary defects results in their placement in diametrally opposed positions in a domain free from inner defects. The resulting texture has lost the rotational symmetry of the parent domains, as illustrated by the last frame captured during the random drift of the isolated domain. The third scenario reported here corresponds to the coalescence of two splay-out domains (Fig. 9.C). In this case, coalescence leads to the formation of a single $s = -1$ defect in one of the boundary intersection points. The new defect interacts with one of the original $s = +1$ defects, developing a 2π wall between them. This is a string-like structure across which the azimuth field performs a full turn. The inner $s = +1$ defect moves toward its negative counterpart until they eventually annihilate. It is not clear why the $s = -1$ defect remains pinned on the boundary since topological arguments would allow it to be inside the domain.

We have also observed other scenarios in which a single $s = -1$ defect forms inside the merged domain, eventually annihilating with one of the original $s = +1$ defects (Fig. 10.A), or even the formation of two boundary $s = -1/2$ defects (Fig. 10.B). In the latter case, the combination of one of the new boundary defects with one of the original inner defects leads to a metastable configuration, with a single $s = +1$ defects in the domain center and two semi-integer defects of opposite charge on the domain boundary. Unlike the case reported in Fig. 9.C, where the same type of defects are formed, attraction force between these boundary defects is apparently weak, and their annihilation is seldom observed.

3.2. *Defect-mediated formation of chiral domains*

As explained above, the different morphologies of isolated domains can be achieved by controlling the thermodynamic conditions of the monolayer. They are also influenced by domain size, which can not really be controlled. For instance, small domains always feature a bend inner texture. It is their coalescence above room temperature that leads to the formation of splay domains provided the resulting domain size is large enough. An intriguing observation in this system is the spontaneous formation of domains with a bend inner texture. Such a texture has a well-defined orientational chirality. Given that the used surfactants are achiral molecules, and that no chiral force is present in the system, monolayers naturally feature an equal amount of bend domains of the two chiralities. In the next section, we will explain that this symmetry can be broken with the application of an external chiral force.

The formation of bend domains can actually be traced back to the coalescence of pairs or achiral domains, and to a process of defect formation and annihilation. This is put into evidence if a 8Az3COOH monolayer is spread in the isotropic *cis* form, and slowly forced into the *trans* form. Phase separation of the two isomers leads to the formation of birefringent domains, which have elliptical shape and display a rather uniform reflectivity. BAM resolution limits our ability to resolve the shape and inner texture of condensed *trans* domains to sizes larger than about 20 μm.

An analysis of the inner azimuthal field of the elliptical domains shows that field lines are symmetrical with respect to the long axis of the ellipse, and originate at a point defect on one of the poles. Field lines converge towards the opposed pole, although their sink is a virtual defect: an imaginary point located outside of the domain (See Fig. 11.A). In consistency with the

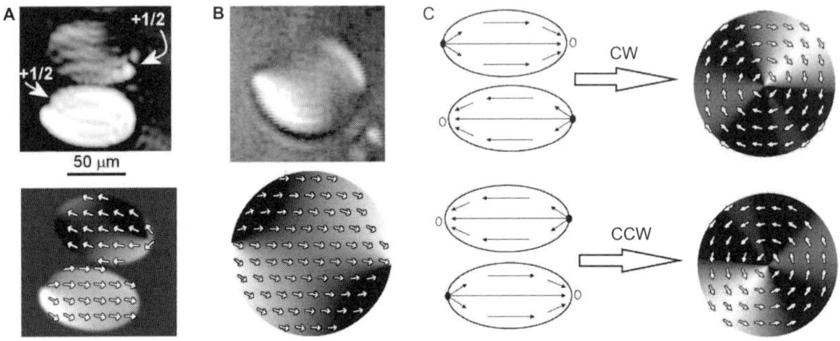

Fig. 11. Ellipsoidal *trans*-rich domains in a 8Az3COOH monolayer evolving from the *cis* to the *trans* form (A). Conditions are T=25.0°C, $\pi \simeq 0$ mN m^{-1}. Their inner texture is characterized by two $s = +1/2$ defects at the ellipse poles, one real and the other virtual. The latter may become real by heating the sample from room temperature up to 30°C (B). Bend domains arise from the coalescence of a doublet of antiparallel ellipsoidal domains (C). The orientational chirality of the bend domain is determined by the relative spatial placement of the members of the doublet, which is random.

topological arguments raised above (total topological charge $s = +1$), we assign a charge $s = +1/2$ to each of these defects. The virtual defect may become real if temperature is raised from room temperature until 30°C. The domain shape becomes more circular and the virtual defect eventually attaches to the domain boundary. The resulting domains has the double boojum configuration described above (Fig. 11.B), which is consistent with the charge assignation to the topological defects in the ellipsoidal domains proposed above.

The tendency of these domains to attract, presumably due to the existence of long-range alectrostatic attractions between condensates, favors the lateral coalescence of drifting elliptic nuclei to give rise to larger aggregates whose textures depend on the relative orientation of the parent domains. We observe that the fusion of domains with parallel azimuthal fields leads, in a simple process, to larger elliptical domains that keep the common orientational distribution of the parents. Conversely, if domains with antiparallel azimuthal distributions merge, the shape of the resulting domain becomes circular and the bend texture appears (11C). Actually, the handedness of the resulting bending distortion depends on the individual azimuthal distributions of the parent couple: a configuration of eastwards (resp. westwards)-top/ westwards (resp. eastwards)-bottom domains gives rise to a clockwise (resp. counter clockwise) texture. The

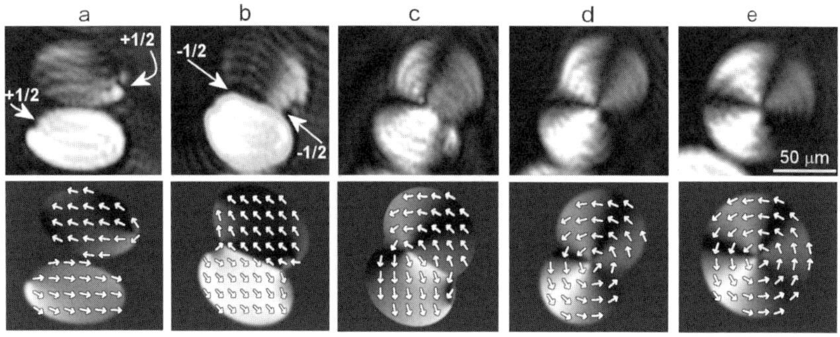

Fig. 12. Sequence of BAM images showing the fusion of two antiparallel elliptical domains (a) that leads to the formation of a circular domain with CCW orientational chirality (e). A sketch of the in-plane orientational field along with the corresponding simulated BAM image is shown below each snapshot. Elapsed time between frames (b) and (d) is 74 s.

remarkable change in both the shape and collective orientational field accompanying this coalescence process is a complex, defect-mediated, phenomenon sketched in Fig. 12, and rationalized as follows. First a pair of elliptical domains must approach each other and collide with an appropriately aligned configuration (panel 12a). When they laterally contact, a pair of $s = -1/2$ defects form at the intersection points (panel 12b). A disclination line stretches between the two semi-integer defects just created. Although this should incur a high elastic energy cost, fusion between antiparallel ellipsoids is a ubiquitous event. Notice at this point, that the total topological charge of the aggregated domain is still $s = +1$, a conserved quantity throughout the remainder of the process similarly to what we have reported for all fusion events. The structure of the azimuth field and the location of the different point defects is clearly sketched in Fig. 13. Each of the newly created defects approaches and annihilates with the respectively closest $s = +1/2$ real defect located at the poles of the parent domains (panel 12b). Subsequently, the remaining virtual ($s = +1/2$) defects move towards the fused domain from their respective escaped poles (panel 12c) and enter into the domain (panel 12d), pulled by a disclination line. The latter finally relaxes (Fig. 12e) by shrinking until collapse, leading to the single $s = +1$ defect, central to the resulting domain, whose outer boundary progressively rounds up to a circular contour (panel 12.f).

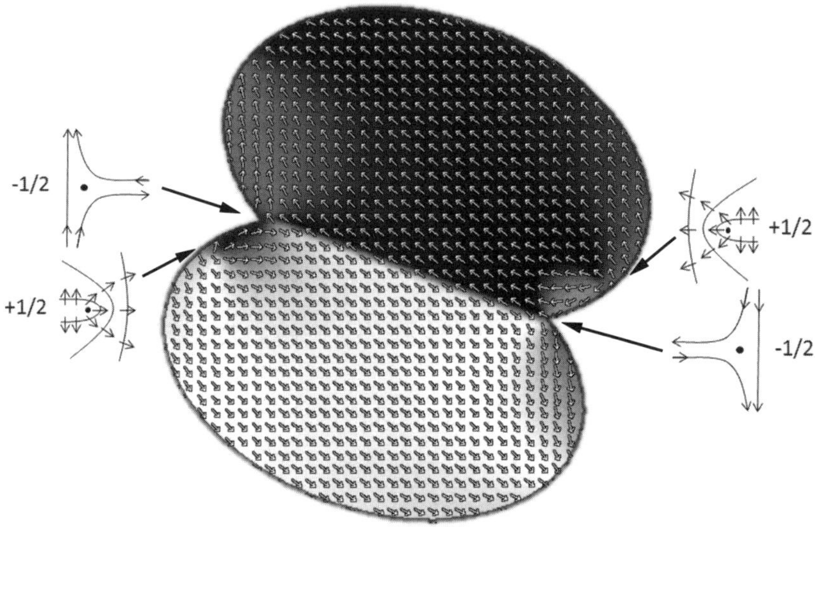

Fig. 13. Sketch of the azimuth field and the structure and position of the different point defects present in the domains that results from the coalescence of two elliptical domains in monolayers of 8Az3COOH (Fig. 12b).

An interesting observation, ubiquitous in defect motion in soft materials, is the enhanced mobility of positive defects with respect to their negative counterpart. Defect motion implies the local reorientation of molecules in the ordered phase. Its dynamics will be a balance between rotational viscosity, elasticity and eventual hydrodynamic (backflow) effects (molecular rotation may lead to mass displacement). Numerical simulations of the annihilation of two planar disclination lines in two dimensions demonstrated that both elastic anisotropy and backflow effects alone could account for this asymmetry, either with integer[33] or semi-integer[34,35] disclinations. Experimentally, backflow effects have been demonstrated to account for asymmetric motion in nematic liquid crystals even with isotropic elasticity.[36,37] The conclusion is that backflow effects can not be decoupled from defect motion in bulk liquid crystals, so the role of elastic anisotropy cannot be quantiatively assessed. The situation is different in Langmuir monolayers, in particular during the asymmetric defect motion that arisess during the coalescence of bend domains (Fig. 9A and defect tracking in Fig. 14A). If we compare the order of magnitude of the rotational viscosity with the effective translational viscosity that arises from coupling between monolayer and subphase, we conclude that rotation of the amphiphilic molecules cannot induce significant hydrodynamic effects.[38] This is consistent with recent experiments were even complex reorientation effects of molecules in the isolated domains can take place without any detectable backflow effects.[29] As a result, observed differential mobility between semi-integer defects of either sign arises solely due to elastic anisotropy.

We have recently proposed a model that relates the observed motion asymmetry during boundary defect anihilation in coalescing bend domains (Fig. 9A) with elastic anisotropy.[38] In two dimensions, only the splay and bend distortion modes will be relevant, so the elastic anisotropy will be given by the ratio between the splay (K_1) and bend (K_3) constants. Defects move due to the elastic attraction with their counterpart, so they both feel the same force. As a result, a ratio between instantaneous velocities can be related to a ratio between the effective viscosity felt by each defect. Experimentally, we have observed a constant terminal velocity during the last seconds prior to defect anihilation, so we will have a well-defined ratio of effective viscosities and, through our model, a ratio of elastic constants. Actually, one can find in the literature measurements by Feder *et al.*[39] for the geometric mean of K_1 and K_3 in this system. As a result, the analysis of defect motion can yield the value of the two elastic constants, and not only their ratio. Moreover, Langmuir monolayers are compressible systems, so monolayer elasticity changes with lateral pressure. By studying the coalescence of bend domains at different lateral pressures we are able to obtain rather elusive equilibrium properties, such as the dependence of elastic constants on lateral pressure, out of relatively simple dynamical measurements of defect anihilation (Fig. 14B).

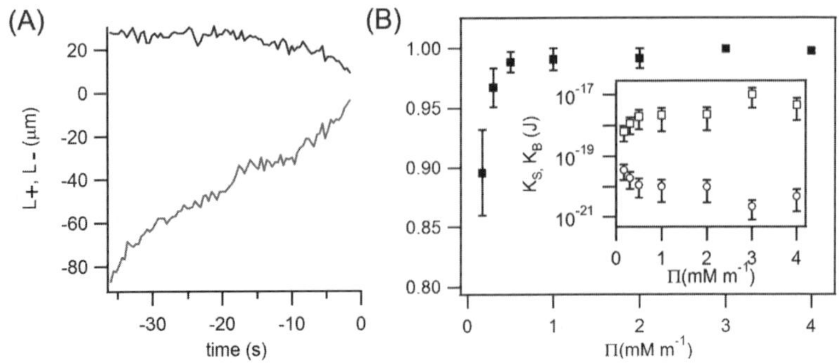

Fig. 14. (A) Position (arclength) of boundary defects in Fig. 9.A with respect to their merging point. Anihilation takes place at t = 0. Defects with charge $s = +1/2$ move up to twice faster than $s = -1/2$ defects. The straight lines are linear fits to the last stage of defect motion, and the ratio between the average velocities is related to the ratio of K_1/K_3 by our model. (B) Anisotropy parameter $\alpha = 1 - K_3/K_1$ and elastic constants as a function of lateral pressure can be obtained from the study of defect dynamics.

3.3. *Quantitative analysis of defect dynamics*

The study of defect dynamics in quasi-two dimensional soft matter is not new, and a significant amount of work has been performed in freestanding Smectic-C films.[40–43] In these systems, however, thickness effects are typically relevant, and affect defect dynamics in a nontrivial way. In comparison, Langmuir monolayers of 8Az3COOH provide a simple and robust experimental system, where material parameters can be directly extracted from an analysis of defect dynamics in films with Smectic-C order. The same kind of analysis can be performed in other scenarios where defects can be clearly tracked towards their annihilation, such as the experiments reported in Fig. 9C.[44] The late stages in the dynamics consist in an inner $s = +1$ defect and a $s = -1$ boundary defect, linked by a string distortion. The distance between defects as a function of time is shown in Fig. 15 where data correspond to the average of three independent measurements. Dispersion across different experiments increases with the distance between defects, presumably because of interaction with the remaining inner $s = +1$ defect or with the boundary, where the molecules adopt a nearly parallel orientation with respect to the normal to the contour.

Data on Fig. 15 reveal that the distance between defect pairs changes linearly with time at large defect separation while it changes as the square root of time in the last part of the collapse dynamics. Qualitatively similar behavior has been reported by Pargellis *et al.*[42] in the context of freely suspended Smectic-C films, where the authors describe the formation of string-like 2π-walls with opposed wedge disclinations at their ends. The confinement along a line of the strain field caused by the defect pair and the observed dynamics could be reproduced by a free energy that included the coupling between an effective electric field and molecular orientation. In the case of Langmuir monolayers, however, the broken mirror symmetry about the air/water interface allows extra terms in the elastic free-energy that can naturally account for the formation of stripe structures.[45–47] This dynamics can be analyzed in terms of the competition of an effective attraction force and an effective viscosity, similarly to what has been reported in the context of defect dynamics in freestanding Smectic-C films[42] and in the recombination of defects split by a heat pulse in hexatic Langmuir monolayers.[48]

The contribution to the droplet elastic free energy of the defect pair grows logarithmically with their distance, d.[49] As a result, the inner $s = +1$ defect is pulled towards its boundary $s = -1$ counterpart with a force that is inversely proportional to d. Moreover, we will assume that there is an additional constant attractive force between the defects arising from the line energy of the string. Consistently with what we have explained above, we will further assume that defect motion results in a pure in-place rotation of the molecular field, without hydrodynamics (backflow effects can be neglected). Rotational viscosity introduces a viscous torque that limits the rate of rotation of molecules thus slowing down the overall defect dynamics. In a purely dissipative regime, this results in a friction force proportional to the defect velocity through an effective friction coefficient. These effects combine into an equation of motion,

$$\mu \dot{d} = \frac{A}{d} + \lambda, \tag{7}$$

where μ is the effective friction coefficient, A is related to the Frank elastic constants, and λ is the line energy per unit string length.

Integrating Eq. 7 by considering that defect collapse takes place at time $t = t_0$, and using the above parameter definitions, yields

$$t_0 - t = d\frac{d_\infty}{D} - \frac{d_\infty^2}{D} \ln\left(1 + \frac{d}{d_\infty}\right). \tag{8}$$

We have combined the three parameters in this problem to give an effective diffusivity for the moving $s = +1$ defects, $D = A/\mu$, and a characteristic length scale $d_\infty = A/\lambda$.[44] This equation can be fitted to data in

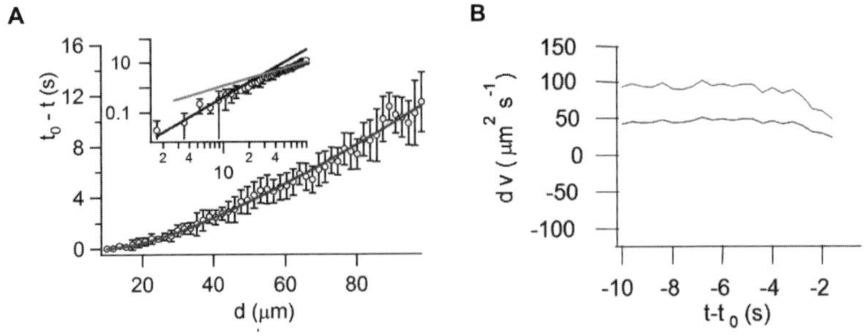

Fig. 15. (A) Anihilation of an inner $s = +1$ and a boundary $s = -1$ point defect linked by a string, corresponding to data in Fig. 9.C. A simple model, discussed in the text, is able to reproduce the observed dynamics, and reveals two limiting regimes, as illustrated in the inset. (B) Data from the dynamics of boundary semi-integer defects corresponding to experiments in Fig. 9.A enables to obtain a value for the viscous diffusivity, which is consistent with the value determined from (A).

Fig. 15A, from where values $d_\infty = 27 \pm 14\mu$m and $D = 120 \pm 40\mu\text{m}^2\text{s}^{-1}$ are extracted. The value for D can be obtained directly from data in Fig. 14A, where boundary semi-integer defects approach due to their attraction, in the absence of any string structure. In that case, the effective dynamics can be approximated as $\mu\dot{d} \simeq A/d$, so that $d\dot{d} \simeq A/\mu = D$ 15.B. We obtain this way a value $D \simeq 100\mu^2 s^{-1}$, consistent with the above estimations.

Equation 8 has two well-defined limiting regimes, determined by the characteristic length scale d_∞. At small defect distances ($d/d_\infty << 1$) diffusive relaxation dominates, and distance between defects varies as the square root of time. This regime can be observed in the inset in Fig. 15A for relative defect distances less than about 20 μm. Data are represented there in a log-log plot after subtracting the fitted d_0 from d. At the opposite limit ($d/d_\infty >> 1$) Eq. 8 predicts a linear regime, which can be observed in our data for the largest defect distances (Fig. 15A). These results can be compared with those presented by Pargellis *et al.*[42] for thin free-standing Smectic-C films.

Even though the above analysis only allows to obtain ratios between characteristic parameters, it is possible to estimate the string line energy per unit length, λ', from a knowledge of the azimuth profile across the string using the usual expression for the elastic energy involving the Frank elastic constants, which we have measured for this system.[38] Notice that we allow λ' to be different from λ since the configuration of the linear structures where they are defined will likely be different. An example of

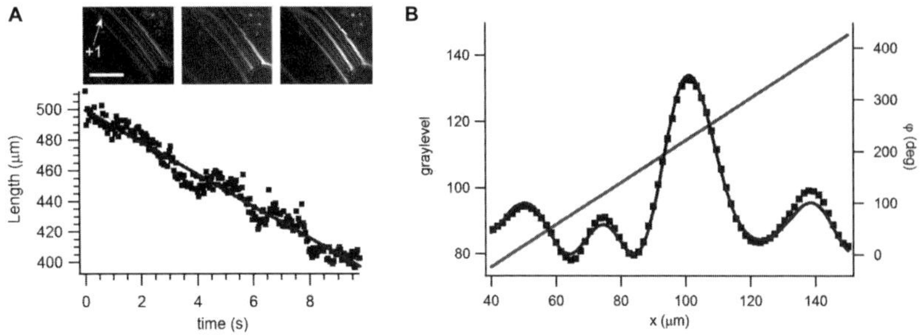

Fig. 16. (A) Distance between the end defects in a shrinking stripe in 8Az3COOH as a function of time. Experimental conditions are T=35.0°C, π = 1.0 mN m^{-1}. Time between the three snap-shots in the inset are (from left to right) 5.3 s, and 4.3 s. The solid line is a fit to the data, as explained in the text. (B) An analysis of the reflectivity profile across the string (symbols and black line) allows to extract the linear azimuth profile (red line), from which a line energy can be estimated.

such an analysis is shown in Fig. 16A, where we analyze a shrinking stripe structure that forms spontaneously in extended regions of *trans* isomer in 8Az3COOH. monolayers. In this example, the stripe features a $s = +1$ singularity at the moving end and a $s = -1$ singularity pinned to a pool of *cis* isomer. Since defects are far apart, their interaction can be neglected. As a result, their dynamics is determined by the stripe line energy and by a viscous drag. This results in a regime of constant velocity (Fig. 16A). The slope yields λ'/μ. The BAM reflectivity profile across the stripe can be deconvoluted to obtain the azimuth profile,[44] which turns out to be linear (Fig. 16B). This results in a estimated value $\lambda' = 0.1$ pN. Combined with data in Fig. 16.A results in a estimated viscous drag $\mu \simeq 10^{-8}$ Kg s^{-1}, which is consistent with theoretical expectations.[44]

4. Mechanical Chiral Selection

The characteristics and shape of amphiphilic assemblies, like Langmuir monolayers, depend on the intermolecular interactions inside the aggregates but also on the interactions between them and the molecules of the domain's environment. Among the different kind of interactions, the chirality of a particular amphiphile plays a decisive role in the features and properties of the aggregates, in particular, their shape and texture. It is well known that mirror-image isomers of a given amphiphile can lead to the formation of two-dimensional aggregates displaying both handednesses.[50] In most cases, this different handedness manifests in the opposite, clockwise (CW) or counterclockwise (CCW), curvature of different parts of the aggregates.

Two dimensional chiral domains can also be formed after spreading achiral amphiphiles on a water subphase. This is what happens in a Langmuir monolayer of the azobenzene derivative 8Az3COOH. As mentioned previously in this chapter, spreading a photostationary solution of a mixture of *cis* and *trans* isomers of this compounds on a water subphase, leads to a monolayer displaying a sort of biphasic two-dimensional emulsion.[12] This is composed of circular *trans*-enriched domains embedded in an isotropic and nearly pure *cis*-matrix. BAM observations show the bend texture of these circular *trans* domains,[13] with equilibrated amounts of CW or CCW orientations (Fig. 17). Then, the monolayer displays a racemic mixture, i.e., equal amounts of CW and CCW domains.

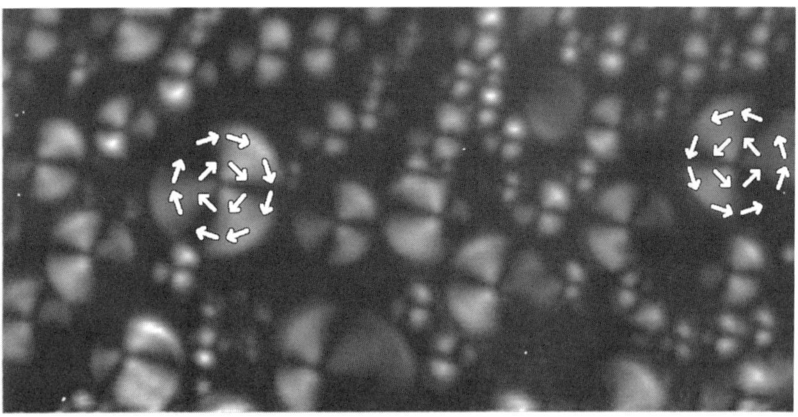

Fig. 17. Typical image of a mixed 8Az3COOH monolayer at low Π values. CW and CCW domains are characterized by the brightest first and third quadrant, respectively. Qualitatively, BAM textures appear as three black brushes, with either two upwards brushes (CCW orientation) or two downwards brushes (CW orientation). The maximum reflectivity appears in the first and third quadrants of the domains for CW and CCW orientation, respectively. The width of the image is 546 μm.

The control of chirality sign of soft assembled materials and crystal structures is a fundamental issue in basic and applied chemistry.[51,52] In particular, the possibility to break the chiral symmetry of solid materials when crystallized under stirring has been widely recognized since the first observation was reported for supersaturated solutions of sodium chlorate.[53,54] Similar results for saturated suspensions ground with glass beads have been published more recently.[55–57] When referring to soft materials the question is still more intriguing. Observations that a dominant chirality of assemblies of achiral molecules could be induced by stirring have appeared during this last decade.[58,59] An even more striking report published by some of us points out to the possibility not only to induce but to select the handedness of labile condensed mesophases.[60,61] The latter phenomenon has been revisited in recent years, extending the discussion to its reversibility or "dynamic nature".[62–66] Interpretation of these observations, based on the usual circular dichroism (CD) measures, have been, however, subject to controversy 34due to the need to discern intrinsic chirality in the condensates from instrumental artifacts.[65,67,68]

A different perspective of such chiral selection process induced by vortexes can also be observed in 8Az3COOH Langmuir monolayers when stirring the aqueous subphase. In contrast to the previous examples, our scenario is different in that the bulk-like vortex force, in spite of being genuinely three-dimensional, is able to imprint a preselected dominant chirality into our two-dimensional condensates. In these experiments, monolayers are prepared by depositing a chloroform solution of a photostationary mixture with the room light of the *cis* and *trans* isomers of 8Az3COOH while stirring the water subphase at a constant rate, Ω ($0 < \Omega < 1600$ rpm) and low surface pressures, Π ($0 < \Pi < 1$ mN m^{-1}). Stirred is continued for 2 min. Then, BAM images of the resulting monolayers are recorded with the analyzer set at 60° counterclockwise with respect to the plane of incidence, which contains the y-axis of the images. Moreover, our BAM images directly reflect molecular arrangement inside the aggregates without any further interpretation of the measurements. As a conclusion we show that the orientational chirality of mesoscopic conglomerates of achiral molecules, forming interfacial sub-millimeter circular domains, is decided by the chiral sign of the underneath vortical flow.

Fig. 18 shows typical BAM images of a mixed 8Az3COOH Langmuir monolayer after stirring the aqueous subphase. A significant excess of CW (resp. CCW) bend domains is evidenced when the monolayer is prepared on a water subphase under CW (resp. CCW) stirring at 1000 rpm. Dominance of a particular handedness is measured in terms of the enantiomeric excess of a particular sign (here CW)

$$ee_{CW} = \frac{2n_{CW} - n_T}{n_T} \times 100$$

where, n_{CW} stands for the number of CW bend-like aggregates with respect

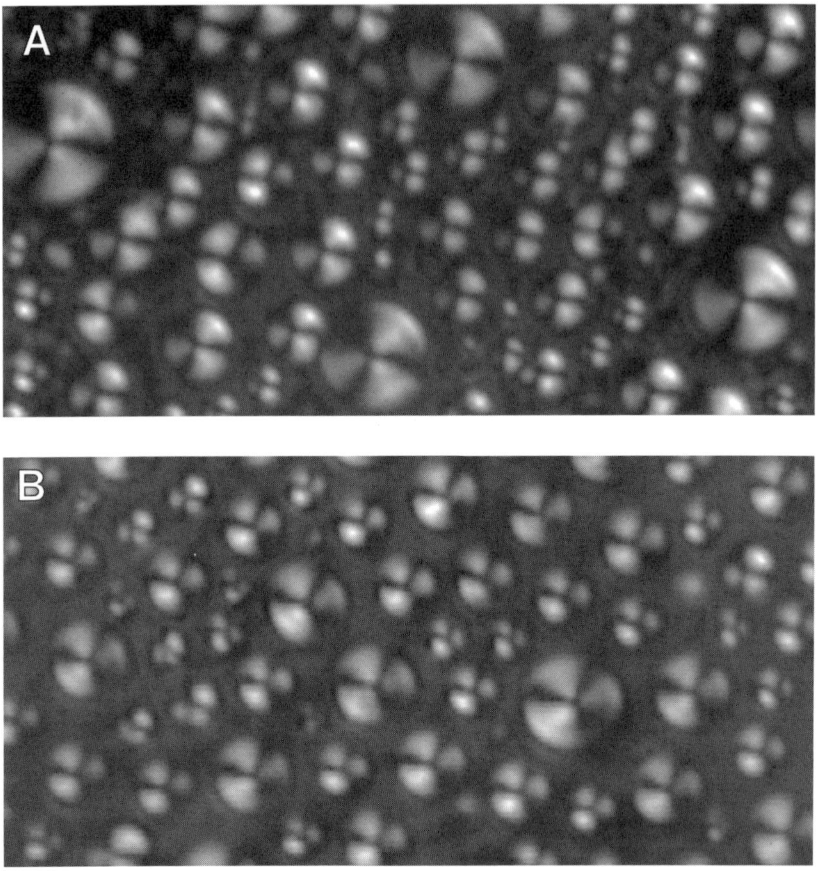

Fig. 18. Typical images of mixed *cis*-8Az3COOH monolayers at low Π values after CW(A) and CCW(B) stirring at 1000 rpm. CW and CCW domains are characterized by the brightest first and third quadrant, respectively. The width of both images is 546 μm.

to the total number n_T of examined domains. In a typical experiment n_T is close to 700. A systematic analysis for different stirring rates is summarized in Fig. 19. It is worth emphasizing that a threshold stirring rate ($\simeq$ 500 rpm) is necessary to observe a significant chirality sign selection and that the enantiomeric excess saturates at a value slightly above 40% at roughly 1000 rpm.

To improve the experimental conditions at which this phenomenon is better observed, we have considered different initial conditions starting from

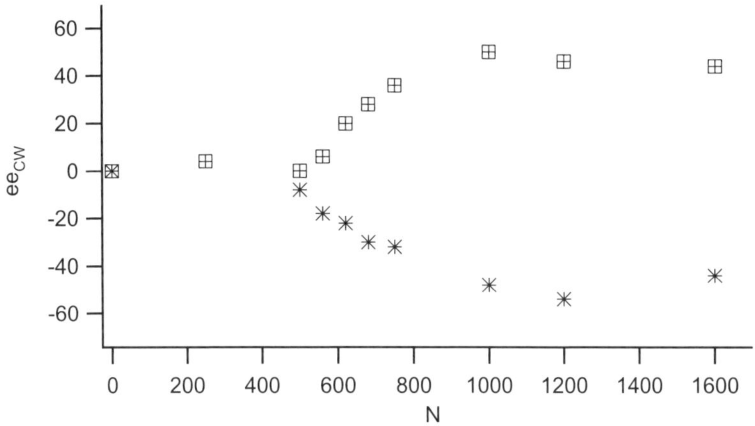

Fig. 19. Dependence of ee_{CW} on stirring rate, Ω in CW (squares) and CCW (stars) sense.

either a pure *cis* or a pure *trans*-monolayer. In the first case, chiral selection takes place similarly to what we have reported but an important difference must be mentioned. Differently from the results in Fig. 19, the handedness dominance increases monotonously with the stirring rate reaching the same upper bound at saturation that was detected for the mixed *cis-trans* monolayer. Moreover, when starting with the completely *trans*-state the phenomenon is completely absent and no signature of enantiomeric excess is evidenced within the entire range of stirring rates investigated.

The latter observations strongly suggest that the chiral selection under vortical stirring is induced by the *cis*-phase content in the monolayer. This is true at low temperature (24–26°C), but when T increases above 30°C, the handedness selection is not observed. This result suggests that it is not the *cis*-isomer itself but its temporal evolution what is responsible of the chiral selection.

Previous results show that mixed *cis-trans* and isotropic *cis*-8Az3COOH monolayers slowly but continuously evolve to a nearly pure *trans*-state.[13,69] This mostly follows from domain coalescence together with the unavoidable conversion of the *cis*-isomer to its *trans*-configuration. This process leads to a monolayer texture consisting of large patches (Smectic-C-like) with stripped reflectivity patterns, totally different from the domain-based textures just reported.

Concerning the evolution of a *cis* monolayer, we conjecture that the slow *cis* → *trans* conversion allows face-to-face assembly of the azobenzene group of the mesogen units (H-aggregation),[13,16] leading to growing nuclei that evolve into domains whose shape depends on temperature. Above 30°C, nucleation proceeds through circular, typical symmetric double-boojum-like, domains whereas at 25°C these domains are elliptical with two different topological defects at the poles (Fig. 11). This asymmetry is responsible of the handedness preference through a complex process of defect annihilation during the coalescence of antiparallel elliptical domains under stirring.[69] The kinetics of this annihilation depends on the nature of the defects involved in the process (see Sec. 3.2), playing these dependencies a crucial role in the chiral selection, and also in the limit value of the enantiomeric excess shown in Fig. 19. The latter means that annihilation of different defects is roughly three times faster than that of similar ones.

The evolution of the monolayer to a *trans*(Smectic C) anticipates a progressive decrease in the enantiomeric excess. This is observed either keeping the monolayer at rest for tens of minutes before stirring or when the subphase is subjected to prolonged stirring after spreading the amphiphilic mixture. Any signature of chiral prevalence vanishes for stirring times larger than 30 min. This process is sketched and interpreted in Fig. 20. Under vigorous stirring, CW and CCW balanced bend domains coming from previously formed *trans*-8Az3COOH patches mix the unbalanced domains formed from the *cis*-counterpart through ellipse fusion,[69] progressively evolving to larger Smectic-C patches. The latter texture is completely unable to accommodate any signature of chirality and thus if maintained gives no chiral signal or, if ruptured by shearing, delivers equal proportion of CW and CCW bend domains. Any enantiomeric excess is in this way eventually eliminated from the monolayer.

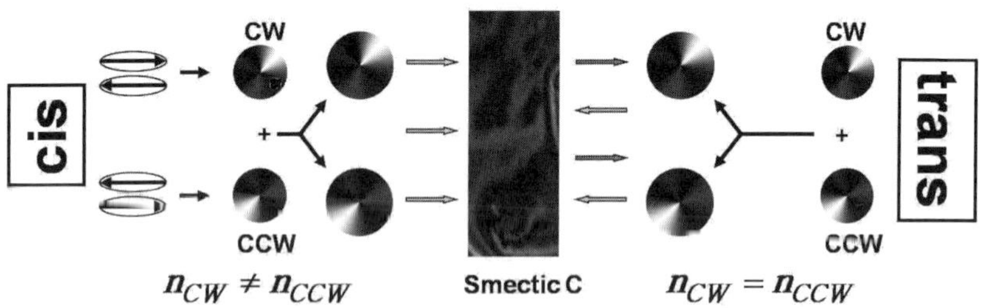

Fig. 20. Sketch of the different processes and evolution of the mixed 8Az3COOH monolayer under vortical stirring (see text).

Our results reinforce the role of interfacial flows of primordial achiral components on the origin of homochiral terrestrial life,[70,71] based on their ability to mould templates to potentially elicit a particular handedness in biopolymers.[72] We also anticipate possibilities for enantioselective catalysis, either directly on aqueous interfaces,[73] or by further transfer of the monolayer to a solid support.[74] Concerning this point, we have successfully transferred these domains by the Langmuir-Blodgett method on a glass slide, with the aim of building liquid crystal θ cells[75] with selected chirality preference. Preliminary results show the ability of these templates to orient thin liquid crystal layers.[76]

5. Two-Dimensional Microfluidics

The accurate control of a network of single or multiphasic fluid flows at the nanoliter scale has led into an emerging and promising new field known as microfluidics.[77,78] A large variety of devices have been developed that allow to control the size, shape and chemical composition of droplets separated at the micron scale.[79,80] At the same time, they have made also possible to investigate physicochemical phenomena in nanoliter-sized fluid volumes at low Reynolds number, among them transport of immersed entities, mixing processes, advection or reaction of dispersed species, etc.

Here we want to adapt some of these features, typical of conventional microfluidics, to a two-dimensional environment. In particular, we will focus on two basic aspects: the dynamics of a fluid flowing trough a single microchannel, and the development of an experimental protocol allowing for the construction of a network of two-dimensional paths and the independent control of monolayers flowing through them. As a first application of this "two-dimensional microfluidics technique" we will measure the diffusion coefficient of a fluorescent probe dispersed in a monolayer advancing through a Y-junction.

5.1. *Dynamics of a two-dimensional fluid flowing trough a microchannel. The bottleneck effect*

Previous studies have shown that the rheological behavior of Langmuir monolayers is complex.[81] Investigations on different monolayers flowing through a narrow channel have been performed to extract steady state flow properties like viscoelastic parameters,[82] unusual velocity profiles,[83–86] or to probe non-Newtonian features in the contraction-expansion flows of

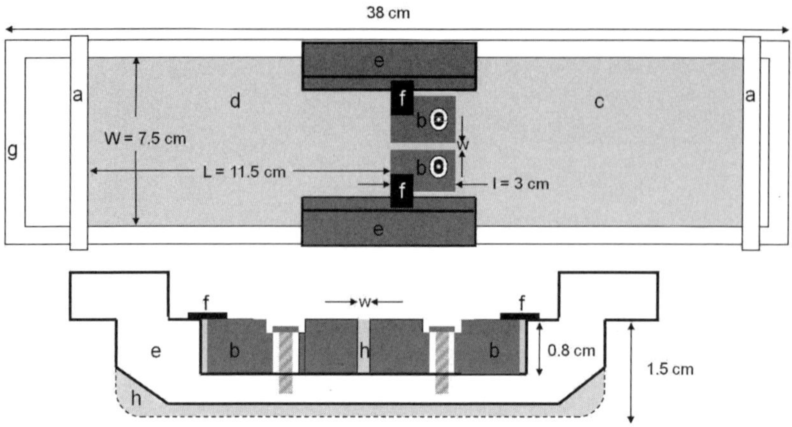

Fig. 21. Scheme of the Langmuir trough used in the bottleneck experiments and section of the channel assembly. a) movable Delrin barrier, b) black Delrin blocks, c) expansion chamber, d) compression chamber, e) black Delrin platform, f) black Delrin pieces to block the lateral channels, g) Teflon Langmuir trough and h) water subphase.

Langmuir monolayers.[87]

In this section, we will focus on a transient phenomenon typically observed at the beginning of the compression of a monolayer forced to flow through a microchannel under constant mechanical driving. This is known as the "bottleneck effect",[77,80] and basically consists in an anomalous and long transient in the surface pressure before attaining its steady value. In conventional microfluidics this effect is commonly masked by the elasticity of the confining tubes[77,80] and this could justify that explicit references to this topic are scarce.[88,89]

In a typical experiment, an expanded monolayer of an amphiphile (elaidic acid) is spread over a water subphase contained in a Langmuir trough divided in two compartments by a microchannel of controllable width (see Fig. 21 for a experimental set-up scheme). When the monolayer is forced to flow through the channel at a constant velocity of the movable barrier, the surface pressure steeply increases while the barrier is moving. When this is stopped, Π decreases slowly to reach a constant value. A similar compression experiment without contraction at the same compression rate and time leads to a slower increase of the surface pressure up to attaining the same constant value as in the compression through the microchannel (Fig. 22).

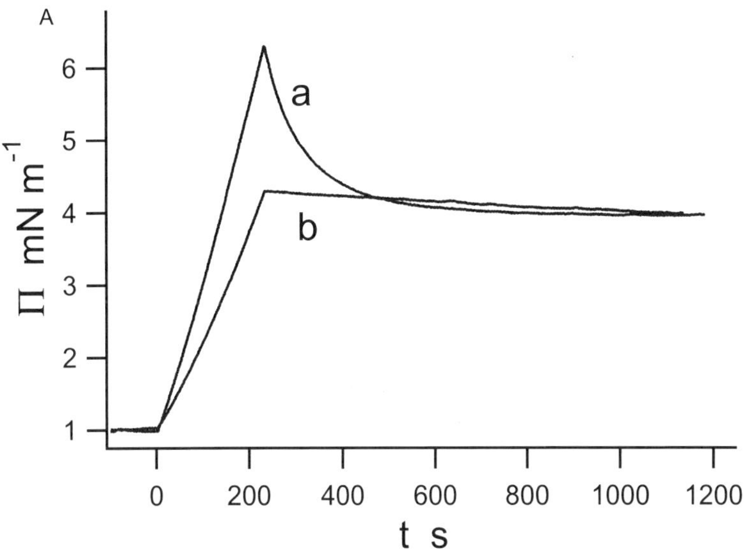

Fig. 22. Evolution of the surface pressure in the compression chamber during a compression of an elaidic acid Langmuir monolayer. a) Flow through a channel 0.4 mm wide and 30 mm long. b) Flow without channel. At $t = 200$ s, the moving barrier, $v_b = 10$ mm min^{-1}, is suddenly stopped. $T = 20°$C.

This distinctive long Π transient depends on the material parameters and geometrical factors of the constrained flow. We can understand this phenomenon as a slow response of the squeezed material to a sudden change of its driving regime. In what follows, we will comment the results of a basic theoretical analysis of the particular case commented above, in which a monolayer is suddenly set to motion from rest, since it admits a simple theoretical description and a better analysis of experimental data.[90]

According to the Langmuir trough scheme depicted in Fig. 21, we consider a compression chamber of Length x Width (LxW) area, limited on one side by a barrier, and contracted at the other end by a narrow channel of length l and width w, that defines a contraction factor α,

$$\alpha = \frac{W}{w}$$

At $t = 0$, when Π has reached an uniform value, Π_0, all over the subphase after spreading, the barrier is abruptly set to motion with a fixed velocity v_b. Assuming a planar, essentially unidirectional, flow of the monolayer at the compression chamber, and restricting our analysis to the early stages of compression when the surface pressure in the expansion area can be still considered unchanged and equal to Π_0 (Fig. 23), the temporal evolutions

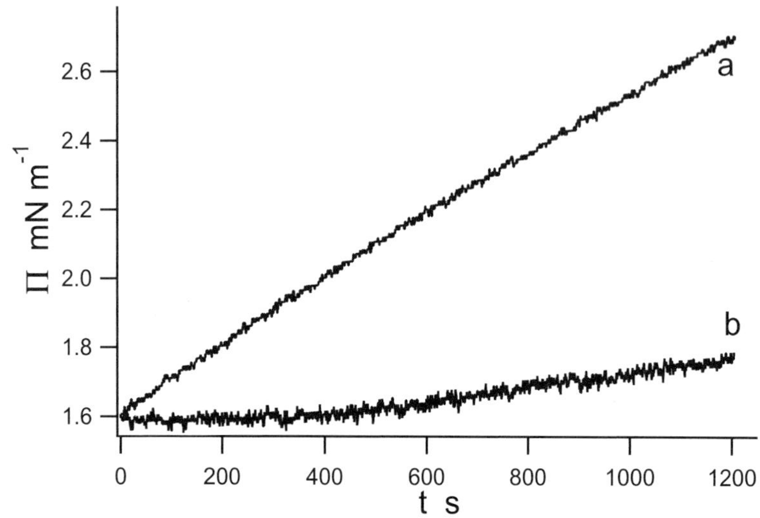

Fig. 23. Temporal evolution of surface pressure in a) the compression chamber, and b) the expansion chamber. $w = 300\ \mu$m, $v_b = 0.5$ mm min^{-1}.

of the velocity inside the channel, v_c and surface pressure, Π, are[90]

$$\Pi(t) = \Pi_0 + \alpha\frac{v_b}{\lambda}(1 - \exp(-t/\tau)) \tag{9}$$

$$v_c(t) = \alpha v_b(1 - \exp(-t/\tau)) \tag{10}$$

in terms of a time constant, τ

$$\tau = \alpha\frac{\kappa L}{\lambda} \tag{11}$$

which in turn depends on the permeability constant, λ

$$\lambda = \frac{v_c(t)}{\Pi(t) - \Pi_0}$$

and the compressibility of the monolayer, κ.

In order to study this permeability constant, λ, the velocity profile of the monolayer flow has been measured across the microchannel. For this purpose, flow visualization has been achieved by means of particle-tracking with a CCD camera, seeding the monolayer with sulfur powder[87,91] and illuminating with an halogen light source. The measurements show that λ is proportional to the channel width, w (see the inset in Fig. 24), and that the monolayer velocity profile is semi-elliptical instead of a parabolic velocity profile expected for a Poiseuille flow regime limited by the viscosity of the monolayer (Fig. 24). This suggests that monolayer flow is governed by the viscosity of the subphase[84] since its in-plane rheology is ultimately coupled to the subphase hydrodynamics. In this case, the relationship between λ and w is given by[84]

$$\lambda = \frac{w}{2\mu_s l} \qquad (12)$$

being μ_s the viscosity of the subphase. The extracted value, 1.07 mPa s, independent on the surface pressure in the investigated range, is in close agreement with the viscosity value corresponding to pure water within the experimental error.

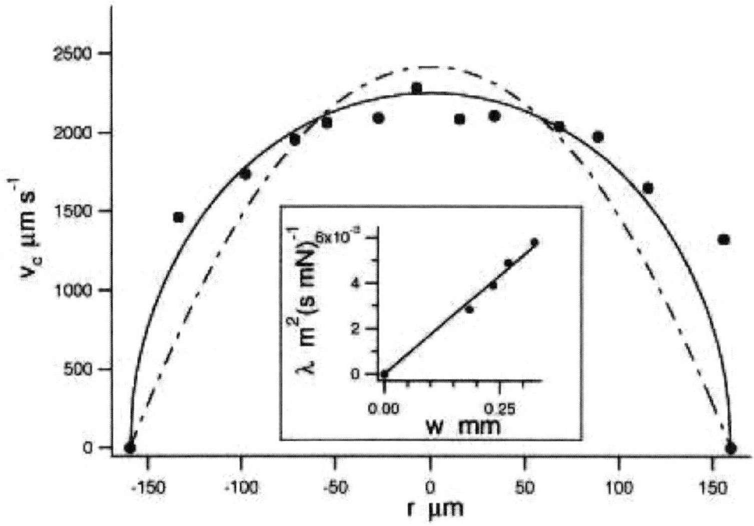

Fig. 24. Velocity profile during monolayer flow in a channel 0.35 mm wide. r represents the distance from the center of the channel. The solid line represents the semi-elliptical profile (see text). The broken line shows the parabolic profile, typical of a Poiseuille flow limited by the viscosity of the monolayer.[84] $v_b = 1$ mm min^{-1}. Inset: Dependence of λ vs w.

The temporal evolution of the surface pressure (Eq. 9) can be easily tested and interesting conclusions can be drawn if we focus at the initial stage of the monolayer compression. According to Eqs. 9 and 11, we derive

$$\left|\frac{d\Pi}{dt}\right|_{t=0} = \frac{1}{\kappa L}\, v_b$$

being this relationship independent on geometrical (α) and viscous (λ) parameters. After a systematic set of experiments varying w and v_b, the linear dependence of this initial slope on v_b was confirmed for the investigated channel widths. However, the proportionality factor, $(\kappa L)^{-1}$, is constant only for $w \leq 400$ μm (see Fig. 25) being the extracted compress-

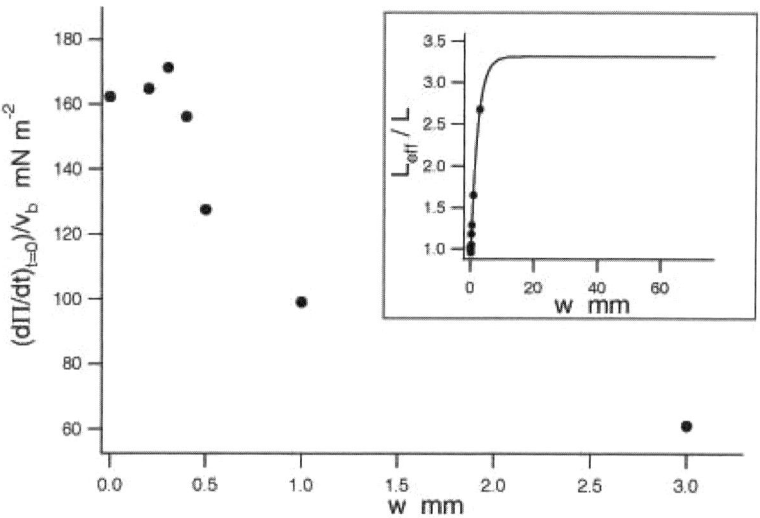

Fig. 25. Dependence of the initial surface pressure temporal slope on the compression rate when varying the channel width, w. Inset shows the L_{eff} dependence on w (see text).

ibility, $\kappa = 5.32\ 10^{-2}$ m mN^{-1}, in fully agreement with the value calculated from the Π *vs* molecular area isotherm of the elaidic acid monolayer at this range of surface pressures. For larger widths, $(\kappa L)^{-1}$ values depend on the geometry of the cell. Instead, we can compute an effective length of the compression chamber, $L_{eff}(w)$, for each channel width using the previous value of κ. Exponentially fitted (see inset in Fig. 25), one finds indeed that L_{eff} tends to 38 cm for large w values, ≥ 1 cm, a value within only a few per cent deviation of the total length of the Langmuir through. This can be interpreted considering that, in spite that the monolayer is a continuous two-dimensional medium extended all over the available area of the Langmuir trough, a microchannel up to 400 μm totally blocks the initial stroke of the compression of the whole monolayer over all the available area of Langmuir trough. This introduces a delay in attaining the steady surface pressure value that decreases by increasing w, and disappears for a channel width only slightly above one tenth of the width of the Langmuir trough.

Dynamical aspects can be discussed by means of Eq. 10. Fig. 26 shows the temporal evolution of the velocity of the monolayer inside the channel, v_c, for various barrier speeds v_b at a fixed channel width w, (w = 400 μm). Numerical fit of these dependences allows to evaluate the

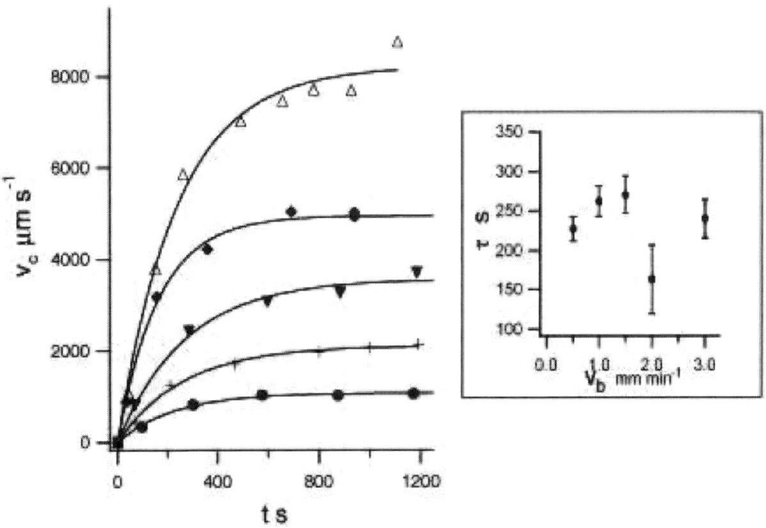

Fig. 26. Velocity transients inside a channel of nominal width 0.4 mm at different barrier velocities: 0.5, 1, 1.5, 2 and 3 mm min^{-1} (from bottom to top). Inset shows the calculated time constant, τ.

time constant τ which is confirmed to remain constant independent of the compression rate (see inset in Fig. 26). In addition, τ values can be also obtained from the experimental values of κ, λ and α using Eq. 11. For experiments conducted at a channel width of 400 μm, we have found τ = 240 $\pm$ 20 s, fully compatible with the results in Fig. 26. A similar level of agreement is found for narrower microchannels, revealing a w^{-2} dependence of τ according to Eqs. 11 and 12. Finally, these measured τ values are also in good agreement with the transient time of the relaxation experiment to rest shown in Fig. 22, reinforcing the validity of the previous theoretical analysis (Eqs. 10 and 11) supported on basic interfacial hydrodynamics rationale.

Finally, just to mention that this analysis of the so-called bottleneck effect, ubiquitous in miniaturized two-dimensional flowing systems, shows that it is directly rooted to the intrinsic material compressibility of the monolayer. This result has to be considered in the development of two-dimensional microfluidic methods, which will certainly be based on the pre-existing knowledge gained from experiments of constrained flows in monolayers, in particular, in those measurements involving the use of constant mechanical pumping, given the largely delayed response inside microchannels.

5.2. *Monolayers flowing trough different microchannels. Two-dimensional microfluidics*

After the analysis of different aspects of the rheology of a monolayer on a water subphase flowing through a microchannel,[81,85,87,90] we propose an experimental set-up to gain control on confined flows of monolayers in a more complex network of microchannels (paths) and, if possible, on circuits patterned at will. Our aim is to stimulate developments parallel to those in conventional microfluidics but now applied to two-dimensional environments.

For this purpose, we will take advantage of the wide variety of available techniques to impress surfaces with motifs of different sizes, shapes and aspect ratios. In particular, we will focus on printing patterns with an important hydrophilic/hydrophobic contrast. This will allow the water subphase to wet only the hydrophilic parts of the circuits and, while monitoring the surface pressures of the different compartments that the hydrophilic paths connect, to control monolayer flows among them. The widths of the hydrophilic paths of the circuits are at least millimetre-wide tracks, one-order of magnitude larger than the standards in the usual closed- or open-microfluidic devices.[92]

The first step of the experimental procedure[93] is the design and preparation of hydrophilic tracks surrounded in a superhydrophobic environment. The substrate of the circuit is a clean and polished brass plate (panel a in

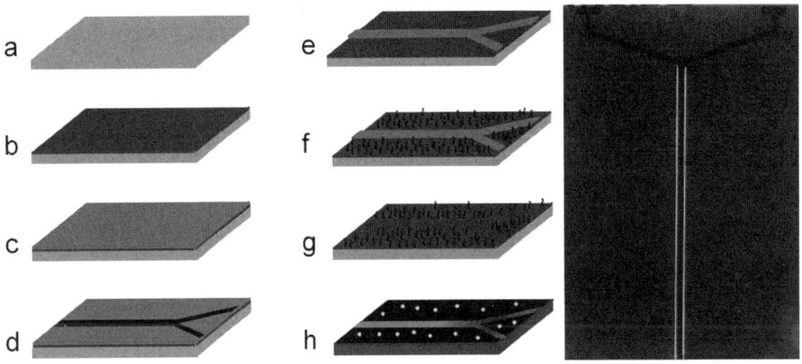

Fig. 27. Left side: scheme of the photolitography procedure to prepare a Y-shaped hydrophilic pattern (see text). Right side: image of a Y-junction followed by a rectilinear path. The width of the resulting channel after merging is 2 mm.

Fig. 27), above which a silver layer is formed by spontaneous deposition during the immersion in an aqueous $AgNO_3$ solution (panel b). Then, the hydrophilic pattern is printed on the silver covered plate by means of a soft lithography technique. After spin coating a layer of a positive photoresist (panel c), the acetate mask with the designed motif is aligned (panel d) and irradiated by UV light. After detaching the mask, the irradiated photoresist is removed by immersion in a developer solution (panel e). The superhydrophobic environment is achieved by means of a self-assembled monolayer of a highly fluorinated aliphatic thiol (panel f) on the exposed silver.[94] The removal of the non irradiated photoresist allows to recover the hydrophilic silver track with the designed circuit shape (panels g and h). Fig. 28 shows a more complex motif fabricated using the same procedure and the high hydrophobic character of the thiolated silver. Contact angle values less than 5° have been measured on free deposited silver surface.

Then, the patterned plate is set into the appropriate Langmuir trough. In Fig. 29, a plate with a Y-junction followed by a rectilinear path connects the three compartments of a thermostated Teflon Langmuir trough. Each compartment is equipped with a barrier and a Wilhelmy plate balance in order to independently monitor and control the three barrier velocities and surface pressures. The brass plate lays on a block which permits a precise connection between each compartment and the appropriate hydrophilic channel (see Fig. 29, panel b).

This "two-dimensional microfluidics" device has been used to measure

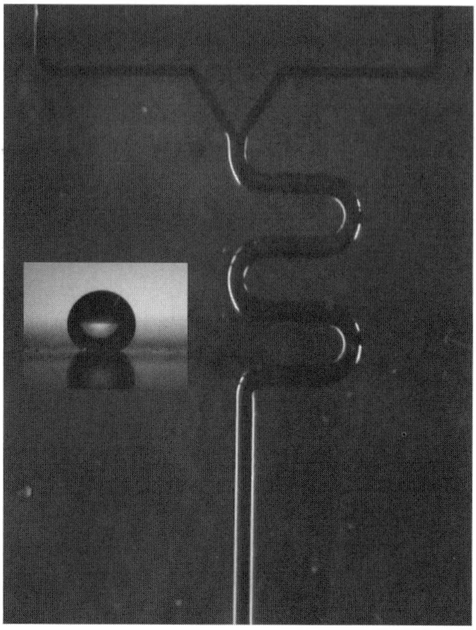

Fig. 28. Patterned plate showing a Y-junction followed by a three loop serpentine. Notice the contact angle ($\sim 140°$) of a 5 μl water droplet deposited on the thiolated silver surface. The width of the resulting channel after merging is 2 mm.

the lateral diffusion coefficient, D, of an amphiphile, a fluorescence probe, at the air-water interface. This parameter is usually determined by fluorescence recovery after photobleaching (FRAP),[95–97] fluorescence correlation spectroscopy (FCS)[98] or two-dimensional voltammetry.[99] In our case, we will directly measure this parameter after the analysis of directly measured concentration profiles at different diffusion times. The fluorescence probe is (N-(Texas Red sulfonyl)-1,2-dihexadecanoyl-snglycero-3-phosphoethanolamine, triethylammonium salt) (TR-DHPE), dispersed in a dimyristoyl-phosphatidyl-choline (DMPC) medium. The diffusion of TR-DHPE has been followed by epi-fluorescence microscopy (FM). This measurement is performed while coflowing a pure and a doped DMPC monolayer through a Y-junction followed by a straight path. When the steady state is attained after adjusting barriers speed, the monolayer flow velocities are measured by Particle Image Velocimetry using polyamide microparticles as tracers and images are recorded at different locations of the straight path. Then diffusion times are computed for each image and the

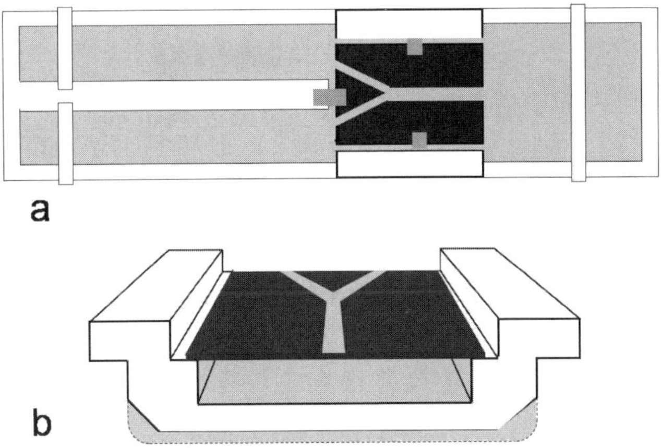

Fig. 29. The three-compartments Langmuir trough (see text).

corresponding concentration profiles of TR-DHPE, perpendicular to the flow direction, are extracted by conventional image analysis (Fig. 30). After fitting the concentration profile at each position (diffusion time) to the diffusion equation,

$$\frac{c}{c^{\circ}} = \frac{1}{2} erfc\left(\frac{x - x_{\circ}}{2\sqrt{Dt}}\right) \Rightarrow I = I^{\circ} + B \;\; erfc[a(x - x_{\circ})]$$

assuming that the recorded intensity of the emitted light, I, is linear with the surface concentration, c, of the probe, the value of D can be calculated from the relationship between a and t, $a^{-1} = 2\sqrt{Dt}$ (Fig. 30). The obtained mean value of the lateral diffusion coefficient of TR-DHPE is $(6 \pm 1)\ 10^{-7}$ $cm^2\ s^{-1}$. This value agrees with that of similar compounds dissolved in DMPC.[98]

Finally, our experimental innovation envisages a plethora of future developments in the field of monolayer flows. Among them we mention two aspects. First to look for miniaturization of the designed circuit networks. This would favor integrability and high throughput surfactant-based chemistry. Second to take advantage of the design variability gained with lithographic techniques to look for different mobility-mediated phenomena for species dispersed in single or multiphase monolayers, particularly sorting, mixing or chemical reactions in restricted 2d interfacial flows.

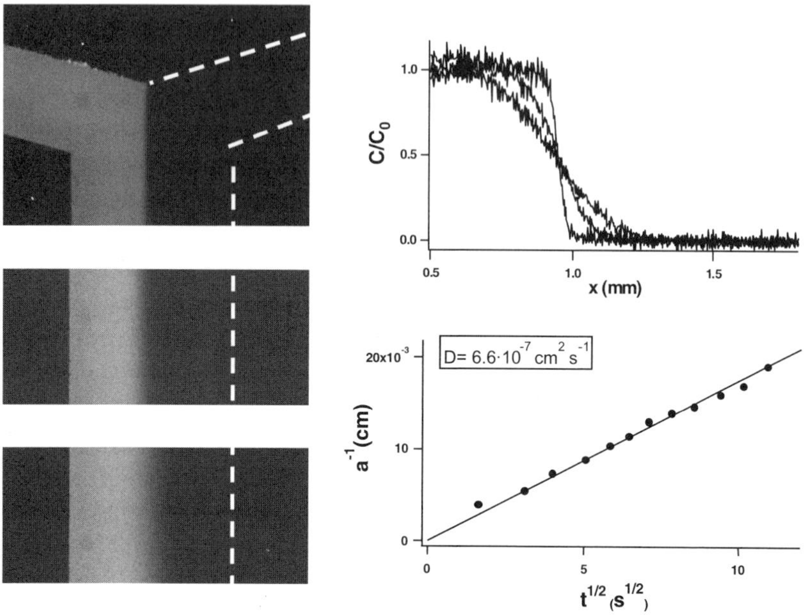

Fig. 30. Evolution of the TR-DHPE concentration profile along the channel. Surface pressures in each compartment are comprised between 1–2 mN m^{-1}. TR-DHPE concentration in the left side compartment of the Langmuir trough is $\sim$ 2%. Temperature is maintained at 25°C. Diffusion times from top to bottom are 2.4, 36.6 and 129.5 s. Bottom right: dependence of a on the diffusion time t.

References

1. P. G. de Gennes, Soft Matter, *Rev. Mod. Phys.* **64**, 645–648 (1992).
2. A. Ullman, *Introduction to Ultrathin Organic Films.* (Academic Press, San Diego, 1991).
3. J. C. Love, L. A. Estroff, J. K. Kriebel, R. G. Nuzzo, and G. M. Whitesides, Self-assembled monolayers of thiolates on metals as a form of nanotechnology, *Chem. Rev.* **105**, 1103–1169 (2005).
4. G. L. Gaines Jr., *Insoluble Monolayers at Liquid-Gas Interfaces.* (Interscience, New York, 1966).
5. V. M. Kaganer, H. Möhwald, and P. Dutta, Structure and phase transitions in Langmuir monolayers, *Rev. Mod. Phys.* **71**, 779–819 (1999).
6. F. Davis and S. Higson, Structured thin films as functional components within biosensors, *Biosens. Bioelec.* **21**, 1–20 (2005).
7. G. Brezesinski and H. Möhwald, Langmuir monolayers to study interac-

tions at model membrane surfaces, *Adv. Coll. Interf. Sci.* **100-102**, 563–584 (2003).
8. S. Hénon and J. Meunier, Microscope at the Brewster-angle: Direct observation of 1st-order phase-transitions in monolayers, *Rev. Sci. Inst.* **62**(4), 936–939 (1991).
9. D. Hönig and D. Möbius, Direct visualization of monolayers at the air-water-interface by Brewster-angle microscopy, *J. Phys. Chem.* **95**(12), 4590–4592 (1991).
10. P. Dynarowicz-Latka, A. Dhanabalan, and O. Oliveira Jr., Modern physcochemical research on Langmuir monolayers, *Adv. Colloid Interface Sci.* **91**, 221–293 (2001).
11. H. Rau. Photoisomerization of azobenzenes. In ed. J. F. Rabek, *Photochemistry and Photophysics*, vol. II, pp. 119–141. CRC Press, Boca Raton (Florida) (1990).
12. J. Crusats, R. Albalat, J. Claret, J. Ignés-Mullol, and F. Sagués, Influence of temperature and composition on the mesoscopic textures of azobenzene Langmuir monolayers, *Langmuir.* **20**, 8668–8674 (2004).
13. J. Ignés-Mullol, J. Claret, R. Albalat, J. Crusats, R. Reigada, M. T. Martín-Romero, and F. Sagués, Texture changes inside smectic-c droplets in azobenzene Langmuir monolayers, *Langmuir.* **21**, 2948–2955 (2005).
14. T. Sato, Y. Ozaki, and K. Iriyama, Molecular aggregation and photoisomerization of Langmuir-Blodgett-films of azobenzene-containing long-chain fatty-acids and their salts studied by Ultraviolet-Visible and Infrared spectroscopies, *Langmuir.* **10**, 2363–2369 (1994).
15. J. Maack, R. C. Ahuja, and H. Tachibana, Resonant and nonresonant investigations of amphiphilic azobenzene derivatives in solution and in monolayers at the air-water-interface, *J. Phys. Chem.* **99**, 9210–9220 (1995).
16. J. M. Pedrosa, M. T. Martín-Romero, L. Camacho, and D. Möbius, Organization of an amphiphilic azobenzene derivative in monolayers at the air-water interface, *J. Phys. Chem. B.* **106**, 2583–2591 (2002).
17. R. Reigada, E. Abad, J. Crusats, J. Claret, J. Ignés-Mullol, and F. Sagués, Orientational structures in confined smectic-c domains in Langmuir monolayers, *J. Chem. Phys.* **121**, 9066–9076 (2004).
18. R. Reigada, F. Sagués, and A. S. Mikhailov, Traveling waves and nonequilibrium stationary patterns in two-component reactive Langmuir monolayers, *Phys. Rev. Lett.* **89**, 038301 (2002).
19. A. Mikhailov and G. Ertl, Nonequilibrium structures in condensed systems, *Science.* **272**, 1596–1597 (1996).
20. M. C. Cross and P. C. Hohenberg, Pattern formation outside of equilibrium, *Rev. Mod. Phys.* **65**, 851–1112 (1993).
21. Y. Tabe and H. Yokoyama, Two dimensional dynamic patterns in illuminated Langmuir monolayers, *Langmuir.* **11**, 4609–4613 (1995).
22. F. Sagués, R. Albalat, R. Reigada, J. Crusats, J. Ignés-Mullol, and J. Claret, Photoswitchable orientational patterns of confined domains in monolayers, *J. Am. Chem. Soc.* **127**, 5296–5297 (2005).
23. J. Crusats, R. Albalat, J. Claret, J. Ignés-Mullol, R. Reigada, and F. Sagués,

Photoinduced molecular reorientation dynamics in confined domains of Langmuir monolayers, *J. Chem. Phys.* **122**, 244722 (2005).
24. J. Ignés-Mullol, J. Claret, and F. Sagués, Star deffects, boojums, and cardioiddroplet shapes in condensed dimyristoylphosphatidylethanolamine monolayers, *J. Phys. Chem. B.* **108**, 612–619 (2004).
25. J. Claret, J. Ignés-Mullol, R. Reigada, F. Sagués, and J. Crusats, Self-organizing propagation waves from dynamic self-assembly in monolayers, *Phys. Rev. E.* **73**, 026225 (2006).
26. J. Claret, R. Crusats, J. Albalat, J. Ignés-Mullol, H. Martinez-Seara, R. Reigada, and F. Sagués, Travelling waves in two-dimensional smectic-c domains, *Eur. Phys. J. E.* **21**, 111–116 (2006).
27. T. Kawai, J. Umemura, and T. Takenaka, UV absorption spectra of azobenzene-containing long-chain fatty acids and their barium salts in spread monolayers and Langmuir-Blodgett films., *Langmuir.* **5**, 1378–1383 (1989).
28. M. K. Durbin, A. Malik, A. G. Richter, C. J. Yu, R. Eisenhower, and P. Dutta, Ordered phases in Langmuir monolayers of an azobenzene derivative, *Langmuir.* **14**, 899–903 (1998).
29. P. Burriel, J. Ignés-Mullol, R. Reigada, and F. Sagués, Collective molecular precession induced by rotating illumination in photosensitive Langmuir monolayers, *Langmuir.* **22**, 187–193 (2006).
30. P. M. Chaikin and T. C. Lubensky, *Principles of Condensed Matter Physics* (Cambridge University Press, 1995).
31. M. Kleman and O. Lavrentovich, *Soft Matter Physics — An Introduction* (Springer, 2003).
32. T. M. Fischer, R. F. Bruinsma, and C. M. Knobler, Textures of surfactant monolayers, *Phys. Rev. E.* **50**, 413–428 (1994).
33. D. Svensek and S. Zumer, Hydrodynamics of pair-annihilating disclinations in smc films, *Phys. Rev. Lett.* **90**(15), 155501 (2003).
34. G. Toth, C. Denniston, and J. M. Yeomans, Hydrodynamics of topological defects in nematic liquid crystals, *Phys. Rev. Lett.* **88**(10), 105504 (2002).
35. G. Toth, C. Denniston, and J. M. Yeomans, Hydrodynamics of domain growth in nematic liquid crystals, *Phys. Rev. E.* **67**, 051705 (2003).
36. P. Oswald, *Rhéophysique. Ou comment coule la matière* (Editions Belin, Paris, 2005).
37. C. Blanc, D. Svensek, S. Zumer, and M. Nobili, Dynamics of nematic liquid crystal disclinations: The role of the backflow, *Phys. Rev. Lett.* **95**(9), (2005).
38. J. Brugués, J. Ignés-Mullol, J. Casademunt, and F. Sagués, Probing elastic anisotropy from defect dynamics in Langmuir monolayers, *Phy. Rev. Lett.* **100**(3) (2008).
39. A. Feder, Y. Tabe, and E. Mazur, Orientational fluctuations in a two dimensional smectic-c liquid crystal with variable density, *Phys. Rev. Lett.* **79**, 1682–1685 (1997).
40. P. V. Dolganov, H. T. Nguyen, G. Joly, V. K. Dolganov, and P. Cluzeau, Shape of nematic droplets in smectic membranes, *Europhys. Lett.* **78**, 66001 (2007).
41. P. V. Dolganov, H. T. Nguyen, G. Joly, V. K. Dolganov, and P. Cluzeau,

Different mechanisms of nucleation and self-organization of droplets in ferroelectric smectic membranes, *Eur. Phys. J. E.* **25**, 31–37 (2008).
42. A. N. Pargellis, P. Finn, J. W. Goodby, P. Panizza, B. Yurke, and P. E. Cladis, Defect dynamics and coarsening dynamics in smectic-c films, *Phys. Rev. A.* **46**, 7765–7776 (1992).
43. J. Z. Pang, C. D. Muzny, and N. A. Clark, String defects in freely suspended liquid-crystal films, *Phys. Rev. Lett.* **69**, 2783–2786 (1992).
44. M. A. Vallvé and J. Ignés-Mullol, Dynamics of point defects and stripe textures in smectic-c Langmuir monolayers, *European Phys. J. E.* **30**, 403–409 (2009).
45. G. A. Hinshaw and R. G. Petschek, Transitions and modulated phases in chiral tilted smectic liquid crystals, *Phys. Rev. A.* **39**, 5914–5926 (1988).
46. J. V. Selinger, Z. G. Wang, R. F. Bruinsma, and C. M. Knobler, Chiral symmetry breaking in Langmuir monolayers and smectic films, *Phys. Rev. Lett.* **70**, 1139–1142 (1993).
47. Y. Tabe, N. Shen, E. Mazur, and H. Yokoyama, Simultaneous observation of molecular tilt and azimuthal angle distributions in spontaneously modulated liquid-crystalline Langmuir monolayers, *Phys. Rev. Lett.* **82**, 759–762 (1999).
48. E. Hatta and T. M. Fischer, String defects connecting pairs of half-integer disclinations and tilting transition of a Langmuir monolayer, *J. Phys. Chem. B.* **109**, 2801–2804 (2005).
49. P. Oswald and P. Pieranski, *Nematic and Cholesteric liquid crystals: concepts and physical properties illustrated by experiments* (Taylor& Francis, 2005).
50. N. Nandi and D. Vollhardt, Effect of molecular chirality on the morphology of biomimetic Langmuir monolayers, *Chem. Rev.* **103**, 4033 (2003).
51. J. J. L. M. Cornelissen, A. E. Rowan, R. J. M. Nolte, and N. A. J. M. Sommerdijk, Chiral architectures from macromolecular building blocks, *Chem. Rev.* **101**, 4039–4070 (2001).
52. L. Pérez-Garcia and D. Amabilino, Spontaneous resolution, whence and whither: from enantiomorphic solids to chiral liquid crystals, monolayers and macro- and supra-molecular polymers and assemblies, *Chem. Soc. Rev.* **36**, 941 (2007).
53. D. P. Kondepudi, R. J. Kaufman, and N. Singh, Chiral symmetry breaking in sodium chlorate crystallizaton, *Science* **250**, 975–976 (1990).
54. J. M. McBride and R. L. Carter, Spontaneous resolution by stirred crystallization, *Angew. Chem. Int. Ed.* **30**, 293–295 (1991).
55. J. M. McBride and J. C. Tully, Did life grind to a start?, *Nature.* **452**, 161–162 (2008).
56. C. Viedma, Chiral symmetry breaking during crystallization: Complete chiral purity induced by nonlinear autocatalysis and recycling, *Phys. Rev. Lett.* **94**, 065504 (2005).
57. W. L. Noorduin, T. Izumi, A. Millemaggi, M. Leeman, H. Meekes, W. J. P. Van Enckevort, R. M. Kellogg, B. Kaptein, E. Vlieg, and D. G. Blackmond, Emergence of a single solid chiral state from a nearly racemic amino acid derivative, *J. Am. Chem. Soc.* **130**, 1158–1159 (2008).
58. W. Dzwolak, A. Loksztejn, A. Galinska-Rakoczy, R. Adachi, Y. Goto, and

L. Rupnicki, Conformational indeterminism in protein misfolding: Chiral amplification on amyloidogenic pathway of insulin, *J. Am. Chem. Soc.* **129**, 7517–7522 (2007).

59. A. Loksztejn and W. Dzwolak, Vortex-induced formation of insulin amyloid superstructures probed by time-lapse atomic force microscopy and circular dichroism spectroscopy, *J. Mol. Biol.* **395**, 643–655 (2010).
60. J. M. Ribó, J. Crusats, F. Sagués, J. Claret, and R. Rubires, Chiral sign induction by vortices during the formation of mesophases in stirred solutions, *Science.* **292**, 2063–2066 (2001).
61. J. Crusats, J. Claret, I. Díez-Perez, Z. El-Hachemi, H. Garcia-Ortega, R. Rubires, F. Sagués, and J. M. Ribó, Chiral shape and enantioselective growth of colloidal particles of self-assembled meso-tetra(phenyl and 4-sulfonatophenyl) porphyrins, *Chem. Comm.* . **13** 1588–1589 (2003).
62. O. Ohno, Y. Kaizu, and H. Kobayashi, J-aggregate formation of a water-soluble porphyrin in acidic aqueous media, *J. Chem. Phys.* **99**, 4128–4139 (1993).
63. T. Yamaguchi, T. Kimura, H. Matsuda, and T. Aida, Macroscopic spinning chirality memorized in spin-coated films of spatially designed dendritic zinc porphyrin J-aggregates, *Angew. Chem. Int. Ed.* **43**, 6350–6355 (2004).
64. A. Tsuda, M. A. Alam, T. Harada, T. Yamaguchi, N. Ishii, and T. Aida, Spectroscopic visualization of vortex flows using dye-containing nanofibers, *Angew. Chem. Int. Ed.* **46**, 8198–8202 (2007).
65. A. D'Urso, R. Randazzo, L. Lo Faro, and R. Purrello, Vortexes and nanoscale chirality, *Angew. Chem. Int. Ed.* **49**, 108–112 (2010).
66. O. Arteaga, A. Canillas, R. Purrello, and J. M. Ribó, Evidence of induced chirality in stirred solutions of supramolecular nanofibers, *Opt. Lett.* **34**, 2177–2179 (2009).
67. M. Wolffs, S. J. George, Z. Tomović, S. C. J. Meskers, A. Schenning, and E. W. Meijer, Macroscopic origin of circular dichroism effects by alignment of self-assembled fibers in solution, *Angew. Chem. Int. Ed.* **46**, 8203–8205 (2007).
68. G. P. Spada, Alignment by the convective and vortex flow of achiral self-assembled fibers induces strong circular dichroism effects, *Angew. Chem. Int. Ed.* **47**, 636–638 (2008).
69. N. Petit-Garrido, J. Ignés-Mullol, J. Claret, and F. Sagués, Chiral selection by interfacial shearing of self-assembled achiral molecules, *Phys. Rev. Lett.* **103**, 237802 (2009).
70. L. D. Barron, Chirality and life, *Space Sci. Rev.* **135**, 187–201 (2008).
71. R. Plasson, D. K. Kondepudi, H. Bersini, A. Commeyras, and K. Asakura, Emergence of homochirality in far from equilibrium systems: Mechanisms and role in prebiotic chemistry, *Chirality.* **19**, 589–600 (2007).
72. R. M. Hazen, Mineral surfaces and the prebiotic selection and organization of biomolecules, *Am. Miner.* **91**, 1715–1729 (2006).
73. H. Zepik, E. Shavit, M. Tang, T. R. Jensen, K. Kjaer, G. Bolbach, L. Leiserowitz, I. Weissbuch, and M. Lahav, Chiral amplification of oligopeptides in two-dimensional crystalline self-assemblies on water, *Science.* **295**, 1266–1269

(2002).
74. K. Katsonis, H. Xu, R. M. Haak, T. Kudernac, Z. Tomovic, S. George, M. V. der Auweraer, A. P. H. J. Schenning, E. W. Meijer, and B. L. Feringa, Emerging solvent-induced homochirality by the confinement of achiral molecules against a solid surface, *Angew. Chem. Int. Ed.* **47**, 4997–5001 (2008).
75. M. Stalder and M. Schadt, Linearly polarized light with axial symmetry generated by liquid-crystal polarization converters, *Optics Lett.* **21**, 1948–1950 (1996).
76. N. Petit-Garrido, J. Ignés-Mullol, J. Claret, R. P. Trivedi, C. Lapointe, F. Sagués, and I. I. Smalyukh, Defects at the interface of 2d and 3d orientationally ordered soft matter systems, in preparation.
77. P. Tabeling, *Introduction to Microfluidics* (Oxford University Press, 2005).
78. T. M. Squires and S. R. Quake, Microfluidics: Fluid physics at the nanoliter scale, *Rev. Mod. Phys.* **77**, 977–1026 (2005).
79. G. M. Whitesides, The origins and the future of microfluidics, *Nature.* **442**, 368–373 (2006).
80. H. A. Stone, A. D. Stroock, and A. Ajdari, Engineering flows in small devices: Microfluidics toward a lab-on-a-chip, *Annu. Rev. Fluid Mech.* **36**, 381–411 (2004).
81. J. Ignés-Mullol, J. Claret, R. Reigada, and F. Sagués, Spread monolayers: Structure, flows and dynamic self-organization phenomena, *Phys. Rep.* **448**, 163–179 (2007).
82. M. Sacchetti, H. Yu, and G. Zografi, Hydrodynamic coupling of monolayers with subphase, *J. Chem. Phys.* **99**, 563–566 (1993).
83. D. K. Schwartz, C. M. Knobler, and R. Bruinsma, Direct observation of Langmuir monolayer flow-through a channel, *Phys. Rev. Lett.* **73**, 2841–2844 (1994).
84. H. A. Stone, Fluid motion of monomolecular films in a channel flow geometry, *Phys. Fluids.* **7**, 2931–2937 (1995).
85. M. L. Kurnaz and D. K. Schwartz, Channel flow in a Langmuir monolayer. unusual velocity profiles in a liquid-crystalline mesophase, *Phys. Rev. E.* **56**, 3378–3384 (1997).
86. A. Ivanova, M. L. Kurnaz, and D. K. Schwartz, Temperature and flow rate dependence of the velocity profile during channel flow of a Langmuir monolayer, *Langmuir.* **15**, 4622–4624 (1999).
87. D. J. Olson and G. G. Fuller, Contraction and expansion flows of Langmuir monolayers, *J. Non-Newtonian Fluid Mech.* **89**, 187–207 (2000).
88. P. Tabeling. Some basic problems of microfluidics. In *14th Australasian Fluid Mechanics Conference*, Adelaide University, Adelaide, Australia (2001).
89. M. Martin, G. Blu, C. Eon, and G. Guiochon, The use of syringe-type pumps in liquid chromatography in order to achieve a constant flow-rate, *J. Chromatogr.* **112**, 399–414 (1975).
90. P. Burriel, J. Claret, J. Ignés-Mullol, and F. Sagués, Bottleneck effect in two-dimensional microfluidics, *Phys. Rev. Lett.* **100**, 134503 (2008).
91. K. S. Yim, C. F. Brooks, G. G. Fuller, D. Winter, and C. D. Eisenbach, Non-newtonian rheology of liquid crystalline polymer monolayers, *Langmuir.*

16, 4325–4332 (2000).
92. R. Seemann, M. Brinkmann, E. J. Kramer, F. F. Lange, and R. Lipowsky, Wetting morphologies at microstructured surfaces, *Proc. Natl. Acad. Sci.* **102**, 1848–1852 (2005).
93. P. Burriel, J. Ignés-Mullol, J. Claret, and F. Sagués, Two-dimensional microfluidics using circuits of wettability contrast, *Langmuir.* **26**, 4613–4615 (2010).
94. I. A. Larmour, S. E. J. Bell, and G. C. Saunders, Remarkably simple fabrication of superhydrophobic surfaces using electroless galvanic deposition, *Angew. Chem. Int. Ed.* **46**, 1710–1712 (2007).
95. J. G. McNally. Quantitative FRAP in analysis of molecular binding dynamics in vivo. In *Fluorescent Proteins*, vol. Methods in Cell Biology, 85, pp. 329–351. Elsevier Academic Press, San Diego (California) (2008).
96. S. Marty, M. Schroeder, K. W. Baker, G. Mazzanti, and A. G. Marangoni, Small-molecule diffusion through polycrystalline triglyceride networks quantified using fluorescence recovery after photobleaching, *Langmuir.* **25**, 8780–8785 (2009).
97. S. Kaufmann, G. Papastavrou, K. Kumar, M. Textor, and E. Reimhult, A detailed investigation of the formation kinetics and layer structure of poly(ethylene glycol) tether supported lipid bilayers, *Soft Matter.* **5**, 2804–2814 (2009).
98. M. Gudmand, M. Fidorra, T. Bjørnholm, and T. Heimburg, Diffusion and partitioning of fluorescent lipid probes in phospholipid monolayers, *Biophys. J.* **96**, 4598–4609 (2009).
99. Y. S. Kang and M. Majda, Headgroup immersion depth and its effect on the lateral diffusion of amphiphiles at the air/water interface, *J. Phys. Chem. B.* **104**, 2082–2089 (2000).

Chapter 5

Hydrodynamic Effects in Multicomponent Fluid Membranes

Shigeyuki Komura[1]*, Sanoop Ramachandran[2] and Masayuki Imai[3]

[1] *Department of Chemistry, Graduate School of Science and Engineering, Tokyo Metropolitan University, Tokyo 192-0397, Japan*

[2] *Physique des Polymères, Université Libre de Bruxelles, Campus Plaine, CP 223, B-1050 Brussels, Belgium*

[3] *Department of Physics, Faculty of Science, Ochanomizu University, Tokyo 112-8610, Japan*

In this chapter, we deal with hydrodynamic effects on multicomponent fluid membranes. Above the miscibility transition temperature, we focus on the hydrodynamic effects on the dynamics of critical concentration fluctuations in two-component fluid membranes, and the wavenumber dependence of the effective diffusion coefficient is shown. Below the miscibility transition temperature, we study the domain growth exponent in a binary fluid membrane using a particle-based simulation method. A change in the growth exponent from two-dimensional to three-dimensional nature with the addition of bulk solvent is observed. Next, using a simplified hydrodynamic theory, we calculate the drag on a liquid domain diffusing in a two-dimensional membrane. The analytical expression for the diffusion coefficient spans the whole spectrum of size ranges. The dynamics of a Gaussian polymer chain embedded in a liquid membrane surrounded by bulk solvent and walls are also discussed. Using the preaveraging approximation, we can circumvent the non-linearity imposed by the hydrodynamics. Within this approximation, the diffusion coefficient of the polymer in the free membrane geometry is obtained for a size range of several decades in order. The polymer relaxation times as well as structure factor are obtained for both confined and free membranes. Finally, we discuss the coupled in-plane dynamics between point particles embedded in stacked fluid membranes.

*komura@tmu.ac.jp

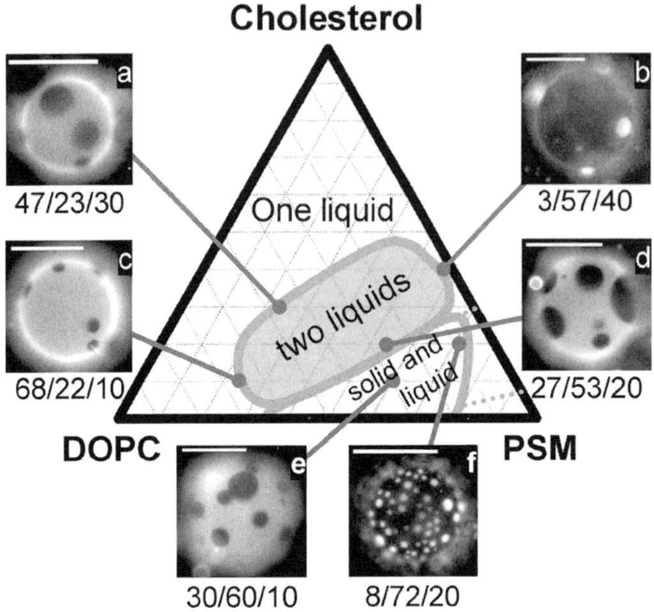

Fig. 1. Phases observed by fluorescence microscopy of giant unilamellar vesicles containing mixture of DOPC, PSM and cholesterol at 25 °C. The dark liquid phase is rich in PSM and cholesterol, while the bright liquid phase is rich in DOPC. Adapted from Ref. 3.

1. Introduction

Biological membranes typically contain various components such as lipid mixtures, sterols, and proteins that are indispensable to cell functions.[1] Rather than being uniformly distributed in the membrane, there is growing evidence that some cellular components are incorporated in domains arising from lateral lipid segregation in membranes. In 1997, Simons and Ikonen proposed a hypothesis which suggests that the lipids organize themselves into sub-micron sized domains termed "lipid rafts".[2] They considered that the lipid rafts serve as platforms for proteins which attribute certain functionality to each domain.

Stimulated by the lipid raft hypothesis, phase separation on vesicles has been extensively investigated using artificial model membranes composed of saturated lipids, unsaturated lipids and cholesterol. Especially, the phase behavior of ternary giant vesicles composed of saturated lipids with high chain melting temperature, unsaturated lipids with low chain melting tem-

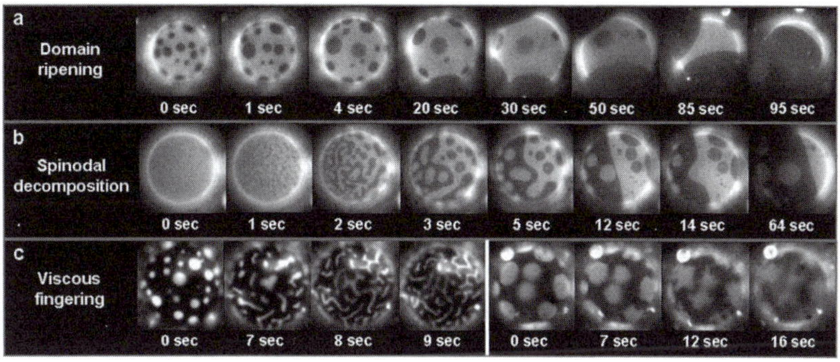

Fig. 2. Phase separation dynamics on ternary giant vesicles. (a) Brownian coagulation observed for 1 : 1 DOPC/DPPC + 25% Chol, (b) spinodal decomposition observed for 1 : 1 DOPC/DPPC + 35% Chol, (c) viscous fingering observed for 1 : 9 DOPC/DPPC + 25% Chol (left series) and 1 : 1 DOPC/DMPC + 25% Chol (right series). Adapted from Ref. 4.

perature, and cholesterol have been investigated in detail for various combinations. Such a ternary membrane is homogeneously mixed in the high temperature region. By decreasing the temperature, the membrane undergoes a phase separation between the coexisting liquid ordered (L_o) and liquid disordered (L_d) phases. As a typical example, we show in Fig. 1 the phase diagram of a ternary vesicle consisting of palmitoyl sphingomyelin (PSM), dioleoylphosphatidylcholine (DOPC) and cholesterol (Chol) together with several fluorescence micrographs.[3] In the pictures, the dark liquid phase is rich in PSM and cholesterol, while the bright liquid phase is rich in DOPC. The minority liquid phase forms circular domains which undergo Brownian motion on the vesicle. A large amount of research has been conducted using various experimental techniques in order to clarify the properties of ternary lipid systems such as to determine the phase diagrams or to identify the domain morphologies.

Among many works, one of the interesting and also important research directions is to investigate the dynamics of the lateral phase separation in fluid membranes. Veatch and Keller observed the kinetics of domain growth on ternary vesicles as presented in Fig. 2.[4] For a vesicle with an off-critical composition, dark circular domains grow by colliding and coalescing with each other rather than through the Ostwald ripening (Fig. 2(a)). When the composition is nearly critical, spinodal decomposition takes place when the temperature is decreased through transition point (Fig. 2(b)). For a highly asymmetric composition as in Fig. 2(c), striped domains are observed when the temperature is raised; possibly viscous fingering. It should be noted that these various morphologies can be found near the miscibility transition temperature.

Yanagisawa *et al.* investigated the domain growth dynamics in ternary vesicle (DOPC/DPPC/Chol).[5] By observing the motion of domains under the optical microscope, they found that each domain is attracted toward the largest one following the flow around it as shown in Fig. 3(a). Figure 3(b) presents the distribution of the moving direction (angle) of domains with respect to the largest domain. Since the resulting distribution has a peak at the center, one finds that the smaller domains are attracted to the largest target domain, and hence their motion is not random. This observation strongly suggests that the hydrodynamic interactions are acting between the domains. Such hydrodynamic interactions can be mediated either by the fluid membrane itself or by the surrounding bulk water.

In this chapter, we shall discuss (mainly theoretically) the dynamics of phase separation in multicomponent membranes, particularly focusing on the effect of bulk solvent such as water. We will repeatedly stress that the hydrodynamic interaction mediated by not only the fluid membrane itself but also the bulk solvent plays an essential role in the dynamical properties of multicomponent membranes at large scales. After providing the general framework of the membrane hydrodynamics in the presence of the bulk solvent (Sec. 2), we shall discuss five related aspects in the dynamics of multicomponent membranes; (i) dynamics of concentration fluctuations above the transition temperature (Sec. 3), (ii) domain growth dynamics below the transition temperature (Sec. 4), (iii) diffusion coefficient of a two-dimensional (2D) liquid domain (Sec. 5), (iv) dynamics of a polymer chain confined in a membrane (Sec. 6), and (v) hydrodynamic coupling between two fluid membranes (Sec. 7). A clear understanding of phase separation dynamics and the diffusion properties in fluid membranes may provide us with a better explanation of the lipid organization in cell membranes.

In Sec. 3, we investigate the hydrodynamic effects on the dynamics of critical concentration fluctuations in two-component fluid membranes. Experimental observations of critical fluctuations point towards the idea that the cell maintains its membrane slightly above the critical point.[6,7] Based on the Ginzburg–Landau approach with full hydrodynamics, we calculate the wavenumber dependence of the effective diffusion coefficient by changing the temperature and/or the thickness of the bulk fluid. We shall also consider the situation when the multicomponent membranes form 2D microemulsions.[8]

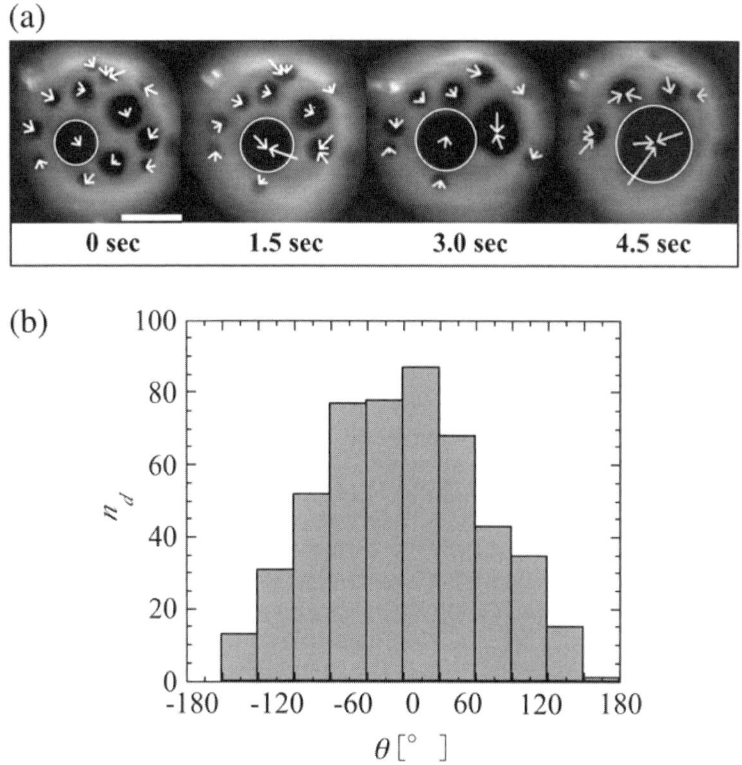

Fig. 3. (a) Time sequence of moving domains on a giant vesicle made of DOPC/DPPC/Chol. Arrow indicates the movement of the domain center during the time interval 1.5 sec. The white circle marks the largest domain to which other domains are attracted. The scale bar corresponds to 5 μm. (b) Distribution of domain migration angle. The vertical axis is the number of domains. Adapted from Ref. 5.

In Sec. 4, we study the effects of an embedding bulk solvent on the phase separation dynamics in a planar fluid membrane using dissipative particle dynamics simulations. We show that the presence of a bulk fluid will alter the domain growth exponent from that of 2D to 3D indicating the significant role played by the membrane-solvent coupling. In order to elucidate the underlying physical mechanism of this effect, we look into the diffusion properties in the membrane by measuring two-particle correlated diffusion. We show that quasi-2D phase separation proceeds by the Brownian coagulation mechanism which reflects the 3D nature of the bulk solvent. Such a behavior is universal as long as the domain size exceeds the hydrodynamic screening length.[9]

In Sec. 5, using a hydrodynamic theory that incorporates a momentum decay mechanism, we discuss the diffusion coefficient of a liquid domain of finite viscosity moving in a fluid membrane. We show an analytical expression for the drag coefficient which covers the whole range of domain sizes. Discussion on several limiting cases are given. The obtained drag coefficient decreases as the domain viscosity becomes smaller with respect to the outer membrane viscosity. This is because the flow induced in the domain acts to transport the fluid in the surrounding matrix more efficiently.

In Sec. 6, we present a Brownian dynamics theory with full hydrodynamics for a Gaussian polymer chain embedded in a liquid membrane which is surrounded by bulk solvent and walls. Within the preaveraging approximation, we obtain the diffusion coefficient of the polymer for the free membrane geometry. We also carry out a Rouse normal mode analysis to obtain the relaxation time and the dynamical structure factor. For large polymer size, both quantities show Zimm-like behavior in the free membrane case, whereas they are Rouse-like for the sandwiched membrane geometry. We use the scaling argument to discuss the effect of excluded volume interactions on the polymer relaxation time.

Finally in the last section of this chapter, we discuss the coupled in-plane diffusion dynamics between point-particles embedded in stacked fluid membranes. We calculate the contributions to the coupling longitudinal and transverse diffusion coefficients due to particle motion within the different as well as the same membranes. We show that the stacked geometry leads to a hydrodynamic coupling between the two membranes.

Before moving to the next section on the membrane hydrodynamics, we shall refer to the seminal theoretical work on which the present chapter is based. In 1975, Saffman and Delbrück (SD) obtained the diffusion coefficient of a protein molecule moving in a membrane under low Reynolds number conditions.[9,10] The protein molecule is assumed to be a rigid disk whereas the membrane is modeled as a 2D thin fluid sheet that is sandwiched in-between a 3D bulk solvent. Solving both the 2D and 3D hydrodynamic equations, they obtained the following expression for the diffusion coefficient of the disk:

$$D = \frac{k_{\mathrm{B}}T}{4\pi\eta}\left(\ln\frac{\eta}{\eta_{\mathrm{s}}R} - \gamma\right). \tag{1}$$

Here R is the radius of the moving disk, η the 2D membrane viscosity, η_{s} the 3D solvent viscosity, and $\gamma = 0.5772\cdots$ is Euler's constant. The above form tells us that the diffusion coefficient depends only logarithmically on the disk radius. However, it should be mentioned that Eq. (1) was derived under the condition such that $R \ll \eta/\eta_s$, i.e., the small disk size limit or the large membrane viscosity limit. This limit is valid for a lipid membrane surrounded by water, and is essentially equivalent to regard the membrane as an isolated pure 2D liquid. The length scale η/η_s, called the SD hydrodynamic screening length, plays an essential role in this problem.

One of the main attempts after the work by Saffman and Delbrück was to go beyond this limit. For example, Hughes *et al.* analyzed the whole range of protein sizes, and obtained the following asymptotic expression in the limit of $R \gg \eta/\eta_s$[11]

$$D = \frac{k_B T}{16\eta_s R}. \tag{2}$$

In this limit, the diffusion coefficient is more strongly dependent on the disk size while it is independent of the membrane viscosity η. Notice that the above expression shows the same scaling as the Stokes-Einstein relation for a rigid sphere in a 3D fluid.

On the other hand, Evans and Sackmann (ES) looked at a slightly different situation, i.e., the diffusion of a protein molecule moving in a supported membrane (instead of a free membrane).[12] The presence of the solid substrate is taken into account through a friction term in the 2D Stokes equation as will be discussed in detail in Secs. 2 and 5. The equivalent hydrodynamic model was independently proposed by Izuyama who suggested that the momentum decay mechanism should generally exist for membranes surrounded by water.[13] One big advantage of this model is that the diffusion coefficient can be analytically obtained over the whole range of the disk sizes. In the small size limit, both the SD model and the ES model give the same logarithmic dependence. In biological systems, the ES model can be more relevant because the cell membranes are strongly anchored to the underlying cytoskeleton, or are tightly adhered to other cells.[14]

2. Membrane Hydrodynamics

In this section, we first establish the governing equations for the fluid membrane and its surrounding environment. Our aim is to derive the membrane mobility tensors which will be used in the later sections.

As shown in Fig. 4, we assume that the membrane is an infinite planar sheet of liquid, and its out-of-plane fluctuations are totally neglected, which

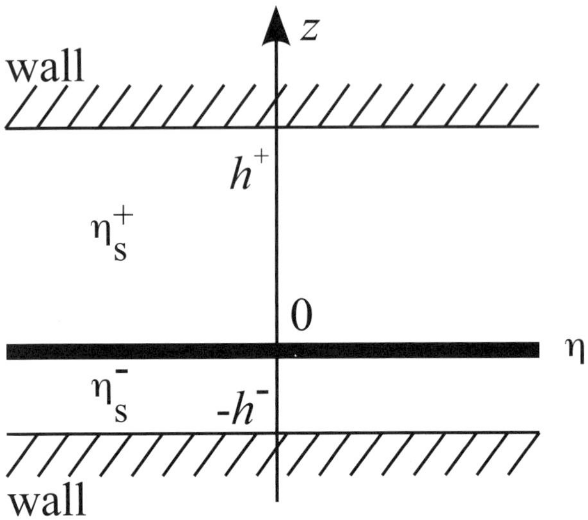

Fig. 4. Schematic picture showing a planar liquid membrane having 2D viscosity η located at $z = 0$. It is sandwiched by a solvent of 3D viscosity $\eta_s^{\pm}$. Two impenetrable walls are located at $z = \pm h^{\pm}$ bounding the solvent.

is justified for typical bending rigidities of bilayers. The liquid membrane is embedded in a bulk fluid such as water or solvent which is bounded by hard walls. Let $\mathbf{v}(\mathbf{r})$ be the 2D velocity of the membrane fluid and the 2D vector $\mathbf{r} = (x, y)$ represents a point in the plane of the membrane. We first assume the membrane to be incompressible

$$\nabla \cdot \mathbf{v} = 0, \tag{3}$$

where ∇ is a 2D differential operator. We work in the low-Reynolds number regime of the membrane hydrodynamics so that the inertial effects can be neglected. This allows us to use the 2D Stokes equation given by

$$\eta \nabla^2 \mathbf{v} - \nabla p + \mathbf{f}_\mathrm{s} + \mathbf{F} = \mathbf{0}, \tag{4}$$

where η is the 2D membrane viscosity, $p(\mathbf{r})$ the 2D in-plane pressure, $\mathbf{f}_\mathrm{s}(\mathbf{r})$ the force exerted on the membrane by the surrounding fluid ("s" stands for the solvent), and $\mathbf{F}(\mathbf{r})$ is any other force acting on the membrane which we shall discuss in later sections.

As presented in Fig. 4, the membrane is fixed in the xy-plane at $z = 0$. The upper ($z > 0$) and the lower ($z < 0$) fluid regions are denoted by "+" and "−", respectively. The velocities and pressures in these regions are written as $\mathbf{v}^{\pm}(\mathbf{r}, z)$ and $p^{\pm}(\mathbf{r}, z)$, respectively. Since the 3D viscosity of the upper and the lower solvent can be different, we denote them as $\eta_{\rm s}^{\pm}$, respectively. Consider the situation in which impenetrable walls are located at $z = \pm h^{\pm}$, where h^{+} and h^{-} can be different in general. Similar to the liquid membrane, the solvent in both regions are taken to be incompressible

$$\tilde{\nabla} \cdot \mathbf{v}^{\pm} = 0, \tag{5}$$

where $\tilde{\nabla}$ represents a 3D differential operator. We also neglect the solvent inertia and hence it obeys the 3D Stokes equations

$$\eta_{\rm s}^{\pm} \tilde{\nabla}^2 \mathbf{v}^{\pm} - \tilde{\nabla} p^{\pm} = 0. \tag{6}$$

The presence of the surrounding solvent is important because it exerts force on the fluid membrane. This force, indicated as $\mathbf{f}_{\rm s}$ in Eq. (4), is given by the projection of $(\boldsymbol{\sigma}^{+} - \boldsymbol{\sigma}^{-})_{z=0} \cdot \hat{\mathbf{e}}_z$ on the xy-plane. Here $\hat{\mathbf{e}}_z$ is the unit vector along the z-axis, and $\boldsymbol{\sigma}^{\pm}$ are the stress tensors due to the solvent

$$\boldsymbol{\sigma}^{\pm} = -p^{\pm}\mathbf{I} + \eta_{\rm s}^{\pm}[\tilde{\nabla}\mathbf{v}^{\pm} + (\tilde{\nabla}\mathbf{v}^{\pm})^{\rm T}]. \tag{7}$$

In the above, $\mathbf{I}$ is the identity tensor and the superscript "T" indicates the transpose.

Using the stick boundary conditions at $z = 0$ and $z = \pm h^{\pm}$, we solve the hydrodynamic equations (5) and (6) to obtain $\mathbf{f}_{\rm s}$. Then we calculate the membrane velocity from Eq. (4) as

$$\mathbf{v}[\mathbf{k}] = \mathbf{G}[\mathbf{k}] \cdot \mathbf{F}[\mathbf{k}], \tag{8}$$

where $\mathbf{v}[\mathbf{k}]$ and $\mathbf{F}[\mathbf{k}]$ are the Fourier components of $\mathbf{v}(\mathbf{r})$ and $\mathbf{F}(\mathbf{r})$ defined by

$$\mathbf{v}(\mathbf{r}) = \int \frac{\mathrm{d}\mathbf{k}}{(2\pi)^2} \mathbf{v}[\mathbf{k}] \exp(i\mathbf{k} \cdot \mathbf{r}), \tag{9}$$

and

$$\mathbf{F}(\mathbf{r}) = \int \frac{\mathrm{d}\mathbf{k}}{(2\pi)^2} \mathbf{F}[\mathbf{k}] \exp(i\mathbf{k} \cdot \mathbf{r}), \tag{10}$$

respectively, and $\mathbf{k} = (k_x, k_y)$. After some calculations shown in Appendix A, one can show that the mobility tensor $\mathbf{G}[\mathbf{k}]$ in Fourier space is given by

$$G_{\alpha\beta}[\mathbf{k}] = \frac{1}{\eta k^2 + k[\eta_{\rm s}^{+} \coth(kh^{+}) + \eta_{\rm s}^{-} \coth(kh^{-})]} \left(\delta_{\alpha\beta} - \frac{k_\alpha k_\beta}{k^2} \right), \tag{11}$$

with $\alpha, \beta = x, y$ and $k = |\mathbf{k}|$. For simplicity, we consider the case when the two walls are located at equal distances from the membrane, i.e., $h^+ = h^- = h$. Then the above mobility tensor becomes

$$G_{\alpha\beta}[\mathbf{k}] = \frac{1}{\eta[k^2 + \nu k \coth(kh)]}\left(\delta_{\alpha\beta} - \frac{k_\alpha k_\beta}{k^2}\right), \tag{12}$$

where $\nu = 2\eta_s/\eta$ with $\eta_s = (\eta_s^+ + \eta_s^-)/2$. An almost equivalent expression to Eq. (12) has also been derived for Langmuir monolayers in which there is only one wall or a substrate.[15,16] In the following, we will use Eq. (12) as the general mobility tensor.

We now discuss the two limiting situations of Eq. (12). Saffman and Delbrück (SD) investigated the case when the two walls are located infinitely away from the membrane.[9,10] This is called as the free membrane case. Taking the limit of $kh \gg 1$ in Eq. (12), the mobility tensor becomes[17–19]

$$G^{\rm SD}_{\alpha\beta}[\mathbf{k}] = \frac{1}{\eta(k^2 + \nu k)}\left(\delta_{\alpha\beta} - \frac{k_\alpha k_\beta}{k^2}\right). \tag{13}$$

We call ν^{-1} as the SD hydrodynamic screening length which was alluded before. The real space expression of this mobility tensor is obtained by the Fourier transform of Eq. (13)

$$\mathbf{G}(\mathbf{r}) = \int \frac{d\mathbf{k}}{(2\pi)^2}\mathbf{G}[\mathbf{k}]\exp(i\mathbf{k}\cdot\mathbf{r}). \tag{14}$$

Performing the calculations presented in Appendix B, we obtain[15,18,19]

$$\begin{aligned} G^{\rm SD}_{\alpha\beta}(\mathbf{r}) &= \frac{1}{4\eta}\left[\mathbf{H}_0(\nu r) - Y_0(\nu r) + \frac{2}{\pi\nu^2 r^2} - \frac{\mathbf{H}_1(\nu r)}{\nu r} + \frac{Y_1(\nu r)}{\nu r}\right]\delta_{\alpha\beta} \\ &+ \frac{1}{4\eta}\left[-\frac{4}{\pi\nu^2 r^2} + \frac{2\mathbf{H}_1(\nu r)}{\nu r} - \frac{2Y_1(\nu r)}{\nu r} - \mathbf{H}_0(\nu r) + Y_0(\nu r)\right]\frac{r_\alpha r_\beta}{r^2}, \end{aligned} \tag{15}$$

where $r = |\mathbf{r}|$. In the above, $\mathbf{H}_n(z)$ are the Struve functions and $Y_n(z)$ are the Neumann functions or the Bessel functions of the second kind.

In the opposite $kh \ll 1$ limit, the membrane is confined between the two walls. Since such a limiting case was considered by Evans and Sackmann (ES),[12] we denote the corresponding quantities with the superscript "ES". In this case, Eq. (12) takes the following form

$$G^{\rm ES}_{\alpha\beta}[\mathbf{k}] = \frac{1}{\eta(k^2 + \kappa^2)}\left(\delta_{\alpha\beta} - \frac{k_\alpha k_\beta}{k^2}\right), \tag{16}$$

where $\kappa = \sqrt{\nu/h}$. This independent length scale κ^{-1} will be called as the ES hydrodynamic screening length. We note that κ^{-1} is the geometric mean of ν^{-1} and h.[20] The above ES mobility tensor was used in a phenomenological membrane hydrodynamic model.[19,21–24] Following the calculations in Appendix B, the real space representation of the ES mobility tensor becomes

$$G^{\mathrm{ES}}_{\alpha\beta}(\mathbf{r}) = \frac{1}{2\pi\eta}\left[K_0(\kappa r) + \frac{K_1(\kappa r)}{\kappa r} - \frac{1}{\kappa^2 r^2}\right]\delta_{\alpha\beta} + \frac{1}{2\pi\eta}\left[-K_0(\kappa r) - \frac{2K_1(\kappa r)}{\kappa r} + \frac{2}{\kappa^2 r^2}\right]\frac{r_\alpha r_\beta}{r^2}, \tag{17}$$

where $K_n(z)$ are the modified Bessel functions of the second kind. In Sec. 3, we shall mainly use the general mobility tensor Eq. (12), whereas either Eq. (13) or (16) is used in Sec. 4 and Sec. 6.

3. Dynamics of Concentration Fluctuations

3.1. *Concentration fluctuations in membranes*

As explained in Introduction, numerous experiments on intact cells and artificial membranes containing saturated lipids, unsaturated lipids and cholesterol have demonstrated the segregation of lipids into liquid-ordered ($\mathrm{L_o}$) and liquid-disordered ($\mathrm{L_d}$) phases below the miscibility transition temperature.[4,25] Below the miscibility transition temperature, the domains undergo coarsening with the largest domain limited by the system size.[25] The primary driving force for the domain coarsening is due to the positive line tension at the domain boundaries. Several people reported on the dynamics of domain coarsening which will be discussed in the next section.[5,26,27]

On the other hand, studies on multicomponent membranes above the transition temperature have also gained much attention. As the critical point is approached from above, one observes composition fluctuations spanning a wide range of length and time scales.[28] Veatch *et al.* made a notable attempt to investigate critical fluctuations in lipid mixtures.[27] Deuterium NMR experiments on model ternary membranes composed of DOPC, DPPC and cholesterol were used to construct the ternary phase diagram. With the determination of the line of miscibility critical points, they observed that the NMR resonances were broadened in the vicinity of the critical points. Such a spectral broadening was attributed to the compositional fluctuations in the membrane having spatial dimensions less than 50 nm.

A more quantitative analysis of critical fluctuations using fluorescence microscopy was addressed by Honerkamp-Smith *et al.* for ternary mixtures

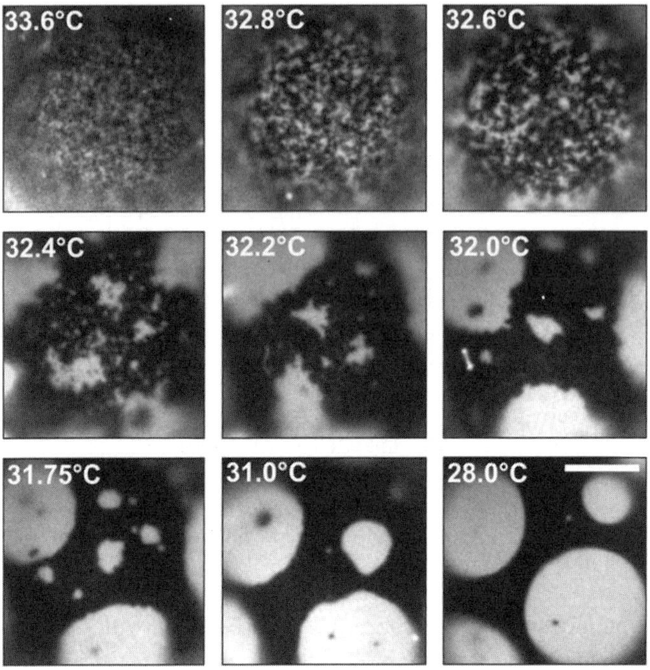

Fig. 5. Giant vesicle (DPPC/diPhyPC/Chol) passing through a critical temperature at $T_c \approx 32.5$ °C. For temperatures above T_c, the vesicle exhibits concentration fluctuations. For temperatures below T_c, the vesicle undergoes macroscopic phase separation. The scale bar corresponds to 20 μm. Adapted from Ref. 6.

of DPPC, diPhyPC and cholesterol.[6] Typical microscope pictures of concentration fluctuations are shown in Fig. 5 for higher temperatures. When the critical temperature T_c is approached from above, the correlation length diverges according to $\xi \approx |T - T_c|^{-\bar{\nu}}$ where $\bar{\nu}$ is the critical exponent.[a] When T_c is approached from below, on the other hand, the order parameter given by the difference in lipid compositions vanishes as $\delta\psi \approx (T_c - T)^{\bar{\beta}}$. From the measurements of the critical exponents $\bar{\nu}$ and $\bar{\beta}$, the authors concluded that the critical behavior in ternary membranes is in the universality class of the 2D Ising model.[29] Based on measurements on the correlation length above T_c and the line tension below T_c, it was shown that giant plasma membrane vesicles extracted from that of living rat basophil leukemia cells also exhibit a critical behavior.[7] This result allows us to speculate that lat-

[a]In order to prevent the confusion, the critical exponents are written with a bar.

eral heterogeneity present in real cell membranes at physiological conditions could correspond to critical fluctuations.[6,7] In other words, concentration fluctuations above (rather than below) the transition temperature can also be responsible for raft structures in cell membranes.

There are several theoretical works on concentration fluctuations in multicomponent membranes. Using renormalization group techniques, Tserkovnyak and Nelson calculated protein diffusion in a multicomponent membrane close to a rigid substrate.[30] They pointed out that, in the vicinity of the critical point, the effective protein diffusion coefficient acquires a power-law behavior. Haataja showed that the effective diffusion coefficient exhibits a crossover from a logarithmic behavior to an algebraic dependence for larger length scales.[31] In his theory, an approximate empirical relation for the diffusion coefficient of a moving object was employed.[32] A rigorous hydrodynamic calculation performed by Inaura and Fujitani showed similar results.[17]

In this section, we use the idea of critical phenomena to calculate the effective diffusion coefficient in multicomponent lipid membranes. Based on Ginzburg–Landau approach with full hydrodynamics, we calculate in particular the decay rate of the concentration fluctuations occurring in membranes.[33] We deal with the general case where the membrane is surrounded by a bulk solvent and two walls as depicted in Fig. 4. As we mentioned before, such a situation is worth considering because biological membranes interact strongly with other cells, substrates or even the underlying cytoskeleton which can affect the structural and transport properties of the membrane.[34] We also study the situation when the multicomponent membranes form 2D microemulsions.[8] This interesting viewpoint of membranes is motivated by a recent work which predicts the reduction of the line tension in membranes containing saturated, unsaturated and hybrid lipids (one tail saturated and the other unsaturated).[35,36] Based on chain entropy arguments, they proposed that hybrid lipids in 2D play an equivalent role to surfactant molecules in 3D. Hence we shall explore the concentration fluctuations in 2D microemulsion within the Ginzburg–Landau approach with full hydrodynamics.

3.2. *Time-dependent Ginzburg–Landau model*

Here we discuss the dynamics of concentration fluctuations in detail. Consider a two-component fluid membrane composed of lipid A and lipid B whose local area fractions are denoted by $\phi_{\rm A}(\mathbf{r})$ and $\phi_{\rm B}(\mathbf{r})$, respectively.

Since the relation $\phi_{\rm A}(\mathbf{r}) + \phi_{\rm B}(\mathbf{r}) = 1$ holds, we introduce a new variable defined by $\psi(\mathbf{r}) = \phi_{\rm A}(\mathbf{r}) - \phi_{\rm B}(\mathbf{r})$. Then the simplest form of the free-energy functional $\mathcal{F}\{\psi\}$ describing the fluctuation around the homogeneous state is

$$\mathcal{F}\{\psi\} = \int d\mathbf{r} \left[\frac{a}{2}\psi^2 + \frac{c}{2}(\nabla\psi)^2 - \mu\psi \right], \tag{18}$$

where $a > 0$ is proportional to the temperature difference from the critical temperature, $c > 0$ is related to the line tension and μ is the chemical potential.

The time evolution of concentration in the presence of hydrodynamic flow is given by the time-dependent Ginzburg–Landau equation for a conserved order parameter[37]

$$\frac{\partial\psi}{\partial t} + \nabla\cdot(\mathbf{v}\psi) = L\nabla^2\frac{\delta\mathcal{F}}{\delta\psi}, \tag{19}$$

where L is the kinetic coefficient. In the membrane hydrodynamic equation (4), we need to incorporate the thermodynamic force due to the concentration fluctuations. Hence we have

$$\mathbf{F} = -\psi\nabla\frac{\delta\mathcal{F}}{\delta\psi}. \tag{20}$$

We implicitly assume that the relaxation of the velocity $\mathbf{v}$ is much faster than that of concentration ψ.[23] The membrane velocity can be formally solved using the appropriate 2D mobility tensor $G_{\alpha\beta}(\mathbf{r},\mathbf{r}')$ derived in the previous section as

$$v_\alpha(\mathbf{r},t) = \int d\mathbf{r}' G_{\alpha\beta}(\mathbf{r},\mathbf{r}')(\nabla'_\beta\psi)\frac{\delta\mathcal{F}}{\delta\psi(\mathbf{r}')}. \tag{21}$$

Since our interest is in the concentration fluctuations around the homogeneous state, we define $\delta\psi(\mathbf{r},t) = \psi(\mathbf{r},t) - \bar{\psi}$, where the bar indicates the spatial average. The free-energy functional expanded in powers of $\delta\psi$ becomes

$$\mathcal{F}\{\delta\psi\} = \int d\mathbf{r} \left[\frac{a}{2}(\delta\psi)^2 + \frac{c}{2}(\nabla\delta\psi)^2 \right]. \tag{22}$$

Substituting Eq. (21) into Eq. (19), we get

$$\begin{aligned}\frac{\partial\delta\psi(\mathbf{r},t)}{\partial t} &= L\nabla^2\frac{\delta\mathcal{F}}{\delta(\delta\psi)} \\ &\quad - \int d\mathbf{r}'(\nabla_\alpha\delta\psi(\mathbf{r}))G_{\alpha\beta}(\mathbf{r},\mathbf{r}')(\nabla'_\beta\delta\psi(\mathbf{r}'))\frac{\delta\mathcal{F}}{\delta(\delta\psi(\mathbf{r}'))}. \end{aligned}\tag{23}$$

We now consider the dynamics of the time-correlation function defined by

$$S(\mathbf{r},t) = \langle \delta\psi(\mathbf{r}_1,t)\delta\psi(\mathbf{r}_2,0)\rangle, \tag{24}$$

where $\mathbf{r} = \mathbf{r}_2 - \mathbf{r}_1$. Within the factorization approximation,[37] the spatial Fourier transform of $S(\mathbf{r},t)$ defined by

$$S(\mathbf{r},t) = \int \frac{d\mathbf{k}}{(2\pi)^2} S[\mathbf{k},t] \exp(i\mathbf{k}\cdot\mathbf{r}), \tag{25}$$

satisfies the following equation

$$\frac{\partial S[\mathbf{k},t]}{\partial t} = -\left(\Gamma^{(1)}[\mathbf{k}] + \Gamma^{(2)}[\mathbf{k}]\right) S[\mathbf{k},t]. \tag{26}$$

The first term in the above equation, $\Gamma^{(1)}[\mathbf{k}]$ denotes the van Hove part of the relaxation rate given by

$$\Gamma^{(1)}[\mathbf{k}] = L k_{\mathrm{B}} T k^2 \chi^{-1}[\mathbf{k}]. \tag{27}$$

Here the static correlation function is defined by

$$\chi[\mathbf{k}] = \langle \delta\psi[\mathbf{k}]\delta\psi[-\mathbf{k}]\rangle = \frac{k_{\mathrm{B}}T}{c(k^2+\xi^{-2})}, \tag{28}$$

where $\xi = (c/a)^{1/2}$ is the correlation length, k_{B} the Boltzmann constant, and T the temperature.

As for the second term in Eq. (26), $\Gamma^{(2)}[\mathbf{k}]$ denotes the hydrodynamic part of the decay rate given by

$$\Gamma^{(2)}[\mathbf{k}] = \frac{1}{\chi[\mathbf{k}]} \int \frac{d\mathbf{q}}{(2\pi)^2} k_\alpha G_{\alpha\beta}[\mathbf{q}] k_\beta \chi[\mathbf{k}+\mathbf{q}]. \tag{29}$$

When we use Eq. (12) for the mobility tensor $G_{\alpha\beta}$, the hydrodynamic part of the decay rate is expressed as

$$\Gamma^{(2)}[\mathbf{k}] = \frac{k_{\mathrm{B}}T}{\eta\chi[\mathbf{k}]} \int \frac{d\mathbf{q}}{(2\pi)^2} \times \frac{\chi[\mathbf{q}]}{|\mathbf{k}-\mathbf{q}|^2 + \nu|\mathbf{k}-\mathbf{q}|\coth(|\mathbf{k}-\mathbf{q}|h)} \frac{k^2q^2 - (\mathbf{k}\cdot\mathbf{q})^2}{|\mathbf{k}-\mathbf{q}|^2}. \tag{30}$$

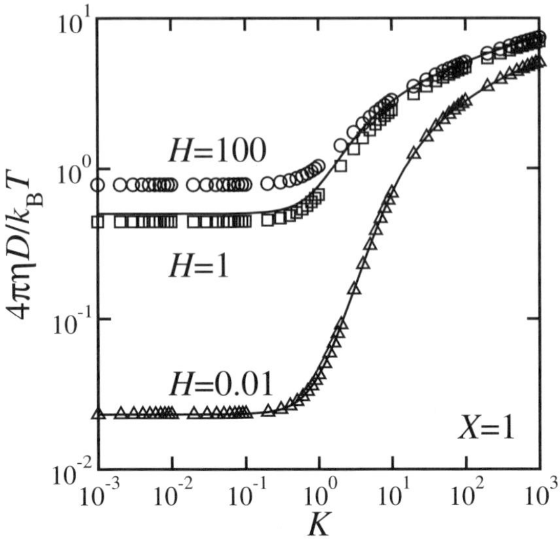

Fig. 6. Scaled effective diffusion coefficient D as a function of K for $H = 0.01, 1, 100$ when $X = 1$ for the confined membrane case. The solid lines are from the analytical expression given in Eq. (33) obtained in the limit of small H.

3.3. *Effective diffusion coefficient*

We now introduce an effective diffusion coefficient (due only to the hydrodynamic effect) $D[\mathbf{k}]$ defined by

$$\Gamma^{(2)}[\mathbf{k}] = k^2 D[\mathbf{k}]. \tag{31}$$

In order to deal with dimensionless quantities, we rescale all the lengths by the SD hydrodynamic screening length $\nu^{-1} = \eta/(2\eta_s)$ such that $K = k/\nu$, $Q = q/\nu$, $X = \xi\nu$ and $H = h\nu$. Then $D[\mathbf{k}]$ can be rewritten as

$$D[K; X, H] = \frac{k_B T}{4\pi^2\eta}(1 + K^2X^2) \times \int_0^\infty dQ \int_0^{2\pi} d\theta \frac{Q^3 \sin^2\theta}{(1 + Q^2X^2)[G^2 + G^{3/2}\coth(\sqrt{G}H)]}, \tag{32}$$

with $G = K^2 + Q^2 - 2KQ\cos\theta$. Since this integral cannot be performed analytically, we evaluate it via a numerical method. We explore the dependencies of D on the variable K, and the parameters X and H.

In Fig. 6, we plot the diffusion coefficient D (scaled by $k_BT/4\pi\eta$) as a function of dimensionless wavenumber K for different solvent thickness H

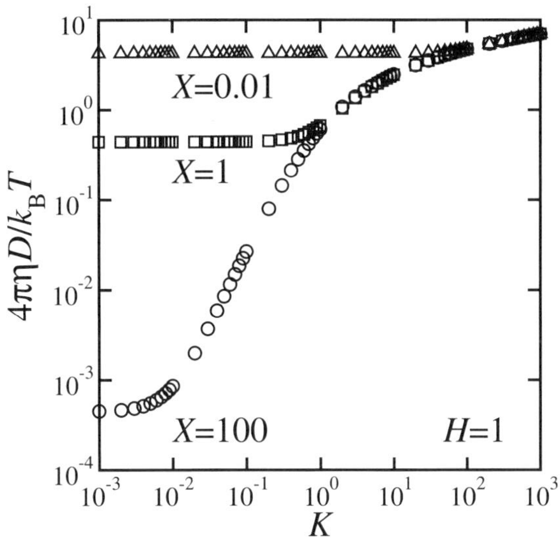

Fig. 7. Scaled effective diffusion coefficient D as a function of K for $X = 0.01, 1, 100$ when $H = 1$ for the confined membrane case.

while the correlation length is fixed to $X = 1$ (i.e., fixed temperature). In the limit of $K \ll 1$, D is almost a constant. The calculated D starts to increase around $K \approx 1$ and a logarithmic behavior (extracted via numerical fitting) is seen for $K \gg 1$. When H is small such as $H = 0.01$, the value of D decreases by one order of magnitude compared to $H = 100$. Figure 7 shows the diffusion coefficient D as a function of wave number K for different X (i.e., different temperature) while the solvent height is fixed to $H = 1$. Again, D is nearly constant for $K \ll 1$, and follows an S-shaped curve with increasing K. Finally, a logarithmic dependence is observed for large K. We note that the this logarithmic behavior for $K \gg 1$ is in contrast to that of 3D critical fluids given by the Kawasaki function which increases linearly with the wavenumber.[38]

In Fig. 8, we explore the effect of the correlation length X on D for different values of H when $K = 10^{-3}$. The quantity X is a measure of an effective size of the correlated region formed transiently in the membrane due to thermal fluctuations. When $X \ll 1$, the diffusion coefficient D decreases only logarithmically, which is typical for a pure 2D system.[9–11] When $X \gg 1$, on the other hand, the behavior of D depends on the value of H. The proximity to the walls results in a loss of momentum from

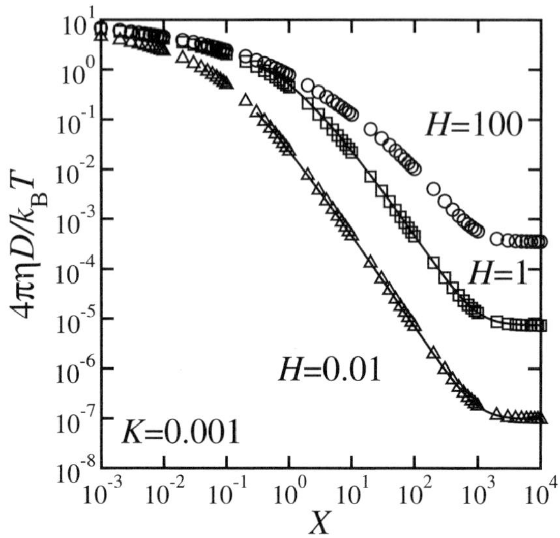

Fig. 8. Scaled effective diffusion coefficient D as a function of X for $H = 0.01, 1, 100$ when $K = 10^{-3}$ for the confined membrane case. The solid lines are from the analytical expression given in Eq. (33) obtained in the limit of small H.

the membrane.[23,39,40] This leads to a rapid suppression of the velocity field within the membrane such that the concentration fluctuations decay slowly. Consequently, the values of D are lower for smaller H. The flattening of the curves for large X is due to the dominance of the X^2 terms in the numerator and denominator in Eq. (32). In Fig. 9, we plot D as a function of H for different values of X when $K = 10^{-3}$. In general, there is a monotonic increase of D followed by a saturation to a constant value for larger H. We see that the effect of H is most prominent for large X (close to the critical point), while there is only a weak dependence for $X \ll 1$ (far from the critical point). The former reflects the fact that the membrane fluid is affected by the the outer environment when the correlation length ξ is larger than the SD screening length ν^{-1} i.e, $X \gg 1$.[39] For correlation lengths smaller than the ν^{-1}, the outer environment surrounding the membrane is less important. The system then behaves as a pure 2D one and the effective diffusion coefficient has only a weak logarithmic size dependence. When the correlation length becomes larger than ν^{-1}, as is the case near the critical point, the outer environment significantly affects the membrane dynamics.

For small h, we showed in Sec. 2 that the general mobility tensor Eq. (12)

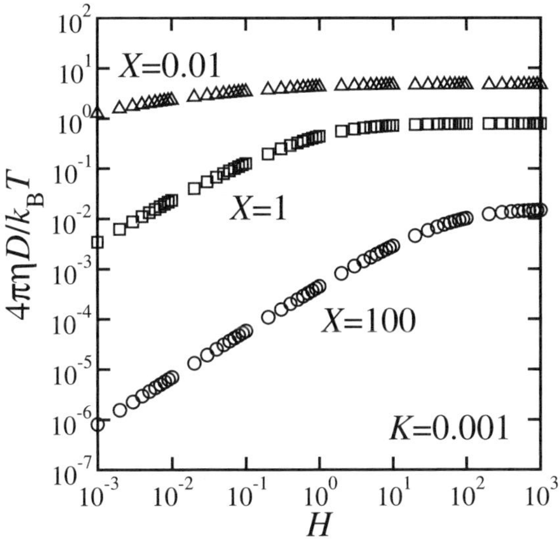

Fig. 9. Scaled effective diffusion coefficient D as a function of H for $X = 0.01, 1, 100$ when $K = 10^{-3}$ for the confined membrane case.

reduces to the ES mobility tensor given by Eq. (16). In this case, one can obtain an analytical expression for the effective diffusion coefficient.[23] In terms of the dimensionless quantities K, X and H, it can be written as

$$D[K; X, H] = \frac{k_{\mathrm{B}}T}{4\pi\eta}\frac{1+K^2X^2}{2K^2X^2}\left[-\ln\left(\frac{X}{\sqrt{H}}\right)\right.$$
$$+\frac{H}{X^2}(1+K^2X^2)\ln\left(\frac{X}{\sqrt{H}(1+K^2X^2)}\right)$$
$$\left.+\frac{H\Omega}{2X^2}\ln\left(\frac{K_+^4+K_-^2+K_+^2\Omega}{\Omega-K_-^2-1}\right)\right], \tag{33}$$

where

$$\Omega = \sqrt{(K^2X^2 + X^2/H - 1)^2 + 4K^2X^2}, \tag{34}$$

and

$$K_\pm = \sqrt{K^2X^2 \pm X^2/H}. \tag{35}$$

Equation (33) is plotted using solid lines in Fig. 6 for $H = 0.01$ and 1 with $X = 1$. For $H = 0.01$, the analytical and numerical data coincide giving credence to accuracy of the numerical solutions. It is seen that even for $H = 1$ the agreement is still acceptable. For $H = 100$, however, a significant deviation is observed (not shown), which is expected as this limit is beyond the valid range of Eq. (33).

The solid lines in Fig. 8 also represent the analytical result of Eq. (33). It is seen that the analytical and the numerical data points almost coincide for $H = 0.01$ and $H = 1$. For $H = 100$, there is significant deviation from the numerical data (not shown). However, the agreement between the numerical result and the analytical expression is beyond the expected range of $H \ll 1$ and reaches up to $H \approx 1$, as pointed out by Stone and Ajdari.[20] Hence Eq. (33) is useful in analyzing the experimental data in many situations.

From the experiments on model multicomponent vesicles, the static critical exponents for the order parameter and correlation length were found to have values close to $\bar{\beta} = 1/8$ and $\bar{\nu} = 1$, respectively.[6] Furthermore, experiments on giant plasma membrane vesicles measured the critical exponent $\bar{\gamma} = 7/4$ which characterizes the critical behavior of the osmotic compressibility.[7] These static exponents coincide with the exact results of the 2D Ising model.[41,42] The description presented in this chapter uses the mean-field approach and therefore the corresponding exponents are $\bar{\beta} = 1/2$, $\bar{\nu} = 1/2$, and $\bar{\gamma} = 1$, respectively. The discrepancies between these values are still under debate.

3.4. *Membrane as a 2D microemulsion*

The role of surfactant molecules in 3D microemulsions is to reduce the surface tension at the interface between oil and water. In an analogy to 3D microemulsions, hybrid lipids (one chain unsaturated and the other saturated) act as lineactant molecules which stabilize finite sized domains in 2D. In other words, hybrid lipids play a similar role to surfactant molecules at the interface between L_o and L_d domains. It should be also noticed that hybrid lipids form a major percentage of all naturally existing lipids.[43,44] Based on a simple model of hybrid lipids, Brewster *et al.* showed that finite sized domains can be formed in equilibrium.[35,36] A subsequent model predicted stabilized domains even in a system of saturated/hybrid/cholesterol lipid membranes.[45] Being motivated by this idea, we calculate the decay rate of concentration fluctuations when the free energy of the multicomponent membrane has the form of a 2D microemulsion.

The free-energy functional for a microemulsion includes a higher order derivative term and is expressed in terms of $\delta\psi$ as[8]

$$\mathcal{F}_{\rm ME}\{\delta\psi\} = \int {\rm d}\mathbf{r} \left[\frac{a}{2}(\delta\psi)^2 + \frac{c}{2}(\nabla\delta\psi)^2 + \frac{g}{2}(\nabla^2\delta\psi)^2\right], \tag{36}$$

with $a, g > 0$ and $c < 0$. The negative value of c creates 2D interfaces, while the term with positive g is a stabilizing term. This form of the free energy has been used previously to study coupled modulated bilayers.[46] As in the previous section, the decay rate of the correlation function can be split into two parts. First, the van Hove part now becomes

$$\Gamma^{(1)}_{\rm ME}[\mathbf{k}] = Lk_{\rm B}Tk^2\chi^{-1}_{\rm ME}[\mathbf{k}], \tag{37}$$

where L is the kinetic coefficient which is assumed to be same as before, and the static correlation function $\chi_{\rm ME}[\mathbf{k}]$ is[47]

$$\chi_{\rm ME}[\mathbf{k}] = \frac{k_{\rm B}T}{gk^4 + ck^2 + a}. \tag{38}$$

By defining

$$k_0^2 = -\frac{c}{2g}, \tag{39}$$

$$\sigma^4 = \frac{a}{g} - \left(\frac{c}{2g}\right)^2, \tag{40}$$

we can write the static correlation function as

$$\chi_{\rm ME}[\mathbf{k}] = \frac{k_{\rm B}T}{g\left[(k^2 - k_0^2)^2 + \sigma^4\right]}. \tag{41}$$

On plotting $\chi_{\rm ME}$ as a function of k, a peak appears at $k = k_0$ followed by a $1/k^4$-decay. The width of the peak is given by σ. A lamellar phase appears when $\sigma = 0$. Notice that $c = 0$ is called the Lifshitz point at which the peak occurs for $k = 0$.[48] Using the form of Eq. (41), we can write Eq. (37) as

$$\Gamma^{(1)}_{\rm ME}[\mathbf{k}] = Lgk^2\left[(k^2 - k_0^2)^2 + \sigma^4\right]. \tag{42}$$

Similar to the previous subsection, we next write the hydrodynamic part of the decay rate in terms of the effective diffusion coefficient $D_{\rm ME}[\mathbf{k}]$ as

$$\Gamma^{(2)}_{\rm ME}[\mathbf{k}] = k^2 D_{\rm ME}[\mathbf{k}]. \tag{43}$$

Using the mobility tensor given by Eq. (12), we can write $D_{\rm ME}$ as

$$D_{\rm ME}[K; K_0, \Sigma, H] = \frac{k_{\rm B}T}{4\pi^2\eta}[(K^2 - K_0^2)^2 + \Sigma^4] \times \int_0^\infty {\rm d}Q \int_0^{2\pi} {\rm d}\theta \frac{Q^3\sin^2\theta}{[(Q^2 - K_0^2)^2 + \Sigma^4][G^2 + G^{3/2}\coth(\sqrt{G}H)]}, \tag{44}$$

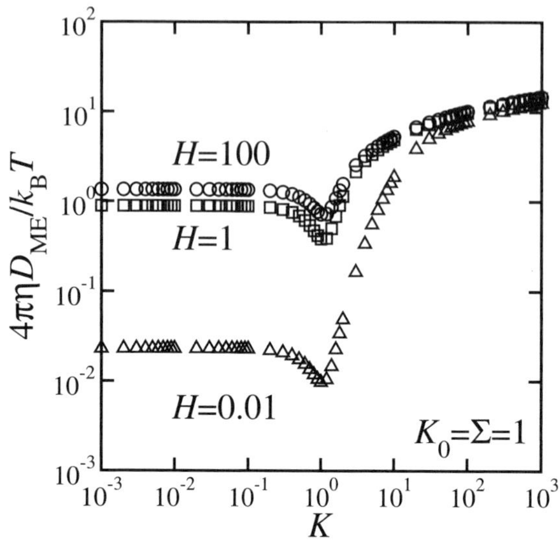

Fig. 10. Scaled effective diffusion coefficient D_{ME} as a function of K for $H = 0.01, 1, 100$ when $K_0 = \Sigma = 1$ for the confined membrane case.

where $K = q/\nu$, $K_0 = q_0/\nu$, $Q = q/\nu$, $\Sigma = \sigma/\nu$, $H = h\nu$, and $G = K^2 + Q^2 - 2KQ\cos\theta$.

In Fig. 10, we plot D_{ME} as a function of K for different values of H when $K_0 = \Sigma = 1$ are fixed. When $K \ll 1$, D shows a constant value. We also observe the 2D characteristic of logarithmic behavior of D for $K \gg 1$. For $K \ll 1$, the effect of the outer environment is felt with the suppression of the diffusion coefficient with smaller H. The curves almost overlap when $K \gg 1$ indicating the negligible effect of the outer environment at large wave numbers. An interesting feature of D_{ME} is the dip occurring at $K \approx K_0$ which does not exist for binary critical fluids. This can be attributed to the peak at $k = k_0$ in $\chi_{\mathrm{ME}}[\mathbf{k}]$.[37,47,49] We also note that, for 3D microemulsions, the effective diffusion coefficient varies linearly with the wavenumber when it is large enough.[37]

Although the analogy between 3D and 2D microemulsions has been invoked, it should be pointed out that a 3D microemulsion formed from an oil-water-surfactant mixture arises predominantly due to the differences in relative affinity between the components. For the 2D case, it is the physical interactions from the hydrocarbon chain packing requirements at the $\mathrm{L_o}/\mathrm{L_d}$ interface that gives rise to the lineactant properties of the hybrid lipid.

3.5. *Biological relevance*

Experiments on real plasma membranes have suggested that the cell maintains the membranes at a critical composition.[7] This leads to nanometer-sized composition fluctuations at the physiological temperatures, although these structures are much smaller than what can be resolved through optical microscopy. We therefore speculate that there is some biological relevance in studying concentration fluctuations. However we have mainly concentrated on the dynamics towards the equilibrium state of lipid bilayer membranes. In real cells, there are many active non-equilibrium cellular processes that are involved in the proper functioning of the cells. It has indeed been proposed that nano-domain formation may be related to the underlying cytoskeleton.[50] There have been several other models which make use of the active non-equilibrium phenomena to explain the existence of finite sized domains in multicomponent membranes. A review of the non-equilibrium models can be found in the review article.[51]

4. Phase Separation Dynamics

4.1. *Macroscopic phase separation in membranes*

Phase separation of binary fluids following a quench has been under study for over forty years.[52] The dynamic scaling hypothesis assumes that there exists a scaling regime characterized by the average domain size R that grows with time t as $R \sim t^{\alpha}$ with an universal exponent α. For 3D off-critical binary fluids, there is an initial growth by the Brownian coagulation process,[53] followed by the Lifshitz-Slyozov (LS) evaporation-condensation process;[54] both mechanisms show a growth exponent $\alpha = 1/3$. This is followed by a late time inertial regime of $\alpha = 2/3$.[55] For critical mixtures, there is an intermediate $\alpha = 1$ regime owing to interface diffusion.[56] The scenario is slightly different for pure 2D systems.[57] For an off-critical mixture, it was predicted that after the initial formation of domains, they grow by the Brownian coagulation mechanism with a different exponent $\alpha = 1/2$ (as will be explained later), followed by a crossover to the LS mechanism which gives $\alpha = 1/3$ even in 2D. For critical mixtures, on the other hand, the initial quench produces an interconnected structure which coarsens and then breaks up due to the interface diffusion with an exponent $\alpha = 1/2$. After the breakup processes, coarsening takes place through Brownian coagulation that is again characterized by the $\alpha = 1/2$ scaling.[53] These predictions were confirmed by molecular dynamics simulations in

2D.[58] The exponent $\alpha = 1/2$ was also observed in 2D lattice-Boltzmann simulations in the presence of thermal noise for a critical mixture.[59]

The domain growth dynamics on ternary vesicles composed of DOPC, DPPC and cholesterol was first investigated by Saeki *et al.*[26] They showed that the average domain size develops according to a power-law behavior with an exponent $\alpha = 0.15$. Subsequently, Yanagisawa *et al.* found that the domain coarsening processes are classified into two types, i.e, normal coarsening and trapped coarsening.[5] In the former case, the domains having flat circular shape grow through Brownian coagulation process, and the growth exponent turns out to be $\alpha = 2/3$. For the trapped coarsening, on the other hand, the domain coarsening is suppressed at a certain domain size because the repulsive inter-domain interactions obstruct the coalescence of domains. The two-color imaging of the trapped domains revealed that the repulsive interactions are induced by the budding of domains.

Although biomembranes composed of lipid bilayers can be regarded as 2D viscous fluids, they are not isolated pure 2D systems since lipids are coupled to the adjacent fluid. Hence it is of great interest to investigate the phase separation dynamics in such a quasi-2D liquid membrane in the presence of hydrodynamic interaction.[b] To address this problem, we consider a 2D binary viscous membrane in contact with a bulk solvent. We employ a simple model in which the membrane is confined to a plane with the bulk fluid particles added above and below. In our model using dissipative particle dynamics (DPD) simulation technique, the exchange of momentum between the membrane and the bulk solvent is naturally taken into account. We particularly focus on the effect of bulk solvent on the quasi-2D phase separation.

4.2. *Model and simulation technique*

We use a structureless model of the 2D fluid membrane within the DPD framework.[60,61] As shown in Fig. 11, the 2D membrane is represented by a single layer of particles confined to a plane. In order to study phase separation, we introduce two species of particles, A and B. The bulk fluid which we call as "solvent" (S) is also represented by single particles of same size as that of the membrane particles. All particles have the same mass m.

In DPD, the interaction between any two particles, within a range r_0, is linearly repulsive. The pairwise interaction leads to full momentum con-

[b]We use the word "quasi-2D" whenever the membrane is coupled to the bulk fluid.

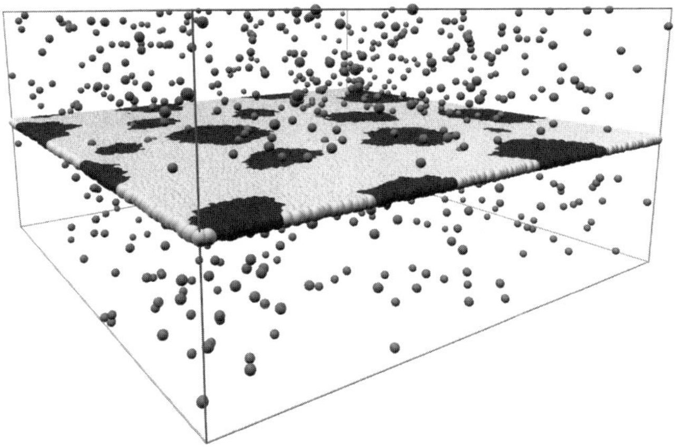

Fig. 11. Image of the fluid membrane with the bulk fluid called solvent. The black (A) and white (B) particles represent the two components constituting the membrane, while gray ones (S) represent the solvent. For clarity, only a fraction of the solvent particles are shown.

servation, which in turn brings out the correct fluid hydrodynamics. The force on a particle i is given by

$$m\frac{\mathrm{d}\mathbf{v}_i}{\mathrm{d}t} = \sum_{j\neq i}\left[\mathbf{F}_{ij}^{\mathrm{C}}(\mathbf{r}_{ij}) + \mathbf{F}_{ij}^{\mathrm{D}}(\mathbf{r}_{ij}, \mathbf{v}_{ij}) + \mathbf{F}_{ij}^{\mathrm{R}}(\mathbf{r}_{ij})\right], \tag{45}$$

where $\mathbf{r}_{ij} = \mathbf{r}_i - \mathbf{r}_j$ and $\mathbf{v}_{ij} = \mathbf{v}_i - \mathbf{v}_j$. Of the three types of forces acting on the particles, the conservative force on particle i due to j is $\mathbf{F}_{ij}^{\mathrm{C}} = a_{ij}\omega(r_{ij})\hat{\mathbf{r}}_{ij}$, where a_{ij} is an interaction strength and $\hat{\mathbf{r}}_{ij} = \mathbf{r}_{ij}/r_{ij}$ with $r_{ij} = |\mathbf{r}_{ij}|$. The second type of force is the dissipative force $\mathbf{F}_{ij}^{\mathrm{D}} = -\Gamma_{ij}\omega^2(r_{ij})(\hat{\mathbf{r}}_{ij}\cdot\mathbf{v}_{ij})\hat{\mathbf{r}}_{ij}$, where Γ_{ij} is the dissipative strength for the pair (i, j). The last is the random force $\mathbf{F}_{ij}^{\mathrm{R}} = \sigma_{ij}(\Delta t)^{-1/2}\omega(r_{ij})\zeta_{ij}\hat{\mathbf{r}}_{ij}$, where σ_{ij} is the amplitude of the random noise for the pair (i, j), and ζ_{ij} is a random variable with zero mean and unit variance which is uncorrelated for different pairs of particles and different time steps. The dissipative and random forces act as a thermostat, provided the fluctuation-dissipation theorem $\sigma_{ij}^2 = 2\Gamma_{ij}k_{\mathrm{B}}T$ is satisfied. The weight factor is chosen as $\omega(r_{ij}) = 1 - r_{ij}/r_0$ up to the cutoff radius r_0 and zero thereafter. The particle trajectories are obtained by solving Eq. (45) using the velocity-Verlet integrator. In the simulation, r_0 and m set the scales for length and mass, respectively, while $k_{\mathrm{B}}T$ sets the energy scale. The time is measured in units of $\tau = (mr_0^2/k_{\mathrm{B}}T)^{1/2}$. The numerical value of the amplitude of the random force is assumed to be the same for all pairs such that $\sigma_{ij} = 3.0[(k_B T)^3 m/r_0^2]^{1/4}$, and the fluid density is set as $\rho = 3.0$. We set $k_B T = 1$ and the integration time step is chosen to be $\Delta t = 0.01\tau$.

The membrane is constructed by placing particles in the xy-plane in the middle of the simulation box (see Fig. 11). Owing to the structureless representation of the constituent particles, we apply an external potential so as to maintain the membrane integrity. This is done by fixing the z-coordinates of all the membrane particles. The work involves the systematic variation of the height of the simulation box starting from the pure 2D case. In the absence of solvent, we work with a 2D-box of dimensions $L_x \times L_y = 80 \times 80$ with 19200 particles constituting the membrane. For the quasi-2D studies, we add solvent particles S above and below the membrane, and increase the height of the box as $L_z = 5, 20$ and 40. For all the cases there are 19200 membrane particles. The largest box size ($L_z = 40$) has 748800 solvent particles. The box with height $L_z = 40$ is found to be sufficiently large enough to prevent the finite size effect which affects the membrane-solvent interaction. The system is then subject to periodic boundary conditions in all the three directions. For phase separation simulations, we introduce two species of membrane particles A and B. The interaction parameter between various particles are given by $a_{AA} = a_{BB} = a_{SS} = a_{AS} = a_{BS} = 25$ and $a_{AB} = 50$. In order to do a quench, the membrane is first equilibrated with a single component, following which a fraction of the particles are instantaneously changed to the B type.

4.3. *Domain growth dynamics*

First we describe the results of the phase separation dynamics. The snapshots for A : B composition set to 70 : 30 (off-critical mixture) are shown in Fig. 12 for both pure 2D case (left column) and quasi-2D case with $L_z = 40$ (right column). Qualitatively, it is seen that the domains for the quasi-2D case are smaller in size when compared at the same time step. We also monitor the average domain size $R(t)$ which can be obtained from the total interface length $L(t)$ between the two components. This is because $R(t)$ and $L(t)$ are related by $L(t) = 2\pi N(t)R(t)$, where $N(t)$ is the number of domains. The area occupied by the B-component is given by $\mathcal{A} = \pi N(t)R^2(t)$ which is a conserved quantity. Then we have

$$R(t) = 2\mathcal{A}/L(t). \tag{46}$$

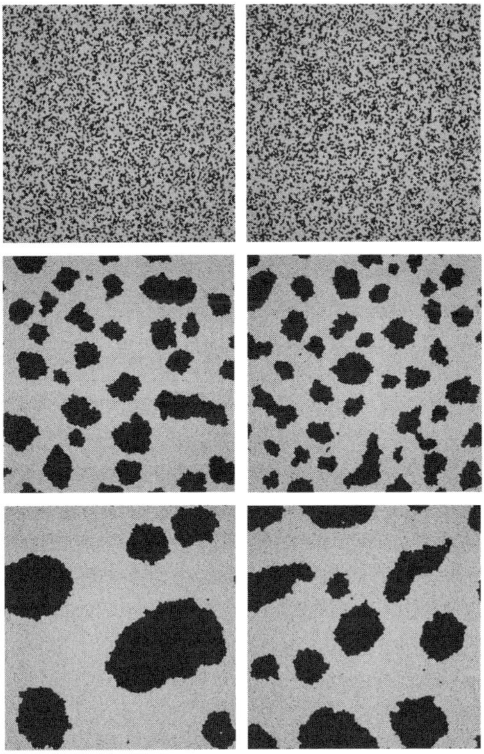

Fig. 12. The snapshots for a 70 : 30 mixture undergoing phase separation at $t = 0, 150$ and 1000 (top to bottom) for a pure 2D (left column) and quasi-2D system with $L_z = 40$ (right column). The above snapshots are from one of the ten independent trials that were conducted.

When the domain size grows as $R \sim t^{\alpha}$, one has $L \sim t^{-\alpha}$ and $N \sim t^{-2\alpha}$. The domain size $R(t)$ for 70 : 30 mixture is shown in Fig. 13. In this plot, an average over 10 independent trials has been taken. It can be seen that the pure 2D case has a growth exponent $\alpha = 1/2$. Upon the addition of solvent, we observe that the exponent shifts to a lower value of $\alpha = 1/3$. This exponent is reminiscent of the phase separation dynamics of an off-critical mixture in 3D. Upon systematically increasing the amount of solvent in the system by changing the height L_z, we can see a clear deviation from the pure 2D behavior. There is no further change if L_z is increased beyond 40. A larger system size $L_x \times L_y = 200 \times 200$ also produced the same scaling for the pure 2D case, which demonstrates that finite-size effects are small.

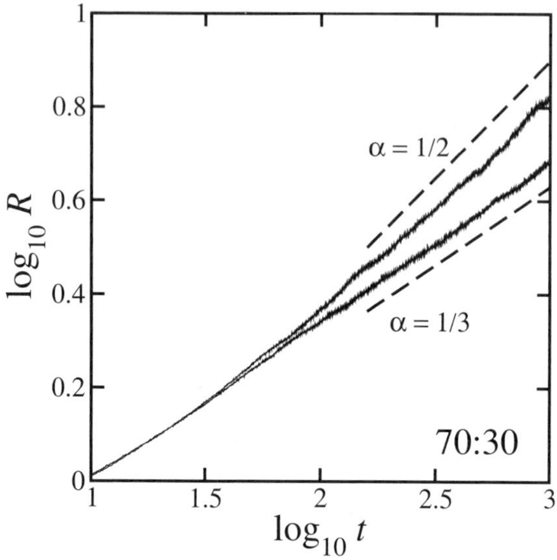

Fig. 13. The average domain size R as a function of time t for a 70 : 30 off-critical mixture. The upper curve is the pure 2D case showing an $\alpha = 1/2$ scaling, and the lower curve is the quasi-2D case when $L_z = 40$ showing a distinct $\alpha = 1/3$ scaling.

In Fig. 14, we show the result for a component ratio of 50 : 50 (critical mixture). In this case, the growth exponent for the pure 2D case is less obvious owing to rapid coarsening of the domains. However, by simulating a bigger system 200×200 with the same areal density, an $\alpha = 1/2$ exponent is indeed obtained. Similar to the off-critical case, the growth of the domains is slowed down by the addition of solvent and the exponent is reduced to $\alpha = 1/3$. These results indicate that solvent is responsible for slowing down the growth dynamics.

The observed exponent $\alpha = 1/2$ in pure 2D systems can be explained in terms of the Brownian coagulation mechanism.[57] From dimensional analysis, the 2D diffusion coefficient of the domain is given by $D_2 \sim k_{\mathrm{B}}T/\eta$, where η is the membrane 2D viscosity. Using the relation

$$R^2 \sim D_2 t \sim (k_{\mathrm{B}}T/\eta)t, \tag{47}$$

we find $R \sim t^{1/2}$. For 3D systems, on the other hand, the diffusion coefficient of the droplet is inversely proportional to its size, $D_3 \sim 1/R$, a well-known Stokes-Einstein relation. Hence the Brownian coagulation

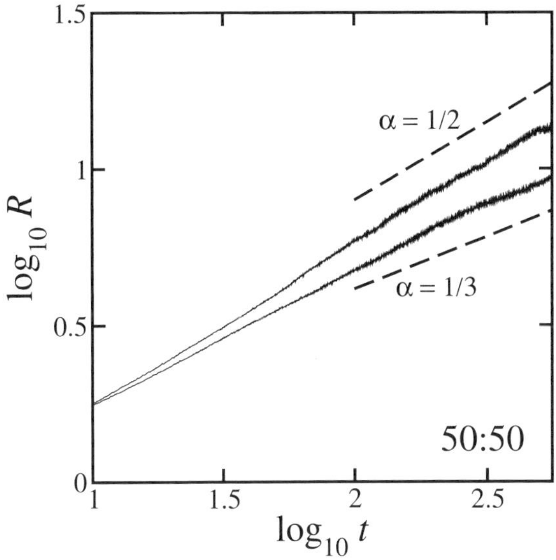

Fig. 14. The average domain size R as a function of time t for a 50 : 50 critical mixture. The upper curve is the pure 2D case showing an $\alpha = 1/2$ scaling, and the lower curve is the quasi-2D case when $L_z = 40$ showing a distinct $\alpha = 1/3$ scaling.

mechanism in 3D gives rise to an exponent $\alpha = 1/3$.[c] The change in the exponent from $\alpha = 1/2$ to $1/3$ due to the addition of solvent implies the crossover from 2D to 3D behaviors of the phase separation dynamics even though the lateral coarsening takes place only within the 2D geometry.[d]

A DPD study by Laradji and Kumar on phase separation dynamics of two-component membranes (both critical and off-critical cases) used a coarse-grained model for the membrane lipids.[62] In their model, the self-assembly of the bilayer in the presence of solvent is naturally taken into account. The exponent for the off-critical case $\alpha = 1/3$ is the same as that obtained in our study, although they attributed this value to the LS mechanism. For critical mixtures in the presence of solvent, they obtained a different value $\alpha = 1/2$.

[c]In general, the exponent is $\alpha = 1/d$, where d is the space dimension.
[d]We note that the LS mechanism shows an exponent of $\alpha = 1/3$ in both 2D and 3D. We thus conclude that our simulations are still in the early time of the coarsening dynamics.

4.4. *Correlated diffusion*

In order to justify our argument in the last subsection, it is necessary to examine the size dependence of the domain diffusion coefficient in quasi-2D systems. This can be calculated by tracking the mean-squared displacement of domains of various radii. The equivalent information can be more efficiently obtained by calculating the two-particle longitudinal coupling diffusion coefficient in a single component membrane rather than in a binary system.

Consider a pair of particles separated by a 2D vector $\mathbf{r}$, undergoing diffusion in the fluid membrane. The two-particle mean squared displacement is given by[18]

$$\langle \Delta r_\alpha^k \Delta r_\beta^l \rangle = 2D_{\alpha\beta}^{kl}(\mathbf{r})t, \tag{48}$$

where Δr_α^k is the displacement of the particle k $(= 1, 2)$ along the axis α $(= x, y)$, $D_{\alpha\beta}^{kl}$ is the diffusion tensor giving self-diffusion when $k = l$ and the coupling between them when $k \neq l$. The x-axis is defined along the line connecting a pair of particles 1 and 2, i.e., $\mathbf{r} = r\hat{x}$. Hence, we have $D_{xy}^{12} = 0$ by symmetry. The longitudinal coupling diffusion coefficient, $D_{\mathrm{L}}(r) = D_{xx}^{12}(r\hat{x})$, gives the coupled diffusion along the line of centers of the particles.

We first describe the analytical expression of $D_{\mathrm{L}}(r)$ based on the SD theory.[9] The longitudinal coupling diffusion coefficient in the SD case can be essentially obtained from Eq. (15) or Eq. (B.4). For over-damped dynamics, we can use the Einstein relation.[18] Then we obtain

$$D_{\mathrm{L}}(r) = \frac{k_{\mathrm{B}}T}{4\pi\eta}\left[-\frac{2}{\nu^2 r^2} + \frac{\pi \mathbf{H}_1(\nu r)}{\nu r} - \frac{\pi Y_1(\nu r)}{\nu r}\right], \tag{49}$$

where $\mathbf{H}_1(z)$ and $Y_1(z)$ are Struve function and Bessel function of the second kind, respectively.

At short distances $r \ll \nu^{-1}$, the asymptotic form of the above expression becomes

$$D_{\mathrm{L}}(r) \approx \frac{k_{\mathrm{B}}T}{4\pi\eta}\left[\ln\left(\frac{2}{\nu r}\right) - \gamma + \frac{1}{2}\right], \tag{50}$$

where $\gamma = 0.5772\cdots$ is Euler's constant. At large inter-particle separations $r \gg \nu^{-1}$, on the other hand, Eq. (49) reduces to

$$D_{\mathrm{L}}(r) \approx \frac{k_{\mathrm{B}}T}{2\pi\eta\nu r} = \frac{k_{\mathrm{B}}T}{4\pi\eta_{\mathrm{s}} r}, \tag{51}$$

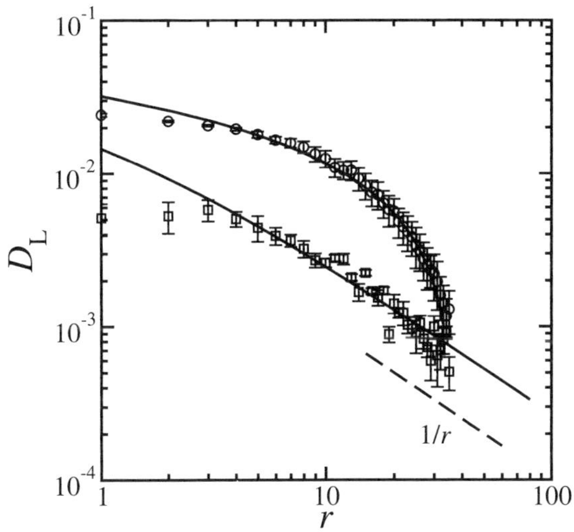

Fig. 15. Longitudinal coupling diffusion D_{L} as a function of particle separation r. The upper circles are data for the pure 2D case. The lower squares correspond to the case with solvent when $L_z = 40$. The upper solid line is the fit by Eq. (50), and the lower solid line is the fit by Eq. (49). The dashed line shows the $1/r$-dependence.

showing the asymptotic $1/r$-decay which reflects the 3D nature of this limit. Notice that Eq. (51) depends only on the solvent viscosity η_{s} but not on the membrane viscosity η any more.

In Fig. 15, we plot the measured longitudinal coupling diffusion coefficient D_{L} as a function of 2D distance r. In these simulations, we have worked with only single component membranes with the same system sizes and number of particles as those used for the phase separation simulations. We have also taken average over 20 independent trials. In the pure 2D case without any solvent, D_{L} shows a logarithmic dependence on r. This is consistent with Eq. (50) obtained when the coupling between the membrane and solvent is very weak so that the membrane can be regarded almost as a pure 2D system. Using Eq. (50) as an approximate expression, we get from the fitting as $k_{\mathrm{B}}T/4\pi\eta \approx 0.89 \times 10^{-2}$ and $\nu^{-1} \approx 20$. In an ideal case, the SD screening length should diverge due to the absence of solvent. The obtained finite value for ν^{-1} is roughly set by the half of the system size in the simulation.

When we add solvent ($L_z = 40$), the D_{L} is decreased and no longer behaves logarithmically. In this case, we use the full expression Eq. (49) for the fitting, and obtained $k_B T/4\pi\eta \approx 1.35 \times 10^{-2}$ and $\nu^{-1} \approx 1$. In the above two fits, we have neglected the first two points as they lie outside the range of validity, $r \gg 1$, of Eq. (49).[18] Since $\nu^{-1} \approx 1$ when the solvent is present, the data shown in Fig. 15 are in the crossover region, $r \gtrsim \nu^{-1}$, showing an approach towards the asymptotic $1/r$-dependence as in Eq. (51). Hence we conclude that the solvent brings in the 3D hydrodynamic property to the diffusion in membranes. This is the reason for the 3D exponent $\alpha = 1/3$ in the phase separation dynamics, and justifies that it is mainly driven by Brownian coagulation mechanism.

In our simulations, the membrane and the solvent have very similar viscosities. This sets the SD length scale to be of the order of unity, which is consistent with the value $\nu^{-1} \approx 1$ obtained from the fitting. As explained above, the fit also provides the 2D membrane viscosity as $\eta \approx 6$, and hence we obtain as $\eta_s \approx 3$. This value is in reasonable agreement with the value $\eta_s \approx 1$ calculated in Ref. 63 by using the reverse Poiseuille flow method. The reason for the slightly higher value of η_s in our simulations is that the tracer particles are of the same size as the membrane particles. This may lead to an underestimation of the correlated diffusion coefficient. In real biomembranes sandwiched by water, however, the value of the SD length is much larger than the lipid size, and is in the order of submicron scale.[9] Hence the 3D nature of hydrodynamics should be observed for large enough domains.[64]

In the SD theory, the bulk fluid is assumed to occupy an infinite space above and below the membrane. As discussed in Sec. 2, the situation is altered when the solvent and the membrane are confined between two solid walls.[17] If the distance h between the membrane and the wall is small enough, the appropriate mobility tensor is given by Eq. (16). Following the same procedure as for the SD case, the longitudinal coupling diffusion coefficient can be obtained as

$$D_L(r) = \frac{k_B T}{2\pi\eta}\left[\frac{1}{\kappa^2 r^2} - \frac{K_1(\kappa r)}{\kappa r}\right], \tag{52}$$

where $K_1(z)$ is modified Bessel function of the second kind (see Eqs. (17) and (B.12)). At short distances $r \ll \kappa^{-1}$, we have

$$D_L(r) \approx \frac{k_B T}{4\pi\eta}\left[\ln\left(\frac{2}{\kappa r}\right) - \gamma + \frac{1}{2}\right], \tag{53}$$

which is almost identical to Eq. (50) except ν is replaced now by κ. At long distances $r \gg \kappa^{-1}$, on the other hand, we get

$$D_{\mathrm{L}}(r) \approx \frac{k_{\mathrm{B}}T}{2\pi\eta\kappa^2 r^2} = \frac{k_{\mathrm{B}}Th}{4\pi\eta_{\mathrm{s}} r^2}, \tag{54}$$

which exhibits a $1/r^2$-dependence. This is in contrast to Eq. (51). Following the similar scaling argument, we predict that, in the presence of walls, the domain growth exponent should be $\alpha = 1/4$ within the Brownian coagulation mechanism. In biological systems, the above model with solid walls can be relevant because the cell membranes are strongly anchored to the underlying cytoskeleton, or are tightly adhered to other cells.

As a related experimental work, the diffusion of tracer particles embedded in a soap film was reported.[65] When the diameter of the tracer particles is close to the thickness of the soap film, the system shows a 2D behavior. On the other hand, if the particle diameter is much smaller than the soap film thickness, it executes a 3D motion. On systematically increasing the soap film thickness, they identified a transition from 2D to 3D nature.

In a recent work, the hydrodynamic effects on the spinodal decomposition kinetics in a planar membrane was studied by Fan *et al.*[66] This work used continuum numerical simulations of a convective time-dependent Ginzburg–Landau equation for the membrane which was then coupled to bulk fluid via the Stokes equation. It was reported that dynamical scaling breaks down for critical lipid mixtures. However, one major distinction from the case discussed in this section, is the lack of thermal fluctuations which rules out the Brownian coagulation mechanism.

5. Diffusion Coefficient of a 2D Liquid Domain

5.1. *Diffusion of liquid domains*

Membrane components such as lipids and proteins are subject to Brownian motion, and the resulting diffusive process is an important mechanism for the transport of these materials. Consequently, it is generally believed that many biochemical functions in membranes are diffusion controlled processes. As mentioned in Introduction, the studies of diffusion in biomembranes have mainly concentrated on protein molecules which are assumed to be a rigid disk.

On the other hand, we described that lipid domains appear due to lateral phase separation between the $\mathrm{L_o}$ and $\mathrm{L_d}$ phases.[25] Liquid lipid domains, however, cannot be considered as rigid objects. One should take into account the fluid nature and finite viscosity of the diffusing domain.

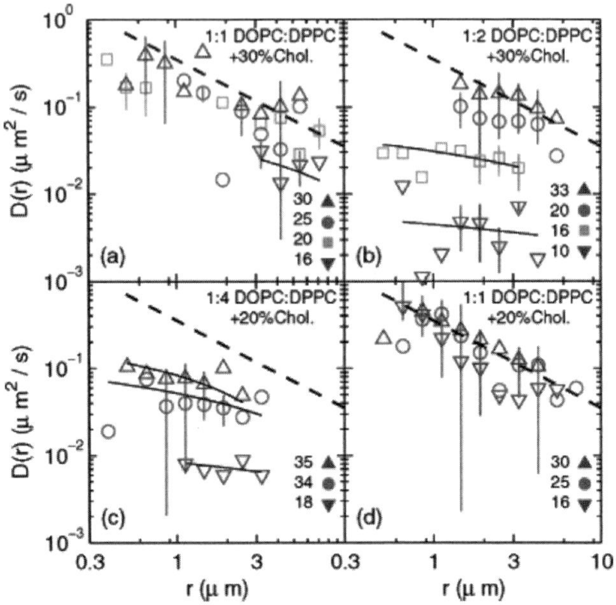

Fig. 16. Diffusion coefficients as a function of domain radius for ternary vesicles composed of DOPC/DPPC/Chol. Both the temperature and composition are varied. Solid lines are fits to a logarithmic dependence on domain size using the SD form Eq. (1). Dashed lines are fits to Eq. (2). Adapted from Ref. 64.

It should be also noted that the viscosity of the lipid domain is different from that of the matrix.[67] In the previous section, we have described that the phase separation in membranes takes place through the Brownian coagulation mechanism. In this section, we discuss the diffusion coefficient (or drag coefficient) of a circular liquid domain by taking into account its finite viscosity using the 2D hydrodynamic equation with momentum decay. We derive an analytical expression for the drag coefficient which covers the whole range of domain sizes. The obtained drag coefficient decreases as the domain viscosity becomes smaller with respect to the outer membrane viscosity.

Experimentally, Cicuta *et al.* reported diffusion coefficients of micron-scale liquid domains in ternary giant unilamellar vesicles.[64] By tracking the trajectory of each domain, they measured the dependence of diffusion coefficient on domain size for different compositions and temperatures as presented in Fig. 16. At high temperatures for which membrane viscosity is low, diffusion coefficients are independent of membrane properties, and domains diffuse with a radial dependence of $D \sim 1/R$ consistent with Eq. (2). In membranes with a higher viscosity, the radial dependence can be fit by the logarithmic relation $D \sim \ln(1/R)$ which is the SD result Eq. (1). We shall reexamine their experimental data by taking into account the difference in the viscosities between domain and matrix.

5.2. *Hydrodynamic model with momentum decay*

The usual method of obtaining the diffusion coefficient D from the drag coefficient ζ is to use the Einstein relation

$$D = \frac{k_{\mathrm{B}}T}{\zeta}. \tag{55}$$

However, for a pure 2D system in the low Reynolds number regime, a linear relation between the velocity and drag force cannot be obtained.[68] This fact is due to the inability to simultaneously satisfy the boundary conditions both at the disk (cylinder) surface and at infinity within the Stokes approximation.[69] Such a problem is known as the Stokes paradox which essentially originates from the constraint of momentum conservation in a pure 2D fluid.

Although a lipid bilayer membrane can be treated as a 2D viscous fluid, it is not an isolated system because the membrane is surrounded by a 3D fluid. Due to the coupling between the membrane and the solvent, the momentum within the membrane can leak away to the outer fluid. Such an effect can be taken into account through a momentum decay term in the equation of motion. Within the Stokes approximation, a hydrodynamic equation which is consistent with the total momentum decay is

$$\eta \nabla^2 \mathbf{v} - \nabla p - \lambda \mathbf{v} = 0, \tag{56}$$

where λ is the momentum decay parameter. A detailed derivation of Eq. (56) in terms of the total momentum decay is given in Ref. 22. On the other hand, the incompressibility condition is given by

$$\nabla \cdot \mathbf{v} = 0, \tag{57}$$

as before. It should be noted that Eq. (56) is the special case of Eq. (4) when $\mathbf{f}_{\mathrm{s}} = -\lambda \mathbf{v}$. Moreover, the ES limit of the mobility tensor Eq. (16) can be directly obtained from Eq. (56) by setting $\lambda = 2\eta_{\mathrm{s}}/h$.[12] Hence Eq. (56) is essentially equivalent to the ES model at least mathematically. Nevertheless we consider here that the parameter λ represents all kinds of frictional

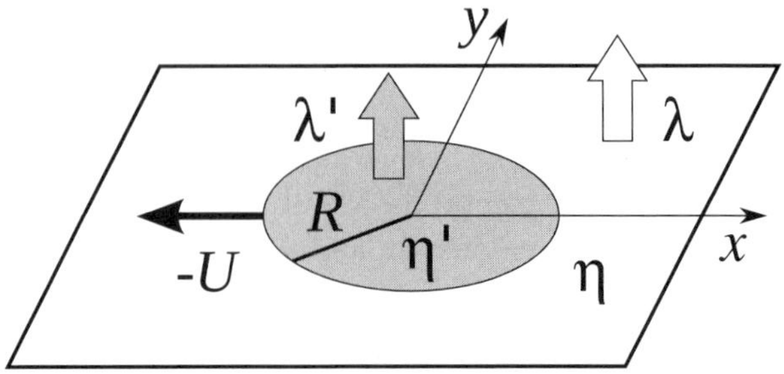

Fig. 17. Schematic picture showing a section of an infinite membrane with viscosity η and momentum decay parameter λ in which a circular liquid domain of radius R with viscosity η' and momentum decay parameter λ' is moving. The liquid domain moves with a velocity $-U$ in the x-direction.

interactions between the lipid head group and the surrounding fluid even for a free membrane, and hence has more general significance. As pointed out in Ref. 18, the translational symmetry along the membrane surface is broken for Eq. (56) due to the friction term. We implicitly assume here that the velocity at infinity vanishes so that the friction term is proportional to $\mathbf{v}$ itself. Due to the presence of the momentum decay term, the Stokes paradox is now eliminated.

As schematically presented in Fig. 17, we consider a circular liquid domain of radius R and viscosity η' moving with a velocity $-\mathbf{U} = (-U, 0)$ in a thin membrane sheet of viscosity η. As a result of the domain motion, a velocity field is induced around it as well as inside the domain. Our purpose is to calculate the drag force experienced by the domain. To generalize our treatment, we allow for the situation where the momentum decay parameters are different between the matrix (λ) and the domain (λ'). For example, sphingomyelin constituting the L_o phases has a large head group which can protrude out into the 3D fluid, whereas the L_d phase are devoid of such structures.[2] It is therefore reasonable to assume that the momentum decay parameter λ experienced by the L_o and L_d domains are different. Hereafter, quantities without prime correspond to those of the matrix, while quantities with prime refer to those of the domain.

The fluid velocity can generally be expressed as a gradient of some potential φ and curl of some vector $\mathbf{A} = (0, 0, A)$ which has only a single component.[69] Then the velocities in the matrix and the domain regions are expressed as

$$\mathbf{v} = -\nabla\varphi + \nabla \times \mathbf{A}, \tag{58}$$

and

$$\mathbf{v}' = -\nabla\varphi' + \nabla \times \mathbf{A}', \tag{59}$$

respectively. Substituting Eqs. (58) and (59) into Eq. (57), we obtain

$$\nabla^2\varphi = 0, \qquad \nabla^2\varphi' = 0, \tag{60}$$

which are the Laplace equations. With the use of Eqs. (58) and (59), one can show that Eq. (56) can be satisfied if the pressures are given by

$$p = \lambda\varphi, \qquad p' = \lambda'\varphi', \tag{61}$$

while A and A' obey the following equations:

$$(\nabla^2 - \kappa^2)A = 0, \qquad (\nabla^2 - \kappa'^2)A' = 0. \tag{62}$$

Here we have defined the (inverse) ES hydrodynamic screening lengths for the matrix and the domain as $\kappa = \sqrt{\lambda/\eta}$ and $\kappa' = \sqrt{\lambda'/\eta'}$, respectively. We remind that this definition is the same as that given in Sec. 2 when $\lambda = 2\eta_{\mathrm{s}}/h$. We shall next seek for the solutions to Eqs. (60) and (62) subject to the appropriate boundary conditions for the translational motion of the domain.

5.3. *Drag coefficient of a liquid domain*

It is convenient to work in the cylindrical polar coordinates (r, θ) with the origin at the center of the circular domain. First we consider the matrix region where $r > R$. Under the condition that the velocity and pressure must approach zero at large distances, we write down the solutions to Eqs. (60) and (62) as follows:

$$\varphi = \frac{C_1}{r}\cos\theta, \qquad A = C_2 K_1(\kappa r)\sin\theta. \tag{63}$$

Here, C_1 and C_2 are unknown coefficients which will be determined from the boundary conditions, and $K_1(z)$ is the modified Bessel function of the second kind of order one. Although the general solutions for φ and A can be expressed as series expansions in terms of r, we have kept only the least number of terms satisfying the requisite pressure and velocity conditions. From Eq. (58), the radial and tangential components of the velocity are given by

$$v_r = \left[\frac{C_1}{r^2} + \frac{C_2}{r}K_1(\kappa r)\right]\cos\theta, \tag{64}$$

$$v_\theta = \left[\frac{C_1}{r^2} + C_2\kappa K_0(\kappa r) + \frac{C_2}{r}K_1(\kappa r)\right]\sin\theta, \tag{65}$$

where we have used the recursion relations among the modified Bessel functions. Then the components of the stress tensor can be obtained as

$$\begin{aligned}\sigma_{rr} &= -p + 2\eta\frac{\partial v_r}{\partial r}\\ &= -\eta\left[C_1\left(\frac{\kappa^2}{r}+\frac{4}{r^3}\right)+C_2\left(\frac{4}{r^2}K_1(\kappa r)+\frac{2\kappa}{r}K_0(\kappa r)\right)\right]\cos\theta, \qquad (66)\\ \sigma_{r\theta} &= \eta\left[\frac{1}{r}\frac{\partial v_r}{\partial\theta}+\frac{\partial v_\theta}{\partial r}-\frac{v_\theta}{r}\right]\\ &= -\eta\left[\frac{4C_1}{r^3}+C_2\left(\frac{4}{r^2}K_1(\kappa r)+\kappa^2K_1(\kappa r)+\frac{2\kappa}{r}K_0(\kappa r)\right)\right]\sin\theta. \qquad (67)\end{aligned}$$

Inside the liquid domain where $r < R$, the proper solutions to Eqs. (60) and (62) subject to the condition that they do not diverge as $r \to 0$ are

$$\varphi' = C_1' r\cos\theta, \qquad A' = C_2' I_1(\kappa' r)\sin\theta, \qquad (68)$$

Here, C_1' and C_2' are unknown coefficients, and $I_1(z)$ is the modified Bessel function of the first kind of order one. Since the corresponding radial and tangential components of the velocity are now

$$v_r' = \left[-C_1' + \frac{C_2'}{r}I_1(\kappa' r)\right]\cos\theta, \qquad (69)$$

$$v_\theta' = \left[C_1' - C_2'\kappa' I_0(\kappa' r) + \frac{C_2'}{r}I_1(\kappa' r)\right]\sin\theta, \qquad (70)$$

the components of the stress tensor can be obtained as

$$\sigma_{rr}' = -\eta'\left[C_1'\kappa'^2 r + C_2'\left(\frac{4}{r^2}I_1(\kappa' r) - \frac{2\kappa'}{r}I_0(\kappa' r)\right)\right]\cos\theta, \qquad (71)$$

$$\sigma_{r\theta}' = -\eta' C_2'\left[\frac{4}{r^2}I_1(\kappa' r) + \kappa'^2 I_1(\kappa' r) - \frac{2\kappa'}{r}I_0(\kappa' r)\right]\sin\theta. \qquad (72)$$

Next we assume that the no-slip condition is satisfied at the domain boundary. This means that, at $r = R$, the radial component of the fluid velocities should be equal to the domain velocity $-U\cos\theta$, the tangential components should be continuous, and so should the components of the stress tensor.[68] These conditions are written as

$$v_r = -U\cos\theta, \qquad (73)$$

$$v_r' = -U\cos\theta, \qquad (74)$$

$$v_\theta = v_\theta', \qquad (75)$$

$$\sigma_{r\theta} = \sigma_{r\theta}'. \qquad (76)$$

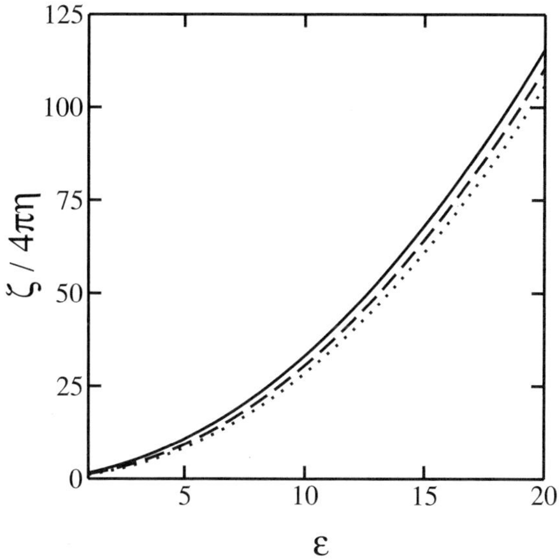

Fig. 18. Scaled drag coefficient ζ as a function of scaled domain radius $\varepsilon = \kappa R$. The different curves are for $E = 0.1$ (solid line), $E = 1$ (dashed line), and $E = 10$ (dotted line).

Since we have kept only the lowest order terms in r for the solutions to Eqs. (60) and (62) (see Eqs. (63) and (68)), we are allowed to neglect the shape deformation of the circular domain while it moves. In other words, we are implicitly assuming that the line tension at the domain boundary exceeds a typical viscous force. A more quantitative argument to justify this condition will be given later.

Using the above four boundary conditions, we can obtain the four unknown coefficients C_1, C_2, C_1' and C_2'. In the following, we introduce the dimensionless parameters to measure the relative viscosities $E = \eta/\eta'$ and the relative decay parameters $L = \lambda/\lambda'$. We also use the notations $\varepsilon = \kappa R$ and $\varepsilon' = \kappa' R = \varepsilon\sqrt{E/L}$ to measure the domain radius. Furthermore, the arguments of the modified Bessel functions will be omitted as $K_n = K_n(\varepsilon)$ and $I_n = I_n(\varepsilon')$ in order to keep the notation compact.

The force exerted on the domain is given by the integral of the stress tensor over the boundary

$$F = R \int_0^{2\pi} \mathrm{d}\theta \, (\sigma_{rr} \cos\theta - \sigma_{r\theta} \sin\theta). \qquad (77)$$

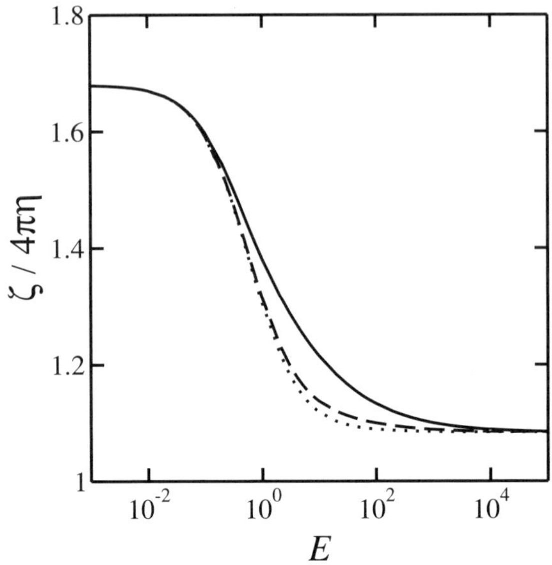

Fig. 19. Scaled drag coefficient ζ as a function of the relative viscosity ratio $E = \eta/\eta'$ when $\varepsilon = 1$. The different curves are for $L = 0.1$ (solid line), $L = 1$ (dashed line) and $L = 10$ (dotted line).

After some calculation, the drag coefficient obtained from $\zeta = F/U$ becomes

$$\frac{\zeta(\varepsilon; E, L)}{4\pi\eta} = \frac{\varepsilon^2}{4} + \frac{\varepsilon K_1[(4+\varepsilon'^2)I_1 - 2\varepsilon' I_0 + 2E(\varepsilon' I_0 - 2I_1)]}{K_0[(4+\varepsilon'^2)I_1 - 2\varepsilon' I_0] + E(2K_0 + \varepsilon K_1)(\varepsilon' I_0 - 2I_1)}. \tag{78}$$

An alternate form can be obtained by using the following recurrence relation

$$I_0(\varepsilon') = \frac{2I_1(\varepsilon')}{\varepsilon'} + I_2(\varepsilon'), \tag{79}$$

so that we have

$$\frac{\zeta(\varepsilon; E, L)}{4\pi\eta} = \frac{\varepsilon^2}{4} + \frac{\varepsilon K_1(\varepsilon' I_1 - 2I_2 + 2EI_2)}{K_0(\varepsilon' I_1 - 2I_2) + E(2K_0 + \varepsilon K_1)I_2}. \tag{80}$$

In Fig. 18, we plot the dimensionless drag coefficient ζ as a function of dimensionless domain size $\varepsilon = \kappa R$ for $E = 0.1, 1, 10$ when $L = 1$. In all these cases, the drag coefficient increases with the domain radius R as it should be. For fixed values of ε, on the other hand, the drag coefficient is smaller when the domain viscosity η' becomes smaller (larger E). Fixing the domain size to $\varepsilon = 1$, we have plotted in Figs. 19 and 20 the drag

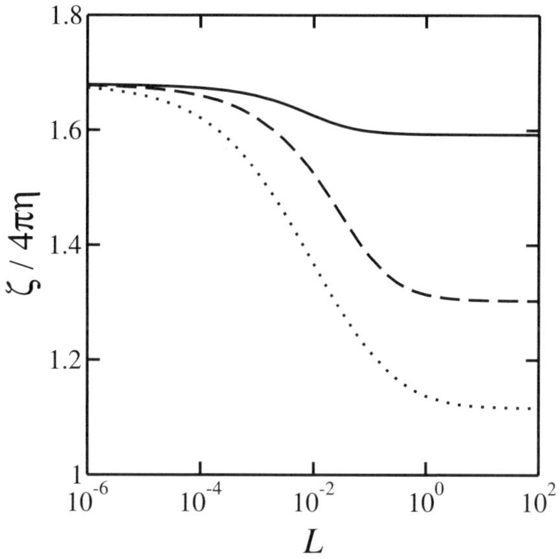

Fig. 20. Scaled drag coefficient ζ as a function of the relative momentum decay parameter ratio $L = \lambda/\lambda'$ when $\varepsilon = 1$. The different curves are for $E = 0.1$ (solid line), $E = 1$ (dashed line) and $E = 10$ (dotted line).

coefficient ζ as a function of E and L, respectively. In the former, we chose various values of L ranging from $L = 0.1$ to 10, while the values of E were changed from $E = 0.1$ to 10 in the latter. From Fig. 19, we see that the drag coefficient monotonically decreases with increasing E (smaller domain viscosity). This can be attributed to the internal flows generated in the domain because they are more efficient in transporting the fluid around it. As a result, a domain with finite viscosity feels a smaller drag force than a rigid disk. The asymptotic values of ζ for $E \to 0$ and $E \to \infty$ converge to respective constants. In Fig. 20, the drag coefficient is a decreasing function of L. The large L values of ζ are dependent on E as discussed below.

Next we examine several asymptotic limits of Eq. (78) or Eq. (80). First we consider arbitrary values of E and L. In the limit of $\varepsilon \to 0$, Eq. (78) gives a logarithmic behavior,

$$\frac{\zeta(\varepsilon \to 0)}{4\pi\eta} \approx \frac{1 + E}{\ln(2/\varepsilon) - \gamma + E[\ln(2/\varepsilon) - \gamma + (1/2)]}, \tag{81}$$

where $\gamma = 0.5772\cdots$ is Euler's constant. In the opposite $\varepsilon \to \infty$ limit, we

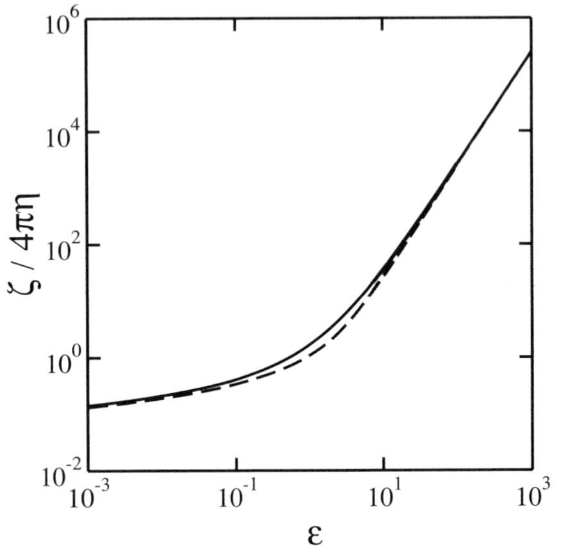

Fig. 21. The limiting scaled drag coefficient ζ_0 (solid line, Eq. (83)) and ζ_∞ (dashed line, Eq. (84)) as a function of scaled domain radius $\varepsilon = \kappa R$ when $L = 1$.

have an algebraic dependence

$$\frac{\zeta(\varepsilon \to \infty)}{4\pi\eta} \approx \frac{\varepsilon^2}{4}, \tag{82}$$

which does not depend on E. We also note that the above asymptotic expressions for small and large domain size limits do not depend on L.

For arbitrary ε, we next consider the rigid disk case ($\eta' \to \infty$) where the limit of $E \to 0$ is taken. Then we obtain

$$\frac{\zeta(\varepsilon; E \to 0)}{4\pi\eta} = \frac{\varepsilon^2}{4} + \frac{\varepsilon K_1}{K_0} \equiv \frac{\zeta_0(\varepsilon)}{4\pi\eta}, \tag{83}$$

which coincides with the result by Evans and Sackmann.[12] The opposite limit of $E \to \infty$ corresponds to a 2D gas bubble ($\eta' \to 0$). In this case, we obtain a different expression:

$$\frac{\zeta(\varepsilon; E \to \infty)}{4\pi\eta} = \frac{\varepsilon^2}{4} + \frac{2\varepsilon K_1}{2K_0 + \varepsilon K_1} \equiv \frac{\zeta_\infty(\varepsilon)}{4\pi\eta}. \tag{84}$$

In Fig. 21, we plot both ζ_0 and ζ_∞ as a function of ε in log-log scales. In the limit of $\varepsilon \to \infty$, the difference between the two curves is the order of

ε which becomes negligibly small compared to the first terms in Eqs. (83) and (84). Upon taking the $\varepsilon \to 0$ limit of Eqs. (83) and (84), we obtain

$$\frac{\zeta_0(\varepsilon \to 0)}{4\pi\eta} \approx \frac{1}{\ln(2/\varepsilon) - \gamma}, \tag{85}$$

$$\frac{\zeta_\infty(\varepsilon \to 0)}{4\pi\eta} \approx \frac{1}{\ln(2/\varepsilon) - \gamma + (1/2)}. \tag{86}$$

These are nothing but the $E \to 0$ and $E \to \infty$ limits of Eq. (81). We note that small ε behavior of ζ_0 and ζ_∞ do not coincide in Fig. 21. The limiting expressions of Eqs. (83) and (84) for $\varepsilon \to \infty$ coincide with Eq. (82).

In the limit of $L \to 0$, which is the case of high momentum dissipation in the domain region, we recover Eq. (83) again;

$$\frac{\zeta(\varepsilon; L \to 0)}{4\pi\eta} = \frac{\varepsilon^2}{4} + \frac{\varepsilon K_1}{K_0}. \tag{87}$$

In the opposite $L \to \infty$ limit, we obtain

$$\frac{\zeta(\varepsilon; L \to \infty)}{4\pi\eta} = \frac{\varepsilon^2}{4} + \frac{2\varepsilon(1+E)K_1}{2(1+E)K_0 + \varepsilon E K_1}. \tag{88}$$

which depends on E as observed in Fig. 20. Equation (88) reduces to Eqs. (83) and (84) for $E \to 0$ and $E \to \infty$, respectively.

When imposing the boundary conditions as given from Eq. (73) to (76), we have assumed that the domain does not undergo any shape deformation. This implies that the line tension should be large enough to overcome the viscous force of deformation. In the case of lipid domains, the line tension was measured to be $\sigma \sim 10^{-12}$ N using domain boundary flicker spectroscopy.[70] For a flow of the order of $U \sim 10^{-6}$ m/s, a typical viscous force turns out be $\eta U \sim 10^{-15}$ N which is much smaller than σ. Hence it is reasonable to assume that the line tension is large enough to maintain the circular shape of the domains unless the temperature is very close to the critical point. Different boundary conditions were used in order to consider the relaxation of deformed domains in a polymer monolayer.[71]

5.4. *Comparison with experiments*

We discuss here the realistic value of $E = \eta/\eta'$ in the case of lipid domains. Applying the pulsed field gradient NMR spectroscopy, Orädd *et al.* measured lateral diffusion coefficients of a single lipid molecule both in the L_o and L_d phases.[67] With the use of the SD logarithmic expression for the diffusion coefficient, the 2D viscosities of these phases are estimated as

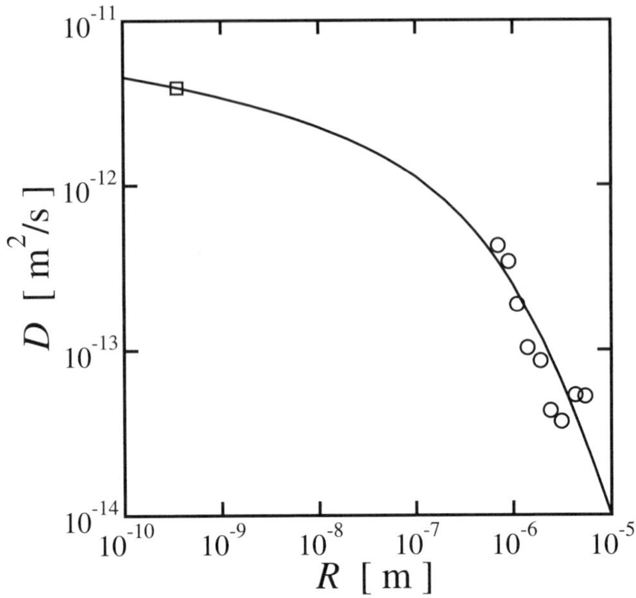

Fig. 22. Diffusion coefficient D as a function of size R. We compare several experimental data obtained by various methods and the analytical expression Eq. (78). The used parameters are $\lambda = 1000$ Ns/m^3, $\eta = 0.63 \times 10^{-9}$ Ns/m, $L = 1$ and $E = 0.072$.

$\eta_o \approx 1.6 \times 10^{-9}$ Ns/m and $\eta_d \approx 0.4 \times 10^{-9}$ Ns/m at 293 K (the membrane thickness is chosen to be $h \approx 3.8 \times 10^{-9}$ m). When the L_o phase forms a domain in the matrix of the L_d phase, the typical viscosity ratio would be $E \approx 0.2$. In the opposite case where the L_d phase forms a domain, the ratio tends to be $E \approx 4$.

In Fig. 22, some known experimental data for the diffusion coefficients of lipid domains as a function of their size R are fitted with Eq. (78). The square data point is the diffusion coefficient of a single lipid molecule obtained from NMR measurements.[67] The circular data points are from Fig. 16(d) obtained from fluorescent microscope measurements in ternary vesicles.[64] From the graph, we can infer that Eq. (78) is a very good fit as the data points span over five decades of domain size range. The membrane viscosity obtained from the fit is in good agreement with the reported values in literature. A somewhat smaller value of $E = 0.072$ is because the experiments here were done at a lower temperature.

6. Dynamics of a Polymer Chain Confined in Membranes

6.1. *Diffusion of 2D polymers*

Integral membrane proteins play a vital role in a variety of cell functions such as solute transport, signal transduction and regulation of membrane composition.[14] Owing to finite temperatures, such proteins along with other membrane components are constantly undergoing Brownian motions. The resulting diffusive motion plays an important role in determining their transport properties. Hence the studies of protein diffusion have been actively performed on model systems[72–76] as well as on living cells.[77–80] As explained in Introduction, the standard approach is to consider them as rigid disks moving in a 2D liquid membrane under low-Reynolds number conditions.

Alternatively, the membrane protein can be regarded as a polymer chain rather than a rigid disk. In this case, the internal degrees of freedom of the polymer should be taken into account, which is the main subject of this section. Apart from the protein analogy, hydrophobically modified polymers which adhere to the membrane could also be described using the present description.[81] In addition, the experiments on DNA molecules embedded on a cationic supported membrane were performed.[82–84] The negative charge of the DNA molecules leads to strong adhesion with the membrane so that only the lateral motions are allowed. The measured diffusion coefficient showed a Rouse-like behavior. More recently, a similar experiment with DNA on a free standing membrane has been conducted.[85] Another related situation is a dilute polymer solution confined between narrow slits. Based on the scaling argument, such a polymer was predicted to show a Rouse-like behavior.[86–88] In an attempt to verify these predictions, experiments on dilute solutions of DNA confined in narrow slits have shown that the exponents for conformation and chain relaxation of DNA is 2.2 which lies between 2D and 3D behaviors.[89]

In this section, we discuss a Brownian dynamics theory for a polymer chain confined in a membrane. As schematically presented in Fig. 23, we consider a polymer chain embedded in a liquid membrane surrounded by a 3D bulk fluid and walls (not shown). We shall obtain the analytical expressions for the polymer diffusion coefficient which are valid for all sizes. A Rouse normal mode analysis is performed to calculate the relaxation time and dynamical structure factor. We also use a scaling theory to discuss the effect of excluded volume interactions on the relaxation times.

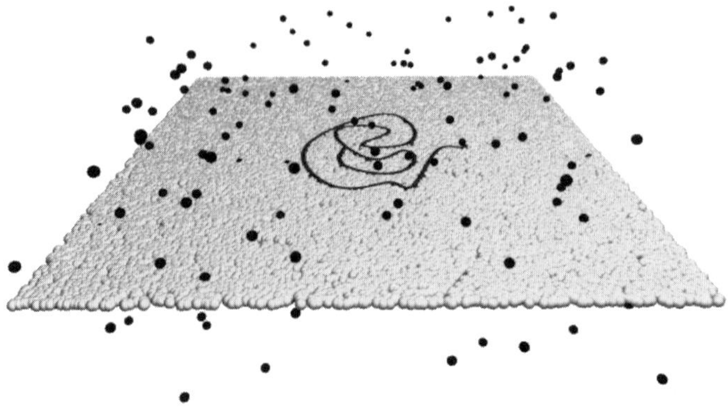

Fig. 23. A transmembrane protein approximated as a polymer chain embedded in a liquid membrane. The membrane itself is surrounded by solvent. Only a few representative solvent particles are shown.

Theoretically, Muthukumar studied the dynamics of a hydrophobic polymer confined in a 2D liquid membrane.[90] In his theory, the membrane itself was treated as an isolated 2D system having an anisotropic viscosity. He showed that the mean squared displacement of a monomer obeys a diffusive law. He also pointed out that the mode dependence of the relaxation time arises from the excluded volume effect.

6.2. *Dynamics of a 2D Gaussian polymer chain*

We now introduce a polymer into the membrane. For simplicity, we first work with a Gaussian polymer chain whose conformation is given by a set of N position vectors denoted as $\{\mathbf{R}_n\} = (\mathbf{R}_1, \ldots, \mathbf{R}_N)$ embedded in the 2D membrane. The excluded volume effects will be discussed later in Sec. 6.5. It is implicitly assumed that the polymer relaxation time is much longer than that for the typical hydrodynamic disturbances so that the membrane can be effectively considered as a 2D liquid. If the polymer consists of monomers which exert a set of point forces $\mathbf{f}_n$ acting at $\mathbf{R}_n$, the external force due to the polymer $\mathbf{F}(\mathbf{r})$ in Eq. (4) can be written as

$$\mathbf{F}(\mathbf{r}) = \sum_n \mathbf{f}_n \, \delta(\mathbf{r} - \mathbf{R}_n). \tag{89}$$

In writing this expression, we have assumed that the superposition principle holds. With the use of the mobility tensor obtained in the previous section,

Eq. (4) can be formally solved as

$$\mathbf{v}(\mathbf{r}) = \sum_n \mathbf{G}(\mathbf{r} - \mathbf{R}_n) \cdot \mathbf{f}_n. \tag{90}$$

Since monomers move with the same velocity as the membrane, their velocities are given by

$$\frac{\partial \mathbf{R}_n(t)}{\partial t} = \mathbf{v}(\mathbf{R}_n) = \sum_m \mathbf{G}_{nm} \cdot \mathbf{f}_m, \tag{91}$$

where we have used the notation $\mathbf{G}_{nm} = \mathbf{G}(\mathbf{R}_n - \mathbf{R}_m)$.

The Langevin equation for a polymer chain embedded in a membrane is written as[91,92]

$$\frac{\partial \mathbf{R}_n(t)}{\partial t} = \sum_m \mathbf{G}_{nm} \cdot \left(-\frac{\partial U}{\partial \mathbf{R}_m} + \zeta_m(t) \right) + \frac{k_B T}{2} \sum_m \frac{\partial}{\partial \mathbf{R}_m} \cdot \mathbf{G}_{nm}, \tag{92}$$

where $\zeta_m(t)$ is the Gaussian random force acting at $\mathbf{R}_m$, k_B the Boltzmann constant, T the temperature. The potential energy of the 2D polymer has the form

$$U = \frac{k_B T}{b^2} \sum_{n=2}^{N} (\mathbf{R}_n - \mathbf{R}_{n-1})^2, \tag{93}$$

where b is the Kuhn length. It can be shown that the mobility tensors Eqs. (15) and (17) satisfy

$$\frac{\partial}{\partial \mathbf{R}_m} \cdot \mathbf{G}_{nm} = 0. \tag{94}$$

Substituting Eq. (93) into Eq. (92) and using Eq. (94), we have

$$\frac{\partial \mathbf{R}_n(t)}{\partial t} = \sum_m \mathbf{G}_{nm} \cdot \left(\frac{2k_B T}{b^2} \frac{\partial^2 \mathbf{R}_m(t)}{\partial m^2} + \zeta_m(t) \right). \tag{95}$$

Due to the hydrodynamic coupling between different parts of the polymer, the above equation is non-linear and difficult to solve analytically. In order to overcome this difficulty, we employ the preaveraging approximation which has been successfully used for a polymer in 3D solvent.[92,93] The validity of this approximation has been studied in some detail for polymers in a 3D bulk fluid, where the preaveraging approximation (Zimm model) yields results which are not very different from more sophisticated calculations.[92] Deviations from the Zimm model have been attributed to non-Gaussian distributions, the slowness to reach asymptotic behaviors or the effect of hydrodynamic fluctuations, although a clear conclusion has not been reached.[94] The success of preaveraging in 3D polymer solutions encourages us to expect the same to hold for 2D systems. Assuming that the polymer is close to its equilibrium, we replace $\mathbf{G}_{nm}$ by its equilibrium value $\langle \mathbf{G}_{nm} \rangle$ such that

$$\begin{aligned}\langle \mathbf{G}_{nm} \rangle &= \int \mathrm{d}\{\mathbf{R}_n\}\, \mathbf{G}_{nm} \Psi(\{\mathbf{R}_n\}) \\ &= \int_0^\infty \mathrm{d}r\, 2\pi r \frac{1}{\pi |n-m| b^2} \exp\left(-\frac{r^2}{|n-m|b^2} \right) \mathbf{G}_{nm}(r) \\ &= g(n-m)\mathbf{I}, \end{aligned} \tag{96}$$

where $\Psi(\{\mathbf{R}_n\})$ is the 2D Gaussian distribution function. Within this approximation, Eq. (95) can be simplified as

$$\frac{\partial \mathbf{R}_n(t)}{\partial t} = \sum_m g(n-m) \left(\frac{2k_\mathrm{B}T}{b^2} \frac{\partial^2 \mathbf{R}_m(t)}{\partial m^2} + \boldsymbol{\zeta}_m(t) \right). \tag{97}$$

The above equation can be rewritten in terms of the Rouse normal coordinates defined by[92]

$$\mathbf{X}_p(t) = \frac{1}{N} \int_0^N \mathrm{d}n \cos\left(\frac{p\pi n}{N} \right) \mathbf{R}_n(t), \tag{98}$$

as

$$\frac{\partial \mathbf{X}_p(t)}{\partial t} = \sum_q g_{pq} [-k_q \mathbf{X}_q(t) + \boldsymbol{\zeta}_q(t)], \tag{99}$$

with

$$k_p = \frac{4\pi^2 k_\mathrm{B} T}{N b^2} p^2. \tag{100}$$

Here $p, q = 0, 1, 2, \ldots$ and g_{pq} is the mobility tensor in terms of the normal coordinates:

$$g_{pq} = \int_0^N \frac{\mathrm{d}n}{N} \int_0^N \frac{\mathrm{d}m}{N} \cos\left(\frac{p\pi n}{N} \right) \cos\left(\frac{q\pi m}{N} \right) g(n-m). \tag{101}$$

If one neglects the contribution from the off-diagonal components of g_{pq}, we finally obtain

$$\frac{\partial \mathbf{X}_p(t)}{\partial t} = g_p \left[-k_p \mathbf{X}_p(t) + \boldsymbol{\zeta}_p(t) \right], \tag{102}$$

with the definition $g_p = g_{pp}$. The Gaussian random forces $\boldsymbol{\zeta}_p(t)$ satisfy the following conditions

$$\langle \zeta_{p\alpha}(t) \rangle = 0, \tag{103}$$

$$\langle \zeta_{p\alpha}(t) \zeta_{q\beta}(t') \rangle = 2\delta_{pq}\delta_{\alpha\beta}(g_p)^{-1} k_\mathrm{B} T \delta(t-t'). \tag{104}$$

Therefore the relaxation time of a polymer in terms of the Rouse modes is given by

$$\tau_p = \frac{1}{g_p k_p}. \tag{105}$$

Furthermore, the polymer diffusion coefficient can be calculated according to the following equation

$$D = k_{\rm B} T g_0 = k_{\rm B} T \int_0^N \frac{\mathrm{d}n}{N} \int_0^N \frac{\mathrm{d}m}{N} g(n-m). \tag{106}$$

Another useful quantity that we calculate is the dynamic structure factor defined by

$$S(\mathbf{k}, t) = \frac{1}{N} \sum_{n,m} \langle \exp[i\mathbf{k} \cdot (\mathbf{R}_n(t) - \mathbf{R}_m(0))] \rangle. \tag{107}$$

Since $\mathbf{R}_n(t) - \mathbf{R}_m(0)$ is a linear function of $\boldsymbol{\zeta}_n(t)$ obeying the Gaussian distribution, the distribution of $\mathbf{R}_n(t) - \mathbf{R}_m(0)$ is also Gaussian.[95] Hence we have

$$\langle \exp(i\mathbf{k} \cdot [\mathbf{R}_n(t) - \mathbf{R}_m(0)]) \rangle = \exp\left(-\frac{k^2}{4} \langle (\mathbf{R}_n(t) - \mathbf{R}_m(0))^2 \rangle \right), \tag{108}$$

in 2D. Denoting the center of mass by $\mathbf{X}_0$ and using the inverse relation of Eq. (98)

$$\mathbf{R}_n = \mathbf{X}_0 + 2 \sum_{p=1}^{\infty} \mathbf{X}_p \cos\left(\frac{p\pi n}{N} \right), \tag{109}$$

we can calculate the dynamical structure factor as

$$S(\mathbf{k}, t) = \frac{1}{N} \sum_{n,m} \exp\Bigg[-k^2 Dt - \frac{1}{4} |n-m| b^2 k^2 - \frac{Nb^2k^2}{\pi^2} \sum_{p=1}^{\infty} \frac{1}{p^2} \cos\left(\frac{p\pi n}{N} \right) \cos\left(\frac{p\pi m}{N} \right) [1 - \exp(-t/\tau_p)] \Bigg]. \tag{110}$$

This completes the general formalism of the 2D polymer dynamics confined in a liquid membrane. In the next section, we consider the free and confined membrane cases separately by using Eqs. (13) or (16) for the mobility tensor.

6.3. *Polymer dynamics: free membrane case*

When the two walls in Fig. 4 are located at infinite distance from the membrane, one can use the SD mobility tensor Eq. (13). We calculate the preaveraged mobility tensor, relaxation time, diffusion coefficient and dynamic structure factor following the recipe described in the previous subsection.

In order to perform the preaveraging of the mobility tensor, we take the configurational average of Eq. (13) by using the equilibrium probability distribution function of a Gaussian polymer in 2D:

$$\langle \mathbf{G}_{nm}^{\mathrm{SD}} \rangle = \left\langle \int \frac{\mathrm{d}\mathbf{k}}{(2\pi)^2} \frac{\mathbf{I} - \hat{\mathbf{k}}\hat{\mathbf{k}}}{\eta(k^2 + \nu k)} \exp[i\mathbf{k} \cdot (\mathbf{R}_n - \mathbf{R}_m)] \right\rangle, \tag{111}$$

where $\hat{\mathbf{k}}$ denotes a unit vector along $\mathbf{k}$. This leads to $\langle \mathbf{G}_{nm}^{\mathrm{SD}} \rangle = g^{\mathrm{SD}}(n-m)\mathbf{I}$ with

$$\begin{aligned} g^{\mathrm{SD}}(n-m) &= \frac{1}{4\pi\eta} \int_0^\infty \mathrm{d}k \, \frac{k}{k^2 + \nu k} \exp\left(-\frac{1}{4} b^2 k^2 |n-m|\right) \\ &= \frac{1}{8\pi\eta} \exp\left(-\frac{1}{4} b^2 \nu^2 |n-m|\right) \\ &\quad \times \left[\pi \mathrm{erfi}\left(\frac{1}{2} b\nu\sqrt{|n-m|}\right) - \mathrm{Ei}\left(\frac{1}{4} b^2 \nu^2 |n-m|\right)\right], \end{aligned} \tag{112}$$

where $\mathrm{erfi}(z)$ is the imaginary error function

$$\mathrm{erfi}(z) = -i\mathrm{erf}(iz), \tag{113}$$

with $\mathrm{erf}(z)$ being the error function

$$\mathrm{erf}(z) = \frac{2}{\sqrt{\pi}} \int_0^z \mathrm{d}u \, e^{-u^2}, \tag{114}$$

whereas $\mathrm{Ei}(-z)$ is the exponential integral given by[96]

$$\mathrm{Ei}(-z) = -\int_z^\infty \mathrm{d}u \, \frac{e^{-u}}{u}. \tag{115}$$

It should be noted that the obtained $g^{\mathrm{SD}}(n-m)$ is real despite the presence of complex functions.

In order to obtain the polymer relaxation time, we first substitute Eq. (112) into Eq. (101) to express the mobility tensor in terms of the Rouse normal coordinates

$$\begin{aligned} g_p^{\mathrm{SD}} &= \frac{1}{\pi\eta N b^2} \int_0^\infty \mathrm{d}k \, \frac{k^3}{(k^2 + \nu k)[k^4 + (4\pi p/Nb^2)^2]} \\ &= \frac{1}{16\pi\eta} \frac{\pi^2 p - \sqrt{2p}\pi^{3/2}\delta + 2\ln(\pi p/\delta^2)\delta^2 + (\sqrt{2\pi/p})\delta^3}{\pi^2 p^2 + \delta^4}, \end{aligned} \tag{116}$$

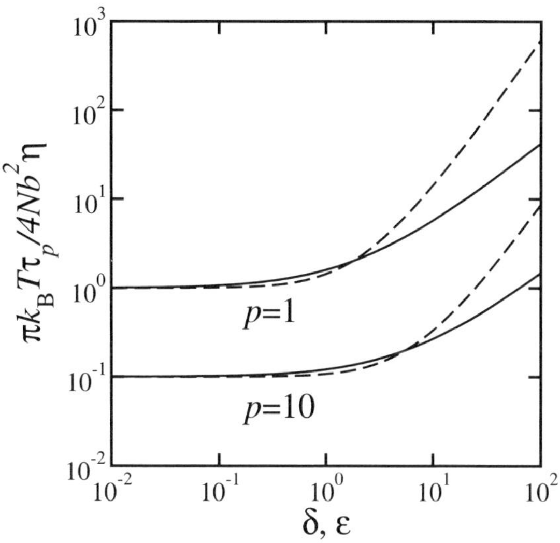

Fig. 24. Scaled relaxation time τ_p as a function of $\delta = R_g\nu$ for free membranes (solid lines) or $\varepsilon = R_g\kappa$ for confined membranes (dashed lines) for $p = 1$ and 10.

(note that $g_p^{\mathrm{SD}} = g_{pp}^{\mathrm{SD}}$). In the above, we have defined the dimensionless polymer size $\delta = \sqrt{N}b\nu/2$. Since the radius of gyration for the 2D Gaussian polymer is $R_g = \sqrt{N}b/2$,[92] δ can be also written as $\delta = R_g\nu$. Using Eq. (105), the relaxation time becomes

$$\tau_p^{\mathrm{SD}} = \frac{4Nb^2\eta}{\pi k_B T}\frac{\pi^2p^2+\delta^4}{p^2[\pi^2p-\sqrt{2p}\pi^{3/2}\delta+2\ln(\pi p/\delta^2)\delta^2+(\sqrt{2\pi/p})\delta^3]}. \quad (117)$$

This expression shows how the presence of the bulk solvent affects the relaxation time.

We consider two asymptotic limits of Eq. (117). For small polymer sizes or $\delta \ll 1$, we have

$$\tau_p^{\mathrm{SD}} \approx \frac{4Nb^2\eta}{\pi k_B T}\frac{1}{p}. \quad (118)$$

For large sizes, the condition $\delta \gg 1$ yields

$$\tau_p^{\mathrm{SD}} \approx \frac{4Nb^2\eta}{\pi k_B T}\frac{\delta}{\sqrt{2\pi}p^{3/2}}. \quad (119)$$

Notice again that Eq. (119) depends only on the solvent viscosity η_s.

In Fig. 24, the scaled relaxation time Eq. (117) is plotted as a function of δ for $p = 1$ and 10 as solid lines. For small δ, the relaxation time

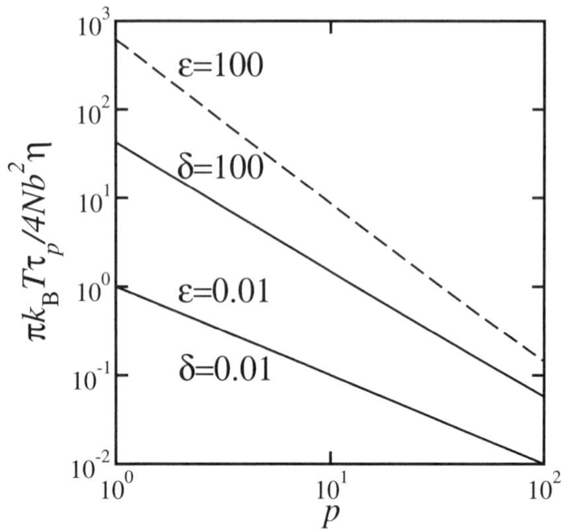

Fig. 25. Scaled relaxation time τ_p as a function of p for different values of $\delta = 0.01, 100$ (solid lines) and $\varepsilon = 0.01, 100$ (dashed lines). The two curves for $\delta = \varepsilon = 0.01$ overlap each other and cannot be distinguished.

is independent of the polymer size, which is consistent with Eq. (118). The relaxation time increases through a crossover regime towards a linear behavior as given by Eq. (119). Such a crossover occurs around the region where the polymer size $R_{\rm g}$ is comparable to the SD hydrodynamic screening length ν^{-1}, i.e., $\delta \sim 1$. In the limit of large δ, the p-dependence of the relaxation time is analogous to that obtained from the Zimm model.[92] The solid lines in Fig. 25 show the relaxation time as a function of the Rouse normal mode p for $\delta = 0.01$ and 100. These lines have slopes -1 and $-3/2$ for $\delta = 0.01$ and 100, respectively. These results indicate the different mode dependencies in the two limiting polymer sizes.

By substituting Eq. (112) into Eq. (106), the diffusion coefficient of the polymer can be obtained as

$$D^{\rm SD} = \frac{k_{\rm B}T}{4\pi\eta}\frac{1}{\delta^4}\left[(\pi\,{\rm erfi}(\delta) - {\rm Ei}(\delta^2))\exp(-\delta^2) + \frac{4\sqrt{\pi}}{3}\delta^3 + \delta^2 - (\ln\delta^2 + \gamma)(\delta^2 - 1) - 2\sqrt{\pi}\delta\right], \quad (120)$$

where $\gamma = 0.5772\cdots$ is Euler's constant. This expression for the polymer diffusion coefficient is valid for all the ranges of δ.

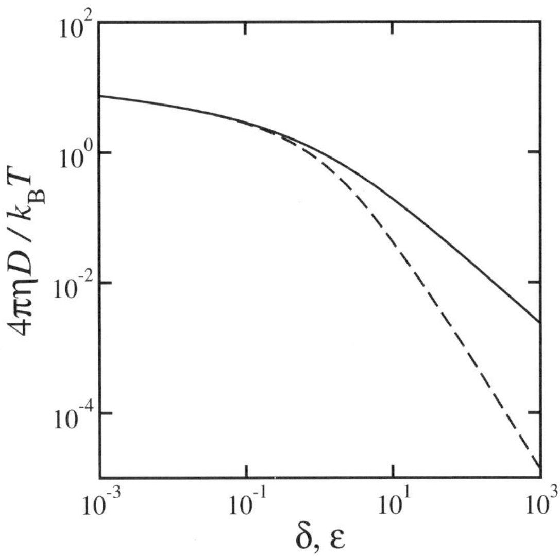

Fig. 26. Scaled diffusion coefficient D as a function of $\delta = R_g\nu$ (solid line) or $\varepsilon = R_g\kappa$ (dashed line).

We now discuss the asymptotic limits of Eq. (120) for small and large polymer sizes. When $\delta \ll 1$, it reduces to

$$D^{\mathrm{SD}} \approx \frac{k_{\mathrm{B}}T}{4\pi\eta}\left(-\ln\delta - \frac{\gamma}{2} + \frac{3}{4}\right). \tag{121}$$

Such a logarithmic behavior is consistent with that of an object in a pure 2D system.[9,10] In the opposite limit of $\delta \gg 1$, we have

$$D^{\mathrm{SD}} \approx \frac{k_{\mathrm{B}}T}{4\pi\eta}\frac{4\sqrt{\pi}}{3\delta} = \frac{k_{\mathrm{B}}T}{6\sqrt{\pi}\eta_{\mathrm{s}}R_{\mathrm{g}}}. \tag{122}$$

Similar to Eq. (119), this expression depends only on η_{s}. The obtained $1/(\eta_{\mathrm{s}}R_{\mathrm{g}})$-dependence is analogous to that of an object moving in 3D fluid as well as the result by Hughes *et al.* (see Eq. (2)).[11] In Fig. 26, we plot the diffusion coefficient D^{SD} as a function of δ (solid curve). With the increase in the polymer size, there is a crossover from logarithmic to algebraic decay indicated by Eqs. (121) and (122), respectively.

The last quantity calculated for the free membrane geometry is the dynamical structure factor $S^{\mathrm{SD}}(\mathbf{k}, t)$ defined in Eq. (107). Since the full expression is rather complicated, we derive several analytical expressions for the limiting cases. For $R_g k \ll 1$, we have

$$S^{\rm SD}(\mathbf{k},t) \approx N \exp(-k^2 D^{\rm SD} t), \tag{123}$$

where $D^{\rm SD}$ is given by Eq. (120). This is reasonable because only the center of mass motion of the polymer is captured in the small angle regime.

For $R_g k \gg 1$, on the other hand, we only consider the time region $t \ll \tau_p^{\rm SD}$ because $S^{\rm SD}(\mathbf{k},t)$ becomes very small for $t \gg \tau_p^{\rm SD}$. In this case, we have

$$\begin{aligned} S^{\rm SD}(\mathbf{k},t) &\approx \frac{1}{N}\sum_{n,m} \exp\left[-\frac{1}{4}|n-m|b^2k^2 \right. \\ &\quad \left. - \frac{Nb^2k^2}{\pi^2}\sum_{p=1}^{\infty}\frac{1}{p^2}\cos\left(\frac{p\pi n}{N}\right)\cos\left(\frac{p\pi m}{N}\right)[1-\exp(-t/\tau_p^{\rm SD})]\right] \\ &= \frac{8}{b^2k^2}\int_0^{\infty} du\, \exp(-u - k^2 I_1(u)), \end{aligned} \tag{124}$$

with

$$I_1(u) = \frac{Nb^2}{2\pi^2}\int_0^{\infty} dp\,\frac{1}{p^2}\cos\left(\frac{4\pi p u}{Nb^2k^2}\right)[1-\exp(-t/\tau_p^{\rm SD})]. \tag{125}$$

When $\delta \gg 1$, we can use the limiting expression for $\tau_p^{\rm SD}$ as obtained in Eq. (119). In this case, the above expression becomes

$$S^{\rm SD}(\mathbf{k},t) \approx \frac{8}{b^2k^2}\int_0^{\infty} du\, \exp\left[-u - (\Gamma^{\rm SD}t)^{2/3} w_1((\Gamma^{\rm SD}t)^{-2/3}u)\right], \tag{126}$$

with the decay rate

$$\Gamma^{\rm SD} = \frac{k_{\rm B}T\sqrt{N}bk^3}{16\eta\delta} = \frac{k_{\rm B}Tk^3}{8\eta\nu}, \tag{127}$$

and

$$w_1(u) = \frac{2}{\pi}\int_0^{\infty} dx\,\frac{\cos(xu)}{x^2}[1-\exp(-x^{3/2}/\sqrt{2})]. \tag{128}$$

Notice the k^3-dependence of the decay rate. For $\Gamma^{\rm SD}t \gg 1$, the above expression is further simplified to

$$S^{\rm SD}(\mathbf{k},t) \approx S^{\rm SD}(\mathbf{k},0)\exp(-1.35(\Gamma^{\rm SD}t)^{2/3}), \tag{129}$$

since $w_1(0) = \Gamma(1/3) \approx 1.35$. This expression is valid in the limit of large polymer sizes such that $R_g k \gg 1$ and $\delta \gg 1$. Large wave vectors (probing the internal motion of the polymer) and hydrodynamic screening lengths (high solvent viscosity) will also lead to the same expression. The applicable time window for Eq. (129) is $1/\Gamma^{\rm SD} \ll t \ll \tau_p^{\rm SD}$. These expressions for large δ are analogous to that obtained from the Zimm model[92] in the $R_{\rm g}k \gg 1$ regime for a polymer in 3D.

6.4. *Polymer dynamics: confined membrane case*

When the thickness of the solvent layers is very small, the membrane is almost confined by the two walls. However, there is a thin lubricating layer between the membrane and the walls so that $h \neq 0$. In this case, we use the ES mobility tensor Eq. (16).

By using Eq. (16), the pre-averaged mobility tensor is calculated from

$$\langle \mathbf{G}_{nm}^{\rm ES} \rangle = \left\langle \int \frac{d\mathbf{k}}{(2\pi)^2} \frac{\mathbf{I} - \hat{\mathbf{k}}\hat{\mathbf{k}}}{\eta(k^2+\kappa^2)} \exp[i\mathbf{k}\cdot(\mathbf{R}_n - \mathbf{R}_m)] \right\rangle. \tag{130}$$

This results in $\langle \mathbf{G}_{nm}^{\rm ES} \rangle = g^{\rm ES}(n-m)\mathbf{I}$ with[21]

$$\begin{aligned} g^{\rm ES}(n-m) &= \frac{1}{4\pi\eta} \int_0^\infty dk \, \frac{k}{k^2+\kappa^2} \exp\left(-\frac{1}{4} b^2 k^2 |n-m|\right) \\ &= -\frac{1}{8\pi\eta} \exp\left(\frac{1}{4} b^2 \kappa^2 |n-m|\right) \mathrm{Ei}\left(-\frac{1}{4} b^2 \kappa^2 |n-m|\right). \end{aligned} \tag{131}$$

The polymer relaxation time can be obtained by substituting Eq. (131) into Eq. (101). Then we have

$$\begin{aligned} g_p^{\rm ES} &= \frac{1}{\pi\eta N b^2} \int_0^\infty dk \, \frac{k^3}{(k^2+\kappa^2)[k^4 + (4\pi p/Nb^2)^2]} \\ &= \frac{1}{16\pi\eta} \frac{\pi^2 p + 2\varepsilon^2 \ln(\varepsilon^2/(\pi p))}{\pi^2 p^2 + \varepsilon^4}. \end{aligned} \tag{132}$$

In the above, we have defined the dimensionless polymer size as $\varepsilon = \sqrt{N} b\kappa/2 = R_{\rm g}\kappa$ which should be distinguished from δ in the previous subsection. Then the relaxation time can be written as

$$\tau_p^{\rm ES} = \frac{4Nb^2\eta}{\pi k_{\rm B}T} \frac{\pi^2 p^2 + \varepsilon^4}{p^2[\pi^2 p + 2\varepsilon^2 \ln(\varepsilon^2/(\pi p))]}. \tag{133}$$

In the limit of $\varepsilon \ll 1$, it reduces to

$$\tau_p^{\rm ES} \approx \frac{4Nb^2\eta}{\pi k_{\rm B}T} \frac{1}{p}, \tag{134}$$

which coincides with Eq. (118). In the opposite limit of $\varepsilon \gg 1$, one gets

$$\tau_p^{\rm ES} \approx \frac{4Nb^2\eta}{\pi k_{\rm B}T} \frac{\varepsilon^2}{2p^2 \ln(\varepsilon^2/(\pi p))}. \tag{135}$$

In Fig. 24, we plot τ_p^{ES} as a function of ε for $p = 1$ and 10 in dashed lines. The algebraic ε^2-dependence in Eq. (135) is seen for large ε. The dashed lines in Fig. 25 are the plots of τ_p^{ES} as a function of p for $\varepsilon = 0.01$ and 100. For $\varepsilon = 100$, the slope -2 is consistent with Eq. (135) neglecting the logarithmic correction. Notice that this p-dependence is in contrast to that for the free membrane case given in Eq. (119).

With the use of Eq. (131), the diffusion coefficient for the confined membrane geometry is written as[21]

$$D^{\mathrm{ES}} = \frac{k_{\mathrm{B}}T}{4\pi\eta}\frac{1}{\varepsilon^4}[(1+\varepsilon^2)(2\ln\varepsilon + \gamma) - \varepsilon^2 - \exp(\varepsilon^2)\mathrm{Ei}(-\varepsilon^2)]. \tag{136}$$

The limiting expression for $\varepsilon \ll 1$ is

$$D^{\mathrm{ES}} \approx \frac{k_{\mathrm{B}}T}{4\pi\eta}\left(-\ln\varepsilon - \frac{\gamma}{2} + \frac{3}{4}\right), \tag{137}$$

which coincides with Eq. (121) as long as ε is replaced by δ. When $\varepsilon \gg 1$, Eq. (136) reduces to

$$D^{\mathrm{ES}} \approx \frac{k_{\mathrm{B}}T}{4\pi\eta}\frac{1}{\varepsilon^2} = \frac{k_{\mathrm{B}}Th}{8\pi\eta_{\mathrm{s}}R_{\mathrm{g}}^2}. \tag{138}$$

This $1/R_{\mathrm{g}}^2$-dependence is a characteristic of a system in which there is momentum loss from the membrane to the surrounding environment.[40] The dashed line in Fig. 26 shows the plot of D^{ES} as a function of ε, showing the logarithmic and algebraic behaviors as derived above.

The dynamic structure factor can be calculated in the same manner as before. For $R_{\mathrm{g}}k \ll 1$, we have

$$S^{\mathrm{ES}}(\mathbf{k},t) \approx N\exp(-k^2 D^{\mathrm{ES}} t). \tag{139}$$

For $R_{\mathrm{g}}k \gg 1$ and $t \ll \tau_p^{\mathrm{ES}}$, we get

$$S^{\mathrm{ES}}(\mathbf{k},t) \approx \frac{8}{b^2k^2}\int_0^\infty du\, \exp(-u - k^2 I_2(u)), \tag{140}$$

with

$$I_2(u) = \frac{Nb^2}{2\pi^2}\int_0^\infty dp\, \frac{1}{p^2}\cos\left(\frac{4\pi pu}{Nb^2k^2}\right)[1 - \exp(-t/\tau_p^{\mathrm{ES}})]. \tag{141}$$

By considering $\varepsilon \gg 1$ and neglecting the logarithmic dependence in Eq. (135), the above expression becomes

$$S^{\mathrm{ES}}(\mathbf{k},t) \approx \frac{8}{b^2k^2}\int_0^\infty du\, \exp\left[-u - (\Gamma^{\mathrm{ES}}t)^{1/2} w_2((\Gamma^{\mathrm{ES}}t)^{-1/2}u)\right], \tag{142}$$

with the decay rate

$$\Gamma^{\mathrm{ES}} = \frac{k_{\mathrm{B}}TNb^2k^4}{32\pi\eta\varepsilon^2} = \frac{k_{\mathrm{B}}Tk^4}{8\pi\eta\kappa^2}, \tag{143}$$

and

$$w_2(u) = \frac{2}{\pi}\int_0^\infty \mathrm{d}x\,\frac{\cos(xu)}{x^2}[1-\exp(-x^2)]. \tag{144}$$

Note that Γ^{ES} is proportional to k^4 as for the Rouse model. For $\Gamma^{\mathrm{ES}}t \gg 1$, it can be further simplified to

$$S^{\mathrm{ES}}(\mathbf{k},t) \approx S^{\mathrm{ES}}(\mathbf{k},0)\exp(-1.13(\Gamma^{\mathrm{ES}}t)^{1/2}), \tag{145}$$

since $w_2(0) = 2/\sqrt{\pi} \approx 1.13$. This expression is valid when $R_{\mathrm{g}}k \gg 1$ and $\varepsilon \gg 1$. The relevant time interval for this expression is $1/\Gamma^{\mathrm{ES}} \ll t \ll \tau_p^{\mathrm{ES}}$.

The dynamic structure factor is a quantity readily accessible through scattering experiments. For small wave numbers $R_{\mathrm{g}}k \ll 1$, only the center of mass motion of the polymer can be captured. When $R_{\mathrm{g}}k \gg 1$ and t is much less than the relaxation times, the screening lengths become important. In the limit of $\delta \gg 1$, $S^{\mathrm{SD}}(\mathbf{k},t)$ shows a stretched exponential decay with an exponent 2/3, and so does $S^{\mathrm{ES}}(\mathbf{k},t)$ for $\varepsilon \gg 1$ with an exponent 1/2. Again the role of the outer environment is reflected in the decay rates Γ^{SD} and Γ^{ES} given by Eqs. (127) and (143), respectively. In the free membrane case, the dependence $\Gamma^{\mathrm{SD}} \sim k^3$ resembles that of the Zimm model in 3D. For confined membranes on the other hand the behavior $\Gamma^{\mathrm{ES}} \sim k^4$ is analogous to that obtained from the Rouse model.

The different large size behaviors of the diffusion coefficient can be better understood by focusing on the mobility tensors. It is essentially described as a consequence of the conservation of mass and momentum principles.[40] The mobility tensor $\mathbf{G}(\mathbf{r})$ acts as a Green's function relating a source of disturbance at the origin to the velocity at a point $\mathbf{r}$. At sufficiently large distances, the source of disturbance in the fluid can be thought of as a force monopole which introduces a source for momentum in the fluid. A moving object also causes a perturbation in the mass density and this can be regarded as a mass dipole (a source and a sink of mass density). We now apply the conservations of momentum and mass to the 3D and 2D cases separately.

For the 3D case, the presence of a force monopole at the origin should cause the momentum flux or stress σ decay as $1/r^2$ in order to conserve the total momentum (r^2 being proportional to the area of a sphere surrounding the force monopole). Since the shear stress is related to the fluid velocity through $\sigma \sim \eta_s v/r$, we have $v \sim 1/\eta_s r$. This implies that the mobility tensor should also scale as $1/\eta_s r$. Concerning the mass conservation principle, a mass monopole would create a flow velocity that decays as $1/r^2$. Since we have a mass dipole, the resulting velocity now decays as $1/r^3$. Comparing these two effects, the contribution to the velocity due to momentum conservation (which varies as $1/r$) always dominates at large distances in 3D. This essentially explains the behavior $D^{\mathrm{SD}} \sim 1/\eta_s R_g$.

Let us now consider the 2D case in which the membrane is in contact with the walls leading to a loss of momentum from the membrane. This implies that the momentum is not conserved, and the only contribution to the velocity is from the mass conservation. In 2D, a mass monopole will create a velocity which decays as $1/r$ (r being the perimeter of a circle surrounding the mass monopole). Hence the velocity and the mobility tensor due to a mass dipole decays as $1/r^2$. This explains the scaling $D^{\mathrm{ES}} \sim 1/R_g^2$. These two different behaviors of the mobility tensors are reflected in the diffusion coefficients for large size polymers. Incidentally, for the pure 2D case, the stress decays as $1/r$ due to the momentum conservation. Since the stress scales as $\sigma \sim \eta v/r$, we have $v \sim 1/\eta$. This explains the logarithmic size dependence of the diffusion coefficient.

At this stage, a rough estimate of the screening lengths would be useful. As reported in Ref. 72, the membrane viscosity of DMPC bilayers at 32 °C (rounded to the nearest order) is approximately 0.1 Ns/m^2 and the viscosity of water is $\eta_s \approx 10^{-3}$ Ns/m^2. For supported membranes we can approximate the height of the intervening solvent region to be $h \approx 10^{-8}$ m.[97] Hence we obtain $\nu^{-1} \approx 2.5 \times 10^{-7}$ m and $\kappa^{-1} \approx 0.5 \times 10^{-7}$ m. It is noted that the experiments with DNA adsorbed on supported membranes point towards a Rouse-like behavior.[82,83] The strong electrostatic attraction between the negatively charged DNA and the positively charged membrane prevents any out-of-plane motions of the polymer. Because the typical length scale of a DNA molecule used in the experiments were of several microns, the scenario should be close to the case of $\varepsilon \gg 1$. Hence the present result is consistent with the experimental observations showing the Rouse-like behavior. A more recent experiment on DNA adsorption on free standing cationic giant vesicles showed that the diffusion coefficient of DNA molecules lies in the crossover region between the logarithmic to the algebraic regimes.[85]

6.5. *Excluded volume effects*

So far, we have treated only a 2D Gaussian polymer chain. Here we briefly discuss the effects of excluded volume on the dynamical quantities. Even if the polymer is not a Gaussian chain, one can still use the preaveraging approximation for the mobility tensor. For the free membrane case, one can generally show that

$$g^{\mathrm{SD}}(n-m) = \frac{1}{8\eta}\langle \mathbf{H}_0(\nu r_{nm}) - Y_0(\nu r_{nm})\rangle, \tag{146}$$

where $r_{nm} = |\mathbf{R}_n - \mathbf{R}_m|$ (see appendix B). Then the diffusion coefficient can be expressed as

$$D^{\mathrm{SD}} = \frac{k_{\mathrm{B}}T}{8\eta}\int_0^N \frac{\mathrm{d}n}{N}\int_0^N \frac{\mathrm{d}m}{N}\langle \mathbf{H}_0(\nu r_{nm}) - Y_0(\nu r_{nm})\rangle. \tag{147}$$

Similarly for the confined membrane case, we find

$$D^{\mathrm{ES}} = \frac{k_{\mathrm{B}}T}{4\pi\eta}\int_0^N \frac{\mathrm{d}n}{N}\int_0^N \frac{\mathrm{d}m}{N}\langle K_0(\kappa r_{nm})\rangle, \tag{148}$$

(see Eq. (B.11)). These expressions are the 2D analog of the Kirkwood formula.[92] It should be emphasized that they are rigorous even in the presence of excluded volume effect.

For excluded volume chains, however, the appropriate equilibrium distribution function $\Psi(\{\mathbf{R}_n\})$ needed to calculate averages is not known.[92] Therefore, we cannot obtain rigorous forms of diffusion coefficient. Instead, we shall make use of scaling arguments to infer the effects of excluded volume interactions. Here, we limit our discussion to small and large polymer sizes. A simple argument is that the excluded volume effects lead to a rescaled polymer radius of gyration R_{g}. Within the Flory theory, the radius of gyration scales as $R_{\mathrm{g}} \sim bN^{\nu_{\mathrm{F}}}$ using the Flory exponent ν_{F}. Up to a numerical factor, the diffusion coefficient in the limiting cases will still show the same size dependence as in Eqs. (121), (122), (137) and (138) in which R_{g} is now replaced with that of excluded volume chains. In other words, we replace N with $N^{2\nu_{\mathrm{F}}}$ in these equations. On the other hand, the relaxation time in the presence of the excluded volume effects can be obtained by replacing N/p with $(N/p)^{2\nu_{\mathrm{F}}}$ in the Gaussian chain cases. In the small size limits, i.e, $\delta, \varepsilon \ll 1$, we have

$$\tau^{\mathrm{SD}} \sim \frac{\eta b^2}{k_{\mathrm{B}}T}\left(\frac{N}{p}\right)^{2\nu_{\mathrm{F}}}, \tag{149}$$

limits	ideal chain ($\nu_F = 1/2$)	real chain ($\nu_F = 3/4$)
$\delta \ll 1$	$\tau^{SD} \sim p^{-1}$	$\tau^{SD} \sim p^{-3/2}$
$\varepsilon \ll 1$	$\tau^{ES} \sim p^{-1}$	$\tau^{ES} \sim p^{-3/2}$
$\delta \gg 1$	$\tau^{SD} \sim p^{-3/2}$	$\tau^{SD} \sim p^{-9/4}$
$\varepsilon \gg 1$	$\tau^{ES} \sim p^{-2}$	$\tau^{ES} \sim p^{-3}$

and

$$\tau^{ES} \sim \frac{\eta b^2}{k_B T} \left(\frac{N}{p}\right)^{2\nu_F}. \tag{150}$$

In the opposite of large polymer size limits, i.e., $\delta, \varepsilon \gg 1$, we get

$$\tau^{SD} \sim \frac{\eta b^3 \nu}{k_B T} \left(\frac{N}{p}\right)^{3\nu_F}, \tag{151}$$

and

$$\tau^{ES} \sim \frac{\eta b^4 \kappa^2}{k_B T} \left(\frac{N}{p}\right)^{4\nu_F}. \tag{152}$$

These are the scaling predictions. Notice that the above results can also be obtained by using the relation $\tau \sim R_g^2/D$ where R_g includes the excluded volume effect.[98]

Table 1 shows the comparison between a Gaussian chain polymer ($\nu_F = 1/2$) and a chain with excluded volume interactions ($\nu_F = 3/4$) in 2D. The former Gaussian case recovers all the relations in the previous sections (see Eqs. (118), (134), (119) and (135)). For the chain with $\nu_F = 3/4$, the relaxation time shows a $p^{-3/2}$-dependence for both the free and confined membrane cases when $\delta, \varepsilon \ll 1$. This is consistent with the result in Ref. 90. For $\delta \gg 1$ and $\varepsilon \gg 1$, we have $\tau^{SD} \sim p^{-9/4}$ and $\tau^{ES} \sim p^{-3}$, respectively. These exponents are unique for polymers confined in a membrane.

6.6. *Related works*

The dynamics of a hydrophobic polymer embedded in a 2D membrane was previously investigated by Muthukumar both for a Gaussian polymer and a polymer with excluded volume effects.[90] In his treatment, the membrane was regarded as an isolated entity without any couplings to the outer environment. The membrane 2D nature was taken into account through an anisotropic viscosity. It was shown that the longest relaxation time is proportional to p^{-1} for a Gaussian chain. This agrees with our limiting expressions of the relaxation times in Eqs. (118) and (134) obtained for small δ and ε, respectively. In his treatment, the excluded volume effects were taken into account by a mode-dependent polymer blob size, showing that the relaxation time scales as $p^{-3/2}$. In Sec. 6.5, we showed that the presence of excluded volume interactions would indeed result in the scaling $p^{-3/2}$ in the small size limit both for the free and confined membrane cases (see table 1). Notice again that the polymer dynamics in this limit remains unaffected by the outer environment. For large polymer sizes, we obtain either $\tau^{\mathrm{SD}} \sim p^{-9/4}$ or $\tau^{\mathrm{ES}} \sim p^{-3}$ for real polymer chains.

A related situation to a polymer in a confined membrane is a dilute polymer solution trapped in a slit-like geometry whose width is much smaller than the polymer blob size. Using scaling arguments, Brochard calculated the polymer relaxation time scales as $p^{-5/2}$.[87] An experimental realization of such a geometry was done by Lin *et al.* who confined dilute DNA solution in quasi-2D 110 nm wide slits.[89] The relaxation times measured in this case was found to scale as $p^{-2.2}$. However, it should be emphasized that this scenario is different from the model discussed here. In our model, the polymer chain is strictly confined to the 2D plane of the membrane which itself is embedded in a 3D bulk fluid.

7. Hydrodynamic Coupling Between Two Fluid Membranes

A recurring theme in this chapter has been the induced Brownian motion of membrane constituents owing to the finite temperature at normal physiological conditions. So far we have concentrated on diffusion of objects in single membranes. We now explore the coupling of diffusion dynamics between two fluid membranes through an intervening bulk fluid.[99] It was reported that model experimental systems in which two lipid membranes are stacked on a substrate exhibit correlated dynamics.[97] This planar geometry has become favorable to study the membrane dynamics which are otherwise not possible in vesicles.[100] The coupling effect between two membranes can be important in biological systems with large concentration of cells such as in tissues. Other examples are Gram-negative bacteria which enclose a periplasmic space with an approximate width of about 15 to 20 nm between their inner and outer lipid bilayers.[14] Highly folded membranous organelles such as Golgi apparatus also correspond to a situation in which membranes come in close proximity to each other. In all these cases, it is very relevant to consider the hydrodynamic coupling between two biomembranes.

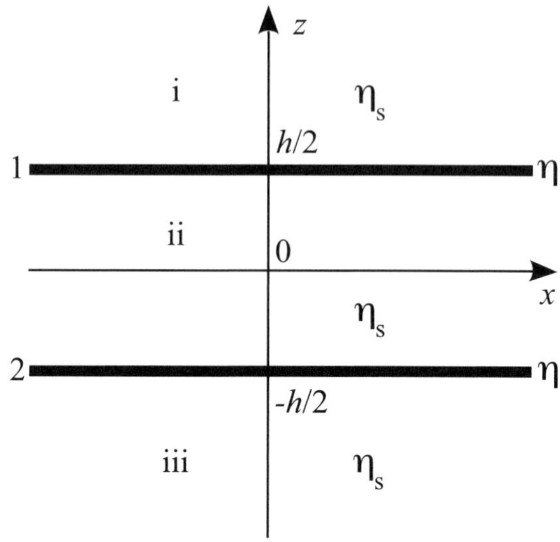

Fig. 27. Schematic picture showing a set of stacked fluid membranes (solid thick lines) having 2D viscosities η located at $z = \pm h/2$. The two membranes are denoted by the labels 1 and 2. The solvent of 3D viscosities η_{s} are labeled by the regions i, ii and iii.

As shown in Fig 27, the membranes are fixed in the xy-plane at $z = \pm h/2$. Let $\mathbf{v}^{(i)}(\mathbf{r})$ be the 2D velocity of the membrane fluids. Here the index $i = 1, 2$ represents the two membranes, and the 2D vector $\mathbf{r} = (x, y)$ represents a point in the planes of the membranes. We let η be the 2D membrane viscosity (same for both the membranes), $p^{(i)}(\mathbf{r})$ the 2D in-plane pressure, $\mathbf{f}_{\mathrm{s}}^{(i)}(\mathbf{r})$ the force exerted on the membrane by the surrounding fluid, and $\mathbf{F}^{(i)}(\mathbf{r})$ be any other force acting on the membrane.

The solvent regions are denoted by the index $j =$ i, ii, iii. The velocities and pressures in these regions are written as $\mathbf{v}^{(j)}(\mathbf{r}, z)$ and $p^{(j)}(\mathbf{r}, z)$, respectively. We assume that the solvent in the three regions have the same 3D viscosity η_{s}. The solvent inertia is neglected and hence it also obeys the 3D Stokes equations. Similar to the fluid membrane, the solvent in all the regions are considered to be incompressible. The presence of the surrounding solvent is important because it exerts force on the liquid membranes. The force on membrane 1, indicated as $\mathbf{f}_{\mathrm{s}}^{(1)}$ is given by the projection of the fluid stress tensor onto the xy-plane of the membrane. Following Sec. 2, the general procedure is to first resolve Eq. (4) into components along $\mathbf{k}$ and perpendicular to it, where $\mathbf{k}$ is the 2D wave vector in Fourier space.

We then solve the resulting differential equations for the velocities. Stick boundary conditions at the membrane-solvent interfaces are imposed. It is also assumed that the solvent velocities decay to zero at sufficiently large distances from the membranes.

Owing to the linearity of governing Stokes equations, the in-plane velocity in membrane 1 can be obtained in Fourier space as

$$v_\alpha^{(1)}[\mathbf{k}] = G_{\alpha\beta}^{(11)}[\mathbf{k}]F_\beta^{(1)}[\mathbf{k}] + G_{\alpha\beta}^{(12)}[\mathbf{k}]F_\beta^{(2)}[\mathbf{k}]. \tag{153}$$

Here $G_{\alpha\beta}^{(11)}$ and $G_{\alpha\beta}^{(12)}$ $(\alpha, \beta = x, y)$ are the mobility tensors given by

$$G_{\alpha\beta}^{(11)}[\mathbf{k}] = \frac{1}{\eta\nu^2}\frac{1 + 2K + \coth(KH)}{Kg(K,H)}\left(\delta_{\alpha\beta} - \frac{k_\alpha k_\beta}{k^2}\right), \tag{154}$$

$$G_{\alpha\beta}^{(12)}[\mathbf{k}] = \frac{1}{\eta\nu^2}\frac{\operatorname{cosech}(KH)}{Kg(K,H)}\left(\delta_{\alpha\beta} - \frac{k_\alpha k_\beta}{k^2}\right), \tag{155}$$

with

$$g(K,H) = 1 + 2K(1+K) + (1+2K)\coth(KH), \tag{156}$$

and $\nu = 2\eta_s/\eta$ as before. The definitions are $H = h\nu$ and $K = k/\nu$ with $k = |\mathbf{k}|$. By symmetry, a similar set of expressions can be written down for the membrane 2 also. The above mobility tensors represent the hydrodynamic coupling between the two membranes.

As in Sec. 4.4, we consider a pair of particles embedded in the membrane undergoing Brownian motion separated by $\mathbf{r}$. For sufficiently large enough times, the particle displacements obey $\langle \Delta r_\alpha \Delta r'_\beta\rangle = 2D_{\alpha\beta}(\mathbf{r})t$ where Δr_α represents the displacement of the first particle and $\Delta r'_\beta$ represents that of the second particle (see also Eq. (48)). In the above, the diffusion tensor for over-damped dynamics is given by the Einstein relation $D_{\alpha\beta} = k_B T G_{\alpha\beta}$. We now apply these definitions to a double-membrane system. The line of centers connecting any two particles in the membranes after projection on to the 2D plane can be taken to be along the x-axis without loss of generality. Then we obtain the coupling longitudinal diffusion coefficients as $D_L^{(11)}(\mathbf{r}) = k_B T G_{xx}^{(11)}(r\hat{e}_x)$ of two particles within the same membrane, and $D_L^{(12)}(\mathbf{r}) = k_B T G_{xx}^{(12)}(r\hat{e}_x)$ of two particles in different membranes. The coupling transverse diffusion coefficients are $D_T^{(11)}(\mathbf{r}) = k_B T G_{yy}^{(11)}(r\hat{e}_x)$ and $D_T^{(12)}(\mathbf{r}) = k_B T G_{yy}^{(12)}(r\hat{e}_x)$. The longitudinal coupling diffusion coefficient is associated with Brownian motion along the line of centers, while the transverse one is associated with motion perpendicular to the line of centers.[18,19,101] In the following, we discuss the above four diffusion coefficients sequentially.

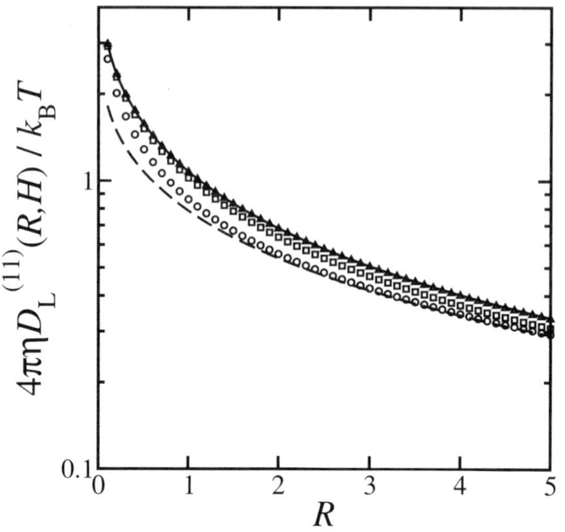

Fig. 28. Scaled $D_{\rm L}^{(11)}(R, H)$ as a function of R for various values of H. The circles, squares and triangles correspond to $H = 0.1$, 1 and 10, respectively. The solid line corresponds to Eq. (158) representing the analytical expression for large H limit. The dashed line corresponds to the small H limit.

The real-space expressions for the mobility tensors can be obtained by inverse Fourier transform. Using the notation $R = r\nu$, we obtain the longitudinal coupling diffusion coefficient within the same membrane as

$$\frac{4\pi\eta D_{\rm L}^{(11)}(R,H)}{k_{\rm B}T} = 2\int_0^\infty {\rm d}K\, \frac{1+2K+\coth(KH)}{g(K,H)}\,\frac{J_1(KR)}{KR}, \tag{157}$$

where $J_n(z)$ are the Bessel function of the first kind of order n. In Fig. 28, we plot the scaled $D_{\rm L}^{(11)}(R, H)$ as a function of R for various H following numerical integration. It is a monotonically decreasing function of R. The circles ($H = 0.1$), squares ($H = 1$) and triangles ($H = 10$) represent the intermediate membrane separations respectively. These symbols represent the same H values in the rest of the figures in this section.

For $H \gg 1$, we can approximate $\coth(KH) \approx 1$ so that the H-dependence drops out. In this limit, the integral can be analytically performed to yield

$$\frac{4\pi\eta D_{\rm L}^{(11)}(R)}{k_{\rm B}T} = -\frac{2}{R^2} + \frac{\pi}{R}\left[\mathbf{H}_1(R) - Y_1(R)\right], \tag{158}$$

which is the same with Eq. (49). This is because the membranes are effectively isolated from each other for large H. For $R \ll 1$, the r.h.s. of

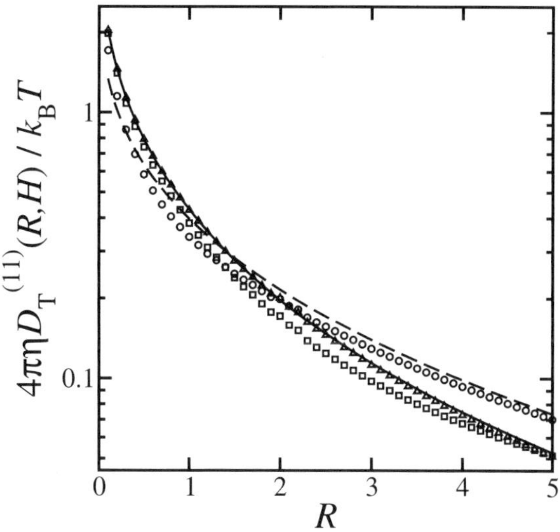

Fig. 29. Scaled $D_{\rm T}^{(11)}(R,H)$ as a function of R for various values of H. The circles, squares and triangles correspond to $H = 0.1$, 1 and 10, respectively. The solid line corresponds to Eq. (160) representing the analytical expression for large H limit. The dashed line corresponds to the small H limit.

Eq. (158) behaves logarithmically, i.e., $\ln(2/R) - \gamma + 0.5$, while in the opposite $R \gg 1$ limit, it decays algebraically as $2/R$ (see also Eqs. (50) and (51)). When $H \ll 1$, on the other hand, Eq. (157) becomes $D_{\rm L}^{(11)}(R/2)/2$. The vanishing thickness of the solvent region ii results in a rescaling of ν^{-1} by a factor of two, and hence the resultant expression. The solid and dashed lines in Fig. 28 represent the above limiting cases of large and small H limits, respectively. It is thus observed that the presence of the second membrane also has a finite contribution to the coupling longitudinal diffusion coefficient within the same membrane.

Following the same argument, the transverse coupling diffusion coefficient between two particles within the same membrane is calculated according to

$$\frac{4\pi\eta D_{\rm T}^{(11)}(R,H)}{k_{\rm B}T} = 2\int_0^\infty {\rm d}K\, \frac{1+2K+\coth(KH)}{g(K,H)} \times \left[J_0(KR) - \frac{J_1(KR)}{KR}\right]. \tag{159}$$

In Fig. 29, the variation of the scaled $D_{\rm T}^{(11)}(R,H)$ as a function of R for various H is plotted using numerical integration. Similar to the longitudinal

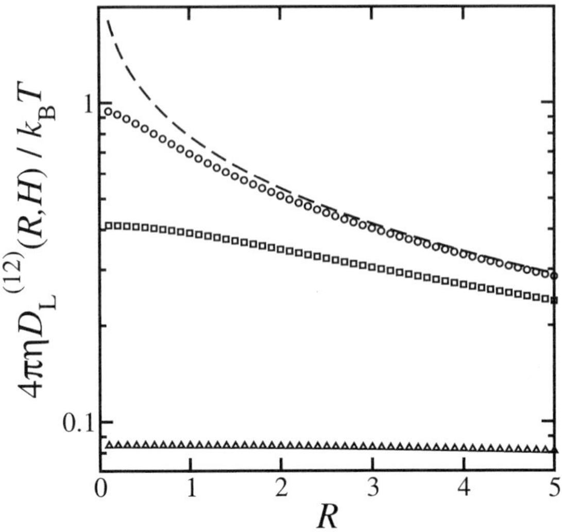

Fig. 30. Scaled $D_{\rm L}^{(12)}(R,H)$ as a function of R for various values of H. The circles, squares and triangles correspond to $H = 0.1$, 1 and 10, respectively. The dashed line corresponds to the small H limit.

case, it is also a monotonically decreasing function of R. For $H \gg 1$, Eq. (159) can be analytically treated to have the form

$$\frac{4\pi\eta D_{\rm T}^{(11)}(R)}{k_{\rm B}T} = \frac{2}{R^2} - \frac{\pi}{R}\left[\mathbf{H}_1(R) - Y_1(R)\right] + \pi\left[\mathbf{H}_0(R) - Y_0(R)\right], \quad (160)$$

which also coincides with the expression for the transverse coupling diffusion coefficient in a single membrane.[18] In this case, the asymptotic behaviors of the r.h.s. of Eq. (160) for small and large R are $\ln(2/R) - \gamma - 0.5$ and $2/R^2$, respectively. In the opposite limit of $H \ll 1$, Eq. (159) becomes $D_{\rm T}^{(11)}(R/2)/2$. As before, the solid and dashed lines in Fig. 29 show the limiting cases of large and small H, respectively.

Now we proceed to calculate the coupling diffusion coefficients of two particles in different membranes. In this case, the longitudinal coupling diffusion coefficient is

$$\frac{4\pi\eta D_{\rm L}^{(12)}(R,H)}{k_{\rm B}T} = 2\int_0^\infty {\rm d}K\, \frac{{\rm cosech}(KH)}{g(K,H)}\, \frac{J_1(KR)}{KR}. \quad (161)$$

The functional dependence of $D_{\rm L}^{(12)}(R,H)$ on R for various H is shown in Fig. 30. Since ${\rm cosech}(KH) \approx 0$ for $H \gg 1$, the above integral vanishes for large inter-membrane distances as seen by the triangles ($H = 10$) in Fig. 30.

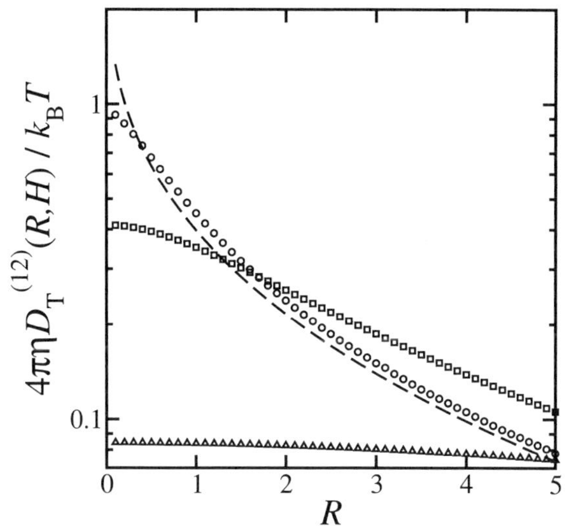

Fig. 31. Scaled $D_{\rm T}^{(12)}(R, H)$ as a function of R for various values of H. The circles, squares and triangles correspond to $H = 0.1$, 1 and 10, respectively. The dashed line corresponds to the small H limit.

This is a reasonable result as the membranes are effectively independent of each other. In the limit of $H \ll 1$, Eq. (161) results in $D_{\rm L}^{(11)}(R/2)/2$ which is plotted by the dashed line in Fig. 30. As mentioned earlier, this function has an initial logarithmic behavior followed by an asymptotic $1/R$-decay.

The transverse coupling diffusion coefficient between two particles in different membranes is given by

$$\frac{4\pi\eta D_{\rm T}^{(12)}(R,H)}{k_{\rm B}T} = 2\int_0^{\infty} {\rm d}K\, \frac{{\rm cosech}(KH)}{g(K,H)} \left[J_0(KR) - \frac{J_1(KR)}{KR}\right]. \quad (162)$$

In Fig. 31, the variation of $D_{\rm T}^{(12)}(R, H)$ as a function of R for various H is plotted. The above integral also vanishes when $H \gg 1$ as expected for the decoupled membranes. In the opposite limit of $H \ll 1$, Eq. (162) results in $D_{\rm T}^{(11)}(R/2)/2$ as plotted by the dashed line in Fig. 31.

From Figs. 30 and 31, it can be observed that for large H, the coupling diffusion coefficients vanish. This is due to the exponential decay of the cosech(KH) term. Up to $H = 1$, the proximity to the second membrane leads to additional contributions to the coupling diffusion coefficients. As estimated before, the hydrodynamic screening length ν^{-1} can be estimated to be of the order of μm. This implies that an adjacent membrane within this distance ($H < 1$) can have a strong bearing on the diffusion dynamics such as for typical stacked supported membrane experiments.[97] Qualitatively, the presence of the second membrane can enhance the effective coupling diffusion.

8. Conclusions

This chapter dealt with hydrodynamic effects on multicomponent fluid membranes. In Sec. 3, the hydrodynamic effects on the dynamics of critical concentration fluctuations in two-component fluid membranes were investigated. Based on the Ginzburg–Landau approach with full hydrodynamics, the wavenumber dependence of the effective diffusion coefficient by changing the temperature and/or the thickness of the bulk fluid was calculated.

In Sec. 4, the effects of an embedding bulk solvent on the phase separation dynamics in a planar fluid membrane using dissipative particle dynamics simulations was presented. We showed that the presence of a bulk fluid will alter the domain growth exponent from that of 2D to 3D indicating the significant role played by the membrane-solvent coupling. We showed that quasi-2D phase separation proceeds by the Brownian coagulation mechanism which reflects the 3D nature of the bulk solvent.

In Sec. 5, using a hydrodynamic theory that incorporated a momentum decay mechanism, we discussed the diffusion coefficient of a liquid domain of finite viscosity moving in a fluid membrane. An analytical expression for the drag coefficient covering the whole range of domain sizes was derived.

In Sec. 6, a Brownian dynamics theory with full hydrodynamics for a Gaussian polymer chain embedded in a liquid membrane which is surrounded by bulk solvent and walls was presented. Within the preaveraging approximation, we obtained the diffusion coefficient of the polymer for the free membrane geometry. A Rouse normal mode analysis was carried out to obtain the relaxation time and the dynamical structure factor. For large polymer size, both quantities show Zimm-like behavior in the free membrane case, whereas they are Rouse-like for the sandwiched membrane geometry.

Finally we discussed the coupled in-plane diffusion dynamics between point-particles embedded in stacked fluid membranes. The contributions to the coupling longitudinal and transverse diffusion coefficients due to particle motion within the different as well as the same membranes was calculated. We showed that the stacked geometry leads to a hydrodynamic coupling between the two membranes.

Even though the models presented in this chapter capture the essential physics, they are simplistic in several respects. The bending stiffness of typical membranes is of the order of $10k_BT$ which is sufficiently large enough to neglect the out-of-plane displacements of the membrane itself. The dynamics of the out-of-plane fluctuations have been previously studied by Levine and MacKintosh.[102] It has also been shown that the out-of-plane fluctuations of the membrane lead to a reduction in the diffusion coefficient of proteins in the single membranes.[103,104] However, it is known that the presence of a substrate or the second membrane would suppress the out-of-plane membrane fluctuations.[105,106] In our treatment, we have also neglected the effects of membrane curvature which can be significant when the radius of curvature becomes close to the hydrodynamic screening length. Recent calculations by Henle *et al.* considered the diffusion of a point object on a spherically closed membrane.[107,108] The extension of this work to finite sized objects is a particularly difficult proposition. We have neglected the finite size effect of the membrane inclusions which are known to modify the membrane response at small inter-particle distances.[18,19] At distances much larger than the inclusion size, however, these effects become unimportant. Overall, we expect that fluctuations and the presence of a substrate will not qualitatively affect our results.

Acknowledgments

We thank D. Andelman, H. Diamant, Y. Fujitani, G. Gompper, T. Hamada, Y. Hirose, T. Kato, S. L. Keller, N. Oppenheimer, Y. Sakuma and K. Seki for useful discussions. This work was supported by KAKENHI (Grant-in-Aid for Scientific Research) on Priority Area "Soft matter physics" and Grant No. 21540420 from the Ministry of Education, Culture, Sports, Science and Technology of Japan. This work was also supported by the JSPS Core-to-Core Program "International research network for non-equilibrium dynamics of soft matter".

Appendix A. Derivation of the General Mobility Tensor

As presented in Fig. 4, we consider a general case where the two walls are located at different distances from the membrane, i.e., $h^+ \neq h^-$. The membrane is assumed to be impermeable.

Our purpose is to derive the in-plane force $\mathbf{f}_s$ on the membrane due to the bulk solvent and walls (see Eq. (4)). We first take the Fourier transform of $\mathbf{v}^{\pm}(\mathbf{r}, z)$ by

$$\mathbf{v}^{\pm}[\mathbf{k}, z] = \int \mathrm{d}\mathbf{r}\, \mathbf{v}^{\pm}(\mathbf{r}, z) \exp(-i\mathbf{k} \cdot \mathbf{r}), \tag{A.1}$$

where $\mathbf{r} = (x, y)$ and $\mathbf{k} = (k_x, k_y)$. The projection of the vector $\mathbf{v}^{\pm}[\mathbf{k}, z]$ on the xy-plane can be expressed as $v_{\parallel}^{\pm}[\mathbf{k}, z]\hat{\mathbf{k}} + v_{\perp}^{\pm}[\mathbf{k}, z]\bar{\mathbf{k}}$ where $\hat{\mathbf{k}} = (k_x/k, k_y/k)$ and $\bar{\mathbf{k}} = (-k_y/k, k_x/k)$ with $k = |\mathbf{k}|$. The subscripts $\parallel$ and $\perp$ indicate the components parallel and perpendicular to $\hat{\mathbf{k}}$, respectively. From Eq. (6), the vertical component $v_{\perp}^{\pm}[\mathbf{k}, z]$ obeys the equation

$$\left(-k^2 + \frac{\partial^2}{\partial z^2}\right) v_{\perp}^{\pm}[\mathbf{k}, z] = 0. \tag{A.2}$$

The solution to the above equation can be written as

$$v_{\perp}^{\pm}[\mathbf{k}, z] = A_1 e^{-kz} + A_2 e^{kz}, \tag{A.3}$$

with unknown coefficients A_1 and A_2. These coefficients are determined by the stick boundary conditions imposed at the membrane-solvent and solvent-wall boundaries:

$$v_{\perp}^{\pm}[\mathbf{k}, z] = \begin{cases} v_{\perp}[\mathbf{k}] & \text{at} \quad z = 0, \\ 0 & \text{at} \quad z = \pm h^{\pm}, \end{cases} \tag{A.4}$$

where $v_{\perp}[\mathbf{k}] = \mathbf{v}[\mathbf{k}] \cdot \bar{\mathbf{k}}$ and see Eq.(9) for the definition of $\mathbf{v}[\mathbf{k}]$. Thus we have

$$v_{\perp}^{\pm}[\mathbf{k}, z] = \mp v_{\perp}[\mathbf{k}] \frac{\sinh(k(z \mp h^{\pm}))}{\sinh(kh^{\pm})}. \tag{A.5}$$

From Eq. (A.5), the in-plane force on the membrane due to the outer solvent and walls becomes

$$\begin{aligned} f_{\mathrm{s}\perp}[\mathbf{k}] &= \mathbf{f}_{\mathrm{s}}[\mathbf{k}] \cdot \bar{\mathbf{k}} \\ &= \frac{\partial}{\partial z}\left(\eta_{\mathrm{s}}^{+} v_{\perp}^{+}[\mathbf{k}, z] - \eta_{\mathrm{s}}^{-} v_{\perp}^{-}[\mathbf{k}, z]\right)_{z=0} \\ &= -\eta_{\mathrm{s}}^{+} v_{\perp} k \coth(kh^{+}) - \eta_{\mathrm{s}}^{-} v_{\perp} k \coth(kh^{-}). \end{aligned} \tag{A.6}$$

Since the incompressibility condition of the membrane fluid implies $v_{\parallel}[\mathbf{k}] = \mathbf{v}[\mathbf{k}] \cdot \mathbf{k} = 0$, the perpendicular component of the Fourier transform of Eq. (4) gives

$$-\eta k^2 v_{\perp}[\mathbf{k}] + f_{\mathrm{s}\perp}[\mathbf{k}] + F_{\perp}[\mathbf{k}] = 0, \tag{A.7}$$

with $F_{\perp}[\mathbf{k}] = \mathbf{F}[\mathbf{k}] \cdot \bar{\mathbf{k}}$. Hence the mobility tensor defined by $\mathbf{v}[\mathbf{k}] = \mathbf{G}[\mathbf{k}] \cdot \mathbf{F}[\mathbf{k}]$ is given by Eq. (11).

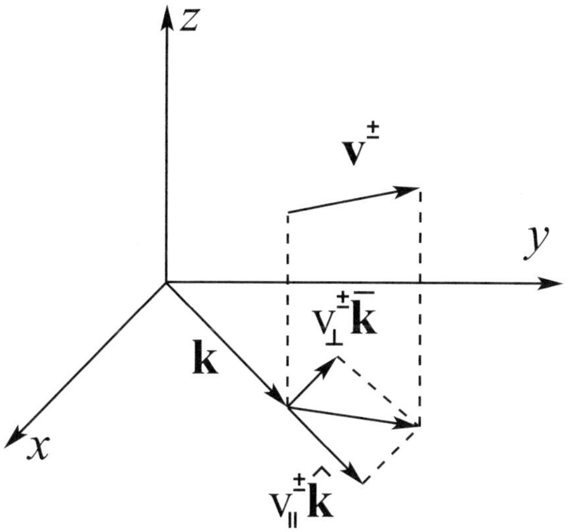

Fig. A.1. Projection of $\mathbf{v}^{\pm}[\mathbf{k}, z]$ onto the xy-plane which is decomposed as $v_{\parallel}^{\pm}\hat{\mathbf{k}} + v_{\perp}^{\pm}\bar{\mathbf{k}}$.

Appendix B. Mobility Tensors in Real Space

B.1. Free membrane case

The mobility tensor in Fourier space is given by Eq. (13). The expression for $G_{\alpha\beta}^{\mathrm{SD}}(\mathbf{r})$ can be found by assuming

$$G_{\alpha\beta}^{\mathrm{SD}}(\mathbf{r}) = B_1\delta_{\alpha\beta} + B_2\frac{r_\alpha r_\beta}{r^2}, \tag{B.1}$$

with two coefficients B_1 and B_2. By considering the diagonal and off-diagonal parts of Eq. (B.1) separately, we have

$$2B_1 + B_2 = \frac{1}{2\pi\eta}\int_0^\infty \mathrm{d}k\,\frac{J_0(kr)}{k+\nu} = \frac{1}{4\eta}\left[\mathbf{H}_0(\nu r) - Y_0(\nu r)\right], \tag{B.2}$$

and

$$B_1 + B_2 = \frac{1}{2\pi\eta}\int_0^\infty \mathrm{d}k\,\frac{J_1(kr)}{kr(k+\nu)} \tag{B.3}$$

$$= \frac{1}{4\eta}\left[-\frac{2}{\pi\nu^2 r^2} + \frac{\mathbf{H}_1(\nu r)}{\nu r} - \frac{Y_1(\nu r)}{\nu r}\right]. \tag{B.4}$$

See Ref. 109 for the integral of Eq. (B.2). In evaluating the integral of Eq. (B.4), we have made use of the following relations;

$$\frac{J_1(z)}{z} = J_0(z) + \frac{\mathrm{d}^2}{\mathrm{d}z^2} J_0(z), \tag{B.5}$$

$$\int_0^\infty \mathrm{d}z\, J_0(z) \frac{\mathrm{d}^2}{\mathrm{d}z^2} \frac{1}{z+a} = \frac{\mathrm{d}^2}{\mathrm{d}a^2} \int_0^\infty \mathrm{d}z\, \frac{J_0(z)}{z+a}. \tag{B.6}$$

Solving Eqs. (B.2) and (B.4) for B_1 and B_2, we get

$$B_1 = \frac{1}{4\eta}\left[\mathbf{H}_0(\nu r) - Y_0(\nu r) + \frac{2}{\pi\nu^2 r^2} - \frac{\mathbf{H}_1(\nu r)}{\nu r} + \frac{Y_1(\nu r)}{\nu r}\right], \tag{B.7}$$

$$B_2 = \frac{1}{4\eta}\left[-\frac{4}{\pi\nu^2 r^2} + \frac{2\mathbf{H}_1(\nu r)}{\nu r} - \frac{2Y_1(\nu r)}{\nu r} - \mathbf{H}_0(\nu r) + Y_0(\nu r)\right]. \tag{B.8}$$

The preaveraging of Eq. (B.1) yields

$$g^{\mathrm{SD}}(n-m) = \frac{1}{2}\langle 2B_1 + B_2 \rangle. \tag{B.9}$$

Hence we obtain Eq. (146) using Eq. (B.2).

B.2. Confined membrane case

Now the mobility tensor in the Fourier space is Eq. (16). Similar to the free membrane case, its real space expression can be written as

$$G^{\mathrm{ES}}_{\alpha\beta}(\mathbf{r}) = C_1 \delta_{\alpha\beta} + C_2 \frac{r_\alpha r_\beta}{r^2}, \tag{B.10}$$

with two coefficients C_1 and C_2. Then it follows that

$$2C_1 + C_2 = \frac{1}{2\pi\eta} \int_0^\infty \mathrm{d}k\, \frac{kJ_0(kr)}{k^2+\kappa^2} = \frac{1}{2\pi\eta} K_0(\kappa r), \tag{B.11}$$

and

$$C_1 + C_2 = \frac{1}{2\pi\eta} \int_0^\infty \mathrm{d}k\, \frac{J_1(kr)}{r(k^2+\kappa^2)} = \frac{1}{2\pi\eta}\left[\frac{1}{\kappa^2 r^2} - \frac{K_1(\kappa r)}{\kappa r}\right]. \tag{B.12}$$

Solving these equations, we obtain

$$C_1 = \frac{1}{2\pi\eta}\left[K_0(\kappa r) + \frac{K_1(\kappa r)}{\kappa r} - \frac{1}{\kappa^2 r^2}\right], \tag{B.13}$$

$$C_2 = \frac{1}{2\pi\eta}\left[-K_0(\kappa r) - \frac{2K_1(\kappa r)}{\kappa r} + \frac{2}{\kappa^2 r^2}\right]. \tag{B.14}$$

References

1. R. Lipowsky and E. Sackmann, *Structure and Dynamics of Membranes* (Elsevier, Amsterdam, 1995).
2. K. Simons and E. Ikonen, Functional rafts in cell membranes, *Nature* **387**, 569–572 (1997).
3. S. L. Veatch and S. L. Keller, Miscibility phase diagrams of giant vesicles containing sphingomyelin, *Phys. Rev. Lett.* **94**, 148101 (2005).
4. S. L. Veatch and S. L. Keller, Separation of liquid phases in giant vesicles of ternary mixtures of phospholipids and cholesterol, *Biophys. J.* **85**, 3074–3083 (2003).
5. M. Yanagisawa, M. Imai, T. Masui, S. Komura, and T. Ohta, Growth dynamics of domains in ternary fluid vesicles, *Biophys. J.* **92**, 115–125 (2007).
6. A. R. Honerkamp-Smith, P. Cicuta, M. Collins, S. L. Veatch, M. Schick, M. den Nijs, and S. L. Keller, Line tensions, correlation lengths and critical exponents in lipd membranes near critical points, *Biophys. J.* **95**, 236–246 (2008).
7. S. L. Veatch, P. Cicuta, P. Sengupta, A. Honerkamp-Smith, D. Holowka, and B. Baird, Critical fluctuations in plasma membrane vesicles, *ACS Chem. Biol.* **3**, 287–293 (2008).
8. G. Gompper and M. Schick, *Phase Transitions and Critical Phenomena* Vol. 16 (Academic, New York, 1994).
9. P. G. Saffman, Brownian motion of thin sheets of viscous fluid, *J. Fluid Mech.* **73**, 593–602 (1976).
10. P. G. Saffman and M. Delbrück, Brownian motion in biological membranes, *Proc. Natl. Acad. Sci. USA* **72**, 3111–3113 (1975).
11. B. D. Hughes, B. A. Pailthorpe, and L. R. White, The translational and rotational drag on a cylinder moving in a membrane, *J. Fluid Mech.* **110**, 349–372 (1981).
12. E. Evans and E. Sackmann, Translational and rotational drag coefficients for a disk moving in a liquid membrane associated with a rigid substrate, *J. Fluid Mech.* **194**, 553–561 (1988).
13. T. Izuyama, *Dynamics of Ordering Processes in Condensed Matter* (Plenum, New York, 1988).
14. B. Alberts, A. Johnson, P. Walter, J. Lewis, and M. Raff, *Molecular Biology of the Cell* (Garland Science, New York, 2008).
15. Th. M. Fischer, The drag on needles moving in a Langmuir monolayer, *J. Fluid Mech.* **498**, 123–137 (2004).
16. D. K. Lubensky and R. E. Goldstein, Hydrodynamics of monolayer domains at the air-water interface, *Phys. Fluids* **8**, 843–854 (1996).
17. K. Inaura and Y. Fujitani, Concentration fluctuation in a two-component fluid membrane surrounded with three-dimensional fluids, *J. Phys. Soc. Jpn.* **77**, 114603 (2008).
18. N. Oppenheimer and H. Diamant, Correlated diffusion of membrane proteins and their effect on membrane viscosity, *Biophys. J.* **96**, 3041–3049 (2009).

19. N. Oppenheimer and H. Diamant, Correlated dynamics of inclusions in a supported membrane, *Phys. Rev. E* **82**, 041912 (2010).
20. H. Stone and A. Ajdari, Hydrodynamics of particles embedded in a flat surfactant layer overlying a subphase of finite depth, *J. Fluid Mech.* **369**, 151–173 (1998).
21. S. Komura and K. Seki, Diffusion constant of a polymer chain in biomembranes, *J. Phys. II* **5**, 5–9 (1995).
22. K. Seki and S. Komura, Brownian dynamics in a thin sheet with momentum decay, *Phys. Rev. E* **47**, 2377–2383 (1993).
23. K. Seki, S. Komura, and M. Imai, Concentration fluctuations in binary fluid membranes, *J. Phys.: Condens. Matter* **19**, 072101 (2007).
24. S. Ramachandran, S. Komura, M. Imai, and K. Seki, Drag coefficient of a liquid domain in a two-dimensional membrane, *Eur. Phys. J. E* **31**, 303–310 (2010).
25. S. L. Veatch and S. L. Keller, Seeing spots: Complex phase behavior in simple membranes, *Biochim. Biophys. Acta, Mol. Cell Res.* **1746**, 172–185 (2005).
26. D. Saeki, T. Hamada, and K. Yoshikawa, Domain-growth kinetics in a cell-sized liposome, *J. Phys. Soc. Jpn.* **75**, 013602 (2006).
27. S. L. Veatch, O. Soubias, S. L. Keller, and K. Gawrisch, Critical fluctuations in domain forming lipid mixtures, *Proc. Natl. Acad. Sci. USA* **104**, 17650–17655 (2007).
28. J. V. Sengers and J. M. H. L. Sengers, Thermodynamic behavior of fluids near the critical point, *Annu. Rev. Phys. Chem.* **37**, 189–222 (1986).
29. L. Onsager, Crystal statistics: A two-dimensional model with an order-disorder transition, *Phys. Rev* **65**, 117–149 (1944).
30. Y. Tserkovnyak and D. R. Nelson, Conditions for extreme sensitivity of protein diffusion in membranes to cell environments, *Proc. Natl. Acad. Sci. USA* **103**, 15002–15007 (2006).
31. M. Haataja, Critical dynamics in multicomponent lipid membranes, *Phys. Rev. E* **80**, 020902 (2009).
32. E. P. Petrov and P. Schwille, Translational diffusion in lipid membranes beyond the Saffman-Delbrück approximation, *Biophys. J.* **94**, L41–L43 (2008).
33. S. Ramachandran, S. Komura, K. Seki, and M. Imai, Hydrodynamic effects on concentration fluctuations in multicomponent membranes, *Soft Matter* **7**, 1524–1531 (2011).
34. K. Simons and W. L. C. Vaz, Model systems, lipid rafts and cell membranes, *Annu. Rev. Biophys. Biomol. Struct.* **33**, 269–295 (2004).
35. R. Brewster, P. A. Pincus, and S. A. Safran, Hybrid lipids as a biological surface-active component, *Biophys. J.* **97**, 1087–1094 (2009).
36. R. Brewster and S. A. Safran, Line active hybrid lipids determine domain size in phase separation of saturated and unsaturated lipids, *Biophys. J.* **98**, L21–L23 (2010).
37. M. Nonomura and T. Ohta, Decay rate of concentration fluctuations in microemulsions, *J. Chem. Phys.* **110**, 7516–7523 (1999).
38. K. Kawasaki, Kinetic equations and time correlation functions of critical

fluctuations, *Ann. Phys.* **61**, 1–56 (1970).
39. S. Ramachandran, S. Komura, K. Seki, and G. Gompper, Dynamics of a polymer chain confined in a membrane, *Eur. Phys. J. E* **34**, 46 (2011).
40. H. Diamant, Hydroydnamic interaction in confined geometries, *J. Phys. Soc. Jpn.* **78**, 041002 (2009).
41. J. S. Rowlinson and B. Widom, *Molecular Theory of Capillarity* (Dover, New York, 1982).
42. N. Goldenfeld, *Lectures on Phase Transitions and the Renormalization Group* (Addison-Wesley, New York, 1993).
43. L. L. M. van Deenen, Chemistry of phospholipids in relation to biological membranes, *Pure Appl. Chem.* **25**, 25–56 (1971).
44. G. van Meer, D. Voelker, and G. Feigenson, Membrane lipids: Where they are and how they behave, *Nature Rev. Mol. Cell. Biol.* **9**, 112–124 (2008).
45. T. Yamamoto, R. Brewster, and S. A. Safran, Chain ordering of hybrid lipids can stabilize domains in saturated/hybrid/cholesterol lipid membranes, *Europhys. Lett.* **91**, 28002 (2010).
46. Y. Hirose, S. Komura, and D. Andelman, Coupled modulated bilayers: A phenomenological model, *ChemPhysChem* **10**, 2839–2846 (2009).
47. M. Teubner and R. Strey, Origin of the scattering peak in microemulsions, *J. Chem. Phys.* **87**, 3195–3200 (1987).
48. G. Gompper and M. Schick, Lattice model of microemulsions, *Phys. Rev. B* **41**, 9148–9162 (1990).
49. M. Hennes and G. Gompper, Dynamical behavior of microemulsion and sponge phases in thermal equilibrium, *Phys. Rev. E* **54**, 3811–3831 (1994).
50. S. Mayor and M. Rao, Rafts: Scale dependent, active lipid organization at the cell surface, *Traffic* **5**, 231–240 (2004).
51. J. Fan, M. Sammalkorpi, and M. Haataja, Formation and regulation of lipid microdomains in cell membranes: Theory, modeling and speculation, *FEBS Lett.* **584**, 1678–1684 (2010).
52. A. J. Bray, Theory of phase-ordering kinetics, *Adv. Phys.* **51**, 481–587 (2002).
53. K. Binder and D. Stauffer, Theory for the slowing down of the relaxation and spinodal decomposition of binary mixtures, *Phys. Rev. Lett.* **33**, 1006–1009 (1974).
54. E. M. Lifshitz and L. P. Pitaevskii, *Physical Kinetics* (Pergamon Press, Oxford, 1981).
55. H. Furukawa, Role of inertia in the late stage of the phase separation of a fluid, *Physica A* **204**, 237–245 (1994).
56. E. D. Siggia, Late stages of spinodal decomposition in binary mixtures, *Phys. Rev. A* **20**, 595–605 (1979).
57. M. S. Miguel, M. Grant, and J. D. Gunton, Phase separation in two-dimensional binary fluids, *Phys. Rev. A* **31**, 1001–1005 (1985).
58. P. Ossadnik, M. F. Gyure, H. E. Stanley, and S. C. Glotzer, Molecular dynamics simulation of spinodal decomposition in a two-dimensional binary fluid mixture, *Phys. Rev. Lett.* **72**, 2498 (1994).
59. G. Gonnella, E. Orlandini, and J. M. Yeomans, Phase separation in two-

dimensional fluids: The role of noise, *Phys. Rev. E* **59**, R4741–R4744 (1999).
60. R. D. Groot and P. B. Warren, Dissipative particle dynamics: Bridging the gap between atomistic and mesoscopic simulation, *J. Chem. Phys.* **107**, 4423–4435 (1997).
61. P. Espanol and P. Warren, Statistical mechanics of dissipative particle dynamics, *Europhys. Lett.* **30**, 191 (1995).
62. M. Laradji and P. B. S. Kumar, Domain growth, budding and fission in phase-separating self-assembled fluid bilayers, *J. Chem. Phys.* **123**, 224902 (2005).
63. W. Pan, D. A. Fedosov, G. E. Karniadakis, and B. Caswell, Hydrodynamic interactions for single dissipative particle dynamics particles and their clusters and filaments, *Phys. Rev. E* **78**, 046706 (2008).
64. P. Cicuta, S. L. Keller, and S. L. Veatch, Diffusion of liquid domains in lipid bilayer membranes, *J. Phys. Chem. B* **111**, 3328–3331 (2007).
65. V. Prasad and E. R. Weeks, Two-dimensional to three-dimensional transition in soap films demonstrated by microrheology, *Phys. Rev. Lett.* **102**, 178302 (2009).
66. J. Fan, T. Han, and M. Haataja, Hydrodynamic effects on spinodal decomposition kinetics in planar lipid bilayer membranes, *J. Chem. Phys.* **133**, 235101 (2010).
67. G. Orädd, P. W. Westerman, and G. Lindblom, Lateral diffusion coefficients of separate lipid species in a ternary raft-forming bilayer: A PFG-NMR multinuclear study, *Biophys. J.* **89**, 315–320 (2005).
68. L. D. Landau and E. M. Lifshitz, *Fluid Mechanics* (Pergamon Press, Oxford, 1987).
69. H. Lamb, *Hydrodynamics* (Cambridge University Press, New York, 1975).
70. C. Esposito, A. Tian, S. Melamed, C. Johnson, S.-Y. Tee, and T. Baumgart, Flicker spectroscopy of thermal lipid bilayer domain boundary fluctuations, *Biophys. J.* **93**, 3169–3181 (2007).
71. E. K. Mann, S. Hénon, D. Langevin, J. Meunier, and L. Léger, Hydrodynamics of domain relaxation in a polymer monolayer, *Phys. Rev. E* **51**, 5708–5720 (1995).
72. R. Peters and R. J. Cherry, Lateral and rotational diffusion of bacteriorhodopsin in lipid bilayers: Experimental test of Saffman-Delbrück equations, *Proc. Natl. Acad. Sci. USA* **79**, 4317–4321 (1982).
73. E. A. J. Reitz and J. J. Neefjes, From fixed to FRAP: Measuring protein mobility and activity in living cells, *Nat. Cell Biol.* **3**, E145–E147 (2001).
74. N. Kahya, E.-I. Pécheur, W. P. de Boeij, D. A. Wiersam, and D. Hoekstra, Reconstitution of membrane proteins into giant unilamellar vesicles via peptide-induced fusion, *Biophys. J.* **81**, 1464–1474 (2001).
75. N. Tsapis, F. Reiss-Husson, R. Ober, M. Genest, R. S. Hodges, and W. Urbach, Self diffusion and spectral modifications of a membrane protein, the rubrivivax gelatinosus LH2 complex, incorporated into a monoolein cubic phase, *Biophys. J.* **81**, 1613–1623 (2001).
76. Y. Gambin, R. Lopez-Esparza, M. Reffay, E. Sierecki, N. S. Gov, M. Genest, R. S. Hodges, and W. Urbach, Lateral mobility of proteins in liquid

membranes revisited, *Proc. Natl. Acad. Sci. USA* **103**, 2098–2102 (2007).
77. D. F. Kucik, E. L. Olson, and M. P. Sheetz, Weak dependence of mobility of membrane protein aggregates on aggregate size supports a viscous model of retardation of diffusion, *Biophys. J.* **76**, 314–322 (1999).
78. J. Lippincott-Schwartz, E. Snapp, and A. Kenworthy, Studying protein dynamics in living cells, *Nat. Rev. Mol. Cell Biol.* **2**, 444–456 (2001).
79. M. Vrljic, Y. Nishimura, S. Brasselet, W. E. Moerner, and H. M. McConnell, Translational diffusion of individual class II MHC membrane proteins in cells, *Biophys. J.* **83**, 2681–2692 (2002).
80. A. K. Kenworthy, B. J. Nichols, C. L. Remmert, G. M. Hendrix, M. Kumar, J. Zimmerberg, and J. Lippincott-Schwartz, Dynamics of putative raft-associated proteins at the cell surface, *J. Cell. Biol.* **165**, 735–746 (2004).
81. Y. Yang, R. Prudhomme, K. M. McGrath, P. Richetti, and C. M. Marques, Confinement of polysoaps in membrane lyotropic phases, *Phys. Rev. Lett.* **12**, 2729–2732 (1998).
82. B. Maier and J. O. Rädler, Conformation and self-diffusion of single DNA molecules confined to two-dimensions, *Phys. Rev. Lett.* **82**, 1911–1914 (1999).
83. B. Maier and J. O. Rädler, DNA on fluid membranes: A model polymer in two dimensions, *Macromolecules* **33**, 7185–7194 (2000).
84. B. Maier and J. O. Rädler, Shape of self-avoiding walks in two dimensions, *Macromolecules* **34**, 5723–5724 (2001).
85. C. Herold, P. Schwille, and E. P. Petrov, DNA condensation at freestanding cationic lipid bilayers, *Phys. Rev. Lett.* **104**, 148102 (2010).
86. M. Daoud and P. G. de Gennes, Statistics of macromolecular solutions trapped in small pores, *J. Phys.* **38**, 85–93 (1977).
87. F. Brochard, Dynamics of polymer chains trapped in a slit, *J. Phys.* **38**, 1285–1291 (1977).
88. P. G. de Gennes and F. Brochard, Dynamics of confined polymer chains, *J. Chem. Phys.* **67**, 52–56 (1977).
89. P.-K. Lin, C.-C. Fu, Y.-R. Chen, P.-K. Wei, C. H. Kuan, and W. S. Fann, Static conformations and dynamics of a single DNA molecules confined in nanoslits, *Phys. Rev. E* **76**, 011806 (2007).
90. M. Muthukumar, Brownian dynamics of polymer chains in membranes, *J. Chem. Phys.* **82**, 5696–5706 (1985).
91. D. L. Ermak and J. A. McCammon, Brownian dynamics with hydrodynamic interactions, *J. Chem. Phys.* **69**, 1352–1360 (1978).
92. M. Doi and S. F. Edwards, *The Theory of Polymer Dynamics* (Clarendon Press, Oxford, 1986).
93. B. H. Zimm, Dynamics of polymer molecules in dilute solution: Viscoelasticity, flow birefringence and dielectric loss, *J. Chem. Phys.* **24**, 269–279 (1956).
94. W. H. Stockmayer and B. Hammouda, Quasi-elastic light scattering as a diagnostic of single chain dynamics, *Pure Appl. Chem.* **56**, 1373–1378 (1984).
95. W. Feller, *An Introduction to Probability Theory and its Applications* (Wiley, New York, 1968).

96. M. Abramowitz and I. A. Stegun, *Handbook of Mathematical Functions* (Dover, New York, 1972).
97. Y. Kaizuka and J. T. Groves, Structure and dynamics of supported intermembrane junctions, *Biophys. J.* **86**, 905–912 (2004).
98. M. Rubinstein and R. H. Colby, *Polymer Physics* (University Press, Oxford, 2004).
99. S. Ramachandran and S. Komura, Hydrodynamic coupling of two fluid membranes, *J. Phys.: Condens. Matter* **23**, 072205 (2011).
100. J. T. Groves, Bending mechanics and molecular organization in biological membranes, *Annu. Rev. Phys. Chem.* **58**, 697–717 (2007).
101. S. Ramachandran, S. Komura, and G. Gompper, Effects of an embedding bulk fluid on phase separation dynamics in a thin liquid film, *Europhys. Lett.* **89**, 56001 (2010).
102. A. J. Levine and F. C. MacKintosh, Dynamics of viscoelastic membranes, *Phys. Rev. E* **66**, 061606 (2002).
103. A. Naji, P. J. Atzberger, and F. L. H. Brown, Hybrid elastic and discrete-particle approach to biomembrane dynamics with application to the mobility of curved integral membrane proteins, *Phys. Rev. Lett.* **102**, 138102 (2009).
104. E. Reister-Gottfried, S. M. Leitenberger, and U. Seifert, Diffusing proteins on a fluctuating membrane: Analytical theory and simulations, *Phys. Rev. E* **81**(3), 031903 (2010).
105. S. Sankararaman, G. I. Menon, and P. B. S. Kumar, Two-component fluid membranes near repulsive walls: Linearized hydrodynamics of equilibrium and nonequilibrium states, *Phys. Rev. E* **66**, 031914 (2002).
106. N. Gov, A. G. Zilman, and S. Safran, Hydrodynamics of confined geometries, *Phys. Rev. E* **70**, 011104 (2004).
107. M. L. Henle, R. McGotry, A. B. Schofield, A. D. Dinsmore, and A. J. Levine, The effect of curvature and topology on membrane hydrodynamics, *Europhys. Lett.* **84**, 48001 (2008).
108. M. L. Henle and A. J. Levine, Hydrodynamics in curved membranes: The effect of geometry on particulate mobility, *Phys. Rev. E* **81**, 011905 (2010).
109. I. S. Gradshteyn and I. M. Ryzhik, *Table of Integrals, Series and Products* (Academic Press, London, 1994).

Chapter 6

Actively Twisted Polymers and Filaments in Biology

Hirofumi Wada[1*] and Roland R. Netz[2†]

[1]*Yukawa Institute for Theoretical Physics, Kyoto University, Kyoto 606-8502, Japan*

[2]*Fachbereich Physik, Freie Universität Berlin, Arnimallee 14, 14195 Berlin, Germany*

In this chapter, we study nonequilibrium twist dynamics of rotationally driven semiflexible polymers and filaments in a viscous fluid using analytical and numerical approaches. A specific yet versatile system that we study is a uniform elastic rod or polymer that is forced to axially rotate at one end at a given frequency with the other end free. We begin this chapter by presenting an overview of the kinematics and linear elasticity theory for thin elastic filaments, and briefly describe the hydrodynamic Brownian dynamics simulation method which allows us to study the nonlinear dynamics of such driven filaments and polymers. Various interesting properties, including the twirling-whirling transition for stiff rods and the plectoneme transition for semiflexible polymers, are discussed and analyzed in terms of their geometry and energetics by combining exact analysis, scaling arguments and numerical simulations. Biological implications are extensively discussed in relation to in-vivo DNA dynamics during transcription and nuclesome transformation.

1. Introduction

1.1. *Physical properties of semiflexible polymers*

Polymers that possess finite mechanical resistance against bending and twisting (as well as stretching) perturbations are known as semiflexible or stiff chains. Physical properties of semiflexible polymers are distinct from those of their flexible counterparts such as freely-jointed chains.[1–3]

*hwada@yukawa.kyoto-u.ac.jp

†netz@ph.tum.de

Competition between mechanical and entropic effects naturally leads to the introduction of bend and twist persistence lengths, fundamental length scales that characterize stiffness of the polymer in a thermal environment. For example, the bend persistence length ℓ_P of a polymer with bending modulus A is defined as the length at which the mechanical bending moment of curvature $\sim 1/\ell_P$ becomes comparable to k_BT, i.e., $A/\ell_P = k_BT$, giving[1–3]

$$\ell_P = \frac{A}{k_BT}. \tag{1}$$

Similarly, the twist persistence length ℓ_T of the polymer with twisting modulus C may be defined by

$$\ell_T = \frac{C}{k_BT}. \tag{2}$$

This indicates that an aixal torque to twist the polymer over the distance of ℓ_T is equal to k_BT. For typical synthetic and biological polymers, those persistence lengths are widely distributed from several nano to even few millimeters: For double-stranded DNA polymers, $\ell_P \approx 50$ nm, and $\ell_T \approx 75 - 100$ nm at physiological salt concentrations (10-100 mM NaCl). For filamentous protein assemblies such as microtubules, on the other hand, bend persistence length can reach a few millimeters.[4] At length scales smaller than the bend persistence length ℓ_P, a polymer chain can be viewed as a rather stiff rod. At length scales much larger than ℓ_P, chain conformations are fully randomized by thermal fluctuations, and the statistical behavior of the chain is well described by that of flexible chains. For the particular example of ds DNA filaments, many important processes in cellular environments, such as DNA compaction into hierarchical chromosomal order and various interactions with motor- and enzymatic proteins, involve length scales comparable to the DNA persistence length.[5] Therefore, the semiflexibility of DNA polymers is indeed crucial for understanding regulatory mechanisms of those vital processes from a physical point of view.

Motivated by those general considerations, physical properties of semiflexible polymers have received continuous interest and attention from biophysicists for the last two decades. In fact, it has been confirmed many times that semiflexible polymers constitute a minimal yet realistic mechanical model for biological macromolecules such as DNA, filamentous proteins, and their supramolecular assembly like bacterial flagellar filaments.[4–7] Various equilibrium properties of DNA as a semiflexible polymer, such as the scaling properties of radius of gyration,[2,3,8] correlation properties characterizing chain conformations,[9,10] first-order nature of collapsing transitions,[11] and unique folding morphologies driven by multivalent ions[12–16] or confinements ("DNA packaging"),[17–19] as well as mechanical properties like the force-extension relationship in the random coil state[20–23] and in the condensed state,[24,25] have been revealed by experimental, analytical, and various numerical approaches. Theoretical approaches are usually based on the Kratky-Porod wormlike chain model that describes a locally inextensible and fluctuating curve with bending stiffness,[26,27] i.e., a minimal theoretical model of semiflexible polymers. Various static and dynamic properties predicted from the worm like chain model have been successfully tested through detailed comparison with experimental data, obtained from newly emerged micromanipulation techniques[28,29] employing optical tweezers[30,31], magnetic beads[32] and atomic force microscopy (AFM).[33,34] Theories were also generalized in several ways including extensibility,[35,36] chains with multiple stable states in bond lengths,[37] ribbon-shaped chains,[38–40] and chains with "subelastic" bending energy.[41]

The reason for the success of the simple worm like chain model is, that bending elasticity typically dominates the shape dynamics of biopolymers, as has been recognized in detailed theoretical and experimental studies.[42–46] On the other hand, the importance of twist elasticity in DNA filaments has been realized in the context of DNA supercoiling, that plays an important role in many cellular processes including regulation of gene expressions.[47] Figure 1 shows the AFM images of plasmid DNA that is originally negatively supercoiled but displays a positively supercoiled structure in presence of intercalating anticancer drugs.[48] It is also possible today to induce and control supercoiling of DNA mechanically in single molecule experiments.[49–52] Driven by those experiments where both stretching and twisting of single biopolymers could be controlled, theories were generalized to include twisting effects as well and provide quantitative agreement on the static level.[53–61]

The double helical nature of ds-DNA backbone provides a number of unique mechanical features such as twist-bend[62] and twist-stretch coupling.[63–65] However, even when the DNA polymer is simply modeled as a *uniform* elastic rod neglecting its backbone double helical structure, a number of interesting features still arise from the geometric and entropic interplay between bending and twisting.[66] At length scales relevant to biopolymers (typically nano to micro meters), inertial effects of the surrounding solvent fluids can be ignored, and the motion of a polymer is governed by

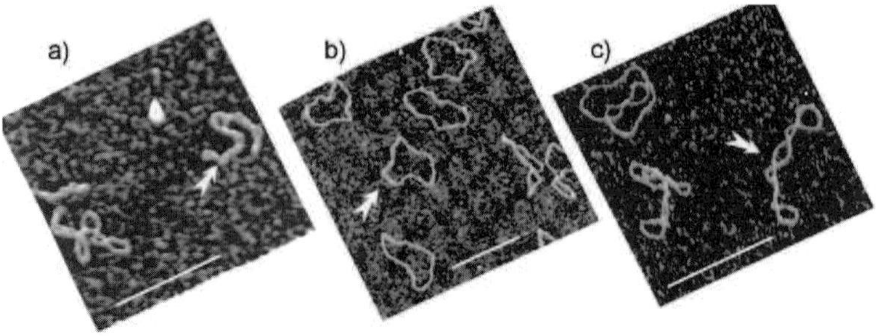

Fig. 1. Atomic force microscopy images (in air) of pUC19 plasmid DNA (2686 bp, total length 913 nm), deposited at 5 mM of $MgCl_2$ in 10 mM Tris/Trizma, pH 7.2 at room temperature, in presence of Daunorubicin (Dau) (an anticancer drug that intercalates with DNA and reduces the twist) at increasing concentrations from (a) to (c). (a): No Dau, all DNA are negatively supercoiled. (b) Dau of 3.6 μM: DNA in the relaxed state or with just a few crossings due to the reduction of twist. (c) 37 μM Dau induced DNA positive supercoiling. The scale bar represents 500 nm. V. Víglasky, F. Valle, J. Adamcik, I. Joab, D. Podhradsky and G. Dietler, *Electrophoresis* (2003) **24**, 1703-1711. (Ref.[48]) Copyright Wiley-VCH Verlag GmbH & Co. KGaA. Reproduced with permission.

the balance between elastic forces and viscous drag forces from the fluid.[1] Linear equations for the viscous bend and twist dynamics of semiflexible polymers have been derived and analyzed previously in detail.[43,67–70] Fluorescence correlation spectroscopy results on DNA monomer kinetics[71] or fluorescent images of isolated actin filaments in solution[42] have confirmed those theoretical predictions.

1.2. *Twist dynamics in biology: Transcription driven instability in DNA*

The in-vivo functioning of DNA often involves non-equilibrium and nonlinear dynamic twist effects that mostly have to do with the activity of various DNA-processing proteins.[47,72–74] In replication, the process by which a DNA chain is duplicated into two identical daughter strands, the helical nature of DNA requires daughter strands to unwind and thus the mother strand to rotate around its axis. Whereas this large scale motion has been considered a conceptual obstacle, it was resolved by Levinthal and Crane who demonstrated by a simple calculation that the viscous torque associated with simple axial-spinning of DNA (like in a speedometer cable) is rather small compared to typical biological free energies.[75]

In transcription, the process by which the DNA informational content is copied into a continuously growing RNA chain, it was suggested that a long nascent RNA chain might (either due to its own hydrodynamic friction or via anchoring to some other cellular component) provide enough rotational resistance to force the DNA strand to rotate around its own axis (see Fig. 2). As the transcription complex including the RNA polymerase (RNAP) moves along the DNA, positive torsional stress builds up ahead of the complex, and negative torsional stress behind the complex. This "twin supercoiled-domain model" was proposed by Liu and Wang.[76] Single-molecule experiments confirmed DNA rotation consistent with its helical pitch during transcription by *E. Coli* RNA polymerase.[77] The step size of RNA polymerase is thought to be nearly 1 bp, so DNA rotates by a full 2π turn in 10.5 steps of RNAP. Since transcription of one full turn is completed in 0.1 sec, thus 100 bp is transcribed in 1 sec, axial rotation rate (driving frequency) of DNA at the cranking point of RNAP amounts to $\omega \approx 2\pi \times 10/\text{sec} \approx 60$ rad/sec. Liu and Wang pointed out that, for DNA supercoiling to happen, the template DNA has to be anchored to some other cellular machinery and has to be prevented to rotate, because for an unanchored DNA, the hydrodynamic rotational torque seems to be too small to induce any structural change in DNA, as considered in the context of DNA replications. To be more precise, when a rigid rod of length L and of diameter a is rotated about its long axis at frequency ω in a fluid of viscosity η, the viscous drag torque acting on this rod is given by

$$M = \pi \eta a^2 \omega L, \tag{3}$$

where $a = 2$ nm is the diameter of DNA, and $\eta = 10^{-3}$ Pa·s is the water viscosity. For a linear DNA strand of length $L \approx 1\ \mu$m that is rotated at one end with frequency $\omega \approx 60$ rad/sec, the maximal viscous torque estimated from Eq. (3) is $M \approx 10^{-4} k_B T$, much smaller than free energies to induce any structural change in DNA, as pointed out by Nelson before.[78] Previous experimental in-vivo and in-vitro studies did confirm large degrees of DNA supercoiling upon transcription, being positive in front and negative behind the transcriptional complex.[76,79–81]

However, this cannot be the whole story. While many in-vivo and in-vitro experiments observed the transcription-driven DNA supercoiling, they did so not only for anchored DNA template, but also for topologically open DNA template (i.e., even when the two ends of the DNA chain were free to rotate).[82–84] This finding is clearly at odds with the above-mentioned simple axial-spinning scenario,[75] since the rotational friction of bare DNA cannot be large enough to develop sufficient torsional stress needed to induce supercoiling. In an effort to reconcile the conflicting experimental results, Phil Nelson argued that this axial spinning scenario may be misleading, and proposed the idea that DNA intrinsic bends can increase the rotational friction enough to induce supercoiling.[78] However, an experiment by Stupina and Wang[85] concluded that DNA several kb in length cannot provide enough rotational friction to prevent the diffusional merging of oppositely supercoiled domains, even when controlled DNA static bends were introduced. A single molecule DNA unzipping experiment looked at the rotational friction of double-stranded DNA up to a very rapid rate of 2,000 turns per sec,[86] but failed to observe the static-bend-induced anomaly of the rotational friction. Although previous experimental tests did not find any direct evidence of the static-kink induced supercoiling, it is suggested that a protein stabilized bend can serve as a significant barrier against DNA rotational diffusion. So far, the issue of the rotational friction of a semiflexible chain is still a matter of interest and debate. Several studies did not provide a clear-cut answer as to what the influence of static DNA bends is and whether other factors present in in-vivo studies are needed to induce substantial supercoiling of DNA under twist injection.[78–88]

1.3. *Aims and organization of this chapter*

Even apart from the biological complexity involved in transcription-driven instability in DNA, a fundamental scenario on twist propagation in semiflexible polymers remained largely elusive, except the well-established linear response regime. Motivated by this observation, the purpose of this chapter is to introduce the fundamental scenario of nonlinear twist propagation dynamics in semiflexible polymers far from equilibrium. Excluding any structural disorder such as sequence-dependent static kinks, we concentrate on the simplest model, i.e., a uniform straight elastic rod or a polymer chain that is forced to rotate at one end at frequency ω_0 with the other end free to move and rotate in a zero Reynolds number fluid, see Fig. 3. At exactly zero temperature, there are no thermally-driven shape fluctuations (persistence lengths are infinite), and the dynamics of the rod is determined purely by the balance between the rod elasticity and viscous hydrodynamic forces. This model was first proposed and analyzed by Wolgemuth, Powers and Goldstein.[89] They showed that there is an elastohydrodynamic instability that occurs at a critical driving frequency,[89]

$$\omega_c \sim \frac{A}{\eta a^2 L^2} \sim \frac{k_B T \ell_P}{\eta a^2 L^2}. \tag{4}$$

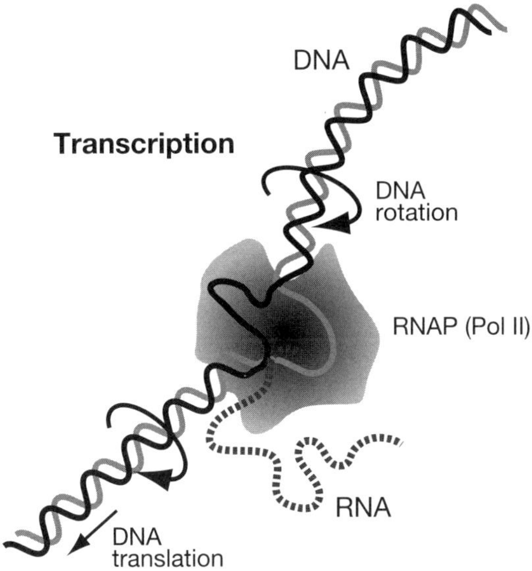

Fig. 2. Schematic illustration of DNA transcription: RNA polymerase II (PoLII) unwinds the DNA double helix in order to read out sequence information, translating along the DNA backbone. Torsional stress is continuously injected at the cranking point of PoLII, which forces DNA strand to rotate about its long axis and may induce certain structural changes in DNA.

For $\omega_0 < \omega_c$ the rod stays straight and undergoes simple axial spinning (*twirling*), while for $\omega_0 > \omega_c$ the rod buckles and displays a combination of axial spinning and rigid-body rotation (*whirling*), see also Fig. 12. Subsequently, a large amplitude whirling state for $\omega_0 > \omega_c$ was found numerically, and the transition from twirling to whirling was shown to be discontinuous[90,91] and to depend on initial rod configurations.[91] Because of the subcritical nature of the bifurcation, at finite temperatures, thermally-driven shape fluctuations of the rod round off the transition and shift the instability to lower frequency.[91]

In the first half of the chapter, we will review these elastohydrodynamic aspects. To get a feeling for the dynamics of elastic filaments driven in a viscous fluid, we begin with scaling arguments on characteristic time scales for stretch, bend and twist relaxations in a thin elastic rod. We then briefly review kinematics of a curve in three-dimensional space, and derive equations of motion for an elastic rod. For this aim, we first define the elastic energy of a slender rod with bend and twist degrees of freedom, and use a variational argument to calculate the elastic translational force and

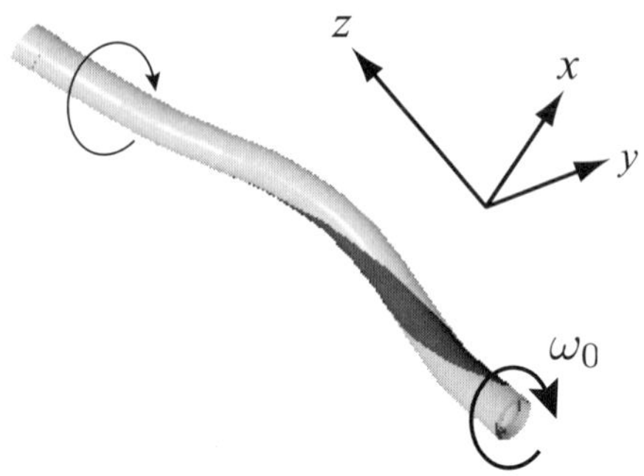

Fig. 3. Schematic picture of our model system: An isotropic elastic rod is forced to rotate at its clamped base at frequency ω_0 with the other end free in a Stokes fluid of viscosity η.

moment about the local tangent per unit length of the rod. Identifying viscous drag force and torque acting on a slender rod in a zero-Reynolds number fluid (there are a few different ways to do this depending on how precisely we describe Stokes flow around the slender rod), we can obtain the equations of motion for the moving rod.

Our arguments in this chapter combine analytical and numerical results. To make this chapter self-contained, we introduce an equivalent discrete chain model using a body-fixed Euler angle frame, and describe quickly our Brownian hydrodynamic simulation method. Using these analytical and numerical tools, we discuss the linear stability analysis for the twirling-whirling shape transition, and numerically construct a stability diagram for this transition. Our numerical approach also reveals a large amplitude whirling motion after the shape transition, where twist is largely relieved via out-of-plane bending (writhing). In this regime, twist transport is strongly nonlinear. We examine geometrical aspects of this phenomenon and show a nontrivial role of writhing on the energetics of this dynamical buckling of the rotating rod. Those geometric arguments are not mathematically new, but help us to intuitively understand how the writhing mechanism transports torsional stress in forced elastic filaments.

In the latter half of this chapter, we consider the opposite limit, where the chain contour length L is much larger than bend and twist persistence lengths, i.e., $L \gg \ell_P, \ell_T$. Remember that the critical torque for twirlingwhiring transition at zero temperature is given by

$$M_c \sim \eta a^2 \omega_c L \approx 8.9 \frac{k_B T \ell_P}{L}, \tag{5}$$

where the numerical prefactor in ω_c given in Eq. (98) was taken into account. For DNA of $L \approx 1\mu$m long, this buckling torque is about $M_c \approx 0.5 k_B T$. Thus chain conformations are dominated by entropic effects at ω_c, and the twirling-whirling instability is washed out by thermal fluctuations for the specific case of DNA. As our main result, we find in addition to the well-known axial-spinning regime, realized for low driving frequency $\omega_0 < \omega_*$, a novel plectoneme dominated regime at high frequencies $\omega_0 > \omega_*$. In this regime, twist is converted locally into writhe (in the form of plectonemes) close to the driven end and then diffuses out to the free end. For sufficiently long chains, the crossover frequency ω_* is much larger than the twirling-whirling threshold ω_c. Quite surprisingly, in the plectoneme regime the overall rotational friction is significantly reduced as compared to the axial-spinning scenario, i.e., the plectoneme dissipation channel transforms the injected twist into writhing motion at very low frictional cost.[92] This mechanism could be biologically relevant since it shows how transcription might produce positively supercoiled DNA structures that could favorably interact with negatively supercoiled nucleosomal structure ahead of the RNA polymerase[74,83,84] even in topologically open systems and at very small energy expenditure.

Our entire discussion presented in this chapter neglects one important feature found in many biological polymers, namely, electrostatic effects. Considering that many biological macromolecules are water soluble (which is important in order to exercise their biological functions in living cells), electrostatic effects often dominate statics and dynamics of biopolymer systems. For example, DNA is a highly charged polyelectrolyte, carrying two negative elementary charges per $\lambda \approx 0.34$ nm. In eukaryotic cells, DNA has a length of about $L \approx 1$ cm, which amounts to a total charge of nearly $Q = (2e)\lambda/L \approx 10^9 e$, and is tightly compacted into chromosomal structures inside the cell nucleus. The fundamental unit of this compacted state is the so-called *nucleosome*, in which a short DNA strand of 146 base pairs is wrapped roughly twice around a positively charged protein (i.e., histon octamer). Therefore, the first step of hierarchical compaction of genomic DNA is mostly driven by attractive Coulombic interactions between oppositely charged DNA and histon proteins.[93] It is also well known that thermodynamic stability of DNA-histon complex structure is crucially affected by the salt concentration of the surrounding medium.[94–96] Whereas numerous interesting phenomena arise from charge effects in biopolymer systems,[3,93–98] we will leave these important issues aside in this chapter and focus only on mechanical aspects in forced DNA dynamics.

2. Viscous Relaxations of Stretch, Bend, and Twist in Elastic Rods

At length and time scales relevant to our discussion (typically nm to several tenth μm), inertial effects are much smaller than viscous effects, and the dynamics of elastic filaments are completely determined by the balance between internal elastic stresses and hydrodynamic viscous drag from the surrounding fluid. This regime is usually called "low-Reynolds number" regime, in which fluid inertia is entirely neglected and the Reynolds number is set exactly zero, Re = 0. In this zero Reynolds number limit, the creeping flow produced by moving rods is described by[99,100]

$$0 = -\nabla p + \eta \nabla^2 \mathbf{v}, \tag{6}$$

where η is the solvent viscosity and the pressure p is determined so that the velocity field $\mathbf{v}$ satisfies the incompressibility condition $\nabla \cdot \mathbf{v} = 0$.

To get a feeling for the dynamic characteristics of moving elastic rods and polymers in such a zero-Reynolds number fluid, let us compare characteristic time scales for stretching, bending, and twisting.[101] Consider a uniform straight rod of radius a and of total length L, with the Young's modulus E, the bending and twisting modulus A and C, embedded in a viscous Newtonian fluid.

Stretching. Let us assume that the rod is initially aligned along the z direction. If it is now subjected to small stretching or compression along its long axis direction, the elastic internal force per unit length is $\sim \pi a^2 E \partial_z^2 u$, where u is the displacement of the cross section of the rod at z and E is the Young's modulus. In the inertialess limit, this elastic restoring force balances with the drag force per unit length of the rod moving with the velocity $v \sim \dot{u}$, where the dot indicates the time derivative $\dot{u} = \partial u / \partial t$ in this chapter. We thus obtain $\eta v \sim \pi a^2 E \partial_z^2 u$, which suggests the characteristic relaxation time for longitudinal stretching or compression of the rod to be given by

$$\tau_{stretch} \sim \left(\frac{L}{a}\right)^2 \frac{\eta}{E}, \tag{7}$$

where we assumed that L defines the length scale of typical stretching deformations.

Twisting. For a straight rod, we can define a rotational angle of its cross section $\phi(z)$ (relative to the equilibrium state) at arbitrary z. When subjected to a certain twisting deformation, an elastic torque through a cross section is $C\partial_z\phi$, thus the resulting elastic force is $C\partial_z^2\phi$. As the rod rotates with a rotational rate $\omega \sim \dot{\phi}$, it experiences a viscous rotational drag per length, $\sim \pi a^2\eta\omega$, which balances with the local elastic force: $\pi a^2\eta\omega \sim C\partial_z^2\phi$. This sets the characteristic time scale for twisting of the rod as

$$\tau_{twist} \sim \frac{\eta a^2 L^2}{C}. \tag{8}$$

Our local balance argument suggests that twist in a slender elastic rod relaxes diffusively.

Bending. In contrast to stretching and twisting, bending or shape deformation of a thin rod relaxes much more slowly. The bending elastic energy of a rod depends on its curvature, its restoring force is proportional to $\sim A\partial_z^4 u_\perp$, where $u_\perp(z)$ describes small deflections of a rod from its straight configuration. Balancing this with the viscous drag force per length, we obtain $\eta\dot{u}_\perp \sim A\partial_z^4 u_\perp$, thus the characteristic bending relaxation time of a fluctuating rod is

$$\tau_{bend} \sim \frac{\eta L^4}{A} \sim \left(\frac{L}{a}\right)^4 \frac{\eta}{E}, \tag{9}$$

where for the second expression, we have used that the bending modulus for a thin elastic rod of radius a is given by $A = \pi a^4 E/4$.[102,103]

Let us now compare these time scales. Noting that the ratio of bending to twisting modulus is related to the Poisson's ratio σ of a material that makes up the rod as $A/C = 1 + \sigma$ and is order of one for typical materials,[102,103] we see

$$\frac{\tau_{stretch}}{\tau_{bend}} \sim \left(\frac{a}{L}\right)^2, \qquad \frac{\tau_{twist}}{\tau_{bend}} \sim \left(\frac{a}{L}\right)^2. \tag{10}$$

These quantities are thus typically very small: for ds-DNA of length 1 μm, we have $(a/L)^2 \sim 10^{-6}$. Therefore, for a freely relaxing rod or polymer in a viscous fluid, extensional and (axial) twisting perturbations relax much faster than bending modulations. This explains on physical grounds why many biopolymers such as DNA are often modeled as an inextensible polymer chain without twisting but with bending elasticity when one is interested in shape relaxation dynamics of those biopolymers. One may notice that the twist relaxation time is comparable to the stretch relaxation time. As a matter of fact, rods and filaments with bending and twisting elasticity are often assumed to be inextensible. This is particularly reasonable when one is interested in the dynamics of rotationally driven rods, because no stretching is induced by this process and effects from stretch-twist couplings are sufficiently small compared to twist-bend coupling effects.[101] Throughout this chapter, we will consider inextensible rods and polymers in our analytical arugments, while we may allow only small chain extensibility in numerical treatments by choosing a sufficiently large stretching modulus.

3. Thin Rod Kinematics

To proceed, we begin with a brief review of rod kinematics to see how one can mathematically parameterize the shape of a thin rod with twist in three dimensional space. When the total length of a rod is much longer than its cross section, the configuration of the curve may be described as a spatial curve with its centerline parameterized by the arclength s, where $0 \leq s \leq L$. A material orthonormal frame (i.e., the generalized Frenet basis) $\{\hat{\mathbf{e}}_1, \hat{\mathbf{e}}_2, \hat{\mathbf{e}}_3\}$ is defined at each point along the centerline of the moving curve, $\mathbf{r}(s,t)$, where $\hat{\mathbf{e}}_3 = \partial \mathbf{r}/\partial s$ points along the tangent and $\hat{\mathbf{e}}_1$, $\hat{\mathbf{e}}_2$ correspond to the principle axes of the cross section (see Fig. 4).[101,104,105] We take a right-handed triad of unit vectors (directors) $\{\hat{\mathbf{e}}_\alpha\}$, where $\alpha = 1, 2, 3$, then $\hat{\mathbf{e}}_2 = \hat{\mathbf{e}}_3 \times \hat{\mathbf{e}}_1$. If one knows how this material frame (body-fixed frame of directors) is rotated as it moves along the arclength, the configuration of the rod is completely specified. Corresponding rotational matrices have all information about the local change of the rod shape, and are given in terms of the strain components for the rod deformations:[101,104,105]

$$\partial_s \hat{\mathbf{e}}_\alpha = \sum_{\beta=1}^{3} K_{\alpha\beta} \cdot \hat{\mathbf{e}}_\beta = -\sum_{\beta=1}^{3} \hat{\mathbf{e}}_\beta \cdot K_{\alpha\beta}, \tag{11}$$

where ∂_s denotes the partial derivative with respect to s. $\mathbf{K}$ is an antisymmetric 3×3 matrix, i.e., $\mathbf{K}^t = -\mathbf{K}$, and is given by

$$\mathbf{K} = \begin{bmatrix} 0 & \Omega_3 & -\Omega_2 \\ -\Omega_3 & 0 & \Omega_1 \\ \Omega_2 & -\Omega_1 & 0 \end{bmatrix}. \tag{12}$$

Introducing the strain vector

$$\boldsymbol{\Omega} = \Omega_1 \hat{\mathbf{e}}_1 + \Omega_2 \hat{\mathbf{e}}_2 + \Omega_3 \hat{\mathbf{e}}_3, \tag{13}$$

the kinematic relation (11) can be expressed in a more familiar form:

$$\partial_s \hat{\mathbf{e}}_\alpha = \boldsymbol{\Omega} \times \hat{\mathbf{e}}_\alpha. \tag{14}$$

The strains are obtained from spatial changes of the directors, that is, $\Omega_1 = \hat{\mathbf{e}}_3 \cdot \partial_s \hat{\mathbf{e}}_2$ and its cyclic permutations of the indices for Ω_2 and Ω_3. Ω_1 and Ω_2 are the components of curvature of the centerline with respect to the two principle axes of the cross section, and Ω_3 represents the twist density. Note that the strains Ω_α have the dimension of $[\text{length}]^{-1}$, unlike strains in a three dimensional elastic body which are non-dimensional. (In this sense, Ω_α might be rather considered as the rate of strains as s increases.) The rotation angle about the tangent, $\phi(s,t)$, may be introduced through the variational relation

$$\delta\phi = \hat{\mathbf{e}}_2 \cdot \delta\hat{\mathbf{e}}_1 = -\hat{\mathbf{e}}_1 \cdot \delta\hat{\mathbf{e}}_2. \tag{15}$$

One should note, however, that there is in general no globally defined angle $\phi(s)$ that satisfies $\partial\phi(s)/\partial s = \Omega_3(s)$.[101] Twist density Ω_3 also receives a change due to the variations in $\mathbf{r}(s)$, see Eq. (34) given below. Therefore, *when and only when* the rod shape is "frozen", i.e., $\delta\mathbf{r}(s) = 0$ for all s, there can be a globally defined variable $\phi(s)$ that satisfies $\Omega_3(s) = \partial\phi/\partial s$.[101,106] For example, we can clearly define an excess twist angle $\phi(s)$ that satisfies $\Omega(s) = d\phi(s)/ds$ for one-dimensional twist diffusion process in a slender rod. In this case, the rod shape is assumed to be straight irrespective of the twist profile, but in general the situation is more complex. We also remark here that Eq. (C15) in Ref.[107] is thus misleading as a general statement and is correct only under the variational constraint $\delta\mathbf{r}(s) = 0$.

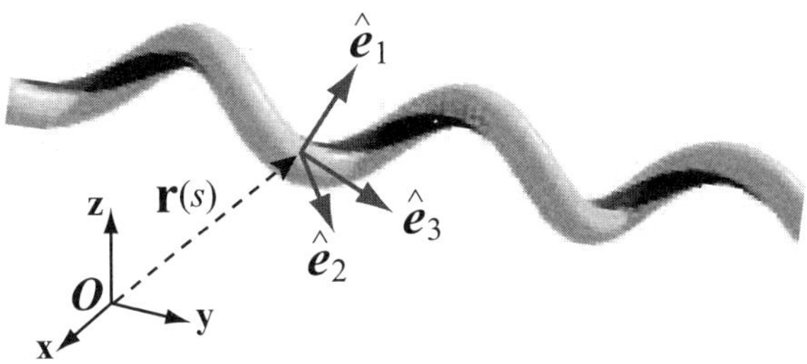

Fig. 4. Centerline of the elastic curve is specified as position vector $\mathbf{r}(s)$, and the material frame $\{\hat{\mathbf{e}}_\alpha\}$ is defined at each point of the centerline. The director $\hat{\mathbf{e}}_1$ and $\hat{\mathbf{e}}_2$ point towards two principal directions in the cross section.

The filament shape is alternatively described by the original Frenet formulation of space curves in terms of the unit tangent $\hat{\mathbf{t}} \equiv \hat{\mathbf{e}}_3$, normal

$\hat{\mathbf{n}} = \partial_s^2\mathbf{r}/|\partial_s^2\mathbf{r}|$ and binormal vector $\hat{\mathbf{b}} = \hat{\mathbf{t}} \times \hat{\mathbf{n}}$ (the Frenet-Serret frame). They satisfy the Frenet-Serret equations,[104,105]

$$\partial_s \hat{\mathbf{t}} = \kappa \hat{\mathbf{n}}, \tag{16}$$

$$\partial_s \hat{\mathbf{n}} = -\kappa \hat{\mathbf{t}} + \tau \hat{\mathbf{b}}, \tag{17}$$

$$\partial_s \hat{\mathbf{b}} = -\tau \hat{\mathbf{n}}, \tag{18}$$

where κ is the curvature related to (Ω_1, Ω_2) via $\kappa = (\Omega_1^2 + \Omega_2^2)^{1/2}$, and τ is the geometric torsion. Transformation from one description to the other is obtained via the rotation by an angle φ about the common tangent $\mathbf{t} = \hat{\mathbf{e}}_3$, that is,

$$\hat{\mathbf{e}}_1 + i\hat{\mathbf{e}}_2 = e^{-i\varphi(s)}(\mathbf{n} + i\mathbf{b}). \tag{19}$$

Plugging this into equations (16)-(18), and comparing with equation (14), we obtain the relation between the strains $\boldsymbol{\Omega}$ and the curvature κ and torsion τ as $\Omega_1 = \kappa \sin\varphi$, $\Omega_2 = \kappa \cos\varphi$, and

$$\Omega_3 = \tau + \frac{d\varphi}{ds}. \tag{20}$$

This relation highlights the fact that the twist density Ω_3 is the axial twist plus geometric torsion determined from the rod centerline shape. This fact suggests that specifying only the curvature κ and twist Ω_3 does not determine the shape uniquely. To illustrate this situation clearly, Powers gives a useful example,[101] see Fig. 5. The left structure in Fig. 5 is a twisted circle with curvature $\kappa = 1/R$ and twist $\Omega_3 = d\varphi/ds = 1/R$:

$$\boldsymbol{\Omega}_{\text{circ}} = \frac{\cos(s/R)}{R}\hat{\mathbf{e}}_1 + \frac{\sin(s/R)}{R}\hat{\mathbf{e}}_2 + \frac{1}{R}\hat{\mathbf{e}}_3. \tag{21}$$

On the other hand, the right structure in Fig. 5 shows a helix that has the same curvature $\kappa = 1/R$ and the same twist $\Omega_3 = \tau = 1/R$ (with $d\varphi/ds = 0$!), but whose strain vector is given by

$$\boldsymbol{\Omega}_{\text{helix}} = \frac{1}{R}\hat{\mathbf{e}}_2 + \frac{1}{R}\hat{\mathbf{e}}_3. \tag{22}$$

This example clearly shows that in general all three variables Ω_1, Ω_2 and Ω_3, are necessary to uniquely specify the shape of a rod.

4. Thin Rod Mechanics

To describe the time evolution of an elastic rod, we need to know the force and torque fields acting on each element of the rod. This can be made

Fig. 5. Two different shapes, circle and helix, of a rod with the same curvature κ, the same twist Ω_3, and of the same length. The red strip traces the path of the unit vector $\hat{\mathbf{e}}_1$ for each configuration. This interesting demonstration was given in Ref. 101.

explicitly once an elastic energy functional is given. Because the energetic penalty is imposed only against rod shape deformations, the elastic energy is given in terms of strains Ω_α. Within linear elasticity theory, the elastic energy functional is given by[102,103,108]

$$E = \int_0^L e(\Omega_1, \Omega_2, \Omega_3) ds - \int_0^L \Lambda(s) ds, \tag{23}$$

where L is the total contour length of rod, and

$$e = \frac{A_1}{2}\Omega_1^2 + \frac{A_2}{2}\Omega_2^2 + \frac{C}{2}\Omega_3^2. \tag{24}$$

For rods with non-circular cross section, the two bending moduli in two principal directions, A_1 and A_2, can be different. It is suggested that such bending anisotropies in DNA can have interesting consequences on the structure of short DNA fragments.[109] For rods with circular cross section of diameter a (that we mostly study in this chapter), the bending becomes isotropic, i.e., $A_1 = A_2 = A$, and we have

$$e = \frac{A}{2}\left(\Omega_1^2 + \Omega_2^2\right) + \frac{C}{2}\Omega_3^2 = \frac{A}{2}\left(\partial_s \hat{\mathbf{t}}\right)^2 + \frac{C}{2}\Omega^2, \tag{25}$$

In what follows, Ω_3 is abbreviated as Ω when it cannot be confused with either Ω_1 or Ω_2. The Lagrange multiplier $\Lambda(s)$ in Eq. (23) can be interpreted as a local tension (with negative sign) generated in the rod that enforces the local inextensibility, $\hat{\mathbf{t}} \cdot \partial_s \dot{\mathbf{r}} = 0$.[6,108]

In order to obtain the force density $\mathbf{f}(s)$ and torque density $\mathbf{m}(s)$ following from the energy functional Eq. (23), we rely on the principle of virtual work, instead of the classical approach based on the Kirchhoff rod equations with constitutive relations.[101–104] The final form of the mechanical balance equations for the rod is of course the same, but our present approach, originally given by Goldstein et al.,[108] appears to be more straightforward and easily allows generalizations to elastic filaments with more complex architectures.[7,110] A formulation of the Kirchhoff rod equations using the framework of variational approach is given in the recent review paper by Powers.[101] In the next subsections, we look at how the variations of the strains are given in terms of $\delta\mathbf{r}(s)$ and $\delta\phi(s)$, and subsequently apply them to derive the force and torque densities.

4.1. *Variational relations*

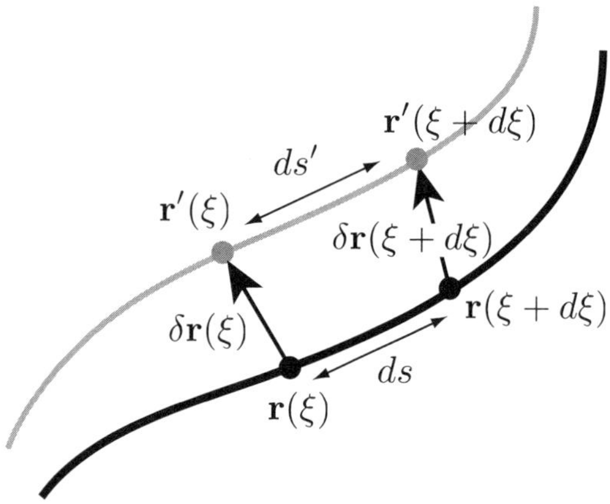

Fig. 6. Schematics on the variation of the filament shape. A hypothetically displaced configuration of the curve is described by the gray line. The configurations of both curves are parameterized by the common material coordinate ξ. Since the variation of a curve generally stretches the curve, the line element ds is also subject to variation.

Suppose taking the variation of the centerline position of the curve $\mathbf{r}$, as shown in Fig. 6:

$$\mathbf{r} \to \mathbf{r}' = \mathbf{r} + \delta\mathbf{r} \tag{26}$$

Note that the arclength s is also subjected to the variational change, since it is intrinsic to $\mathbf{r}$. The centerline of a rod $\mathbf{r}(\xi)$ is parameterized by ξ fixed to a given material point (material coordinate). Note that ξ is a label assigned at each point of the rod centerline, and is convected upon any deformation of the rod centerline, thus is free from any variational procedure, unlike the arclength s. The arclength element in the (hypothetically) deformed curve is

$$(ds')^2 = |\mathbf{r}'(\xi + d\xi) - \mathbf{r}'(\xi)|^2 = |\partial_\xi \mathbf{r} + \partial_\xi(\delta\mathbf{r})|^2 (d\xi)^2. \tag{27}$$

Retaining the terms up to first order in the variation δ, and noting that $ds = |\partial_\xi \mathbf{r}| d\xi$, we obtain $ds' = [1 + \hat{\mathbf{e}}_3 \cdot \partial_s(\delta\mathbf{r})]ds$. The variation of arclength ds and the local tangent $\hat{\mathbf{e}}_3$ are found to be

$$\delta(ds) = \hat{\mathbf{e}}_3 \cdot \partial_s(\delta\mathbf{r}) ds, \tag{28}$$

and

$$\delta\hat{\mathbf{e}}_3 = \partial_s(\delta\mathbf{r}) - [\hat{\mathbf{e}}_3 \cdot \partial_s(\delta\mathbf{r})]\hat{\mathbf{e}}_3. \tag{29}$$

In general, the variation δ and the derivative with respect to s do not commute, that is, $\delta\partial_s - \partial_s\delta = -[\hat{\mathbf{e}}_3 \cdot \partial_s(\delta\mathbf{r})]\partial_s$. However, when the rod is inextensible, i.e., $\delta(ds) = \hat{\mathbf{e}}_3 \cdot \partial_s(\delta\mathbf{r}) ds = 0$, the order of these two operations becomes immaterial.[101]

Next we consider how the director vectors $\hat{\mathbf{e}}_1$ and $\hat{\mathbf{e}}_2$ change upon the variational operation considered above. Decomposing an infinitesimal variation $\delta\hat{\mathbf{e}}_1$ as $\delta\hat{\mathbf{e}}_1 = (\hat{\mathbf{e}}_2 \cdot \delta\hat{\mathbf{e}}_1)\hat{\mathbf{e}}_2 + (\hat{\mathbf{e}}_3 \cdot \delta\hat{\mathbf{e}}_1)\hat{\mathbf{e}}_3$, and using $\delta\phi = \hat{\mathbf{e}}_2 \cdot \delta\hat{\mathbf{e}}_1$ (see Eq. (15)), we obtain

$$\delta\hat{\mathbf{e}}_1 = (\delta\phi)\hat{\mathbf{e}}_2 - [\hat{\mathbf{e}}_1 \cdot \partial_s(\delta\mathbf{r})]\hat{\mathbf{e}}_3, \tag{30}$$

where Eq. (29) has been made use of. Likewise, we also obtain

$$\delta\hat{\mathbf{e}}_2 = -(\delta\phi)\hat{\mathbf{e}}_1 - [\hat{\mathbf{e}}_2 \cdot \partial_s(\delta\mathbf{r})]\hat{\mathbf{e}}_3. \tag{31}$$

Applying the variation δ to the kinematic relation $\Omega_1 = \hat{\mathbf{e}}_3 \cdot \partial_s\hat{\mathbf{e}}_2$, we have $\delta\Omega_1 = \delta\hat{\mathbf{e}}_3 \cdot \partial_s\hat{\mathbf{e}}_2 + \hat{\mathbf{e}}_3 \cdot \partial_s(\delta\hat{\mathbf{e}}_2) - [\hat{\mathbf{e}}_3 \cdot \partial_s(\delta\mathbf{r})]\Omega_1$, where the last term comes from the commutation relation between δ and ∂_s. Substituting the following relations $\delta\hat{\mathbf{e}}_3 \cdot \partial_s\hat{\mathbf{e}}_2 = -\Omega_3\hat{\mathbf{e}}_1 \cdot \partial_s(\delta\mathbf{r})$ and $\hat{\mathbf{e}}_3 \cdot \partial_s(\delta\hat{\mathbf{e}}_2) = (\delta\phi)\hat{\mathbf{e}}_2 - \partial_s[\hat{\mathbf{e}}_2 \cdot \partial_s(\delta\mathbf{r})]$ into the above equation for $\delta\Omega_1$, we obtain

$$\delta\Omega_1 = (\delta\phi)\Omega_2 - 2\Omega_1\hat{\mathbf{e}}_3 \cdot \partial_s(\delta\mathbf{r}) - \hat{\mathbf{e}}_2 \cdot \partial_s^2(\delta\mathbf{r}). \tag{32}$$

Making similar steps, we also obtain

$$\delta\Omega_2 = -(\delta\phi)\Omega_1 - 2\Omega_2\hat{\mathbf{e}}_3 \cdot \partial_s(\delta\mathbf{r}) + \hat{\mathbf{e}}_1 \cdot \partial_s^2(\delta\mathbf{r}), \tag{33}$$

$$\delta\Omega_3 = \partial_s(\delta\phi) + (\Omega_1\hat{\mathbf{e}}_1 + \Omega_2\hat{\mathbf{e}}_2) \cdot \partial_s(\delta\mathbf{r}) - \Omega_3\hat{\mathbf{e}}_3 \cdot \partial_s(\delta\mathbf{r}). \tag{34}$$

Equations (32)-(34) give the infinitesimal variation of Ω_α in terms of $\delta\mathbf{r}$ and $\delta\phi$.

4.2. *Calculation of the forces and moments*

Our elastic energy density $e = e(\Omega_\alpha)$ is a function of the rate of strain Ω_α only. Therefore, using Eqs. (32)-(34), it is straightforward to write down δE in the form:

$$\delta E = (\text{boundary terms}) - \int_0^L ds\mathbf{f}(s)\delta\mathbf{r}(s) - \int_0^L dsm_3(s)\delta\phi(s). \quad (35)$$

According to the principle of virtual work, the elastic translational force density $\mathbf{f}(s)$ is obtained by changing infinitesimally the centerline of the curve $\delta\mathbf{r}(s)$, while keeping the axial spinning angle $\delta\phi(s)$ unchanged. On the other hand, the axial torque (the moment about the local tangent per unit length) $m_3(s)$ can be obtained by fixing the centerline shape, i.e., $\delta\mathbf{r}(s) = 0$. Note that the constraint $\delta\phi = 0$ is not equivalent to $\delta\Omega_3 = 0$ (twist fixed). In fact, the twist changes upon the variation with respect to $\delta\mathbf{r}$ as seen from Eq. (34), which gives the coupling between twist and writhe (out-of-plane bending).

The moment vector $\mathbf{M}$ is defined as $\mathbf{M} = M_\alpha\hat{\mathbf{e}}_\alpha$, where $M_\alpha = \partial e/\partial\Omega_\alpha$. For the energy density e given in Eq. (24), we have

$$\mathbf{M} = A_1\Omega_1\hat{\mathbf{e}}_1 + A_2\Omega_2\hat{\mathbf{e}}_2 + C\Omega_3\hat{\mathbf{e}}_3. \quad (36)$$

Applying δ to Eq. (23) leads to $\delta E = \int \delta(ds)e(\Omega_\alpha) + \int dsM_\alpha\delta\Omega_\alpha$. Using Eqs. (32), (33) and (34), and integrating the second derivative terms by parts once, we obtain

$$\begin{aligned}\delta E = &-\mathbf{M}(L)\cdot\delta\boldsymbol{\phi}(L) + \mathbf{M}(0)\cdot\delta\boldsymbol{\phi}(0) + \int ds(M_1\Omega_2 - M_2\Omega_1 - \partial_s M_3)\delta\phi \\ &+ \int ds\,[(-\partial_s M_2 + M_1\Omega_3 - M_3\Omega_1)\,\hat{\mathbf{e}}_1 + (\partial_s M_1 + M_2\Omega_3 - M_3\Omega_2)\,\hat{\mathbf{e}}_2 \\ &- (M_1\Omega_1 + M_2\Omega_2 + M_3\Omega_3 - e + \Lambda)\,\hat{\mathbf{e}}_3]\cdot\partial_s(\delta\mathbf{r}), \end{aligned} \quad (37)$$

where the infinitesimal angle variation $\delta\boldsymbol{\phi} = \delta\phi_\alpha\hat{\mathbf{e}}_\alpha$ is given by $\delta\phi_1 = -\hat{\mathbf{e}}_2\cdot\delta\hat{\mathbf{e}}_3$, $\delta\phi_2 = \hat{\mathbf{e}}_1\cdot\delta\hat{\mathbf{e}}_3$ and $\delta\phi_3 = \delta\phi$. The force $\mathbf{F}(s)$ acting on the cross section of the rod at s is defined by $\mathbf{f}(s) = \partial_s\mathbf{F}$. From Eq. (37), those components $F_\alpha = \mathbf{F}(s)\cdot\hat{\mathbf{e}}_\alpha$ are found to be

$$F_1 = -\partial_s M_2 - M_1\Omega_3 + M_3\Omega_1, \quad (38)$$

$$F_2 = \partial_s M_1 - M_2\Omega_3 + M_3\Omega_2, \quad (39)$$

$$F_3 = M_1\Omega_1 + M_2\Omega_2 + M_3\Omega_3 - e(\Omega_\alpha) + \Lambda(s). \quad (40)$$

Integrating Eq. (37) once again by parts, we obtain δE in the desired form:

$$\delta E = -\mathbf{M}(L) \cdot \delta\boldsymbol{\phi}(L) + \mathbf{M}(0) \cdot \delta\boldsymbol{\phi}(0) + \mathbf{F}(L) \cdot \delta\mathbf{r}(L) - \mathbf{F}(0) \cdot \delta\mathbf{r}(0)$$
$$- \int_0^L ds\mathbf{f}(s) \cdot \delta\mathbf{r}(s) + \int ds(M_1\Omega_2 - M_2\Omega_1 - \partial_s M_3)\delta\phi, \qquad (41)$$

The axial torque about its local tangent, m_3, is thus given by

$$m_3(s) = \partial_s M_3 + M_2\Omega_1 - M_1\Omega_2. \qquad (42)$$

Note that the first term on the right hand side, $\partial_s M_3 = C\partial_s\Omega_3$, means that a differential twist density generates axial torque, while the remaining term, $M_2\Omega_1 - M_1\Omega_2 = (A_2 - A_1)\Omega_1\Omega_2$, suggests that an anisotropic bending can exert axial torque, which vanishes for an isotropic rod. In particular, for such an isotropic rod, the force $\mathbf{F}(s)$ given in Eqs. (38)-(40) and the moment $\mathbf{M}(s)$ in Eq. (36) can be written in a rather compact form. Using the relations easily derived from the kinematic equations, $\partial_s^2\hat{\mathbf{e}}_3 = (\partial_s\Omega_2)\hat{\mathbf{e}}_1 - (\partial_s\Omega_1)\hat{\mathbf{e}}_2 - (\Omega_1^2 + \Omega_2^2)\hat{\mathbf{e}}_3 + (\Omega_1\hat{\mathbf{e}}_1 + \Omega_2\hat{\mathbf{e}}_2)\Omega_3$, and $\hat{\mathbf{e}}_3 \times \partial_s\hat{\mathbf{e}}_3 = \Omega_1\hat{\mathbf{e}}_1 + \Omega_2\hat{\mathbf{e}}_2$, we find

$$\mathbf{F} = -A\partial_s^3\mathbf{r} + C\left[\Omega_3(\partial_s\mathbf{r} \times \partial_s^2\mathbf{r})\right] - \Lambda\partial_s\mathbf{r}, \qquad (43)$$
$$\mathbf{M} = A\partial_s\mathbf{r} \times \partial_s^2\mathbf{r} + C\Omega_3\partial_s\mathbf{r}, \qquad (44)$$

where the tangential part was absorbed into the Lagrange multiplier function $\Lambda(s)$. There are a variety of boundary conditions to which the rod can be subjected. In particular, when external forces $\mathbf{F}_{ext}$ and torques $\mathbf{M}_{ext}$ are applied at the rod ends, the boundary conditions read from Eq. (41) as

$$\mathbf{M}(L) = \mathbf{M}_{ext}(L), \quad \mathbf{M}(0) = -\mathbf{M}_{ext}(0) \qquad (45)$$

and

$$\mathbf{F}(L) = \mathbf{F}_{ext}(L), \quad \mathbf{F}(0) = -\mathbf{F}_{ext}(0). \qquad (46)$$

Or, when one end of the rod, say $s = 0$, is "clamped", $\mathbf{r}(0)$ and $\partial_s\mathbf{r}(0)$ are specified instead of $\mathbf{F}_{ext}(0)$ and $\mathbf{M}_{ext}(0)$.

5. Equations of Motion for Rods with Bend and Twist: Rouse Dynamics

Now we are at a position to write down the time evolution equation for a moving rod with bending and twisting elasticity in a viscous fluid. For simplicity, we first consider the so-called Rouse dynamics, in which only a local isotropic drag is provided from the surrounding fluid. The hydrodynamic drag force $\zeta\dot{\mathbf{r}}$ and torque $\zeta_r\dot{\phi}$ are in this case exactly balanced by the elastic force and moment per unit length derived above:

$$\zeta\dot{\mathbf{r}} = \partial_s\mathbf{F} = \mathbf{f}(s), \quad \zeta_r\dot{\phi} = \partial_s M_3 = m_3(s), \tag{47}$$

where $\zeta = 3\pi\eta a$ and $\zeta_r = \pi\eta a^3$ are translational and rotational friction coefficients.[99] On the other hand, if we choose $\delta\Omega_3 = \dot{\Omega}_3\delta t$ in Eq. (34), we obtain the geometric relation

$$\dot{\Omega}_3 = \partial_s(\dot{\phi}) - \Omega_3\partial_s\mathbf{r}\cdot\partial_s(\dot{\mathbf{r}}) + \left(\partial_s\mathbf{r}\times\partial_s^2\mathbf{r}\right)\cdot\partial_s(\dot{\mathbf{r}}). \tag{48}$$

This equation describes how twist is transported in the rod: It is a local conservation law of twist. The first term on the right-hand-side shows twist-change due to differential rates of angular rotation (pure twist about local tangent), the second term describes twist-change due to stretching (e.g., extension of a straight, twisted rod decreases Ω_3) and the third term accounts for out-of-plane bending motion (writhing).[7,101] The last term is explicitly nonlinear, and is usually interpreted as a sink or source in the twist dynamics. Substituting Eq. (43) and (48) into Eqs. (47), we arrive at the coupled equations of motion for bend and twist of the uniform rod of circular cross section as

$$\zeta\dot{\mathbf{r}} = -A\partial_s^4\mathbf{r} + C\partial_s\left[\Omega(\partial_s\mathbf{r}\times\partial_s^2\mathbf{r})\right] - \partial_s(\Lambda\partial_s\mathbf{r}), \tag{49}$$

$$\zeta_r\dot{\Omega} = C\partial_s^2\Omega + \zeta_r(\partial_s\mathbf{r}\times\partial_s^2\mathbf{r})\cdot\partial_s(\dot{\mathbf{r}}), \tag{50}$$

where the abbreviation $\Omega\equiv\Omega_3$ has been used, and strict local inextensibility, $\partial_s\mathbf{r}\cdot\partial_s(\dot{\mathbf{r}}) = 0$, has been assumed. Applying this constraint to Eq. (49), we see that the Lagrange multiplier Λ must satisfy

$$\left(\partial_s^2 - \kappa^2\right)\Lambda(s) = \partial_s\mathbf{r}\cdot\partial_s^2\left[-A\partial_s^3\mathbf{r} + C\Omega(\partial_s\mathbf{r}\times\partial_s^2\mathbf{r})\right], \tag{51}$$

which closes the dynamics.

5.1. *Kirchhoff rod equation*

It is useful to have the mechanical balance equations of a thin rod in terms of the internal elastic force $\mathbf{F}(s)$ and moment $\mathbf{M}(s)$. Such equations are known as Kirchhoff rod equations first given by Gustav Kirchhoff in 1859:[102–104,111–116]

$$\partial_s\mathbf{F} + \mathbf{f}_{ext} = 0, \tag{52}$$

$$\partial_s\mathbf{M} + \hat{\mathbf{t}}\times\mathbf{F} + \mathbf{m}_{ext} = 0, \tag{53}$$

where $\mathbf{f}_{ext}$ and $\mathbf{m}_{ext}$ are external body force and moment per unit length acting on the rod at s, which include viscous drag contributions from the surrounding fluid. Equations (52) and (53) can be obtained by considering the conservation of translational and angular momentum of an infinitesimal element bounded by two adjoining cross-sections of the rod.[103] A more systematic derivation of Eqs. (52) and (53) using the framework of the above-presented variational approach can be found in Ref.[101] The Kirchhoff equations (52) and (53) are closed by giving the constitutive relation of linear elasticity relating the moment $\mathbf{M}$ to the strains $\mathbf{\Omega}$. For an isotropic rod, we can use $\mathbf{F}$ and $\mathbf{M}$ given in Eqs. (43) and (44), and the viscous drag force $\mathbf{f}_{ext} = -\zeta \mathbf{v} = -\zeta \dot{\mathbf{r}}$ and $\mathbf{m}_{ext} = -\zeta_r \omega \hat{\mathbf{t}}$ in Rouse dynamics. It is thus straightforward to check that Eqs. (52)-(53) give the equations of motion of the rod per unit length, Eqs. (49) and (50).

Using the Euler angles representation, one can see that the Kirchhoff rod equations have the same mathematical structure as the Lagrange mechanics of a heavy top (see Appendix A). Kirchhoff himself already addressed this analogy to the dynamics of a spinning top (known as the Kirchhoff kinetic analogy[102]), which is useful to find general analytic solutions for static rod shapes under stresses and different boundary constraints.[115,117–120] The Kirchhoff rod models currently find many applications in biological mechanics, such as plant tendrils[121,122] and growing microbial fibers,[123] as well as biopolymer mechanics.

5.2. *Topological aspects in closed curves*

Here we make some comments on topological aspects of curves moving in three dimensions. For closed curves (such as plasmid DNA), the linking number Lk is an integer topological invariant and is divided into twist Tw and writhe Wr:

$$\mathrm{Lk} = \mathrm{Tw} + \mathrm{Wr}. \tag{54}$$

Twist is defined by

$$\mathrm{Tw} = \frac{1}{2\pi} \oint ds \Omega_3(s) = \frac{1}{2\pi} \oint ds (\hat{\mathbf{e}}_3 \times \hat{\mathbf{e}}_1) \cdot \partial_s \hat{\mathbf{e}}_1, \tag{55}$$

where the second equality follows from $\Omega_3 = \hat{\mathbf{e}}_2 \cdot \partial_s \hat{\mathbf{e}}_1$. Writhe Wr is a measure for how much the curve is winding around itself, and is a function of the centerline $\mathbf{r}(s)$ only:

$$\mathrm{Wr} = \frac{1}{4\pi} \oint \frac{(\mathbf{r}(s) - \mathbf{r}(s')) \cdot (\partial_s \mathbf{r}(s) - \partial_{s'} \mathbf{r}(s'))}{|\mathbf{r}(s) - \mathbf{r}(s')|^3} ds ds'. \tag{56}$$

This double integral form is known as the Călugăreanu-White (CW) formula.[124,125] As long as a curve does not self-intersect, the linking number Lk is invariant under shape deformations of the curve. Because twist Tw changes continuously under any smooth deformations, if the centerline crosses itself once, there is a discontinuity of 2 in Lk, and thus a discontinuous change of 2 in writhing number Wr.

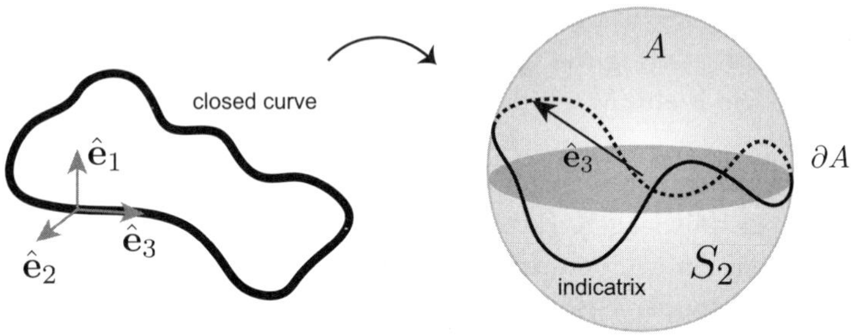

Fig. 7. Schematic illustration of the geometry used to derive Fuller's results.

Fuller showed that the writhe Wr of a closed curve can be given as an integer plus the signed area swept out by the unit tangent vector, $\hat{\mathbf{e}}_3(s)$, on a unit sphere (Poincaré sphere) S_2.[126] We will here reproduce Fuller's result by following the argument detailed by Kamien in Ref.[127,128] Consider a closed curve in 3 dimensional space characterized by a local director frame $\hat{\mathbf{e}}_\alpha$. As we go along the closed curve, the unit tangent $\hat{\mathbf{e}}_3 = \partial_s \mathbf{r}$ traces out a closed curve on the sphere S_2, which is called the tangent indicatrix (see Fig. 7).[129] Let us call the area enclosed by this curve on the sphere as A, and its boundary as ∂A. We also define the unit tangent vector to ∂A as $\hat{\mathbf{T}}(s)$. Noting that $\hat{\mathbf{e}}_1$ is normal to $\hat{\mathbf{e}}_3$, $\hat{\mathbf{e}}_1$ is always on the tangent plane to S_2 together with $\hat{\mathbf{T}}(s)$. Therefore, if we define the orthonormal frame $(\hat{\mathbf{E}}_1, \hat{\mathbf{E}}_2)$ on the tangent plane, $\hat{\mathbf{e}}_1$ and $\hat{\mathbf{T}}$ can be written as

$$\hat{\mathbf{e}}_1(s) = \cos\theta(s)\hat{\mathbf{E}}_1(s) + \sin\theta(s)\hat{\mathbf{E}}_2(s), \tag{57}$$

$$\hat{\mathbf{T}}(s) = \cos\theta_t(s)\hat{\mathbf{E}}_1(s) + \sin\theta_t(s)\hat{\mathbf{E}}_2(s). \tag{58}$$

In general, the local frame $\hat{\mathbf{E}}_\alpha(\xi_1, \xi_2)$ can be defined at any point of the unit sphere in terms of surface coordinates (ξ_1, ξ_2). On the tangent indicatrix ∂A, they are parameterized as $\xi_1 = \hat{\xi}_1(s), \xi_2 = \hat{\xi}_2(s)$, and we have implied $\hat{\mathbf{E}}_1(s) = \hat{E}_1(\hat{\xi}_1(s), \hat{\xi}_2(s))$. The Gauss-Bonnet theorem applied to the region

A ensures

$$\int_A d^2\xi K + \int_{\partial A} ds\kappa_G = 2\pi, \tag{59}$$

where K is the Gaussian curvature and $\kappa_G = \hat{\mathbf{e}}_3 \times \hat{\mathbf{T}} \cdot \partial_s \hat{\mathbf{T}}$ is the geodesic curvature that measures a rate of change of $\hat{\mathbf{T}}(s)$ in the tangent plane. (Note that $\hat{\mathbf{e}}_3$ is the unit normal to this plane.) Substituting Eqs. (57)-(58) into κ_G and twist density $\Omega_3 = (\hat{\mathbf{e}}_3 \times \hat{\mathbf{e}}_1) \cdot \partial_s \hat{\mathbf{e}}_1$, and noting that $(\hat{\mathbf{E}}_1, \hat{\mathbf{E}}_2, \hat{\mathbf{e}}_3)$ consists of the orthonormal frame by construction, we find

$$\kappa_G = \partial_s \theta_t(s) - \hat{\mathbf{E}}_1 \cdot \partial_s \hat{\mathbf{E}}_2, \tag{60}$$

$$\Omega_3 = \partial_s \theta(s) - \hat{\mathbf{E}}_1 \cdot \partial_s \hat{\mathbf{E}}_2. \tag{61}$$

Substituting these results into Eq. (59), and using $K = 1$ for the unit sphere, we arrive at Fuller's result:

$$\frac{A}{2\pi} + \mathrm{Tw} = 0, \quad \mathrm{mod}\, 1, \tag{62}$$

where we have used the fact that $\oint ds \partial_s(\theta - \theta_t)$ is an integer multiplied by 2π for closed curves. Comparing Eq. (62) with Eq. (54), we see that $\mathrm{Wr} = A/2\pi$ modulo 1. Since the solid angle enclosed by the curve is determined only up to 4π (the area of the unit sphere), the writhe Wr of a closed non-self intersecting curve is known up to ± 2 by this method:

$$1 + \mathrm{Wr} = \frac{A}{2\pi} \quad \mathrm{mod}\ 2, \tag{63}$$

which is the Fuller's theorem as established in Ref.[126]

In the same paper,[126] Fuller also gave a useful theorem to calculate Wr of a closed curve. Let Γ_0 and Γ_1 be two closed non-self-intersecting space curves. If Γ_0 can be deformed continuously into Γ_1 through non-self-intersecting curves, Γ_λ, where $0 \leq \lambda \leq 1$, and *if* Γ_0 *and* Γ_λ *never have any points that are oppositely directed (anti-podal)*, then the writhing numbers of two curves are related by

$$\mathrm{Wr}(\Gamma_1) - \mathrm{Wr}(\Gamma_0) = \frac{1}{2\pi} \int \frac{\hat{\mathbf{t}}_0 \times \hat{\mathbf{t}}_1}{1 + \hat{\mathbf{t}}_0 \cdot \hat{\mathbf{t}}_1} \cdot \frac{d}{du} \left(\hat{\mathbf{t}}_0 + \hat{\mathbf{t}}_1 \right) du, \tag{64}$$

where u is a common parameter for both curves which is not necessarily the arclength of Γ_0 or Γ_1, and $\hat{\mathbf{t}}_\lambda(u) = \partial_u \mathbf{r}_\lambda / |\partial_u \mathbf{r}_\lambda|$ is the unit tangent to Γ_λ. While the original paper by Fuller only provides this formula, a concise mathematical proof of Eq. (64) is given by Aldinger *et al.*[130] By choosing the reference curve Γ_0 as a particularly simple one with zero writhe, the computation of Wr comes down to estimating the single integral in

Eq. (64). This sounds simpler compared to the CW double integral formula, and the Fuller's formula has been widely used for calculating supercoiled DNA conformations observed in magnetic tweezer experiments (with appropriate closure operations explained below). However, Neukirch and Starostin pointed out that, the second hypotheses of Fuller's theorem, that the two curves never have antipodal configurations throughout continuous morphing, is not met for plectoneme configurations, and use of Eq. (64) generally leads to incorrect predictions. They show that, once there is any anti-podal configuration, i.e., $\hat{\mathbf{t}}_\lambda(u^*)\cdot\hat{\mathbf{t}}_0(u^*) = -1$ for a certain u^*, it results in a discrepancy in Fuller's formula of ± 2 from the actual (CW) writhe. As a matter of fact, this issue is still a matter of controversy. For example, Bouchiat and Mezard proposed to regularize the model by removing singular configurations, i.e., antipodal points, which, they suppose, ultimately corresponds to taking into account the physical thickness of any polymers.[60,131,132] While Samuel *et al.* clearly distinguish Fuller' formula from the CW writhe formula and warn of the risk for using Fuller's formula as "local writhe formula" as Bouchiat and Mezard did, they nevertheless emphasize the need for regularizing cut-off scale irrespective of the writhe formula used.[133–135] In contrast, both Rossetto-Maggs[136,137] and Neukirch-Starostin[129,138] suggested by using both analytical and numerical methods that a divergence in writhe distribution is an artifact of incorrect use of Fuller's formula and is not present in CW double integral formulation. If DNA is subjected to a sufficiently large stretching force but low twist, there is no antipodal event, and both writhe formulae can give the same correct results (i.e., perturbative regime).[59,139] This is currently the only regime which is not controversial.

The definition of Wr and the concept of Lk conservation can be generalized to an open curve by putting virtual extensions to the top and the base of the curve and by connecting these extensions in an appropriate manner. If the CW double integral formula, Eq. (56), is applied to an open chain (with an appropriate virtual closing), we need to evaluate not only the chain-chain contribution, but also the chain-extension ones in order to obtain a correct writhe that does not change topology. A general closing procedure of an open continuous space curve and a computation of its writhe is proposed by Starostin.[140] Prescriptions to accurately compute the writhe for a polymer chain consisting of *discrete* segments (like in our dynamic simulations, see Sec. 6 below) have been given by Klenin and Langowski.[141] While the original work only dealt with closed curves, we present in the Appendix B our extended formula for open curves with parallel and terminal tangent vectors, which accounts for contributions to Wr from the coupling between the real curve section and virtual extension sections.

5.3. *Intuitive understanding of twist*

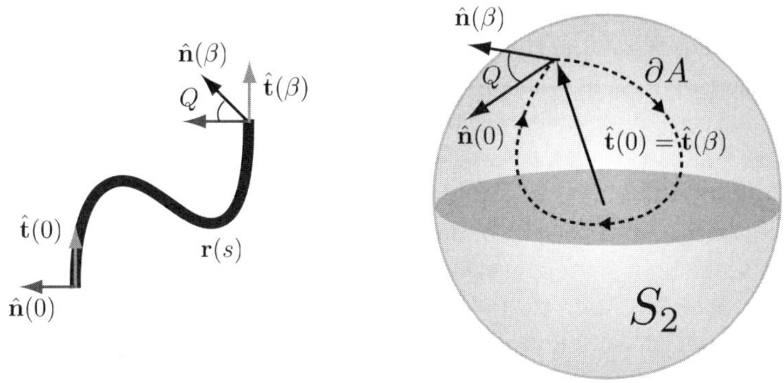

Fig. 8. Parallel transport of the normal vector $\hat{\mathbf{n}}$ on the Poincaré sphere S_2 along a tangent indicatrix ∂A for a section of a curve $\mathbf{r}(s)$, where $\hat{\mathbf{t}}(0) = \hat{\mathbf{t}}(\beta)$. This generally makes a finite angle between $\hat{\mathbf{n}}(0)$ and $\hat{\mathbf{n}}(\beta)$, which corresponds to torsion for the rod section considered.

Equation (20) says that the twist density Ω_3 is the sum of two contributions: a pure twisting about its local tangent and a geometric torsion. To understand its geometric meaning more clearly, we here give a simple plausibility argument: We consider a tangent vector $\hat{\mathbf{t}}(s) \equiv \hat{\mathbf{e}}_3(s)$ together with its normal vector $\hat{\mathbf{n}}(s)$, and monitor how this normal vector $\hat{\mathbf{n}}(s)$ changes via the parallel transport on the sphere S_2 as the tangent $\hat{\mathbf{t}}$ moves along the curve. For simplicity, let us assume that $\hat{\mathbf{t}}$ moves from $s = 0$ to $s = \beta$, where $\hat{\mathbf{t}}(0) = \hat{\mathbf{t}}(\beta)$, see Fig. 8. In this case, the tangent $\mathbf{t}$ comes back to the starting point on S_2 and thus its tip sweeps a closed curve ∂A. Since $\hat{\mathbf{n}}$ is normal to $\hat{\mathbf{t}}$, it is always staying on the tangent plane to S_2, so $\hat{\mathbf{n}}$ undergoes a parallel transport along ∂A. However, $\hat{\mathbf{n}}$ does not necessarily come back to the starting configuration, even when $\hat{\mathbf{t}}$ does. The initial and final vectors, $\hat{\mathbf{n}}(0)$ and $\hat{\mathbf{n}}(\beta)$, generally make a finite angle Q. This means that the vector $\hat{\mathbf{n}}$ rotates by an angle Q about its tangent $\hat{\mathbf{t}}$ as it moves along the rod contour $\mathbf{r}(s)$ from $s = 0$ to $s = \beta$, and gives a twisting of the rod within that rod section. The value of Q depends solely on the shape of the rod, $\mathbf{r}(s)$, which thus corresponds to torsion. There is of course another degree of freedom: at each point, $\hat{\mathbf{n}}$ can be rotated at any angle about its local tangent $\hat{\mathbf{t}}$, which corresponds to spinning degree of freedom. Total rotation of $\mathbf{n}$ between $s = 0$ and $s = \beta$ is the sum of these rotations, and it gives $\int_0^\beta ds\Omega_3(s)$.

5.4. *Link flow conservation for open curves*

The above topological argument can in fact be extended to the dynamical case. When we look at changes in writhe, but not the absolute value of Wr itself, it satisfies a local conservation law. This concept is applied not only to closed curves but also to general open curves in motion, for which conventionally the linking number Lk or writhe Wr are not defined. This geometrically inspired conservation law was first pointed out by Kamien[127,128] and by Goldstein.[7] A key equation is the geometric relation for twist, Eq. (48). First, we note that changes of Wr of a deforming curve is shown by Aldinger et al. to follow[130]

$$\frac{d}{dt}\mathrm{Wr} = \frac{1}{2\pi}\oint ds\,\hat{\mathbf{e}}_3 \cdot \dot{\hat{\mathbf{e}}}_3 \times \partial_s \hat{\mathbf{e}}_3. \tag{65}$$

Assuming that continuous deformation of the curve does not lead to discontinuous integer change in Wr, we can introduce a "writhe density" wr (which is the non-integer contribution of the true writhe) whose time derivative satisfies $\partial(wr)/\partial t = \hat{\mathbf{e}}_3 \cdot \dot{\hat{\mathbf{e}}}_3 \times \partial_s \hat{\mathbf{e}}_3/2\pi$. We thus can rearrange Eq. (48) in the form $\partial_t(\Omega_3 + 2\pi\, wr) = \partial_s \omega$. Defining a "link density" $\rho = \Omega_3 + 2\pi wr$, we obtain a local link conservation law given by $\partial_t \rho + \partial_s j_{Link} = 0$, where $j_{Link} = -\omega = -(C/\zeta_r)\partial_s \Omega$ (for Rouse dynamics). As noted by Goldstein,[108] there is in general no true link density as it is not a single integral of a local quantity unlike Tw. However, changes in link can be written as a single integral over a density, i.e.,

$$\int ds\partial_t \rho - \frac{d}{dt}\mathrm{Lk} = \int ds \left(\dot{\Omega}_3 + \hat{\mathbf{e}}_3 \cdot \dot{\hat{\mathbf{e}}}_3 \times \partial_s \hat{\mathbf{e}}_3\right), \tag{66}$$

which is obtained by integrating the above link conservation over the curve contour. The present argument is applied to an open curve, so changes in writhe and link still have meaning for open polymers, in contrast to the global variables Wr and Lk.[101,108,128]

6. Numerical Method: Brownian Dynamics for Twist-Storing Polymers

The obtained equations of motion in the continuum limit, Eqs. (49) and (50), are highly nonlinear, and it is impossible to integrate those equations analytically in general cases. To study the driven dynamics of an elastic rod, therefore, we have to rely on numerical approaches. We are selectively interested in slow dynamics of driven polymers, whose typical time scales would be $10^{-6} - 10^{-4}$ sec. For example, the typical diffusion time of a sphere of size $a = 10$ nm in water is $\tau \sim \eta a^3/k_BT \sim 10^{-6}$ sec. On this time scale, any fast dynamics such as vibrational modes of solvent dynamics and wave propagations in elastic media die out. Molecular dynamics simulations with explicit solvent dynamics are therefore obviously inefficient for our aim. We thus employ Brownian hydrodynamics simulation methods, in which a polymer chain is coarse-grained on nano-scales and solvent is assumed to be a uniform background medium. Effects of solvent flow are implicitly taken into account as position-dependent mobilities of spheres.

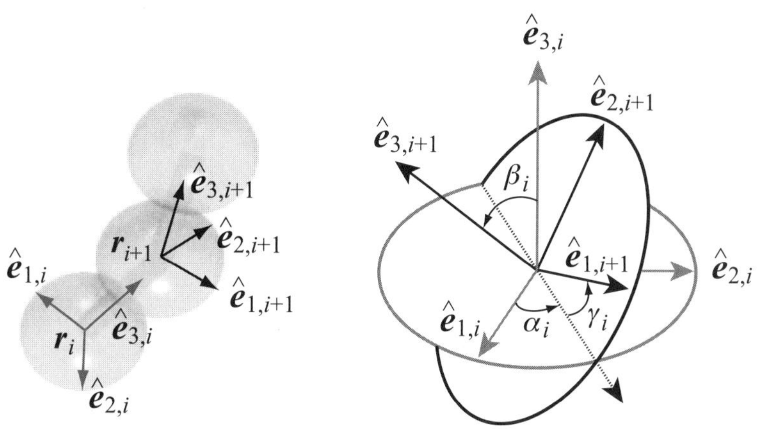

Fig. 9. Transformation of the coordinate Σ_i to Σ_{i+1} parameterized by the Euler angles $(\alpha_i, \beta_i, \gamma_i)$.

There are several different numerical methods proposed for dynamic simulations of elastic filaments with bend and twist.[91,106,142–146] Our method is largely based on the early work by Allison *et al.*[142] and its significantly improved version by Chirico and Langowski.[106,143] In our simulations,[107] an elastic rod or polymer is modeled as a chain of $N + 1$ connected spheres of diameter a subjected to external driving force and torque, as well as inter-sphere potential interactions. Each sphere is specified by its position $\mathbf{r}_i(t)$ and a body-fixed frame $\Sigma_i = (\hat{\mathbf{e}}_{1i}, \hat{\mathbf{e}}_{2i}, \hat{\mathbf{e}}_{3i})$, which corresponds to the local orthogonal director frame $\{\hat{\mathbf{e}}_\alpha\}$ in the continuum limit discussed above. A *finite* angle Euler transformation matrix transforms Σ_i into Σ_{i+1}, where three Euler angles $(\alpha_i, \beta_i, \gamma_i)$ are related to Σ vectors as described previously:[106,142,143] see also Fig. 9. The rate of strains in this discrete model, $\mathbf{\Omega}_i$, is given in terms of the Euler angles as[106,107]

$$\Omega_{1i}a = \beta_i \sin\alpha_i, \quad \Omega_{2i}a = \beta_i \cos\alpha_i, \quad \Omega_{3i}a = \alpha_i + \gamma_i, \tag{67}$$

which are understood by considering the generator matrix of the Euler transformation matrix. Note the difference from Eqs. (A.4)-(A.6) in Appendix A, in which the strains $\mathbf{\Omega}$ in the continuum description are given in terms of infinitesimal changes of Euler angles that directly transform the space-fixd reference frame to the local director frame $\{\hat{\mathbf{e}}_\alpha\}$ at s. Note also that the local bending curvature is given by the bending angle between neighboring bonds: $\kappa = (\Omega_1^2 + \Omega_2^2)^{1/2} = \beta/a$. The total energy of the system includes three contributions, $E_{tot} = E_{st} + E_{el} + E_{LJ}$. The first one is a stretching energy ensuring connectivity of spheres,

$$E_{st} = \frac{K}{2}\sum_{i=1}^{N}(|\mathbf{r}_{i+1} - \mathbf{r}_i| - a)^2. \tag{68}$$

The stretching modulus K is set sufficiently large to keep bond length fluctuations negligible, which is almost equivalent to the inextensible limit in our analytic models. The second contribution arises from the bending and twisting deformations of the rod:

$$E_{el} = \sum_{i=1}^{N}\left[\frac{A}{2}\beta_i^2 + \frac{C}{2}(\alpha_i + \gamma_i)^2\right]. \tag{69}$$

The third contribution is the truncated Lennard-Jones potential to account for chain self-avoidance:

$$E_{LJ} = \epsilon_{LJ}\sum_{i<j}\left(\frac{a^{12}}{|\mathbf{r}_i - \mathbf{r}_j|^{12}} - 2\frac{a^6}{|\mathbf{r}_i - \mathbf{r}_j|^6}\right), \tag{70}$$

which is applied only for $|\mathbf{r}_i - \mathbf{r}_j| < a$. The way to calculate elastic force $\mathbf{F}_i$ and torque about the local tangent M_i acting on each sphere are detailed in Refs.[106,107,143] In the zero Reynolds number limit, inertia is unimportant, and the elastic force and torque balance with viscous drag forces and torques, leading to the coupled Langevin equations:

$$\frac{\partial}{\partial t}\mathbf{r}_i = \sum_{j=1}^{N+1} \boldsymbol{\mu}_{ij} \cdot \mathbf{F}_j + \boldsymbol{\xi}_i(t), \tag{71}$$

$$\pi\eta a^3 \frac{\partial}{\partial t}\phi_i = M_i + \Xi_i(t), \tag{72}$$

for $i = 1, 2, \cdots N + 1$. Long-ranged (non-local) hydrodynamic interactions (HIs) between two spheres i and j can be included via the Rotne-Prager mobility tensor $\boldsymbol{\mu}_{ij}$ given by

$$\boldsymbol{\mu}_{ij} = \frac{1}{8\pi\eta r_{ij}} \left[\mathbf{1} + \frac{\mathbf{r}_{ij}\mathbf{r}_{ij}}{r_{ij}^2} + \frac{a^2}{2r_{ij}^2}\left(\frac{\mathbf{1}}{3} - \frac{\mathbf{r}_{ij}\mathbf{r}_{ij}}{r_{ij}^2} \right) \right], \tag{73}$$

for non-overlapping spheres $(r_{ij} > a)$[147,148] and for overlapping spheres $(r_{ij} < a)$[149]

$$\boldsymbol{\mu}_{ij} = \frac{1}{8\pi\eta a} \left[\left(\frac{8}{3} - \frac{3r_{ij}}{2a} \right) \mathbf{1} + \frac{\mathbf{r}_{ij}\mathbf{r}_{ij}}{2ar_{ij}} \right], \tag{74}$$

where $\mathbf{r}_{ij} = \mathbf{r}_i - \mathbf{r}_j$. This mobility tensor includes the lowest corrections in powers of the sphere size over the Oseen tensor, and correctly describes the flow field between spheres separated by a distance larger than a. For the translational self-mobility of the spherical monomers with diameter a, we use the standard Stokes formula $\boldsymbol{\mu}_{ii} = \mathbf{1}/(3\pi\eta a) \equiv \mu_0\mathbf{1}$.[99] The vectorial random displacements $\boldsymbol{\xi}(t)$ and $\Xi(t)$ model the coupling to a heat bath and obey the fluctuation-dissipation relations

$$\langle \boldsymbol{\xi}_i(t)\boldsymbol{\xi}_j(t') \rangle = 2k_BT\boldsymbol{\mu}_{ij}\delta(t - t'), \tag{75}$$

$$\langle \Xi_i(t)\Xi_j(t') \rangle = 2\pi\eta a^3 k_BT\delta(t - t'), \tag{76}$$

which are implemented numerically by a Cholesky factorization.[147] For further details on the low Reynolds number hydrodynamics a recent review exists.[100]

For the numerical integrations we discretize the Langevin equations (71) and (72) with a time step Δ and rescale all lengths, times and energies, leading to the dimensionless parameter $\tilde{\Delta} = \Delta k_BT\mu_0/a^2$. We typically set the twist-bend rigidity ratio to $C/A = 1$ (i.e., the Poisson ratio $\sigma = 0$: the incompressible limit) and the stretching modulus to $K/k_BTa^2 = 10^3$, which gives negligible bond length fluctuations. The self-crossing of the chain is entirely prevented by setting $\epsilon_{LJ}/k_BT = 5 - 10$. For sufficient numerical accuracy we typically choose a time step $\tilde{\Delta} = 0.0004$ (or smaller depending on elastic moduli or driving field strength used). Output values are calculated every 10^3-10^4 steps, total simulation times are over 10^8 steps, the first $10^6 - 10^7$ steps are not included in the data analysis. In most simulations presented below, we fix a bond vector at one end of the filament towards $\hat{\mathbf{z}}$ direction (i.e., the clamped condition) and rotate it at a given rotational velocity ω_0 about its axis. Such end constraints, $\partial_s \mathbf{r}(0) = \hat{\mathbf{z}}$, are realized by fixing the first two monomers in space by applying virtual forces, which also act (via the mobility tensor) on the rest of the filament. The rotational driving at the base is imposed with $\dot{\phi}_1 = \omega_0$. The number of beads is changed in the range $L/a = N = 30 - 100$ when HIs are included, but is increased up to $N = 200$ when HI is switched off.

7. Equilibrium Twist Diffusion in Semiflexible Polymers

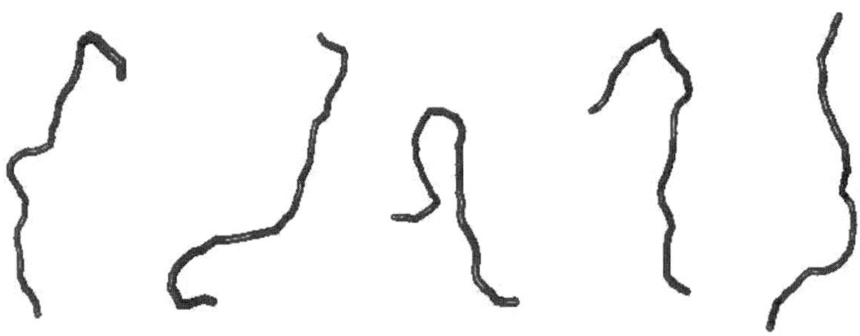

Fig. 10. Typical chain configuration for $N = 50$, $\ell_P = 10a$ and $C/A = 2$ generated from the Brownian dynamics simulations with hydrodynamic interactions.

In this section, we consider twist fluctuations in single semiflexible polymers at thermal equilibrium using analytical and numerical approaches. Statistical and dynamical properties of thermally excited bending fluctuations have been studied in detail before, including hydrodynamic effects.[45,46] For a freely fluctuating polymer in solution, twist fluctuations relax much faster than bend fluctuations as we have discussed above. The polymer chain is thermally bent beyond its bend persistence length ℓ_P, and the shape of a sufficiently long chain of length $L \gg \ell_P$ significantly differs from a straight configuration (see Fig. 10). In fact, axial spinning of a randomly bent rod accompanies flexings of such bent structure. For an isotropic rod, the bending energy depends only on its curvature, $\kappa = (\Omega_1^2 + \Omega_2^2)^{1/2}$, and not its direction. Thus flexing can occur without generating any axial torque, that is, a thermally-bent rod has no ener-

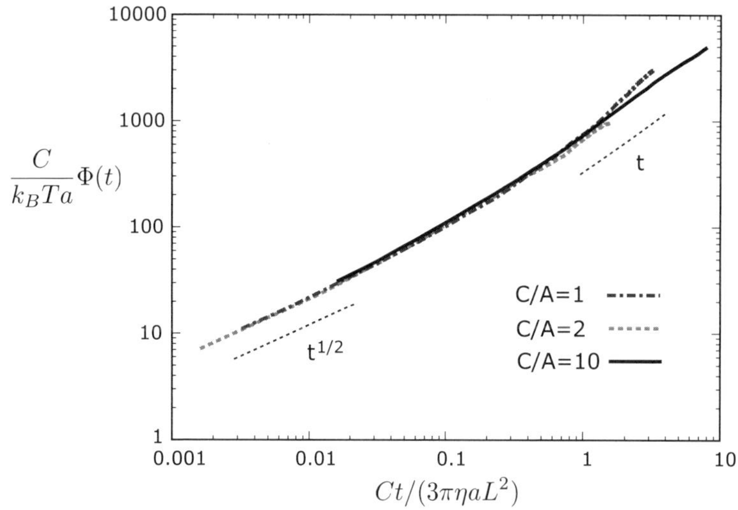

Fig. 11. Time correlation function of $\phi(s=0,t)$, $\Phi(t)$, defined in Eq. (80), obtained from numerical simulations for $N = 50$ and for $C/A = 1, 2, 10$, plotted in the scaling form predicted from Eq. (81).

getic barrier to spinning. For a freely fluctuating rod at thermal equilibrium, bending and twisting fluctuations are expected to be independent, each obeying standard Gaussian statistics ensuring fluctuation-dissipation relationships. Note that bending and twisting couple at the linear level only for rods with anisotropic bending elasticity $A_1 \neq A_2$, as can be seen from Eq. (42). Linearizing the equation of motion for twist Ω about the straight rod configuration for an isotropic rod, and at the same time setting $\Omega = \partial\phi/\partial s$, we see that the axial spinning angle $\phi(s,t)$ obeys a diffusion equation

$$\pi\eta a^2 \frac{\partial\phi}{\partial t} = C\frac{\partial^2\phi}{\partial s^2} + \xi(s,t), \tag{77}$$

where the random thermal noise of zero mean satisfies

$$\langle\xi(s,t)\xi(s',t')\rangle = 2\pi\eta a^2 k_B T\delta(s-s')\delta(t-t'). \tag{78}$$

Both ends of the polymer are free to rotate, thus torsional stress must vanish at $s = 0$ and $s = L$:

$$\frac{\partial\phi}{\partial s}(s=0,t) = 0 \qquad \frac{\partial\phi}{\partial s}(s=L,t) = 0. \tag{79}$$

The above equation can be solved using the standard normal-mode analysis.[1] For example, a temporal correlation function of ϕ at the end located at $s = 0$, defined by

$$\Phi(t) = \langle[\phi(s=0,t) - \phi(s=0,0)]^2\rangle, \tag{80}$$

can be calculated as

$$\Phi(t) = \frac{k_B T L}{\pi^2 C}\left(\frac{t}{\tau_r} + c_0\sqrt{\frac{t}{\tau_r}}\right), \tag{81}$$

where τ_r is the characteristic twist relaxation time given by

$$\tau_r = \frac{\eta a^2 L^2}{2\pi C}. \tag{82}$$

The numerical constant c_0 is given by $c_0 = \int_{-\infty}^{\infty} du(1-e^{-u^2})/u^2 = 3.54...$ The correlation function thus behaves as $\Phi(t) \sim t^{1/2}$ for short times $t < \tau_r$, and as $\Phi(t) \sim t$ at long times $t \gg \tau_r$.[69]

To see whether there is any influence of shape fluctuations upon twist relaxation dynamics in a semiflexilbe polymer, we performed free diffusion simulations of a semiflexible chain using our hydrodynamic simulation that includes full coupling between bending and twisting modes. In Fig. 11, we compare the function $\Phi(t)$ obtained from our simulations for a chain of length $L = 50a$, bend persistence length $\ell_P = 10a$, and for different values of C/A. The agreement between the linearized simple theory and the simulations is good, demonstrating that twist dynamics has little influence on the internal shape dynamics of a freely diffusing semiflexible chain (as shown previously by Maggs).[69] However, as we will see from now, this is not at all true for a rotationally driven polymer, i.e., in the presence of constant twist injection.

8. Twisted Elastic Rods and Buckling Instabiliy

In the remainder of this chapter, we will mainly discuss nonequilibrium viscous dynamics found in elastic rods and in semiflexible polymers. We prepare an elastic rod or polymer with one end clamped (i.e., the position and the tangent of one end is fixed). This clamped end is forced to rotate at frequency ω_0, while the other end is entirely free to relax. We will look at what happens in this polymer chain as the frequency ω_0 is increased up to sufficiently high values.

To disentangle mechanical and entropic effects, we begin with arguments for the zero temperature system. As a result of balance between twist injection, twist diffusion, and writhing, it is known that there are twirling and whirling dynamical steady states in this system.[89] This transition from

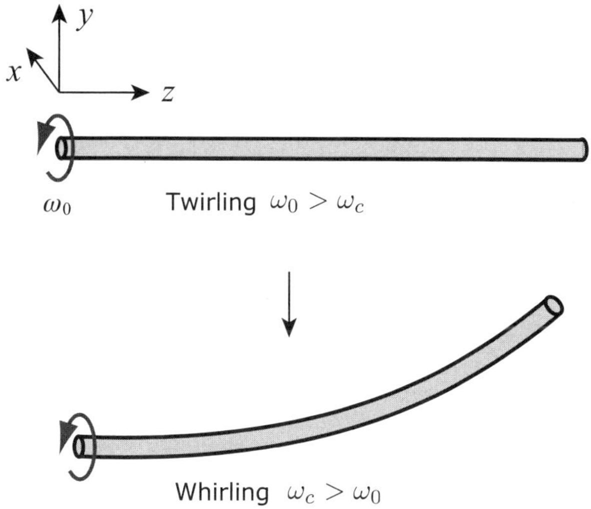

Fig. 12. Schematic picture of the model system: An elastic rod is forced to rotate at its base at frequency ω_0 with the other end free in a fluid of viscosity η. For $\omega_0 < \omega_c$, the rod shows twirling motion, while for $\omega_0 > \omega_c$, the whirling state is realized.

twirling to whirling motion is a dynamical analogue of the classical buckling transition of a uniformly twisted rod with clamped ends. It may therefore be instructive to first recall the argument on *static* buckling of an axially stressed rod.

8.1. *Twisted elastic rod without stretching*

Suppose that an initially straight elastic rod of length L is subjected to a uniform twist of $\Omega_3 = \tau$ by externally applied torques at both ends. As it is axially twisted, τ increases, and at some point the straight rod loses its stability and buckles: see Fig. 13. We analyze the onset of this shape instability by using the results that we have obtained before, that is, we perform a linear stability analysis. The initial loop formation and subsequent plectoneme growth involving self contacts are nonlinear phenomena and require different approaches.[150] There is currently an extensive literature on these classical problems.[115,117–120] For a historical account of the study of twisted elastic rings and its instability called *Michell's instability*[151] or *Zajac's instability*,[152] see Ref.[153] and references therein.

To leading order, the deflection of the rod is perpendicular to the z-axis,

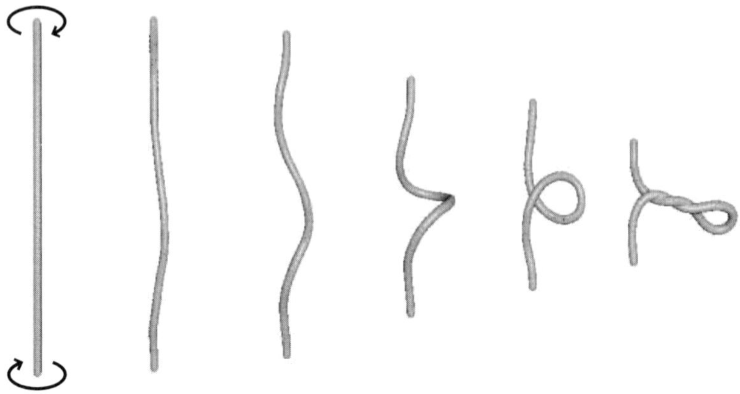

Fig. 13. Example of a classical buckling instability of a twisted elastic rod and subsequent plectoneme formation process. This illustration is obtained from dynamic simulations.

so we set

$$\mathbf{r} = \mathbf{r}_\perp + z\hat{\mathbf{z}}, \quad \mathbf{r}_\perp(z) = (X(z), Y(z), 0). \tag{83}$$

The tension $-\Lambda$, arising from the inextensibility of the rod, is second order in X and Y, and will be neglected in the following linear stability analysis. The final results are different depending on the boundary conditions at the upper and lower ends. We consider two cases, one is the clamped case and the other is the hinged case for both ends. Extensions to mixed cases would be straightforward. We first analyze the clamped boundary condition case.

Clamped case— If the upper and lower ends are clamped, they do not move laterally, and their tangent vectors are also fixed pointing towards $\hat{\mathbf{z}}$.

$$\mathbf{r}_\perp(z=0) = \mathbf{r}_\perp(z=L) = 0, \qquad \partial_z\mathbf{r}(z=0) = \partial_z\mathbf{r}(z=L) = 0. \tag{84}$$

The mechanical equilibrium condition is given by

$$\partial_s\mathbf{F} = -A\partial_z^4\mathbf{r} + C\partial_s\left[\Omega_3(\partial_s\mathbf{r}\times\partial_s^2\mathbf{r})\right] - \partial_s(\Lambda\partial_s\mathbf{r}) = \mathbf{0}. \tag{85}$$

Linearizing this leads to $0 = -A\partial_z^4\mathbf{r}_\perp + C\tau(\hat{\mathbf{z}}\times\partial_z^3\mathbf{r}_\perp)$, or

$$-A\frac{d^4X}{dz^4} - C\tau\frac{d^2Y}{dz^3} = 0, \quad -A\frac{d^4Y}{dz^4} + C\tau\frac{d^2X}{dz^3} = 0. \tag{86}$$

Following Landau-Lifshitz,[103] we introduce a complex variable $\xi = X + iY$, which satisfies $d^4\xi/dz^4 - i\kappa d^3\xi/dz^3 = 0$, where $\kappa = C\tau/A$, with the boundary conditions $\xi(0) = \xi(L) = 0$ and $\partial_z\xi(0) = \partial_z\xi(L) = 0$. Because the eigenvalues for this equation are triply degenerated, the solution is given in the form $\xi(z) = a(1 + i\kappa z - e^{i\kappa z}) + bz^2$, which already satisfies $\xi(0) = \partial_z \xi(0) = 0$. Plugging the solution into the remaining boundary conditions determines (a, b). Non-zero values of (a, b) ensure the appearance of the buckled configuration, which occurs when $\tan \kappa L/2 = \kappa L/2$. The minimum τ that satisfies this condition is the critical twist we are looking for, which is

$$\tau_c = 8.98 \frac{A}{CL}. \tag{87}$$

On the scaling level, there is a simple interpretation of the final result Eq. (87): At the critical point, the rod of length L buckles with a curvature of $\kappa_c \sim 1/L$. The bending moment of this deformation is $\sim A\kappa_c \sim A/L$. When an externally applied torque that balances the twisting moment, $C\tau$, becomes comparable to this bending moment, the buckling can occur. We thus predict $C\tau_c \sim A/L$, which reproduces the critical torque condition Eq. (87) apart from the numerical prefactor.

Hinged case— If the upper and lower ends are hinged, they do not move laterally, but their tangents are free to change. Instead the bending moments at the ends vanish. The corresponding boundary conditions are thus given by

$$\mathbf{r}_\perp(z = 0) = \mathbf{r}_\perp(z = L) = 0, \qquad \partial_z^2 \mathbf{r}(z = 0) = \partial_z^2 \mathbf{r}(z = L) = 0. \tag{88}$$

The solution of the above mechanical equilibrium equation under the hinged boundary conditions is given by $\xi(z) = a(1 - e^{i\kappa z} - \kappa^2 z^2/2) + bz$. Repeating the same argument as before, we obtain the critical twist in the hinged case as $\tau_c = 2\pi A/CL$.

8.2. *Twisted elastic rod under stretching*

When a tension f is applied externally to a rod, the critical torque will be modified, because the stretching force ($f > 0$) will prevent a buckling of the rod. To take into account the effect of tension, we now set $\Lambda = -f$, which leads to the modified linearized equation

$$\frac{d^4\xi}{dz^4} - i\kappa \frac{d^3\xi}{dz^3} - \alpha^2 \frac{d^2\xi}{dz^2} = 0, \tag{89}$$

where we have defined the inverse length scale

$$\alpha = (f/A)^{1/2}. \tag{90}$$

Because the eigenvalues of Eq. (89) are 0 (double-degenerated), and $\lambda_\pm$ where

$$\lambda_\pm = \frac{\kappa}{2} \pm \sqrt{\frac{\kappa^2}{4} - \alpha^2}, \tag{91}$$

the solution of Eq. (89) is of the form $\xi(z) = a(1 + i\lambda_+ z - e^{i\lambda_+ z}) + b(1 + i\lambda_- z - e^{i\lambda_- z})$ for the clamped case in Eq. (84). Repeating the same argument as before, non-zero values (a, b) exist when κ satisfies the equation $F(\kappa L/2, \alpha L) = 0$, where[154]

$$F(x, a) = 2\sqrt{x^2 - a^2}\left(\cos x - \cos\sqrt{x^2 - a^2}\right) + a^2 \sin\sqrt{x^2 - a^2}. \tag{92}$$

Numerical solutions of this equation are given in Fig. 14, where the critical twist under tension f, $\tau_c(f)$ (rescaled by $\tau_c(0)$ given in Eq. (87)), is plotted as a function of $\alpha L = (fL^2/A)^{1/2}$. One can see that the critical twist $\tau_c(f)$ increases with the stretching force f. In particular, for $(fL^2/A)^{1/2} \gg 1$, it approaches the result for the hinged boundary condition case, Eq. (93) and the dotted line in Fig. 14, thus we know $\tau_c \sim (fA)^{1/2}/C$. For $(fL^2/A)^{1/2} \ll 1$, on the other hand, $F(x, a)$ can be expanded as $F(x, a) \approx 2\frac{\cos x}{x^2}\left[\tan(x/2) - x/2\right] a^2 + O(a^4)$. We thus recover Eq. (87) in the limit of vanishing stretching force, as also confirmed by the numerical result in Fig. 14.

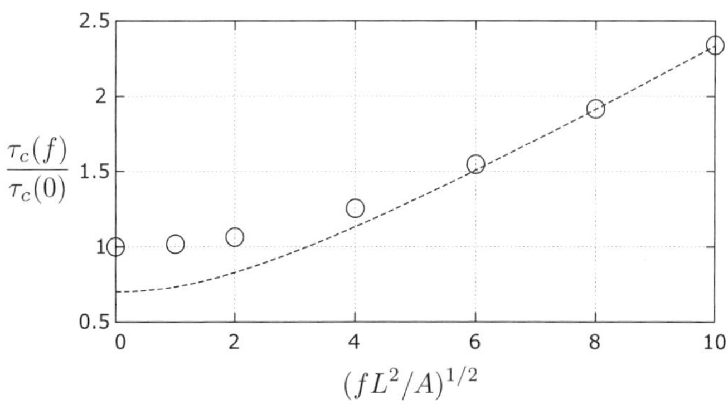

Fig. 14. Critical twist $\tau_c(f)$ rescaled by its zero force value $\tau_c(0) = 8.98A/(CL)$, plotted as a function of rescaled stretching force $(fL^2/A)^{1/2}$. Open circles are the data for the clamped boundary conditions case obtained by solving Eq. (92) numerically. The dotted line is the one for the hinged boundary conditions case, given in Eq. (93).

Hinged case— For the hinged boundary conditions given in Eq. (88), the corresponding solution is given by $\xi(z) = a[\lambda_-^2(e^{i\lambda_+ z} - 1) - \lambda_+^2(e^{i\lambda_- z} -$

$1)] + bz$. Repeating the same argument, the condition for non-zero (a, b) is found to be $\sin\sqrt{(\kappa L)^2/4 - \alpha^2} = 0$. The minimum κ satisfying this (under the assumption that $\kappa L > 2\alpha$) is $\sqrt{(\kappa L)^2/4 - \alpha^2} = \pi/2$. We then obtain the critical twist given by

$$\tau_c = \frac{2A}{C}\sqrt{\frac{f}{A} + \frac{\pi^2}{L^2}}, \tag{93}$$

which recovers the classical result given by Love.[102] For infinitely long rods, $L \to \infty$, the critical torque for the buckling under stretching force f is given as $\tau_c = (4fA)^{1/2}/C$.

9. Rotationally Driven Stiff Filaments: The Twirling-Whirling Instability

Going back to the problem at hand, we can now estimate the critical frequency on the scaling level from analogious arguments. At low rotational frequency, $\omega_0 < \omega_c$, the rod is twisted but remains straight and points towards the $\mathbf{z} = \hat{\mathbf{t}}(0)$ direction. From Eq. (50), the twist Ω obeys the diffusion equation

$$\zeta_r \dot{\Omega} = C\frac{\partial^2 \Omega}{\partial s^2}. \tag{94}$$

For twist boundary conditions, the local torque balance at the driving end suggests $\zeta_r \omega_0 = C\partial_s \Omega(0)$ and the elastic moment vanishes at the free end, $C\Omega(s = L) = 0$. Solving Eq. (94) under these boundary conditions, Wolgemuth et al. found that the steady state twist profile is linear in space ($\dot{\Omega} = 0$):[89]

$$\Omega(s) = \frac{\zeta_r \omega_0}{C}(s - L). \tag{95}$$

From the boundary conditions Eq. (45), the external torque M_{ext} required to apply at the driving end in order to maintain the rod rotating at a given frequency ω_0, is thus obtained as

$$M_{ext} = -C\Omega(0) \;=\; \zeta_r \omega_0 L \;=\; \pi\eta a^2 \omega_0 L, \tag{96}$$

which is proportional to ω_0. This is just a consequence of the linearized viscous dynamics. The rotational friction constant is simply given by $\pi\eta a^2 L$. In this axial-spinning regime, the torque at the base balances the total rotational drag, $\pi\eta a^2 \omega_0 L \sim C\Omega(0)$. This diffusion-dominated axial spinning mode is called twirling. As the driving frequency ω_0 increases, the rod buckles due to dynamical twist/bend coupling. On the scaling level, this elastic instability appears when the twisting external torque $M_{ext} = \pi\eta a^2\omega_0$ becomes comparable to the bending torque, A/L, i.e., giving a critical frequency $\omega_c \sim A/\pi\eta a^2 L^2$, which is independent of the twist rigidity C.[89] This is in contrast to the static case Eq. (87) and is characteristic to this dynamical problem. Linearizing Eq. (49) about the straight configuration as we have done for the static torsional buckling analysis, we find that the transverse displacement variable $\xi(s) = X(s) + iY(s)$ obeys

$$\zeta\dot{\xi} = -A\partial_s^4\xi + i\zeta_r\omega_0\partial_s\left[(s-L)\partial_s^2\xi\right], \tag{97}$$

with the boundary conditions $\xi(0) = 0$ and $\partial_s\xi(0) = 0$ (clamped condition at $z = 0$) and $\partial_s^2\xi(L) = \partial_s^3\xi(L) = 0$ (vanishing force and bend moment at $z = L$). Similar to the previous static case, non-zero solutions only appear in Eq. (97) for special values of ω_0. The lowest value of such ω_0 was shown to be[89,91,101]

$$\omega_c = 8.9\frac{A}{\pi\eta a^2 L^2} = 8.9\frac{k_B T}{\pi\eta a^2 \ell_P}\left(\frac{\ell_P}{L}\right)^2. \tag{98}$$

For $\omega_0 > \omega_c$, twirling is unstable against infinitely small disturbances and the rod starts buckling to relieve twist via writhing, leading to crankshafting motion. At the buckling, the free end traces out a circle about the rotational axis, which grows exponentially in time until the rod bends over. The rod then reaches the stationary state, in which the rod exhibits a hybrid motion, i.e., it shows rigid-body-like rotation (crankshafting) and also spins rapidly about its local tangent vector.

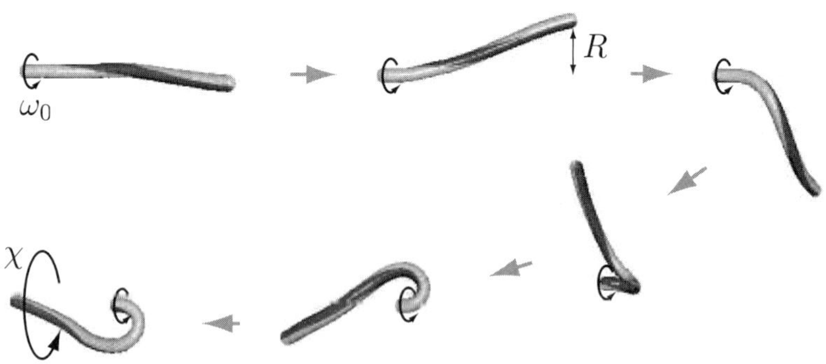

Fig. 15. Sequence of snapshots of a rotating filament of length $N = 30$ and stiffness $\ell_P/L = 10^3$ for $\omega_0/\omega_c = 1.20$.

9.1. *Nonlinear whirling dynamics*

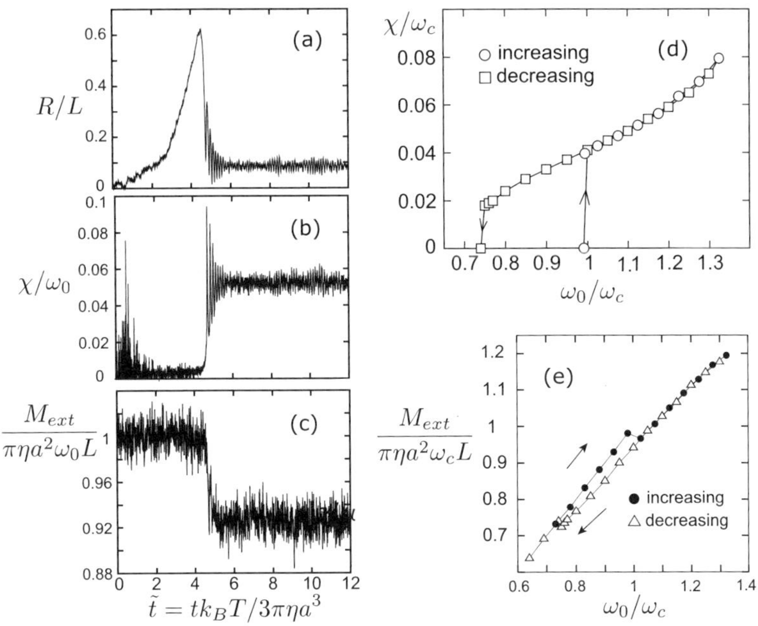

Fig. 16. Time evolution of various quantities corresponding to Fig. 15. (a) the end-point radius R, (b) the crankshafting frequency χ and (c) the torque externally applied at the driving end M_{ext}. (d) Hysteresis loop of χ for a $N = 30$ and $\ell_P/L = 10^3$ rod. (e) same for M_{ext}.

A numerical approach is most suitable to study nonlinear time evolution of the rod after the bifurcation. The time evolution of the rod shape for $N = 30$, persistence length $\ell_P/L = 10^3$ for $\omega_0/\omega_c = 1.20$ (starting with a straight shape) is shown in Fig. 15. Twirling is unstable against thermal disturbance and the rod buckles to relieve twist, leading to crankshafting motion. The radial distance of the free end from the rotational axis, called R, initially increases exponentially in time until the rod bends over (see Fig. 16 (a)) to reach a steady whirling motion.[90,101,155] This large-amplitude whirling motion is a combination of rigid-body-like rotation with frequency χ and much faster axial rotation. The crankshafting frequency χ increases drastically at the transition from a transient small-ampliutde whirling to a steady large-amplitude whirling state, while the torque applied at the base, M_{ext}, shows a small but steep decrease: see Fig. 16 (b) and (c). In other words, *the transition to the whirling state can be viewed as a way to reduce the dissipated power.*

In Figs. 16 (d) and (e), the steady state values of χ and M_{ext} are plotted against ω_0/ω_c across the transition for a rather stiff rod with $\ell_P/L = 10^3$. A pronounced hysteresis is revealed, suggesting that the observed transition is strongly discontinuous and the final state depends not only on the driving frequency ω_0 but also on the initial shape of the rod. As initial configuration of the rod we take part of a circle of radius L/α, and characterize its degree of initial bending by a parameter α (see Fig. 17 (a)). The bifurcation frequencies depend on the rate with which the frequency is changed, but only for very flexible rods does the hysteresis disappear. We construct the stability diagram shown in Fig. 17 (a) in the (ω_0, α) plane at zero temperature (i.e. the deterministic noise-less limit where $\ell_P/L = \infty$) by following the time evolution of a rod with an initial circular shape with bending angle α and the stationary twist profile. Twirling is the final stationary state for a rod with an initial shape bending angle below the separatrics, while above the separatrics the whirling state is obtained. The separatrics terminates for $\alpha = 0$ at $\omega_0 = \omega_c$ as given by Eq. (98),[89,101] indicating that ω_c is an instability point and actually the correct critical frequency at zero-temperature above which the rod is unstable against infinitesimal disturbance. By construction, the stability diagram does not depend on the stiffness of the filament. The cause of the discrepancy to the previous simulation[90] at zero temperatures which yielded a much lower transition frequency[90] is unclear.

9.2. *Influence of thermal fluctuations on the dynamical transition*

At finite temperature (i.e. finite ℓ_P/L), the whirling transition can be observed below ω_c: In Fig. 17 (b) the actual transition frequency ω_c^T is plotted as a function of ℓ_P/L, obtained from simulations where the driving frequency ω_0 is increased in steps of size $\Delta\omega_0 = 10^{-2}\omega_c$ every 10^7 simulation time steps. The time-dependent behavior of χ and ω_0 across the transition is monitored, from which ω_c^T is determined. An even slower frequency increase rate might give a slightly lower ω_c^T, so the results in Fig. 17 (b) correspond to an upper limit of the transition in the stationary limit. A simple scaling argument explains the observed trend: Balancing the bending energy $E \sim A\alpha^2/(2L)$ with $k_BT/2$, one obtains a spontaneous mean

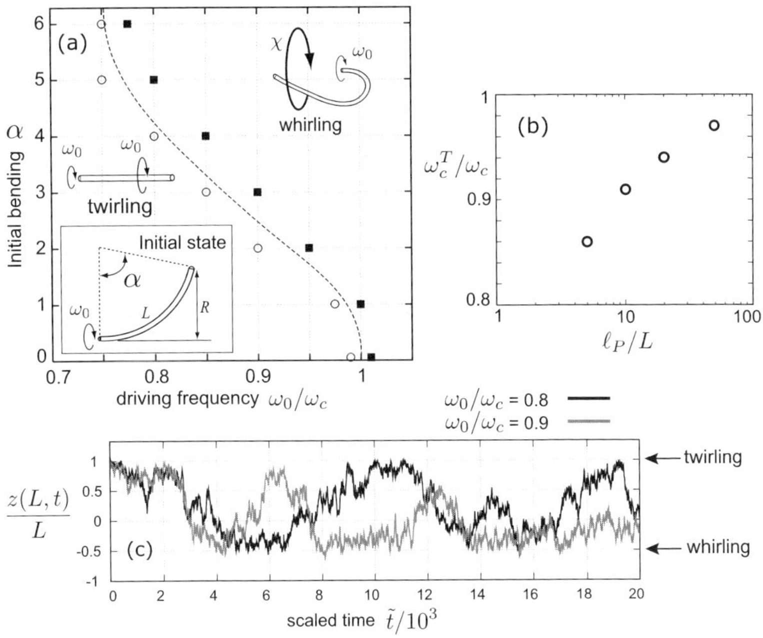

Fig. 17. (a) Zero-temperature stability diagram as a function of initial bending angle α and driving frequency ω for a $N = 30$ rod. (b) Observed transition frequency ω_c^T for varying stiffness ℓ_P/L for $N = 30$ rod. (c) Time evolution of z position of the free end, $z(s = L, t)$, of a flexible rod with $\ell_P/L = 2$ rotating at frequency below ω_c: $\omega_0/\omega_c = 0.8$ (black) and $\omega_0/\omega_c = 0.9$ (gray), showing that the rod stochastically sw itches between twirling and whirling states.

squared bending angle

$$\langle \alpha^2 \rangle \sim L/\ell_P, \tag{99}$$

which is larger for smaller ℓ_P and thus leads to reduced stability (see the stability diagram). Below $\ell_P/L \approx 5$, the elastic rod switches stochastically between the two states, and the transition is washed out, see Fig. 17 (c). The dependence of the twirling-whirling transition on the persistence length could be experimentally tested with actin or tubulin filaments of varying length.[156,157]

9.3. Energetic aspects of writhing dynamics

We plot in Fig. 18 the relationship between the external torque M_{ext} and the driving frequency ω_0 across the twirling-whirling transition obtained from our numerical simulations in a wider range of ω_0/ω_c than Fig. 16 (e). For $\omega_0 < \omega_c$, our numerical data confirms the analytic results for the linearized dynamics, Eqs. (95)-(96). Interestingly, however, the torque M_{ext} grows only sub-linearly above the buckling transition (i.e., for $\omega_0 > \omega_c$). This result further supports the idea that the transition to the whirling state can be viewed as a way to reduce the dissipated power, $P = M_{ext}\omega_0$.

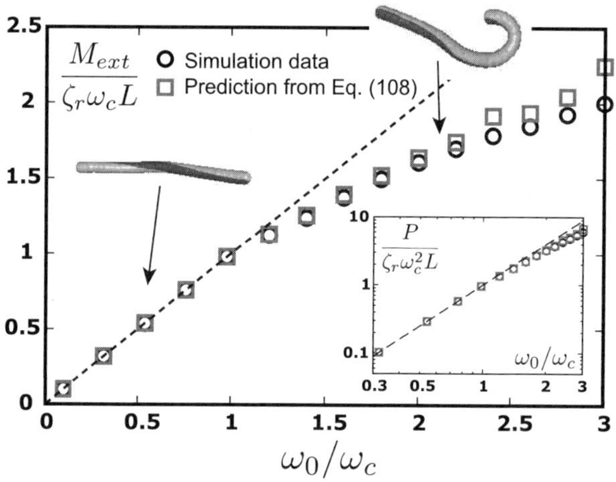

Fig. 18. Relationship between the external torque M_{ext} and the driving freque ncy ω_0 obtained from our dynamic simulations for a rod with its length $N =$ 30 and the stiffness $\ell_p/L = 10^3$. Circle (black) points are M_{ext} "measured" directly at the base, while square (gray) points are the prediction from our theory: Eq. (108) is numerica lly evaluated. Two quantities are determined independently and show good agreement.

However, this finding may be counter-intuitive, considering that viscous dissipation must substantially increase to execute the crankshafting motion. An important key is the geometric relation for twist dynamics, Eq. (66). Note first that the centerline of a whirling rod exhibits rigid-body rotation with a frequency χ independent of time, given by

$$\dot{\mathbf{r}} = \chi \hat{\mathbf{z}} \times \mathbf{r}(s), \tag{100}$$

where $\hat{\mathbf{z}} = \hat{\mathbf{t}}(0)$ is the axial spinning axis. Plugging this equation into Eq. (66) (with the local inextensibility $\mathbf{t} \cdot \partial_s \dot{\mathbf{r}} = 0$), we obtain the local conservation law for twist *without* any source terms, $\dot{\Omega} + \partial_s j = 0$. The corresponding "twist current density" is given by

$$j(s) = -\omega(s) + \chi \cos\theta(s), \tag{101}$$

where we have defined $\hat{\mathbf{z}} \cdot \hat{\mathbf{t}}(s) = \cos\theta(s) = \partial z(s)/\partial s$. In addition to the diffusive current $-\omega$, Eq. (101) has an additional contribution, $\chi\cos\theta$, which describes a twist change due to writhing. Note, however, that it is not generally true that a twist current can be given in such a form as Eq. (101). As emphasized in Eq. (66), twist current should read as $-\omega$ and the writhing term generally works as a source.[89] If we parameterize the tangent as $\hat{\mathbf{t}}(s) = (\sin\theta\cos\psi, \sin\theta\sin\psi, \cos\theta)$, the writhing term in Eq. (66) can be written as $\sin\theta(\dot{\psi}\partial_s\theta - \dot{\theta}\partial_s\psi)$, not a partial derivative of a certain current with respect to s. In our particular problem, however, steady state crankshafting motion has an axial symmetry that ensures $\dot{\theta} = 0$ and $\dot{\psi} = \chi$ and allows us to define the twist current of the form in Eq. (101). Equation (101) is central in our following argument. At the steady state, an injected twist at $s = 0$ has to exit the rod at $s = L$ either in the form of axial spinning or writhing, meaning that the twist current is constant, $j(s) = j(0) = -\omega_0 + \chi$. This condition leads to the following geometric relation

$$\omega_0 = \omega(s) + \chi(1 - \cos\theta(s)). \tag{102}$$

In particular, at the free end $s = L$, we have

$$\omega_0 = \omega(L) + \chi(1 - \cos\theta(L)). \tag{103}$$

We note that the essentially same relation was previously given (but in the weak bending limit) in Ref.[89] One finds immediately that twist transport via writhing is maximized for $\cos\theta(L) = -1$ at a given ω_0.

Our argument so far depends only on geometry, and contains no physics. Now, substituting the viscous torque balance equation (47) into Eq. (102), and integrating over s, we find

$$C\Omega(0) = -\zeta_r\omega_0 L + \zeta_r\chi L(1 - \sigma), \tag{104}$$

where we have used $\Omega(s = L) = 0$ and defined the fractional extension of the rod $\sigma = z(L)/L$. One can note that $\sigma \approx 1$ for twirling while $\sigma < 0$ for whirling, and $\int_0^L ds(1 - \cos\theta) = L(1 - \sigma)$. Despite the linearity of the elastic constitutive relations and viscous dynamics, we find nonlinear dependences on rotational frequencies due to the change of rod shape (see Fig. 18). To see how the torque M_{ext} is related to rod shape changes more quantitatively, we return to the Kirchhoff equations (52) and (53). Integrating Eq. (53) over the rod length, and using the boundary conditions $\mathbf{r}(0) = \mathbf{0}$ and $\mathbf{F}(L) = \mathbf{M}(L) = \mathbf{0}$, we find

$$\mathbf{M}(0) + \int_0^L [\mathbf{r}(s) \times \zeta \mathbf{v}(s)]\, ds + \int_0^L \zeta_r \omega(s)\hat{\mathbf{t}} = 0. \tag{105}$$

For the whirling rod, the backbone shape follows from Eq. (100): $\mathbf{v} = \chi \hat{\mathbf{z}} \times \mathbf{r}$. Plugging this into Eq. (105), and noting $\mathbf{M}(0) = A\hat{\mathbf{z}} \times \partial_s^2 \mathbf{r}(0) + C\Omega(0)\hat{\mathbf{z}}$ from Eq. (44), we obtain

$$C\Omega(0) + \zeta\chi \int_0^L |\mathbf{r}_\perp(s)|^2 ds + \zeta_r \int_0^L \omega(s)\hat{\mathbf{t}} \cdot \hat{\mathbf{z}} = 0, \tag{106}$$

where $\mathbf{r}_\perp(s) = [x(s), y(s), 0]$. This equation describes the overall torque balance between the rod's internal torque and the hydrodynamic friction torque. Plugging Eqs. (102) and (104) into this equation, we find the steady state crankshafting frequency as

$$\chi = \omega_0 \frac{\zeta_r(1-\sigma)L}{\int_0^L ds\, [\zeta_r(1-\cos\theta)^2 + \zeta|\mathbf{r}_\perp|^2]}. \tag{107}$$

Applying this to a uniformly and weakly bent rod at $\omega \approx \omega_c$, we obtain $\chi \approx 30A/(\zeta L^4) \approx \omega_c(a/L)^2$, which is in agreement with the value given in Ref.[101] Finally, the external torque applied at the base, $M_{ext} = \hat{\mathbf{z}} \cdot \mathbf{M}_{ext}$, is from the boundary condition given by $M_{ext} = -C\Omega(0)$. Combining Eq. (104) and (107), we obtain

$$M_{ext}(\omega_0) = \zeta_r \omega_0 L \left[1 - \frac{\zeta_r L(1-\sigma)^2}{\int_0^L ds\, [\zeta_r(1-\cos\theta)^2 + \zeta|\mathbf{r}_\perp|^2]}\right]. \tag{108}$$

Because $M_{ext}^0 = \zeta_r \omega_0 L$ is the torque to twirl a rigid rod of length L at ω_0, this proves

$$\frac{M_{ext} - M_{ext}^0}{M_{ext}^0} = -\frac{\zeta_r L(1-\sigma)^2}{\int_0^L ds\, [\zeta_r(1-\cos\theta)^2 + \zeta|\mathbf{r}_\perp|^2]} < 0. \tag{109}$$

Note that $\zeta_r \omega_0 L$ is the torque to twirl a rigid rod of length L at ω_0. Note also that the second term in the bracket in Eq. (108), the nonlinear writhe correction term, is always positive for $\omega_0 > \omega_c$, and is essentially independent of the viscocity, which indicates that it has a purely geometric origin. Recalling $P = \omega_0 M_{ext}$ (see also inset of Fig. 18), one arrives at a physically important conclusion: The role of the whirling transition is to reduce the viscous power dissipation. This result reveals a dissipative origin of the large amplitude whirling transition, and allows us to view this phenomena as a dynamical non-equilibrium analog of a uniformly over-twisted rod that buckles to reduce its elastic energy.

To obtain an exact value of M_{ext}, we still have to determine the full shape of the rod to evaluate the integral in Eq. (108). We compare in Fig. 18 our prediction from Eq. (108) with the data obtained directly from the numerical simulations. In the comparison, the rod shape was extracted from the numerical simulations to evaluate the integral in Eq. (108). There is very good agreement, which further validates our physical arguments. At higher ω_0, deviations become visible because the assumed axial symmetry is violated due to secondary shape bifurcations.

9.4. *Geometric and topological aspects of writhing dynamics*

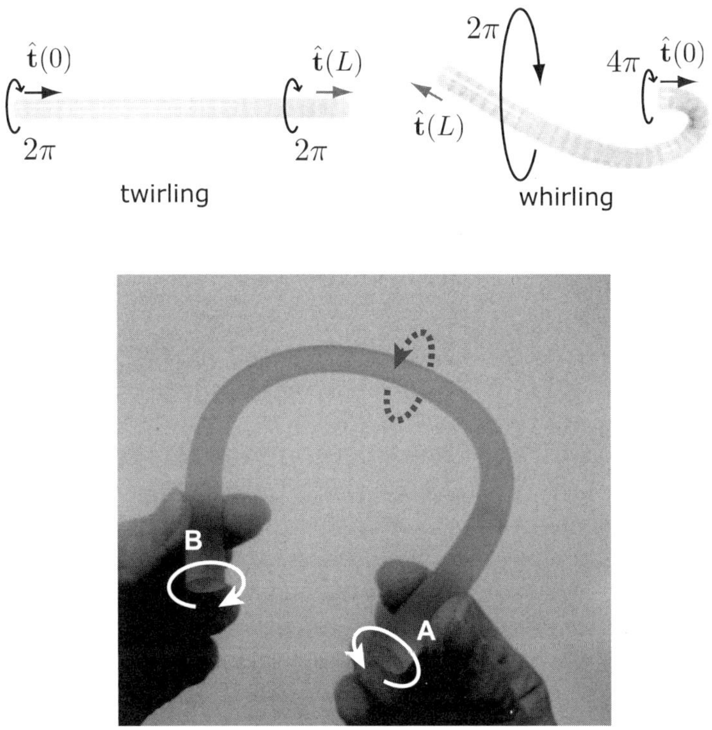

Fig. 19. (Top) Geometric aspect of twist transport in a whirling rod: 2π global rotation relieves 4π axial twist. (Bottom) Simple demonstration with a rod that is bent into a half circle.

The buckled rod undergoes a cycle of global motions that exactly comes back to its original shape while the driving base experiences a rotation of 4π about its axis during it. On this observation, there is a simple geometric interpretation; see Fig. 19. Take an open rod and fold it into a half-circle. After rotating both ends by 2π into opposite directions, the rod executes a 2π rotation about its axis without changing its shape. Looking at this phenomenon in a co-rotating frame at the end A (shown in Fig. 19 below), the other end B executes a rotation of 4π about its tangent, while at the same time the rod centerline undergoes a global rotation of 2π about the tangent at A. Thus, 2π global rotation is equal to 4π axial roation. According to Eq. (112), this amounts to $\omega_B - \omega_A = 2\chi$. This "double frequency" event found in the large amplitude whirling is actually relevant to the Feynman's plate trick in which rotation and translation are done at the same time: see the photos on page 30 in Ref.,[158] and see also Fig. 19 (Top). This global rotation (whirling motion) can relieve twist at the driving end most efficiently, but it costs larger dissipation as a tradeoff. This is why this mode prevails only at elevated frequency ω_0. While not mathematically new, the above simple geometric consideration is helpful to more intuitively understand why a rotating rod adopts such a unique buckled shape upon its transition.

There are geometric phases at play here.[159–161] In polymer physics literature, geometric phase is simply accounted for as writhe, Wr.[137,160,162] According to Fuller's result, Wr is an integer plus $Ar/2\pi$, where Ar is the signed area enclosed by the trace of $\hat{\mathbf{t}}$ on a unit sphere (tangent indicatrix). As we have seen before, the rate of change of this non-integer part of Wr generally satisfies a conservation law for open filaments undergoing smooth deformations.[108,127,128] To see this more directly, let us apply Eq. (65) to our whirling rod, which gives

$$\frac{d}{dt}\mathrm{Wr} = \frac{\chi}{2\pi}\left(1 - \cos\theta(L)\right). \tag{110}$$

Thus the "velocity" of writhe is zero for a twirling rod, $\theta = 0$, and is nearly χ/π for rods executing large-amplitude whirling. For a whirling rod, its tangent indicatrix on a unit sphere is not closed, but its "velocity" dAr/dt is well defined and local. Considering an areal element dA swept out by the unique shorter geodesics connecting the tangent $\hat{\mathbf{t}}(s)$ to the north pole $\hat{\mathbf{z}} = \hat{\mathbf{t}}(0)$ within an infinitely short time dt, it is $dAr = \int_0^{\theta(L)} (\chi dt)\sin\theta d\theta = \chi(1-\cos\theta(L))dt$, which confirms $d\mathrm{Wr}/dt = (1/2\pi)dAr/dt$.[133] Noting $d\mathrm{Tw}/dt = 0$ (which is trivial from the conservation $\dot{\Omega} + \partial_s j = 0$), and still keeping the notation Lk = Wr + Tw, we find

$$\frac{d}{dt}\text{Lk} = \int \frac{ds}{2\pi}\partial_s\omega(s) = \frac{\omega(L) - \omega_0}{2\pi} = \frac{\chi}{2\pi}(1 - \cos\theta(L)) = \frac{d}{dt}\text{Wr}. \quad (111)$$

Thus the link exits at the end solely by writhing, quite similar to the geometric untwisting for a twist free rod with $C/\zeta_r = 0$,[108] but our example holds at a rotational dissipative steady state. Note also that the "velocity" of link, $d\text{Lk}/dt$, has its meaning as a "net" number of rotations of the rod per unit time.

10. Rotationally Driven Semiflexible Polymers: Plectoneme Transitions

Next we proceed to study nonequilibrium twist dynamics of highly fluctuating semiflexible polymers. We consider basically the same system but now the chain total length is much longer than its bend and twist persistence lengths. Thus polymer conformations are considerably modified by entropic effects compared to the low temperature cases we have discussed above. In Fig. 20 we show typical chain snapshots obtained in our dynamic simulations for driving frequencies $\omega_0/\omega_c = 0.3$, 15 and 40 for a chain with $L/\ell_P = 10$. For $\omega_0 < \omega_c$, the chain flexes randomly due to thermal motions and spins about its local axis at frequency $\sim \omega_0$. For $\omega_0 \gg \omega_c$, in contrast, the polymer exhibits continuous generation of plectoneme-like structures and their diffusive transport from the driving end to the free end. Furthermore, those plectonemes are formed and localized close to the driving end, in a highly twisted region nearby the cranking point. Conformational dynamics of a fluctuating polymer is far more complicated than for stiff rods studied above, and an exact analytical approach using elastohydrodynamic equations is impossible. In the following we present a much-simplified scaling theory, valid for long elastic rods $L \gg \ell_P$, that establishes a minimal framework for treating the competitive twist transport due to axial spinning and plectoneme creation (writhing).

10.1. *Three dissipation channels and scaling arguments*

There are three different ways for the polymer to transport the injected twist to its free end, see Fig. 21. The first one is the (A) axial spinning mode, where the polymer rotates about its contour like a speedometer cable, which we show to be the dominant dissipation mode for low enough driving frequencies. The second one is the (B) solid-body rotation mode, where

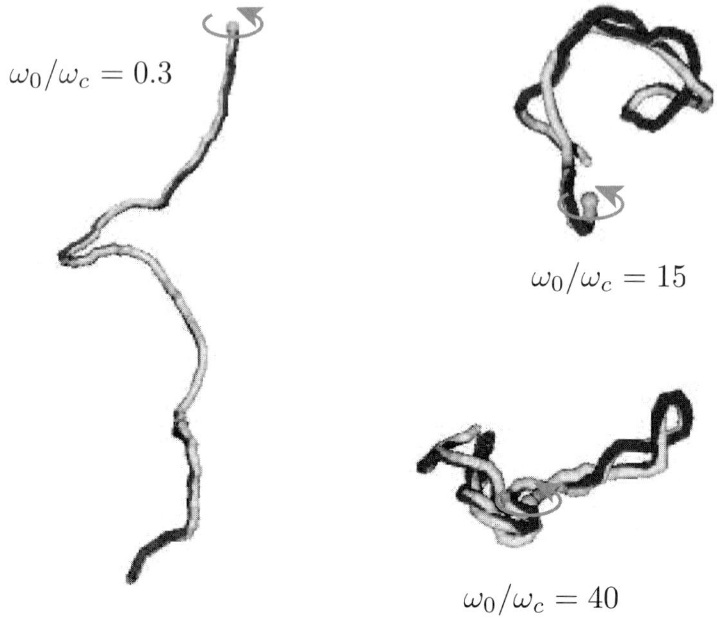

Fig. 20. Typical chain conformations from our hydrodynamic simulations for chain length $L = 100a$ and persistence length $\ell_P = 10a$ and rotational frequency $\omega_0/\omega_c = 0.3, 15$ and 40

the whole polymer coil whirls around the rotational axis at some frequency χ. This was the mode that realized the efficient twist transport in the stiff rod limit. For a flexible polymer, however, there is one more channel, namely the (C) plectoneme creation/diffusion mode, in which plectoneme-like structures are continuously generated at the rotated end and are transported diffusively towards the free end. One should note that this type of motion does not much concert with the solid-body rotation. Thus (B) and (C) should be distinguished as different dissipation mechanisms. As we show below, this dynamic twist-writhe conversion is a highly nonlinear mechanism that provides a very efficient means of relieving torsional stress at elevated driving frequencies.

Note that in a stationary state, the twist that is injected into the polymer has to exit the chain at the free end either in the form of axial spinning, solid-body rotation, or plectoneme creation. In the stationary state, the twist current in Eq. (101) satisfies $j(0) = j(L)$, which on the scaling level

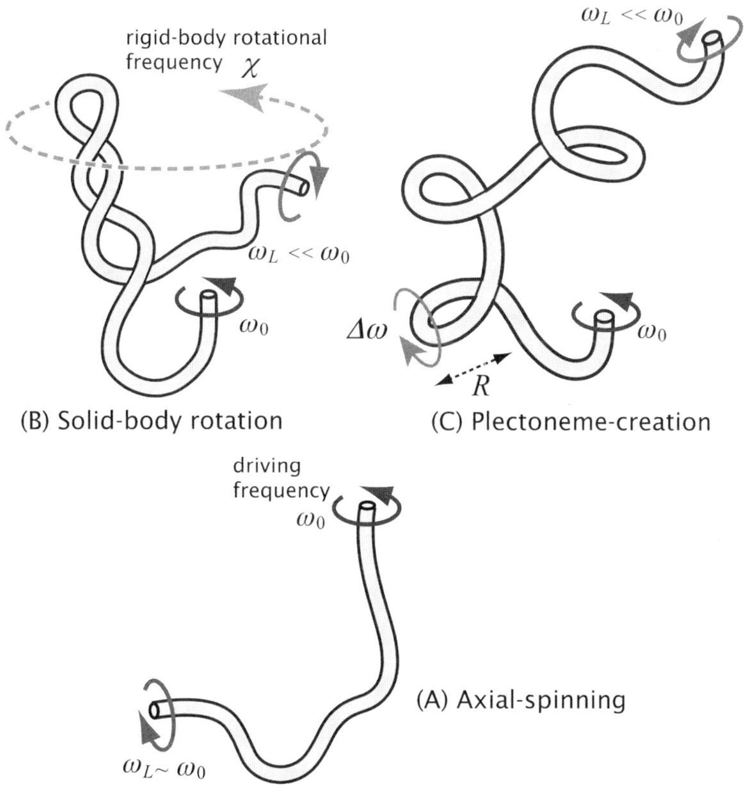

Fig. 21. A semiflexible polymer is rotated at frequency ω_0 at a fixed end and exhibits rotation at frequency ω_L at its free end. The three twist-dissipation channels are (A) axial-spinning, (B) solid-body rotation, and (C) plectoneme-creation.

gives

$$\omega_0 = \omega_L + \chi + \Delta\omega, \tag{112}$$

where $\omega_L = \omega(s = L)$ is the axial spinning frequency of the free end and $\Delta\omega$ denotes the fraction that is converted into plectoneme creations. Equation (112) again gives a key geometric relation for our scaling arguments presented below. In order to decide which of these three channels is in fact realized, we use the concept of minimum entropy production for non-equilibrium stationary states,[163] according to which the state of least dissipation is stable. We will later confirm each of our scaling results by our simulations, which gives further credibility to our scaling approach. We thus have to estimate the power dissipation in each of the modes depicted

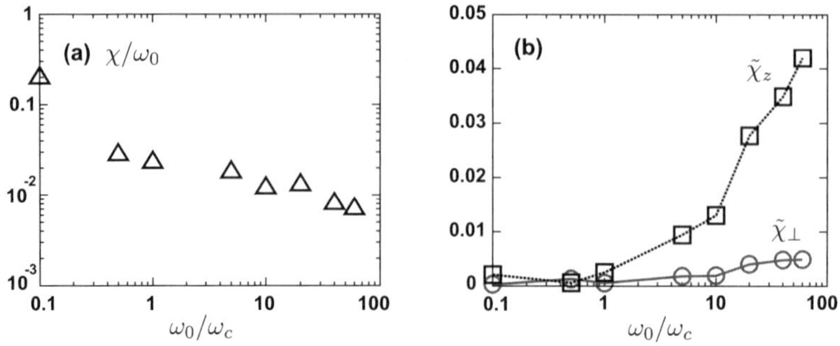

Fig. 22. (a) Relative solid-body rotation frequency, χ/ω_0, as a function of ω_0/ω_c. (b) Rescaled components of the vector $\boldsymbol{\chi}$, $\tilde{\chi}_\perp = \chi_\perp a^2/(\mu_0 k_B T)$, and $\tilde{\chi}_z = \chi_z a^2/(\mu_0 k_B T)$, as a function of ω_0/ω_c. Note that $\chi = (\chi_\perp^2 + \chi_z^2)^{1/2}$. The data are obtained for $L = 50a$ and $\ell_P = \ell_T = 10a$.

in Fig. 21. The power dissipation due to axial spinning is

$$P_{as} \sim \eta L a^2 \omega_L^2. \tag{113}$$

Here we assume that the average axial spinning frequency is ω_L, meaning that in the plectoneme regime the rotational profile $\omega(s)$ decays very quickly (in fact exponentially) along the chain contour to the value $\sim \omega_L$. This is confirmed by our simulations, see Fig. 23 (b). Likewise, the power dissipation due to solid-body rotation is given as $P_{sb} \sim \eta R^3 \chi^2$. Except for a compact globule with $R^3 \sim a^2 L$, we see that axial spinning is a less costly channel for twist transport than solid-body rotation, i.e. $P_{as} \ll P_{sb}$. We therefore neglect solid-body rotation in what follows. To confirm these predictions, we extract the solid-body rotation rate vector, $\boldsymbol{\chi}$, from our hydrodynamic simulations. For rigid-body rotation, i.e, $\dot{\mathbf{r}} \simeq \boldsymbol{\chi} \times \mathbf{r}$, we obtain the vector $\boldsymbol{\chi} = (\chi_\perp, \chi_z)$ via $\boldsymbol{\chi} = \mathbf{I}^{-1} \cdot \mathbf{L}$, where $\mathbf{I} = \sum_{j=1}^{N} (|\mathbf{r}_j|^2 \mathbf{1} - \mathbf{r}_j \mathbf{r}_j)$ is the moment of inertia tensor and $\mathbf{L} = \sum_{j=1}^{N} \mathbf{r} \times d\mathbf{r}/dt$ is the polymer angular momentum (the mass of the polymer beads is set to unity). Figure. 22 (b) shows $\boldsymbol{\chi} = (\chi_\perp, \chi_z)$ as a function of the rescaled driving frequency, ω_0/ω_c; solid body rotation is small for $\omega_0 < \omega_c$, while χ_z grows significantly beyond ω_c. The hydrodynamic shear due to this rotation deforms the plectonemes at high values of ω_0, which is a secondary effect that we neglect in the current version of our scaling arguments. In Fig. 22 (a), we show that $\chi/\omega_0 = |\boldsymbol{\chi}|/\omega_0$ is much smaller than unity, which confirms that solid-body rotation is negligible compared to other dissipation channels.

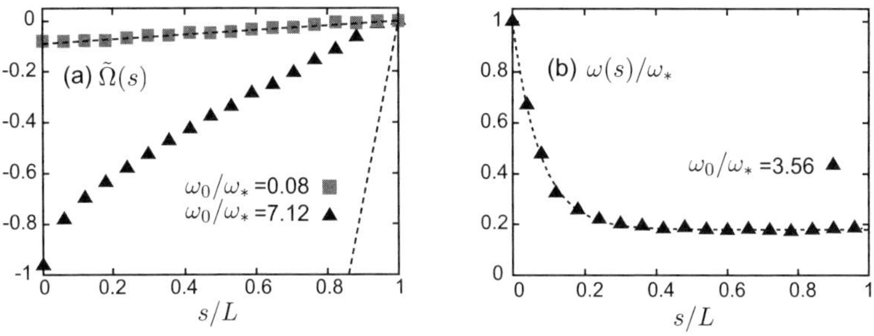

Fig. 23. (a) Steady-state profile of the rescaled twist density, $\tilde{\Omega}(s) = \Omega(s)a$, for two different driving frequencies, $\omega_0/\omega_* = 0.089$ (squares) and 7.12 (triangles). The broken lines are the prediction from linearized theory, see text. (b) Steady-state profile of the rescaled rotational velocity, $\omega(s)/\omega_0$ together with an exponential fit (broken line), for $\omega_0/\omega_* = 3.56$. All data were obtained for $L = 50a$ and $\ell_P = \ell_T = 10a$.

10.2. *Plectoneme creation channel*

The plectoneme-creation channel is more complicated. We consider the creation of a single loop of radius R nearby the driven end, see Fig. 21 (C). Extension to more complex plectoneme structures involving multi-loops (as actually observed in the simulations, see Fig. 20) is straightforward, but does not change the conclusions on the scaling level.

In order to generate a loop of radius R nearby the driven end, there must be a material supply of length $2\pi R$ towards the driven end during the time scale of $(\Delta\omega)^{-1}$. To do so for the polymer, there can be two possible kinematic ways. One possibility is that the whole polymer slides by a distance of $2\pi R$, with the velocity of $v \sim R\Delta\omega$. The other possibility is that the polymer is locally stretched, and the material of length $2\pi R$ is pulled towards the driving end. For sufficiently long chains, the sliding friction becomes indefinitely large, and therefore the polymer coil will be locally stretched to form a loop of R instead of moving the whole chain. In the most general case, a mixed stretching/translation mode would be realized. Let us introduce the factor γ that measures the fraction of excess length due to stretching. Then γR is the material supply towards the driving end due to stretching and $(1-\gamma)R$ is the one due to pure translational motion. In the local stretching mode, the length γR only moves with the speed $v \sim \gamma R\Delta\omega$, its hydrodynamic friction force can be estimated as $\eta(\gamma R)^2\Delta\omega$, where logarithmic hydrodynamic corrections are neglected. For the pure translational motion, the center of mass of the whole chain of length L

translates at a speed $v \sim (1-\gamma)R\Delta\omega$, the associated frictional force thus would be $\eta L(1-\gamma)R(\Delta\omega)$. The viscous power dissipation associated with this mixed stretching/translation mode is therefore given by

$$P_{slide} \sim \eta(\gamma R)^3(\Delta\omega)^2 + \eta L(1-\gamma)^2 R^2(\Delta\omega)^2. \tag{114}$$

Note that the first term from the stretching mode scales as R^3 and the second term from the translation mode scales as R^2L.

Next, we estimate the free energy change due to formation of a loop of size R. The bending energy of one loop is $k_BT\ell_P/R$, the loop creation frequency is $\Delta\omega$, thus the power consumption (the rate of free energy increase) for the loop formation is $\sim k_BT\ell_P\Delta\omega/R$. There is also a free energy increase due to local stretching, which is a bit of a task to evaluate. To simplify our argument, we consider a flexible chain limit since our polymer chain has its persistence lengths much smaller than its total length. See Fig. 24. When the driving end of this chain is started to be pulled at a certain constant speed v at time $t = 0$, then after time t only a part of the chain of length

$$\xi(t) \sim aN(t/\tau_R)^{1/2}, \tag{115}$$

experiences this tension and can be stretched, where τ_R is the longest relaxation time of the chain and is given by the so-called Rouse relaxation time $\tau_R \sim \eta N^2a^3/k_BT$.[1,2] This is valid for $t < \tau_R$ while we obtain $\xi(t) \sim aN$ for $t > \tau_R$. For Zimm dynamics in which hydrodynamic effects are taken into account by the pre-averaging approximation, Eq. (115) will be modified to $\xi(t) \sim aN(t/\tau_Z)^{1/3}$, where τ_Z is the Zimm relaxation time.[1,2] These results are all confirmed by more rigorous calculations using normal-mode analysis.[154] At the driving end, the tension develops as $f(t) \sim \eta\xi(t)v \sim \eta aN(t/\tau_R)^{1/2}v$ for $t < \tau_R$. Since the stretched chain behaves like a flexible chain with stretching modulus $k_BT/(a\xi)$, the streching energy of this chain at time t will be

$$E_{str}(t) \sim \frac{\xi(t)f^2}{k_BT} \sim \left(\frac{\eta^2a\xi^3}{k_BT}\right)v^2 \sim \eta a\left(\frac{k_BT}{\eta a^3}\right)^{1/2} v^2t^{3/2}, \tag{116}$$

apparently independent of chain length N. For the problem at our hand, we may put $v \sim R\Delta\omega$, and estimate the stretching energy E_{str} at time $t \sim (\Delta\omega)^{-1}$. Defining the unit time scale (diffusion time over the monomer length) as $\tau_0 = \eta a^3/k_BT$, we obtain $E_{str} \sim k_BT(R/a)^2(\tau_0\Delta\omega)^{1/2}$. The rate of free energy change associated with the plectoneme creation channel is therefore

$$P_{loop} \sim \frac{k_B T \ell_P}{R} \Delta\omega + k_B T \left(\frac{R}{a}\right)^2 \tau_0^{1/2} (\Delta\omega)^{3/2}, \tag{117}$$

Let us compare the two contributions in Eq. (117). The ratio of the stretching and bending terms is given by $(R/a)^2(R/\ell_P)(\tau_0\Delta\omega)^{1/2}$. From simulation data, a typical loop size R at the transition is seen to be a few times of a, thus $R^2/(\ell_P a) \sim 1$. Since τ_0 sets the shortest time scale involved in our viscous dynamics, any slow dynamics we are interested in now should occur for $\tau_0\Delta\omega < 1$. Under these considerations, we proceed by neglecting the second term in Eq. (117). In fact, with this assumption, we obtain the new transition frequency $\omega_\sharp$ given in Eq. (129). At this transition, we see $\tau_0\Delta\omega < \tau_0\omega_\sharp \sim (a/L)^{3/4} \ll 1$, thus the self-consistency of our argument is ensured. Hereafter we assume

$$P_{loop} \sim \frac{k_B T \ell_P \Delta\omega}{R}. \tag{118}$$

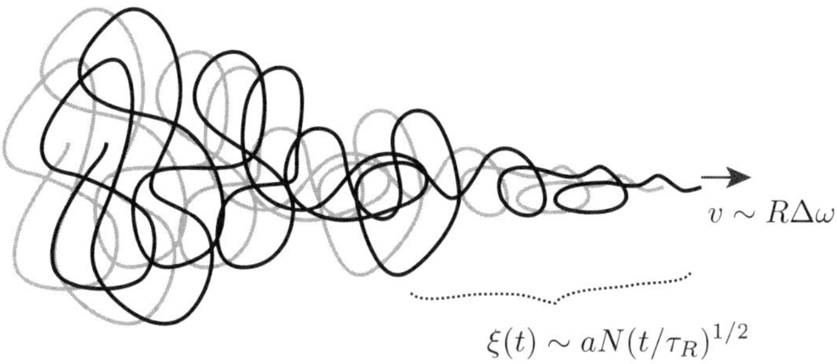

Fig. 24. Cartoon on a dragged flexible polymer chain at one end at a velocity $v \sim R\Delta\omega$. The chain is partially stretched and at the same time its center of mass translates.

There is one remark on the above argument. In the stationary state, the amount of stretching energy will be a constant, and there will be constant motion of excess length from the free end to the rotated end. On the other hand, that excess length is moving back from the rotated end to the free end via the diffusing plectonemes, and there is probably some type of coupling between the two processes. Therefore one should envision some type of cyclic process, where first the chain is stretched (and the excess length is transported to the rotated end), then the plectoneme is produced, which then diffuses back to the free end. Whereas such a dynamical cyclic process can indeed be seen in the simulations, our scaling argument presented above is considerd to be valid after averaging over many such cyclic processes, a process by which any oscillations in physical quantities are well smoothed out.

10.3. *Minimal dissipation hypothesis*

One should here note that in the stiff rod case discussed above, the elastic rod was shown to take the minimum dissipation state at its shape bifurcation. Extending this finding to the current semiflexible case, we assume that the total power consumption for the plectoneme creation channel, $P_{plec} = P_{loop} + P_{slide}$, is also minimized at the stationary state. Unfortunately, however, this minimization cannot be done analytically because P_{slide} given in Eq. (114) is the sum of R^2 and R^3 terms. To proceed, therefore, we consider the two important limiting cases, described by $\gamma = 0$ (translation dominated regime) and $\gamma = 1$ (stretching dominated regime).

10.3.1. *Translation mode dominated case:* $\gamma = 0$

In this limit, we have $P_{slide} \sim \eta L R^2 (\Delta\omega)^2$. Minimizing thus P_{plec} with respect to the loop radius R, we obtain

$$P_{plec} \sim (k_B T \ell_P)^{2/3} (\eta L)^{1/3} (\Delta\omega)^{4/3}, \tag{119}$$

with the loop size given by

$$R \sim \left(\frac{k_B T \ell_P}{\eta L \Delta\omega} \right)^{1/3}. \tag{120}$$

Due to the fractional power law $P_{plec} \sim (\Delta\omega)^{4/3}$, it is easy to see that plectoneme creation is unfavorable at low frequencies but will win over axial spinning at high frequencies.

We now minimize the total dissipation $P \sim P_{as} + P_{plec}$ with respect to the unknown frequency at the free end, ω_L, and use Eq. (66) and $\chi = 0$. For low frequency we obtain pure axial spinning, $\omega_L \sim \omega_0$, i.e. the twist that is injected at one end comes out as axial spinning at the other end. For high frequencies $\omega_0 > \omega_*$, with a crossover frequency defined as

$$\omega_* \equiv \frac{k_B T \ell_P}{\pi \eta a^3 L} \sim \omega_c (L/a), \tag{121}$$

we obtain

$$\omega_L \sim \frac{(k_B T \ell_P)^{2/3}}{\eta^{2/3} a^2 L^{2/3}} \sim \omega_*^{2/3} \omega_0^{1/3}, \tag{122}$$

$$P \sim (\eta L)^{1/3} (k_B T \ell_P)^{2/3} \omega_0^{4/3} \sim \eta L a^2 \omega_*^{2/3} \omega_0^{4/3}. \tag{123}$$

Interestingly, the crossover frequency ω_* is much larger than the zero-temperature twirling-whirling critical frequency ω_c by a factor proportional to the chain contour length, L/a. This is reasonable since conformational fluctuations are more significant for longer chains (for fixed persistence length), and higher driving frequency (or torque injection) is necessary to overcome such entropic barrier to obtain the transition to plectoneme creation mode where the chain gets compact and tightly wound around itself. As we have seen, the twirling-whirling shape transition is completely washed out by thermal fluctuations, and this plectoneme transition is the only shape transition that can be realized for rotationally-driven semiflexible polymers at finite temperatures.

To characterize how much the chain is twisted, we now look at twist Tw defined in Eq. (55). In the axial-spinning regime $\omega_0 < \omega_*$, chain shape fluctuations are decoupled from twisting motion, and the injected twist propagates diffusively along the chain[78] (see also Sec. 7). The thermally averaged twist density $\langle\Omega\rangle$ thus obeys the linear diffusion equation, $\pi\eta a^2\partial_t\langle\Omega\rangle = C\partial_s^2\langle\Omega\rangle$. In steady state under the boundary condition, $C\langle\Omega\rangle(0) = \pi\eta a^2\omega_0 L$, we obtain $\langle\Omega\rangle(s) = (\pi\eta a^2\omega_0/C)(s-L)$, same as for the rigid rod, see Eq. (95). This is in quantitative agreement with simulation for $\omega_0/\omega_* = 0.089$, as seen in Fig. 23 (a) (open squares and broken line). The total twist in this regime is given by

$$\mathrm{Tw} \simeq -\frac{\eta a^2\omega_0 L^2}{4C} = -\frac{1}{4\pi}\frac{A}{C}\frac{L}{a}\frac{\omega_0}{\omega_*}, \tag{124}$$

and thus is linear in the driving frequency ω_0. In the plectoneme regime, on the other hand, the twist is much reduced compared to the linear law (lower broken line in Fig. 23 (a)) and shows a nonlinear spatial profile (open triangles in Fig. 23 (a)). In this regime, the twist density Ω receives an additional contribution from the geometric torsion due to plectoneme formation: remember the discussion in Sec. 4. Plectonemes are however confined close to the driving end as seen from Fig. 23, and their contribution to Tw can be neglected at the leading order. Assuming the twist to be proportional to the average spinning frequency ω_L, we obtain $\mathrm{Tw} \sim A/C(L/a)(\omega_L/\omega_*) \sim (\omega_0/\omega_*)^{1/3}$. Our scaling prediction for Tw is thus given by

$$|\mathrm{Tw}|\frac{C}{A}\left(\frac{a}{L}\right) \sim \begin{cases} \frac{1}{4\pi}(\omega_0/\omega_*) \ (\omega_0 < \omega_*) \\ \\ (\omega_0/\omega_*)^{1/3} \ (\omega_0 > \omega_*). \end{cases} \tag{125}$$

The effective rotational friction coefficient Γ_r is defined through $P \sim \Gamma_r\omega_0^2$, where P is the total dissipation, given by Eq. (123) in the plectoneme regime. We obtain

$$\frac{\Gamma_r}{\pi\eta a^2 L} \sim \begin{cases} 1 & (\omega_0 < \omega_*) \\ (\omega_0/\omega_*)^{-2/3} & (\omega_0 > \omega_*). \end{cases} \tag{126}$$

For $\omega_0 < \omega_*$, the rotational friction corresponds to the axial spinning scenario, while for $\omega_0 > \omega_*$ a drastic decrease is predicted. Note that for $\omega_0 > \omega_*$ the dependence on chain length is $\Gamma_r \sim \omega_0^{-2/3} L^{1/3}$ and thus increases only sub-linearly with polymer length.

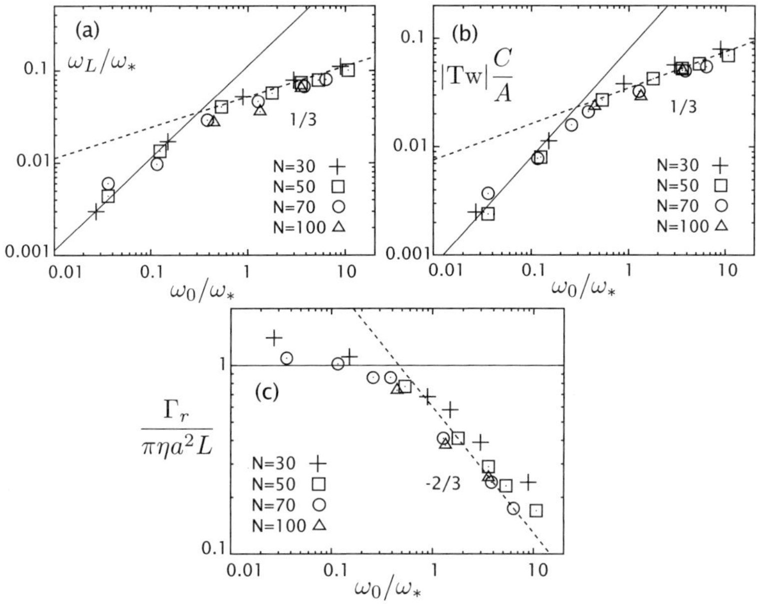

Fig. 25. Scaling plots of hydrodynamic simulation results. (a) Rescaled axial spinning frequency at free end, ω_L/ω_*, (b) total twist divided by the chain length, $|Tw|(a/L))$, and (c) rescaled rotational friction constant, $\Gamma_r/(\pi\eta a^2 L)$, plotted as a function of ω_0/ω_*. Soild lines denote the predictions from linear theory in the axial spinning regime valid for $\omega_0 < \omega_*$, broken lines denote the non-linear results valid in the plectoneme regime as given in Eqs. (122), (125) and (126). The crossover between the regimes occurs around a value of $\omega_0 = c_*\omega_*$ with $c_* = 0.4 \pm 0.1$.

In Fig. 25, ω_L, Tw and Γ_r from simulations are plotted as a function of ω_0/ω_* and confirm the analytical predictions for the non-linear plectoneme regime , Eqs. (122), (125) and (126). In the simulations, the externally applied torque M_{ext} at a given driving frequency ω_0 is measured and averaged to obtain $\Gamma_r = \langle M_{ext}\rangle/\omega_0$. In particular, the crossover frequency between the axial-spinning and the plectoneme regime is quite consistently found to occur at $\omega_0 = c_*\omega_* \sim (L/a)\omega_c$ with $c_* = 0.4 \pm 0.1$ and ω_* defined in Eq. (121). Note that the whirling instability, at $\omega_0 \simeq \omega_c$ and realized only for $L/\ell_P < 1$, is conceptually distinct from the plectoneme transition at $\omega_0 \simeq \omega_*$, which is only observable for long chains $L/\ell_P \gg 1$. In fact, the two transitions do not merge or interconnect at intermediate values of L/ℓ_P: for $L/\ell_P \simeq 1$ a semiflexible chain rather shows a continuous shape and rotational mode evolution with increasing ω_0 without a sharply defined transition. See also Fig. 17 (c).

10.3.2. *Stretching mode dominated case: $\gamma = 1$*

In this opposite limit, we have $P_{slide} \sim \eta R^3(\Delta\omega)^2$. Minimizing then P_{plec} with respect to the loop radius R, this time we obtain

$$P_{plec} \sim (k_B T \ell_P)^{3/4} \eta^{1/4} (\Delta\omega)^{5/4}, \tag{127}$$

with the loop size given by

$$R \sim \left(\frac{k_B T \ell_P}{\eta \Delta\omega}\right)^{1/4}. \tag{128}$$

Again, due to the fractional power law $P_{plec} \sim (\Delta\omega)^{5/4}$, we see that the plectoneme creation is unfavorable at low frequencies but will win over axial spinning at high frequencies. The transition frequency in this case is different from ω_* and is found to be

$$\omega_\sharp \equiv \frac{k_B T \ell_P}{\eta a^{8/3} L^{4/3}} \sim \omega_c (L/a)^{2/3} \sim \omega_* (L/a)^{-1/3}. \tag{129}$$

we obtain

$$\omega_L \sim \frac{(k_B T \ell_P)^{3/4}}{\eta^{3/4} a^2 L} \omega_0^{1/4} \sim \omega_\sharp^{3/4} \omega_0^{1/4}, \tag{130}$$

$$P \sim \eta^{1/4} (k_B T \ell_P)^{3/4} \omega_0^{5/4} \sim \eta L a^2 \omega_\sharp^{3/4} \omega_0^{5/4}. \tag{131}$$

The transition frequency $\omega_\sharp$ is again much larger than the zero-temperature twirling-whirling critical frequency ω_c, but is found to be smaller than ω_* by a factor $(L/a)^{1/3}$. In the stretching dominated scenario that we are discussing now, the dynamics is localized within the chain length of $\sim R$ close to the driving end, whereas in the translation dominated scenario, translation of the whole chain of length L was assumed. This is the reason why

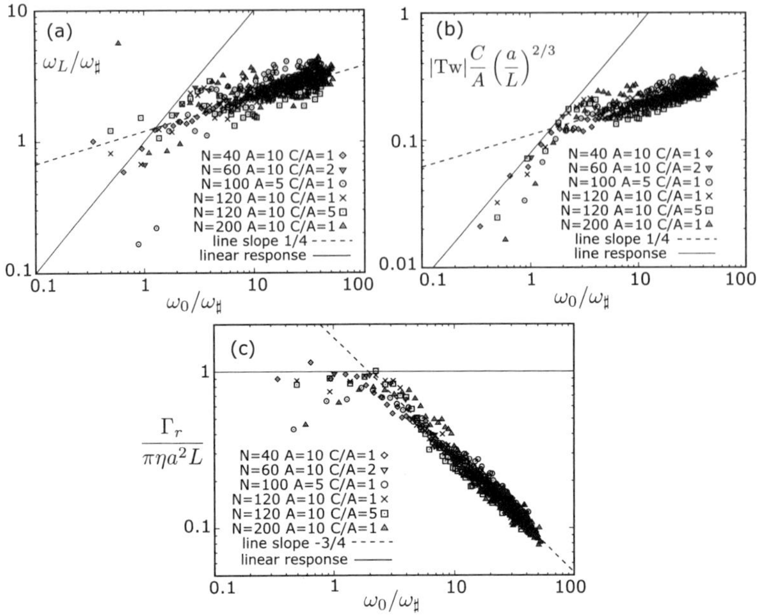

Fig. 26. Scaling plots of dynamic simulation results. (a) Rescaled axial spinning frequency at free end, $\omega_L/\omega_\sharp$, (b) total twist divided by the chain length, $|Tw|(a/L))$, and (c) rescaled rotational friction constant, $\Gamma_r/(\pi\eta a^2 L)$, plotted as a function of $\omega_0/\omega_\sharp$. Soild lines denote the predictions from linear theory in the axial spinning regime valid for $\omega_0 < \omega_\sharp$, broken lines denote the non-linear results valid in the plectoneme regime as given in Eqs. (132), (133) and (134). The crossover between the regimes occurs around a value of $\omega_0 \approx \omega_\sharp$

a lower transition frequency is found in the stretching-dominated scenario for sufficiently long chains.

Taking the similar steps as before, we now give our scaling predictions for axial spin frequency ω_L, twist Tw and the friction constant Γ_r in this stretching dominated scenario (see also Table 1: summary of our scaling predictions):

$$\omega_L/\omega_\sharp \sim \begin{cases} 1 & (\omega_0 < \omega_\sharp) \\ (\omega_0/\omega_\sharp)^{1/4} & (\omega_0 > \omega_\sharp). \end{cases} \tag{132}$$

$$|\mathrm{Tw}|\frac{C}{A}\left(\frac{a}{L}\right)^{2/3} \sim \begin{cases} \frac{1}{4\pi}(\omega_0/\omega_\sharp) \ (\omega_0 < \omega_\sharp) \\ \\ (\omega_0/\omega_\sharp)^{1/4} \ (\omega_0 > \omega_\sharp). \end{cases} \tag{133}$$

$$\frac{\Gamma_r}{\pi\eta a^2 L} \sim \begin{cases} 1 & (\omega_0 < \omega_\sharp) \\ \\ (\omega_0/\omega_\sharp)^{-3/4} & (\omega_0 > \omega_\sharp). \end{cases} \tag{134}$$

To check those predictions, Eqs. (129), (132)-(134), we performed dynamic simulations for chains up to $N = 200$ and for different C/A . Hydrodynamic interactions were switched off for those data, since they have little effects on the scaling behavior as can be seen from our scaling arguments. Comparison between the numerical data and the scaling predictions is shown in Fig. 26, which apparently gives a very good agreement. Although the two scenarios predict different exponents, it is at this point quite difficult to nail down the correct exponents from those numerical data. Simulation data would be best explained by the mixed stretching/translation modes (i.e., at a certain finite γ). Nevertheless, the scaling predictions based on the stretching scenario appear to provide agreement with numerical data for a wider parameter range. This indicates that the effect is important especially for longer chains, since the viscous frictional force becomes indefinitely large as the chain gets longer according to the translation-dominated scenario, so the stretching contribution comes into play for long enough chains.

Translation: $\gamma = 0$ $(\omega_0 > \omega_*)$	Stretching: $\gamma = 1$ $(\omega_0 > \omega_\sharp)$
$\omega_* \sim \omega_c(L/a)$	$\omega_\sharp \sim \omega_c(L/a)^{2/3}$
$P \sim \eta L a^2 \omega_*^{2/3} \omega_0^{4/3}$	$P \sim \eta L a^2 \omega_\sharp^{3/4} \omega_0^{5/4}$
$\omega_L \sim \omega_*^{2/3} \omega_0^{1/3}$	$\omega_L \sim \omega_\sharp^{3/4} \omega_0^{1/4}$
$\mathrm{Tw}\frac{C}{A}(\frac{a}{L}) \sim \omega_*^{2/3} \omega_0^{1/3}$	$\mathrm{Tw}\frac{C}{A}(\frac{a}{L})^{2/3} \sim \omega_\sharp^{3/4} \omega_0^{1/4}$
$\frac{\Gamma_r}{\eta a^2 L} \sim (\omega_0/\omega_*)^{-2/3}$	$\frac{\Gamma_r}{\eta a^2 L} \sim (\omega_0/\omega_\sharp)^{-3/4}$

10.4. *Experimental relevances*

We now produce some numbers relevant for single molecule experiments. Since DNA chains used in those experiments are sufficiently long, we estimate our numbers using the scaling predictions from the stretching scenario, i.e. $\gamma = 1$. For a hydrodynamic diameter of ds-DNA of $a \approx 2$ nm, a bend persistence length $\ell_P \approx 50$ nm and length $L \approx 12\,\mu$m, we obtain according to Eq. (129) a crossover frequency $\omega_\sharp \approx 5 \times 10^4$ rad/s (for comparison, the zero-temperature twirling-whirling transition frequency for these parameters is $\omega_c \approx 600$ rad/sec^{-1}.) The rotational friction of DNA molecules of length $L \approx 12\,\mu$m has been measured for rotational frequencies up to $\omega_0 \approx 12000$ rad/s in DNA unzipping experiments[86,164,165] and showed no detectable non-linear frequency dependence. This is in agreement with our estimate for the threshold $\omega_\sharp$. In recent experiments, the elongation dynamics of supercoiled DNA in response to sudden increase of tension was studied.[88] The dynamics showed almost no effect of the continuous removal of plectonemes, setting an upper bound on the hydrodynamic friction associated with DNA rotation, which turns out to be in agreement with the axial spinning scenario and thus is also consistent with our results.

On the other hand, for longer chains or in a crowded cellular environment with a much elevated viscosity, the crossover frequency $\omega_\sharp$ can be lowered to an experimentally reachable value. The corresponding critical torque at $\omega_\sharp$ scales as $M_c \sim \eta^{1/4}(k_BT\ell_P)^{3/4}\omega_\sharp^{1/4} \sim k_BT(\ell_P/a)(a/L)^{1/3}$, which is thus independent of solution viscosity and is smaller for longer chains. For example, a single *Escherichia coli* RNA polymerase can achieve maximal torque up to $M_{RNAP} \approx$ 1-2 k_BT. By setting $M_c \sim M_{RNAP}$, we obtain critical DNA length of $L_C \approx 3$ kbp. Thus linear DNA templates longer than L_C could induce the plectoneme transitions driven by RNA polymerases, provided that $\omega_\sharp$ is somehow reduced enough (such as due to much higher effective viscosity) so that the rotation can be kinematically possible. Actually, Krebs and Dunaway previously showed *in vivo* that polymerase drives DNA supercoiling during the transcription process for linear open DNA templates if the DNA length is longer than 17 to 19 kbp,[82] which is consistent with our argument.

For chromatin structures, the effective bending persistence length ℓ_P^{chrom} has been shown to stay rather constant while the effective radius a^{chrom} is increased substantially.[74] This means that the critical torque M_*^{chrom} for a chromatin fiber might be of the order of k_BT, quite in reach of typical torques generated by polymerase.

11. Effects of Bending Kinks on Torsional Transport in DNA

We have assumed so far that DNA can be regarded as a homogeneous elastic rod that is straight in its unstressed state (apart from thermally driven flexing). In fact, DNA is known to have many small natural (localized) bends (or "kinks") in its helix backbone. Distribution of these kinks may be characterized by a "structural persistence length" $P \approx 130$ nm, as measured from experiments.[166] This quantity is totally different from a mechanical persistence length that we have defined previously: P is a purely geometric parameter independent of temperature: At zero temperature, the mechanical bend persistence length ℓ_P is infinite, but DNA is still kinked and can be described as a persistent random walk along its contour with a structural persistence length P. The effective (or observable) persistence length of DNA is the mixture of the two lengths, and gives $\ell_P^{eff} \approx 50$ nm. Nelson argued that structural disorder such as bends inherent in DNA, can substantially increase drag force experienced by DNA, which may allow significant development of torsional stress even in linear unanchored DNA upon transcription.[78] Here we briefly review the essential idea of his "hybrid rotation model" presented in Ref. 78.

11.1. *Nelson's hybrid rotation model*

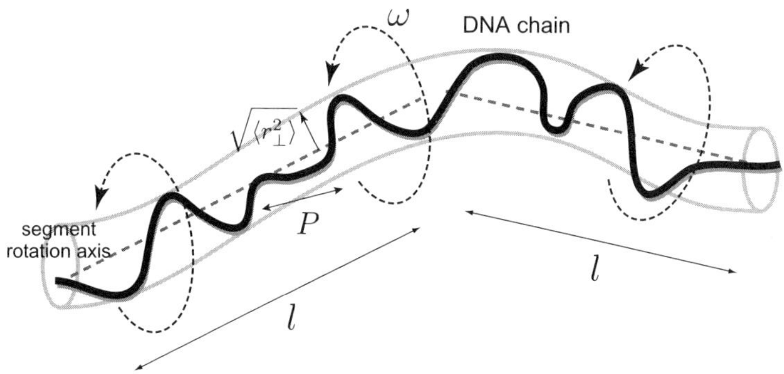

Fig. 27. Schematic illustration of a rotating DNA with many small static bends, envisioned in the scaling argument by P. Nelson. DNA segment of length l (blue line) rotates about its local axis (drawn by dotted lines), in which several kinks with typical distance P are distributed, each executing local crank-shafting motions with a speed $v \sim r_\perp \omega$. Beyond length scale of l, DNA can bend to make connecting segments straight, which costs some bending energy, but reduces viscous torque on each segment in rotation at a frequency ω.

To begin with, we first work out a zero temperature system. We take a long polymer chain in a viscous fluid, and look at a chain segment of length l. In general, when a stiff rod of length l is translating at a speed v in a fluid of viscosity η, the drag force acting on this rod segment is given by $f \sim \eta l v$, where we have neglected hydrodynamic corrections unlike the original argument by Nelson, which changes only numerical factors but not the leading scaling behavior. Now, consider that this segment of length l is kinked many times at random positions, and is rotating about its axis at an angular velocity ω, see Fig. 27. Each element of this segment thus moves at a speed $v \sim r_\perp \omega$, where $r_\perp$ is the distance of the element from the rotational axis. Viscous torque acting over the segment length, M, is thus obtained from $M\omega \sim fv$, i.e.,

$$M \sim \eta l \omega \langle r_\perp^2 \rangle, \tag{135}$$

where $\langle r_\perp^2 \rangle$ represents the average of $r_\perp^2$ over all random distributions of bends. When this segment contains enough numbers of kinks, i.e., $l \gg P$, this average is given by $\langle r_\perp^2 \rangle = Pl/9$,[167] leading to

$$M \sim \eta l^2 P \omega, \tag{136}$$

where we have ignored numerical prefactor. The assumption made here, $P \ll l$, is checked later on to be self-consistent.

Equation (136) suggests that the viscous torque acting on segment increases as l^2. Nelson argues that rotation axes of different segments will be different, because the segment length l is larger than P. In order that those segments are smoothly connected upon rotation, each segment has to be weakly bent over the length of segment l. The bending energy for this is roughly given by $\sim k_B T \ell_P / l$. The actual segment length (or "crossover length" according to Ref.[78]), $l = L_C$, is determined as the length at which the viscous torque and the bending energy cost just balance:

$$\eta L_C^2 P \omega \sim \frac{k_B T \ell_P}{L_C}, \tag{137}$$

which gives $L_C \sim (k_B T \ell_P / \eta \omega P)^{1/3}$, and the viscous torque $M \sim (k_B T \ell_P)^{2/3} (\eta \omega P)^{1/3}$. Including numerical prefactors more carefully, Nelson estimated this crossover length as $L_C \approx 450$ nm for T7 RNA polymerase, which confirms the assumption $l > P$ made above. He concluded that for a DNA polymer longer than L_C, random kink structures almost completely shut down axial spinning of DNA, with a much enhanced rotation friction given by $\Gamma_r \sim \eta P L_C^2$ that is several thousands times larger than the axial spinning one for T7 polymerase.

11.2. *Experimental tests*

Effects of a single large-angle kink on mechanics and dynamics of elastic rods were studied in a macroscale experiment and in classical elastic rod theory.[116,168] These studies confirm that a sharply bent structure does trap axial twist[168] and significantly increases the rotational drag. An elastohydrodynamic shape instability (folding or straightening transition) was also found, which accompanies the nonlinear decrease of effective rotational friction as a function of driving frequency,[116] which is qualitatively consistent with the prediction from Eq. (137). However, in the context of DNA mechanics where thermal effects are dominant, no direct experimental support for the static-kink induced anomaly has so far been reported. Stupina and Wang introduced several DNA segments known to contain stable bends into their plasmid DNA templates, but did not find any evidence that static bends slow down the diffusional merge of oppositely supercoiled domains upon transcription.[85] Thomen and Heslot performed a single molecule DNA unzipping experiment using an optical trap system. They obtained the rotational friction of double-stranded DNA up to a very rapid rate of 2,000 turns per sec by looking at force-velocity curves in unzipping processes, and concluded that the obtained rotational friction is at most a few times larger than expected from simple axial spinning.[86] Predictions from Nelson's hybrid rotation model seem to be consistent with the high rotation speed experimental data, when $\langle r_\perp^2 \rangle \sim L_C^3/P$ instead of $\langle r_\perp^2 \rangle \sim L_C P$ is used, because $L_C \sim P$ at frequencies as high as $\omega \sim 10^4$ sec^{-1}.[164] Nevertheless, a significant disagreement was still pointed out for low rotation speed data for $\omega < 2500$ sec^{-1},[165] which is more relevant to in vivo DNA transcription dynamics.

Although it is not clear why the hybrid rotation model fails to get experimental support, we point out that the above scaling argument might not correctly include effects of thermal (conformational) fluctuations. In estimating the viscous rotational drag on a segment of length l, any effects of thermal flexing were ignored. However, since a typical distance between kinks is P, a bending torque to straightening one kink is roughly $M_b \sim A/P \sim k_B T \ell_P/P$. Using $\ell_P \approx 50$ nm and $P \approx 130$ nm,[78,166] we see $M_b \sim 0.5 k_B T$, which is smaller than $k_B T$. Thus a segment of l would be subject to significant conformational fluctuations and there may be numerous mechanisms to transport torsional stress by executing simple axial rotations. We conclude that a static bend induced enhancement of rotational friction should exist in DNA, but the magnitude of that could be somehow much smaller than expected from the scaling argument reviewed above. A numerical test of the hybrid rotation model is desirable, but it would require a sufficiently long polymer chain with enough statistical averaging over many different distributions of static bends, which makes this a very hard task.

12. Impact of Torsional Stress on Nucleosome Dynamics

Assuming that frozen kinks in double stranded DNA structure are not the origin of the development of torsional stress in DNA as we have discussed above, kink structures in DNA stabilized by bound proteins,[169] such as sharp bends or loops, could provide instead a significant barrier to DNA axial rotational diffusion. In fact, DNA looping is an important mechanism to regulate gene expressions, and superhelical stress in DNA is shown to have significant effects on loop formation of DNA-protein complexes such as DNA-LacI complex (LacI represents a repressor protein that controls the lactose metabolism system in *E. Coli*[5,170–172]). It is also suggested that some single stranded DNA-binding proteins such as Replication protein A can work against external torsional stresses to protect and maintain the double-stranded structure of DNA.[173] It is therefore of great interest to study how injected twist or torsional stress can propagate through a DNA-protein complex and modulate its structural stability.[174]

As a matter of fact, this general question is particularly important for RNA polymerase II -mediated transcription of protein-coding genes in eukaryotic cells.[175] DNA within eukaryotic nucleus is compacted in the form of chromatin, a nucleoprotein complex composed of repeating nucleosomes. The nucleosome is a fundamental unit of chromatin, consisting of 146 bp of DNA tightly wrapped on the surface of a histon protein octamer, which is two copies of each core histon proteins H2A, H2B, H3 and H4. This structure is achieved by the docking of two dimers H2A/H2B onto the tetramer $(H3/H4)_2$, which is stable only when it is mediated by DNA. How transcriptional regulatory proteins can access their DNA recognition sites is not yet known.[176,177] However, increasing number of experiments now suggest that mechanical perturbations generated by transcriptionally active RNA polymerases are intimately coupled to structural modifications of nucleosomes and chromatin, including partial breaking of nucleosome core particles.[175] To be more specific, the positive supercoiling, created in front of the point of twist-injection in the form of plectonemes, may loosen or even drive off histones from their nucleosomal core structures, while the negative supercoiling behind the twist-injection point would form a template for strengthening or reforming the nucleosomal core structure.[178–182] In the last ten years, effects of stretching force on a single nucleosome have been extensively studied both experimentally and theoretically,[93,183–187] but static and dynamic aspects of twist-induced events appear to be yet largely unexplored.

In this section, we consider DNA-protein complexes, especially nucleosomes,[93] and look at effects of positive torsional forces on those structures. Following previous attempts,[94–96,188,189] we construct a simple model for a DNA-protein complexe: A protein is approximated as a rigid sphere of radius R_0 with surface attractive interactions with a semiflexible polymer. A wrapping transition of the semiflexible polymer around the rigid sphere (histon octamer) occurs as the attractive interactions is increased. The fundamental difference in our model from most of the previous ones is that our polymer carries torsional elasticity, which is expected to be very important for events involving symmetry breaking like chiral wrapping transitions that we will study here. In our dynamic simulations, we take the radius of the protein sphere $R_0 = 2a$, and the length of the chain $L = 70a$. The middle part of the chain of length $L_m = 50a$ can interact with the sphere via an attractive potential (specified below), and the rest of the chain (two arms of length $10a$) corresponds to the linker DNA strands extending from the core structure. Taking the DNA diameter $a = 2$ nm, we have $R_0 = 4$ nm, $L = 140$ nm, $L_m = 100$ nm (≈ 300 bp), and the bend persistence length $\ell_P = 50$ nm. These parameter values are comparable to single nucleosome structures determined from X-ray crystallography.[190,191]

12.1. *Model nucleosome: semiflexible polymer-sphere complex*

Attractive interactions between a rigid sphere and DNA strands can be modeled as either isotropic or orientation-dependent ones. When the interaction is dependent on a specific rod orientation, it can be chiral and exerts torques both to the rigid sphere and to the chain. An isotropic potential is obtained simply by switching off this orientational dependence, and in this case there is no torque transmission between the chain and the sphere (i.e., the chain can execute axial rotation almost freely independent of its wrapping configurations). Noting that a polymer chain has a finite thickness, a position of a specific side of the i-th monomer surface, $\mathbf{r}_i^*$, can be expressed as

$$\mathbf{r}_i^* = \mathbf{r}_i + \Delta\hat{\mathbf{e}}_{1i}, \tag{138}$$

where $\mathbf{r}_i$ is the center of the i-th monomer and $\Delta = a/2$ for the orientation-dependent interaction case, whereas $\Delta = 0$ for the isotropic interaction case.

Let the center position of the protein sphere be $\mathbf{R}_0$. The relative position vector between the j-th monomer of the chain and the sphere center is $\boldsymbol{\rho}_i = \mathbf{r}_i^* - \mathbf{R}_0 = (\mathbf{r}_i - \mathbf{R}_0) + \Delta\hat{\mathbf{e}}_{1i}$, and $\rho_i = |\boldsymbol{\rho}_i|$. As a functional form of our interaction potential we take the Mose potential given by

$$U(\rho) = \sigma_0 \left[e^{-2(\rho-\rho_c)/\lambda} - 2e^{-(\rho-\rho_c)/\lambda} \right], \tag{139}$$

where $\sigma_0 = 4k_BT$ controls the strength of attractive interaction, $\rho_c = R_c + a/2 - \Delta$ sets the minimal distance between the sphere center and the DNA rod surface, and $\lambda = a/2 \approx 1$ nm limits the interaction range. For a DNA-histon complex, attraction is mediated by electrostatics, but since the typical Debye screening length is 1 nm at physiological salt concentrations (≈ 100 mM), any short-range attractive potential is sufficient for our present aim. Applying the variational method presented in Sec. 4, we can obtain the forces and torques on the chain and the rigid sphere.

When the DNA chain wraps twice around the histon octamer of radius R_0, the bending energy cost is estimated to be

$$E_{bend} \sim \frac{A}{2R_0^2} \times 4\pi R_0 = \frac{2\pi\ell_P k_BT}{R_0} \approx 75k_BT. \tag{140}$$

On the other hand, the energy gain due to surface binding is $E_{ads} \sim -4\pi R_0\sigma_0 \approx -96k_BT$. Thus the net energy gain per nucleosome core structure is $\Delta E = E_{bend} + E_{ads} \approx -20k_BT$, which is large enough to ensure the wrapped state as an equilibrium structure, and is comparable to experimental values around $-30k_BT$.[185,187] In our simulations, both ends of the polymer chain are constrained to move only along the z direction, and their tangent vectors are fixed to point along the z-direction (see Fig. 28). These boundary constraints mimic typical single molecule manipulation experiments using magnetic tweezers.[185]

12.2. *Wrapping transition driven by an isotropic attractive potential*

We first look at a wrapping transition of a semiflexible chain around a core protein driven by an isotropic interaction potential ($\Delta = 0$). In Fig. 28, the typical time evolution of the wrapping process is shown. In this simulation, the polymer chain can axially rotate at both ends, and there is no conservation of any topological quantity. As is seen, the polymer chain wraps around the spherical core in two turns, selecting the left-handed chirality in this case. There is a spontaneous symmetry breaking event even for the isotropic potential. This wrapping process is typical, and is unchanged when both ends of the chain are blocked to rotate, or when the orientation-dependent potential is used ($\Delta = a/2$).

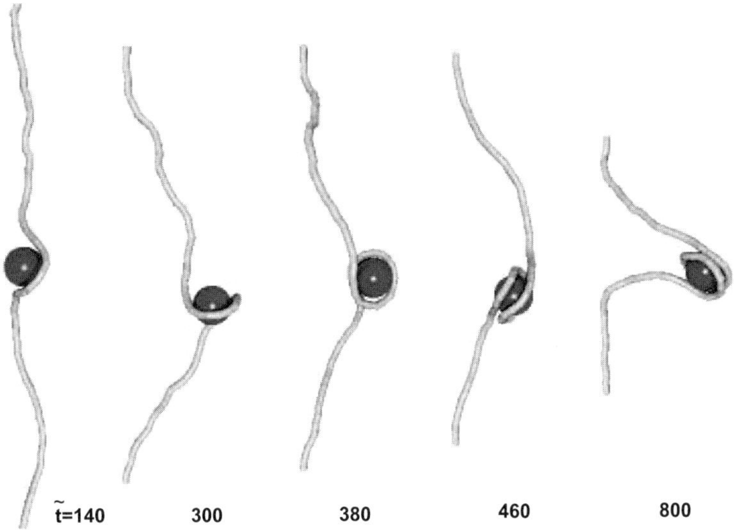

Fig. 28. Time evolution of a semiflexible polymer chain of length $L = 70a$ wrapping around a spherical core protein of radius $R_0 = 2a$. The bend and twist persistence lengths are $\ell_p = \ell_T = 25a$, the adsorption energy per monomer is $\sigma_0 = 4k_BT$. The two chain ends are constrained to the z-axis but otherwise free to rotate. No stretching force along z-direction is applied.

12.3. *Twist propagation into the complex: Enhanced rotational friction*

To understand what happens when the complex is driven by an externally applied torsional stress,[192,193] we performed similar simulations while fixing one end of the chain at the origin and its tangent towards z-direction and enforcing rotation at a given frequency ω_0, with the other end free to rotate but forced to stay on the z aixs. Our results are summarized in Fig. 29. To enhance the visibility of the data, we reduced the magnitude of random thermal noise, in other words, we reduced the effective temperature of the system. The difference between the chiral (orientation-dependent) and isotropic interactions are now remarkable. In the isotropic case ($\Delta = 0$), twist injected at the driving end propagates along the chain without being affected by the histon sphere, because the chain can rotate about its axis as it can "slip" at the sphere surface. Thus the resulting friction constant does not change much compared to the one expected from the simple axial spinning, $\approx \eta a^2 L$. On the other hand, for the orientation-dependent potential ($\Delta = a/2$), the effective rotational friction is found to exceed by far the axial spinning value. Twist propagation in the chain forces the sphere to rotate about z-direction (see the snapshots in Fig. 29), which yields an additional rotational barrier at the driving end. This increased friction can be estimated roughly as follows. A major contribution to this increased rotational friction comes from rotation of the DNA loops of radius R_0. Considering a small segment of this loop of length $dl = R_0 d\theta$, its velocity about z-axis is $v \sim R_0(1+\sin\theta)\omega_0$, and the corresponding viscous dissipation is $dP_{loop} \sim 3\pi\eta v^2 dl \sim 3\pi\eta R_0^3 \omega_0^2 (1+\sin\theta)^2 d\theta$. The total viscous dissipation involved in the loop rotation is thus $P_{loop} \sim \int_{\theta=0}^{\theta=4\pi} dP_{loop} \sim 9\pi^2 \eta R_0^3 \omega_0^2$, which leads to the additional rotational friction $\Delta\Gamma_r \sim 9\pi^2 \eta R_0^3$. Thus the total rotational friction at the driving end amounts to $\Gamma_r/(\pi\eta a^2 L) \sim 1 + 9\pi (R_0/a)^2 R_0/L \approx 7$, in good agreement with the numerical data in Fig. 29.

Those results suggest that various structures formed by DNA and DNA-binding proteins such as loops may indeed act as a significant resistance against diffusive transport of torsional stress in cellular environments in the presence of orientation-dependent interactions. If this scenario does work, diffusive twist transport by DNA axial spinning might be completely shut down in chromatin where nucloesomes are densely packed, which suggests an intimate coupling between DNA supercoiling and chromatin structure upon transcription[84,179,180,194,195] These considerations should be subjected to more detailed studies in future.

13. Conclusion

In order to discuss fundamental scenarios for twist transport dynamics in driven elastic polymers, we studied a particularly simple system: a uniform elastic rod or polymer chain that is forced to axially rotate at one end at frequency ω_0 with the other end free. We began this chapter with the introduction of the kinematics and linear elasticity theory for thin elastic rods, and obtained the equations of motion of the rod in a viscous fluid using the variational approach. The obtained formulation is directly applicable to study semiflexible polymers in contact with a thermal bath by adding Langevin-type random forcing terms. Our hydrodynamic Brownian

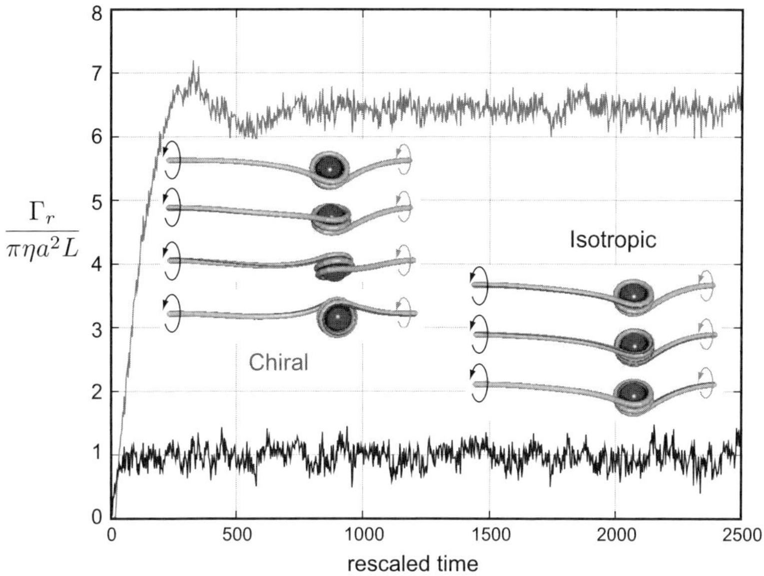

Fig. 29. Twist propagation through the polymer-sphere complexes with orientation-d ependent and isotropic interaction potentials. Effective rotational friction at th e driven end Γ_r, rescaled by that of simple axial spinning, $\pi\eta a^2 L$, as a f unction of the rescaled time for chiral (gray line) and isotropic potential (black line) cases. At both ends, the tangents are fixed towards z-direction and one end is rotated at a given frequency $\omega_0/\omega_c \approx 0.1$ (i.e., close to quasi-static rotation). Simulation parameters: chain length $N = 7\ 0$ with the interacting part of length $N_m = 50$ in the middle, bend persistence length $\ell_P = 25a$, elastic constant ratio $C/A = \ell_T/\ell_P = 1$, rigid sphere radius $R_0 = 2a$, interaction strength $\sigma_0/k_BT = 4$. External tension $f \approx 1.3k_BT/a \approx 2.6$ pN is applied at both ends. To suppress large fluctuations in the data, thermal noise amplitude in the Langevin e quations are set 10^3 times smaller than that at given temperature. Long-ranged hydrodynamic interactions are neglected in this simulation run. Shown together are typical snapshots of the complex. Stripes on the surface of the filament trace the tip of the binormal vector.

dynamics simulation method was also briefly explained. We showed that twisting and bending modes are decoupled on the linear level at thermal equilibrium. This manifests that nonlinear twist-bend coupling (i.e, conversion between twist and writhe) can arise for polymers subjected to an external forcing that at the same time drives the system out of equilibrium.

In the first part, we discussed in detail the twirling-whirling transition dynamics of a rotationally driven elastic rod at zero or very low temperature. We gave evidence for the discontinuous nature of the twirling-whirling transition. We also showed that thermal fluctuations play a significant role in the transition behavior and lead to a decrease of the critical frequency in a range of rod stiffnesses ℓ_P/L that is relevant to biopolymers. Relying on geometric arguments, we elucidated that the transition always occurs to reduce the energy dissipation rate, and calculated a nonlinear relationship between the driving frequency and axial torque at the base, in good agreement with our numerical simulations. We showed that writhing is responsible for the nonlinear response, and discussed those results in connection with geometric phases previously studied in polymers at equilibrium.

In the second part, we turned our interest to the opposite case and considered the high temperature limit. That is, we considered a semiflexible polymer submitted to the same dynamical setup, but where the bend and twist persistence lengths are much smaller than the total contour length of the polymer chain. There, we found a number of features that differ from the stiff rod case. Under significant conformational fluctuations, the zero-temperature twirling-whirling transition is completed washed out, and instead, a novel dynamical regime was distinguished. For small driving frequency, $\omega_0 < \omega_*$, we find the standard axial-spinning regime where the chain flexes randomly and spins about its local axis. For large rotational frequency, $\omega_0 > \omega_*$, twist is locally converted into writhe close to the driven end and then diffuses out to the free end without much concerted solid-body rotation. In this plectoneme regime, the filament exhibits only minimal axial spinning and a significant reduction of the rotational friction as compared to axial spinning is obtained.

In this chapter, we have also included preliminary results on nonequilibrium twist effects on nucleosome structures. We have examined two types of interaction potentials between a polymer and a sphere surface, one is an isotropic attractive potential and the other is a chiral potential that depends on the rod local orientation with respect to the direction from the rod center to the sphere center. We find a wrapping transition of a rigid spherical particle by a twist-storing semiflexible polymer that accompanies a spontaneous left-right symmetry breaking for both chiral and achiral scenarios. Differences appear when we look at the effective rotational friction of the formed complex. For an isotropic complex, the injected torsional stress does not really correlate with the complex, and diffuses out from the free end, leading to small rotational friction similar to a freely spinning rod of the same length. In contrast, transport of external torsional stress is substantially blocked by the formation of the complex for the orientation-dependent interaction case, because rod spinning requires dragging of the core particle. The observed rotational friction is much larger than the one expected from the free axial spinning of the rod, suggesting that structures formed by DNA and DNA-binding proteins such as loops will produce significant resistance against diffusive transport of torsional stress in DNA.

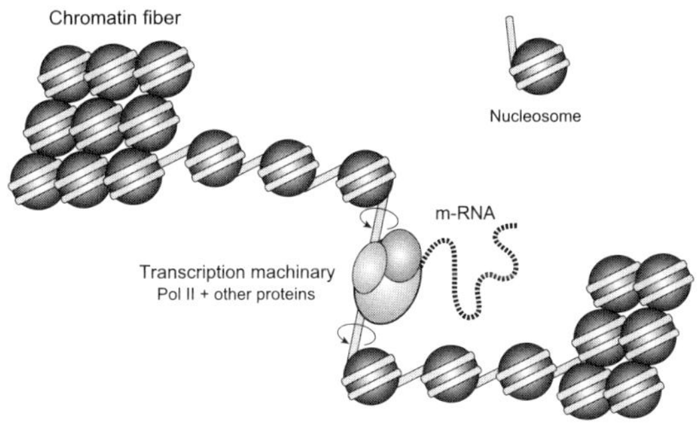

Fig. 30. Cartoon of transcription elongation through chromatin by RNA polymerase II.

An ultimate goal of our study presented in this chapter is to elucidate the roles of torsional stress induced by various DNA-processing proteins upon *in-vivo* functioning of DNA. While the model analyzed in this chapter was designed only to study fundamental aspects of twist transport in a single polymer chain and a number of biophysical features are missing, two biologically relevant conclusions concerning DNA transcription mechanics and dynamics might be extracted. First, the nature of the twist-writhe conversion process leads to a narrow spatial localization of the twist density close to the cranking point by RNA polymerases, which in-vivo might guide and concentrate the activity of twist-sensitive proteins such as topoisomerases to a region close to the polymerase complex.[47] Second, it has been suggested that the positive supercoiling, created in front of the point of twist-injection in the form of plectonemes, might loosen or even drive off histones from their nucleosomale core particles, while the negative supercoiling behind the twist-injection point would form a template for strengthening or reforming the nucleosomal core structure.[84] The detailed physical picture of such transcription dynamics is still a mystery, but our preliminary study clearly suggests that diffusive twist transport by DNA axial spinning might be considerably hindered in chromatin structures, which in turn could induce structural changes of DNA and of chromatin architectures, if torque transmission between the DNA strand and bound proteins is sufficient (i.e., if interactions are orientation-dependent and sufficiently strong). Geometric and mechanical aspects of the transcription dynamics within condensed chromatin fibers are particularly important but challenging issues. These biophysical problems await further studies in future, not only by biological approaches,[196] but also by theoretical physical approaches that can assemble various theoretical tools such as continuum mechanics, statistical mechanics and thermodynamics.[174,195,197]

Acknowledgments

We thank C. Schiller for bringing our attentions to the relationship between geometric phases and twist transport in open polymers, and G. Dietler for kindly allowing us to use Fig. 1 reproduced from Ref.[48] H.W. thanks T. Sakaue, G. Witz, Y. Murayama, and M. Sano for helpful discussions. Financial support from the German Science Foundation (DFG), the German Excellence Initiative, and the Ministry of Education, Culture, Sports, Science, and Technology (MEXT) of Japan (Grant in Aid, No.22740274) is acknowledged. Parts of the numerical calculations were carried out on Altix3700 BX2 at YITP, Kyoto University.

Appendix A.

Introducing the Euler angles (φ, θ, ψ), the local director basis $(\hat{\mathbf{e}}_1, \hat{\mathbf{e}}_2, \hat{\mathbf{e}}_3)$ can be parameterized as

$$\hat{\mathbf{e}}_1 = \begin{pmatrix} \cos\varphi\cos\theta\cos\psi - \sin\varphi\sin\psi \\ \sin\varphi\cos\theta\cos\psi + \cos\varphi\sin\psi \\ -\sin\theta\cos\psi \end{pmatrix}, \tag{A.1}$$

$$\hat{\mathbf{e}}_2 = \begin{pmatrix} -\cos\varphi\cos\theta\sin\psi - \sin\varphi\cos\psi \\ -\sin\varphi\cos\theta\sin\psi + \cos\varphi\cos\psi \\ \sin\theta\sin\psi \end{pmatrix}, \tag{A.2}$$

$$\hat{\mathbf{e}}_3 = \begin{pmatrix} \cos\varphi\sin\theta \\ \sin\varphi\sin\theta \\ \cos\theta \end{pmatrix}. \tag{A.3}$$

The strains are thus expressed in terms of Euler angles as

$$\Omega_1 = \frac{d\theta}{ds}\sin\psi - \frac{d\varphi}{ds}\sin\theta\cos\psi, \tag{A.4}$$

$$\Omega_2 = \frac{d\theta}{ds}\cos\psi + \frac{d\varphi}{ds}\sin\theta\sin\psi, \tag{A.5}$$

$$\Omega_3 = \frac{d\psi}{ds} + \frac{d\varphi}{ds}\cos\theta. \tag{A.6}$$

We can write the elastic energy of an isotropic rod E under an applied tension $\mathbf{F} = F\hat{\mathbf{z}}$ as

$$E = \int_0^L \left[\frac{A}{2}\left(\frac{d\theta}{ds} + \frac{d\varphi}{ds}\sin\theta\right)^2 + \frac{C}{2}\left(\frac{d\psi}{ds} + \frac{d\varphi}{ds}\cos\theta\right)^2 - F\cos\theta \right] ds. \tag{A.7}$$

If the Euler angles $\theta(s), \varphi(s), \psi(s)$ are read as functions of *time* s, Eq. (A.7) formally corresponds to an action functional for a symmetric top in a gravitational field with the principal moments of inertia around the symmetry axis C and perpendicular to it, A. The statics of elastic rods is thus closely related to the dynamics of spinning tops (Kirchhoff kinetic analogy), and solutions of the resulting Euler-Lagrange equations are given explicitly in terms of elliptic functions and integrals.[115]

Here we make some remarks on topological quantities expressed in terms of the Euler angles. Let our polymer be sufficiently stretched along z direction by an external pulling force (or due to its large enough bend persistence length). To evaluate the writhe of this chain, we may use Fuller's formula, Eq. (64), by closing the chain as shown in Fig. B.1. Writhe for any closing segment, where $\Gamma_0 = \Gamma$, are trivially zero. Choosing the reference curve Γ_0 as the z axis, we obtain

$$\mathrm{Wr} = \frac{1}{2\pi}\int_0^L \frac{\hat{\mathbf{e}}_z \times \hat{\mathbf{t}}(s)}{1 + \hat{\mathbf{e}}_z \cdot \hat{\mathbf{t}}(s)} \cdot \frac{d\hat{\mathbf{t}}}{ds} ds = \frac{1}{2\pi}\int_0^L \frac{d\phi}{ds}\left(1 - \cos\theta(s)\right) ds. \tag{A.8}$$

On the other hand, the twist Tw is simply known from Eq. (A.6) as

$$\mathrm{Tw} = \frac{1}{2\pi}\int_0^L \left(\frac{d\psi}{ds} + \frac{d\varphi}{ds}\cos\theta\right). \tag{A.9}$$

These two expressions lead us to

$$\mathrm{Lk} = \mathrm{Tw} + \mathrm{Wr} = \frac{1}{2\pi}\int_0^L \frac{d}{ds}(\phi + \psi) ds. \tag{A.10}$$

These expressions, Eqs. (A.8)-(54), are previously used in the theory of DNA supercoiling, in order to write the elastic free energy under *global* topological constraints in a *local* form. For example, in magnetic tweezer experiments of DNA, the linking number, Lk, is simply a number of turns imposed, i.e., Lk$=n$. Thus, as Bouchiat and Mezard explicitly stated,[60] Eq. (A.10) assumes that the Euler angles are smooth functions of the arclength parameter s for any configurations of DNA chain. However, this assumption may break down at the coordinate singularities of the Euler angles, $\theta = \pi$. Since the reference curve Γ_0 is now the z axis, $\theta = \pi$

corresponds to the anti-podal configuration of the two curves. See also Sec. 5.2. Therefore, as pointed in Refs.,[129,136] Eqs. (A.8)-(A.10) should be used with care. These are safely applied for a fluctuating polymer chain under large pulling force and with low torsional stress where plectonemes are absent.[59,139]

Appendix B.

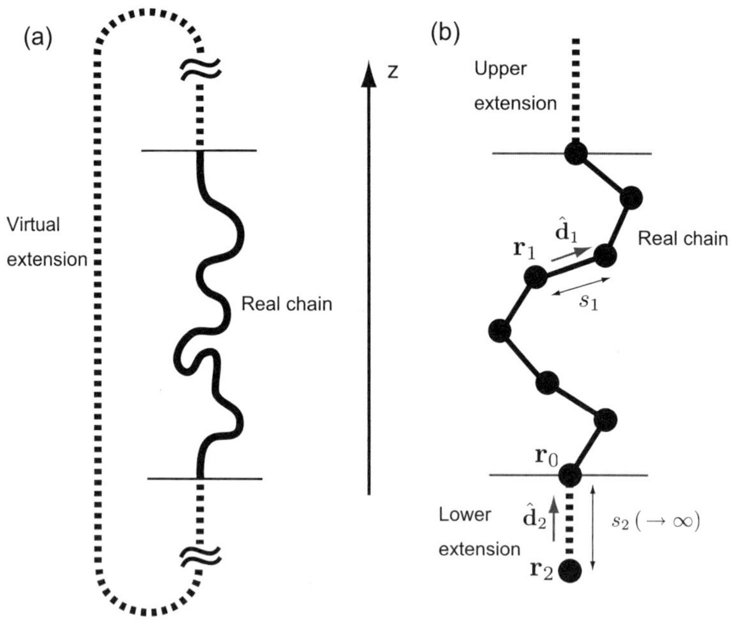

Fig. B.1. Schematic illustration to show the parameterizations used in the calculation of chain/lower-extension contribution to Wr.

To calculate Wr for an open chain in a typical magnetic tweezer experimental setup (see Fig. B.1), we use "Method 1b" by Klenin and Langowski.[141] First, the chain can be closed by attaching two infinitely straight segments to both ends of the chain. The chain/chain contribution to the writhe Wr is already given in Ref.,[141] we here evaluate the additional chain/extension contributions. Specifically, we describe in detail the calculation of the chain/lower extension contribution, Ω_C: The chain/upper extension contribution is obtained in the same way. Take one bead of the chain at position $\mathbf{r}_1$. Its bond vector can be written as $\mathbf{u}_1 = s_1\hat{\mathbf{d}}_1$, where $\hat{\mathbf{d}}_1$ is the unit tangent and s_1 is the bond length. For the moment, we consider an extension of finite length s_2 and along the $\hat{\mathbf{d}}_2 = \hat{\mathbf{z}}$, attached to the end of the chain at $\mathbf{r}_0$, see Fig. B.1. The end of the extension is thus $\mathbf{r}_2 = \mathbf{r}_0 - s_2\hat{\mathbf{d}}_2$. In the final stage, we will take $s_2 \to \infty$. Now, we decompose the vector $\mathbf{r}_{10} = \mathbf{r}_0 - \mathbf{r}_1$ in the (not orthonormal) frame $(-\hat{\mathbf{d}}_1, \hat{\mathbf{d}}_2, \hat{\mathbf{d}}_1 \times \hat{\mathbf{d}}_2)$:

$$\mathbf{r}_{10} = -a_1\hat{\mathbf{e}}_1 + a_2\hat{\mathbf{d}}_2 + a_0(\hat{\mathbf{d}}_1 \times \hat{\mathbf{d}}_2), \tag{B.1}$$

and thus $\mathbf{r}_{12} = -a_1\hat{\mathbf{d}}_1 + (a_2 - s_2)\hat{\mathbf{d}}_2 + a_0(\hat{\mathbf{d}}_1 \times \hat{\mathbf{d}}_2)$. Here, the coefficients a_1, a_2 and a_0 are obtained as

$$a_1 = \mathbf{r}_{10} \cdot (\hat{\mathbf{d}}_2 \cos\beta - \hat{\mathbf{d}}_1)/\sin^2\beta, \tag{B.2}$$

$$a_2 = \mathbf{r}_{10} \cdot (\hat{\mathbf{d}}_2 \cos\beta - \hat{\mathbf{d}}_1)/\sin^2\beta, \tag{B.3}$$

$$a_0 = \mathbf{r}_{10} \cdot (\hat{\mathbf{d}}_1 \times \hat{\mathbf{d}}_2)/\sin^2\beta, \tag{B.4}$$

where $\cos\beta = \hat{\mathbf{d}}_1 \cdot \hat{\mathbf{d}}_2$. and $\sin^2\beta = 1 - (\hat{\mathbf{d}}_1 \cdot \hat{\mathbf{d}}_2)^2$. Note that these quantities are defined by using $\mathbf{r}_{10}$, not by $\mathbf{r}_{12}$ as done in the chain/chain contribution.[141] The segments are represented parametrically as

$$\mathbf{R}_1 = \mathbf{r}_1 + x_1\hat{\mathbf{d}}_1 \ (0 \le x_1 \le s_1), \quad \mathbf{R}_2 = \mathbf{r}_2 + x_2\hat{\mathbf{d}}_2 \ (0 \le x_2 \le s_2). \tag{B.5}$$

We thus obtain

$$\mathbf{R}_{12} = -(a_1 + x_1)\hat{\mathbf{d}}_1 + (a_2 + x_2 - s_2)\hat{\mathbf{d}}_2 + a_0(\hat{\mathbf{d}}_1 \times \hat{\mathbf{d}}_2), \tag{B.6}$$

and $R_{12}^2 = (a_1+x_1)^2+(a_2+x_2-s_2)^2-2(a_1+x_1)(a_2+x_2-s_2)\cos\beta+a_0^2\sin^2\beta$. Introducing the parameters $t_1 = a_1 + x_1$ and $t_2 = a_2 + x_2 - s_2$, we obtain

$$\frac{\Omega_C}{4\pi} = -\frac{1}{4\pi}\int_{a_1}^{a_1+s_1} dt_1 \int_{-s_2+a_2}^{a_2} dt_2 \frac{a_0 \sin^2\beta}{(t_1^2 + t_2^2 - 2t_1t_2\cos\beta + a_0^2\cos^2\beta)^{3/2}}. \tag{B.7}$$

This integral can be performed analytically,[141] which leads to $\Omega_C/4\pi = F(a_1 + s_1, a_2) - F(a_1 + s_1, -s_2 + a_2) - F(a_1, a_2) + F(a_1, -s_2 + a_2)$, where

$$F(y_1, y_2) = -\frac{1}{4\pi}\tan^{-1}\left[\frac{y_1y_2 + a_0^2\cos\beta}{a_0(y_1^2 + y_2^2 - 2y_1y_2\cos\beta + a_0^2\sin^2\beta)^{1/2}}\right]. \tag{B.8}$$

Taking the infinite segment limit, $s_2 \to \infty$, we finally arrive at our final result

$$\frac{\Omega_C}{4\pi} = F(a_1 + s_1, a_2) - F(a_1, a_2) + F_\infty(a_1 + s_1) - F_\infty(a_1), \tag{B.9}$$

where $F_\infty(y_1) = F(y_1, \infty) = -1/(4\pi)\tan^{-1}(y_1/a_0)$. Summing up Ω_C over all segments of the chain, one can obtain the writhe from the coupling between the chain and lower extension.

To confirm that the above formula works for an *open* twisted polymer, we performed a Brownian dynamics simulation, see Fig. B.2. Our polymer chain consists of $N = 70$ monomers, with the bend and twist persistence length $\ell_P = 10a$ and $\ell_T = 15a$ (i.e., the elastic constant ratio $C/A = 1.5$ valid for DNA filaments). First, we constrain our polymer chain along the z axis, and rotate its upper end by six turns, which injects linking number $\mathrm{Lk} = 6$. At $t = 0$, we relax the both ends of the polymer to move only laterally under an external stretching force $F_{ext} = 0.8k_BT/a$, while allowing no axial rotations at either ends. Self intersections are prevented by the truncated Lennerd-Jones potential, and the two infinite side planes at the ends confine the polymer and prevent it from untwisting by looping around its ends. Thus Lk of the chain is invariant throughout the relaxation of the polymer chain. In the simulation shown, the plectonemes are formed at the lower side of the chain: see Fig. B.2 (a). We monitored the writhe Wr (by applying the formula explained above) and twist Tw, which is calculated according to[141]

$$\mathrm{Tw} = \frac{1}{2\pi}\sum_{i=1}^{N}(\alpha_i + \gamma_i). \tag{B.10}$$

The linking number Lk is obtained by summing Wr and Tw. As seen in Fig. B.2 (b), while both Wr and Tw change in time and fluctuate a lot due to thermal effects, Lk is constant throughout the simulation, verifying that the topological variables are correctly computed by the above method.

Appendix C.

In this Appendix, we derive an energy conservation law in a moving elastic rod in a viscous fluid. For this aim, we explain some more about kinematics of *temporal* changes of a rod, in addition to Sec. 3. The temporal change of the directors $\{\hat{\mathbf{e}}_\alpha\}$ can be written as

$$\frac{\partial \hat{\mathbf{e}}_\alpha}{\partial t} = \boldsymbol{\omega} \times \hat{\mathbf{e}}_\alpha, \tag{C.1}$$

where $\boldsymbol{\omega}$ is often called spin vector. The time change of the strain vector $\boldsymbol{\Omega}$ defined in Eq. (13) becomes

$$\frac{\partial \boldsymbol{\Omega}}{\partial t} = \frac{\partial \boldsymbol{\omega}}{\partial s} + \boldsymbol{\omega} \times \boldsymbol{\Omega}. \tag{C.2}$$

In the material frame, it can be written as

$$\frac{\partial \Omega_\alpha}{\partial t}\hat{\mathbf{e}}_\alpha = \frac{\partial \omega_\alpha}{\partial s}\hat{\mathbf{e}}_\alpha + (\boldsymbol{\Omega} \times \boldsymbol{\omega})_\alpha \hat{\mathbf{e}}_\alpha. \tag{C.3}$$

Note that $\boldsymbol{\Omega} \times \boldsymbol{\omega}$ appears on the right-hand side of this equation, instead of $\boldsymbol{\omega} \times \boldsymbol{\Omega}$ as in Eq. (C.2), because of the change of basis directors.[101]

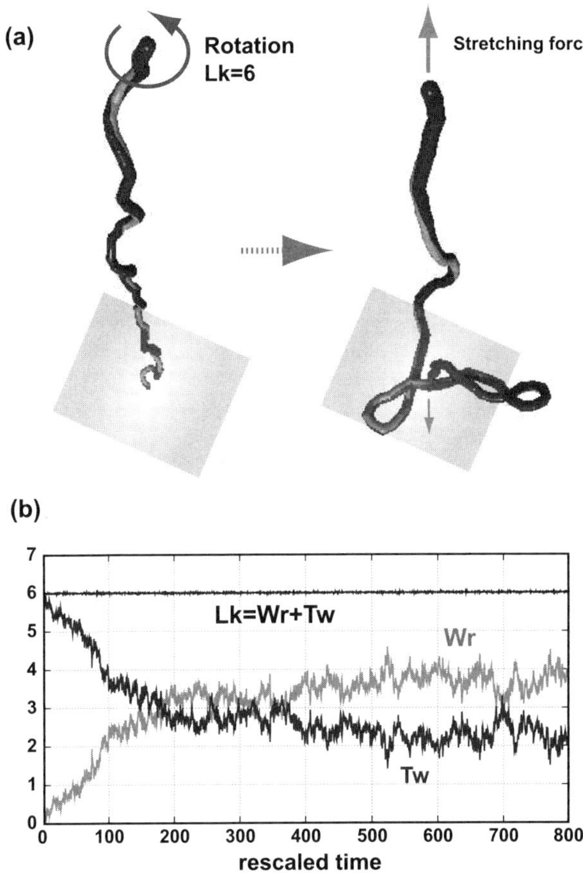

Fig. B.2. (a) Typical snapshots of a twisted polymer chain obtained from the simula tion (left) right after the release of two ends of the polymer chain and (right) a plectoneme configuration at thermal equilibrium. (b) Time changes of the topologi cal variables, writhe Wr (gray) and twist Tw (black) and linking number Lk=Wr+Tw.

We consider an inextensible rod, thus derivatives with respect to arclength s and time t commute, i.e., $\partial_s\partial_t(\cdot) = \partial_t\partial_s(\cdot)$, and $\delta(ds) = 0$. In this case, the variation of the elastic energy is $\delta E = \int_0^L M_\alpha \delta\Omega_\alpha ds$ (see Sec. 4), where the summation is always implied for repeated indices. This suggests

$$\frac{\partial E}{\partial t} = \int_0^L M_\alpha \frac{\partial \Omega_\alpha}{\partial t} ds. \tag{C.4}$$

Using Eq. (C.3), we can obtain

$$M_\alpha \frac{\partial \Omega_\alpha}{\partial t} = \frac{\partial}{\partial s}(M_\alpha \omega_\alpha) - \omega_\alpha \frac{\partial M_\alpha}{\partial s} - \boldsymbol{\omega} \cdot (\mathbf{M} \times \boldsymbol{\Omega}). \quad \text{(C.5)}$$

On the other hand, noting $\boldsymbol{\omega} \cdot \partial_s \mathbf{M} = \omega_\alpha \partial M_\alpha + \boldsymbol{\omega} \cdot (\boldsymbol{\Omega} \times \mathbf{M})$, we find

$$M_\alpha \frac{\partial \Omega_\alpha}{\partial t} = \frac{\partial}{\partial s}(M_\alpha \omega_\alpha) - \boldsymbol{\omega} \cdot \frac{\partial \mathbf{M}}{\partial s}. \quad \text{(C.6)}$$

Here, using the Kirchhoff equation given in Eq. (53) with identifying external moments as viscous drag ones only, i.e., $\mathbf{m}_{ext} = -\mathbf{m}_v$, we have

$$\boldsymbol{\omega} \cdot \frac{\partial \mathbf{M}}{\partial s} = -\mathbf{F} \cdot (\boldsymbol{\omega} \times \hat{\mathbf{e}}_3) + \boldsymbol{\omega} \cdot \mathbf{m}_v. \quad \text{(C.7)}$$

From Eq. (C.1), we find $\boldsymbol{\omega} \times \hat{\mathbf{e}}_3 = \partial_s \mathbf{v}$, where $\mathbf{v} = \dot{\mathbf{r}}$, and then we can further change Eq. (C.7) to obtain

$$\boldsymbol{\omega} \cdot \frac{\partial \mathbf{M}}{\partial s} = -\frac{\partial}{\partial s}(\mathbf{F} \cdot \mathbf{v}) + \mathbf{f}_v \cdot \mathbf{v} + \mathbf{m}_v \cdot \boldsymbol{\omega}, \quad \text{(C.8)}$$

where the force balance condition $\partial_s \mathbf{F} - \mathbf{f}_v = \mathbf{0}$ from Eq. (52), was used. Plugging Eqs. (C.6) and (C.8) into Eq. (C.4), we arrive at

$$\frac{\partial E}{\partial t} = \int_0^L \frac{\partial}{\partial s}(M_\alpha \omega_\alpha + \mathbf{F} \cdot \mathbf{v}) ds - \int_0^L (\mathbf{f}_v \cdot \mathbf{v} + \mathbf{m}_v \cdot \boldsymbol{\omega}) ds. \quad \text{(C.9)}$$

Identifying the viscous power dissipation of this system as

$$P_{diss} = \int_0^L (\mathbf{f}_v \cdot \mathbf{v} + \mathbf{m}_v \cdot \boldsymbol{\omega}) ds, \quad \text{(C.10)}$$

Eq. (C.9) tells us that the work done at the boundaries on this rod per unit time is equal to the viscous power dissipation plus the change of the elastic energy of this rod per unit time. Performing the integral in Eq. (C.9), we obtain

$$P_{diss} + \frac{\partial E}{\partial t} = \mathbf{M}(L) \cdot \boldsymbol{\omega}(L) + \mathbf{F}(L) \cdot \mathbf{v}(L) - \mathbf{M}(0) \cdot \boldsymbol{\omega}(0) - \mathbf{F}(0) \cdot \mathbf{v}(0). \quad \text{(C.11)}$$

For a rotating rod that we are considering, the boundary conditions read $\boldsymbol{\omega}(0) = \omega_0 \hat{\mathbf{t}}(0)$, $\mathbf{v}(0) = \mathbf{0}$ at the driving base, and $\mathbf{M}(L) = \mathbf{0}$ and $\mathbf{F}(L) = \mathbf{0}$ at the free end. In a stationary state, there is no change in time of the elastic energy, $\partial E/\partial t = 0$. Writing the externally applied axial torque as $M_{ext} = \mathbf{M}(0) \cdot \hat{\mathbf{t}}(0)$, we arrive at the simple result

$$P_{diss} = \omega_0 M_{ext}. \quad \text{(C.12)}$$

This says that the work externally done per unit time is consumed as viscous dissipation.

References

1. M. Doi and S. F. Edwards, *The Theory of Polymer Dynamics* (Oxford University Press, New York, 1986).
2. P.-G. D. Gennes, *Scaling Concepts in Polymer Physics* (Cornell University Press, Ithaca, 1979).
3. R. R. Netz and D. Andelman, Neutral and charged polymers at interfaces, *Phys. Rep.* **380**, 1–95 (2003).
4. C. H. Wiggins, Biopolymer mechanics: stability, dynamics, and statistics, *Math. Meth. Appl. Sci.* **24**, 1325 (2001).
5. R. Philips, J. Kondev, and J. Theriot, *Physical Biology of the Cell* (Garland Science, Taylor & Francis Group, New York, 2009).
6. R. E. Goldstein and S. A. Langer, Nonlinear dynamics of stiff polymers, *Phys. Rev. Lett.* **75**, 1094 (1995).
7. R. E. Goldstein, A. Goriely, G. Huber, and C. W. Wolgemuth, Bistable helices, *Phys. Rev. Lett.* **84**, 1631 (2000).
8. F. Valle, M. Favre, P. D. L. Rios, A. Rosa, and G. Dietler, Scaling exponents and probability distributions of DNA end-to-end distance, *Phys. Rev. Lett.* **95**, 158105 (2005).
9. P. A. Wiggins, T. van der Heijden, F. Moreno-Herrero, A. Spakowitz, R. Philips, J. Widom, C. Dekker, and P. C. Nelson, High flexibility of DNA on short length scales probed by atomic force microscopy, *Nat. Nanotech.* **1**, 137–141 (2006).
10. G. Witz, K. Rechendorff, J. Adamcik, and G. Dietler, Conformation of circular DNA in two dimensions, *Phys. Rev. Lett.* **101**, 148103 (2008).
11. K. Yoshikawa, M. Takahashi, V. V. Vasilevskaya, and A. R. Khokhlov, Large discrete transition in a single DNA molecule appears continuous in the ensemble, *Phys. Rev. Lett.* **76**, 3029 (1996).
12. L. Gosule and J. Schellman, Compact form of DNA induced by spermidine, *Nature* **259**, 333–335 (1976).
13. R. W. Wilson and V. A. Bloomfield, Counterion-induced condensation of deoxyribonucleic acid. a light-scattering study, *Biochemistry* **18**, 2192–2196 (1979).
14. J. Widom and R. L. Baldwin, Cation-induced toroidal condensation of DNA: Studies with $Co^{3+}(NH_3)_6$, *J. Mol. Biol.* **144**, 431–453 (1980).
15. V. A. Bloomfield, DNA condensation, *Curr. Opin. Struct. Biol.* **6**, 334–341 (1996).
16. H. Noguchi and K. Yoshikawa, Morphological variation in a collapsed single homopolymer chain, *J. Chem. Phys.* **109**, 5070–5077 (1998).
17. T. Odijk, Statics and dynamics of condensed DNA within phages and globules, *Phil. Trans. R. Soc. Lond. A* **362**, 1497–1517 (2004).
18. D. Marrenduzzo and C. Micheletti, Thermodynamics of DNA packaging inside a viral capsid: The role of DNA intrinsic thickness, *J. Mol. Biol.* **330**, 485–492 (2003).
19. D. Marrenduzzo, Statistics of confined polymers and the melting of a DNA spool, *EPL* **85**, 38005 (2009).

20. A. Vologodskii, DNA extension under the action of an external force, *Macromolecules* **27**, 5623–5625 (1994).
21. C. Bustamante, J. F. Marko, E. D. Siggia, and S. Smith, Entropic elasticity of lambda-phage DNA, *Science* **265**, 1599–1600 (1994).
22. T. T. Perkins, D. E. Smith, R. G. Larson, and S. Chu, Stretching of a single tethered polymer in a uniform flow, *Science* **268**, 83–87 (1995).
23. J. F. Marko and E. D. Siggia, Stretching DNA, *Macromolecules* **28**, 8759–8770 (1995).
24. C. Baumann, V. Bloomfield, S. Smith, C. Bustamante, M. Wang, and S. Block, Stretching of single collapsed DNA molecules, *Biophys. J.* **78**, 1965–1978 (2000).
25. Y. Murayama, Y. Sakamaki, and M. Sano, Elastic response of single DNA molecules exhibits a reentrant collapsing transition, *Phys. Rev. Lett.* **90**, 018102 (2003).
26. O. Kratky and G. Porod, Röntgenuntersuchung gelöster fadenmoleküle, *Recl. Trav. Chim. Pays-Bas* **68**, 1106–1123 (1949).
27. N. Saitô, K. Takahashi, and Y. Yunoki, The statistical mechanical theory of stiff chains, *J. Phys. Soc. Jpn.* **22**, 219–226 (1967).
28. F. Ritort, Single-molecule experiments in biological physics: Methods and applications, *J. Phys.: Condens. Matter* **18**, R531–R583 (2006).
29. K. C. Neuman and A. Nagy, Single-molecule force spectroscopy: optical tweezers, magnetic tweezers and atomic force microscopy, *Nat. Methods* **5**, 491–505 (2008).
30. S. Chu, Laser manipulation of atoms and particles, *Science* **253**, 861–866 (1991).
31. M. D. Wang, H. Yin, R. Landick, J. Gelles, and S. M. Block, Stretching DNA with optical tweezers, *Biophys. J.* **72**, 1335–1346 (1997).
32. S. B. Smith, L. Finzi, and C. Bustamante, Direct mechanical measurements of the elasticity of single DNA molecules by using magnetic beads, *Science* **258**, 1122–1126 (1992).
33. G. Binig, C. G. Quate, and C. Gerber, Atomic force microscope, *Phys. Rev. Lett.* **56**, 930 (1986).
34. B. Cappella and G. Dietler, Force-distance curves by atomic force microscopy, *Sur. Sci. Rep.* **34**, 1–104 (1999).
35. R. A. Harris and J. E. Hearst, On polymer dynamics, *J. Chem. Phys.* **44**, 2595–2602 (1966).
36. R. G. Winkler, P. Reineker, and L. Harnau, Models and equilibrium properties of stiff molecular chains, *J. Chem. Phys.* **101**, 8119–8129 (1994).
37. C. Storm and P. C. Nelson, Theory of high-force DNA stretching and overstretching, *Phys. Rev. E* **67**, 051906 (2003).
38. R. Golestanian and T. B. Liverpool, Statistical mechanics of semiflexible ribbon polymers, *Phys. Rev. E* **62**, 5488 (2000).
39. E. L. Starostin and G. H. M. van der Heijden, Tension-induced multistability in inextensible helical ribbons, *Phys. Rev. Lett.* **101**, 084301 (2008).
40. L. Giomi and L. Mahadevan, Statistical mechanics of developable ribbons, *Phys. Rev. Lett.* **104**, 238104 (2010).

41. P. A. Wiggins and P. C. Nelson, Generalized theory of semiflexible polymers, *Phys. Rev. E* **73**, 031906 (2000).
42. D. Riveline, C. H. Wiggins, R. E. Goldstein, and A. Ott, Elastohydrodynamic study of actin filaments using fluorescence microscopy, *Phys. Rev. E* **56**, R1330 (1997).
43. O. Hallatschek, E. Frey, and K. Kroy, Tension dynamics in semiflexible polymers I: Coarse-grained equations of motion, *Phys. Rev. E* **75**, 031905 (2007).
44. T. Hiraiwa and T. Ohta, Viscoelastic behavior of a single semiflexible polymer chain, *J. Phys. Soc. Jpn.* **77**, 023001 (2008).
45. M. Hinczewski, X. Schlagberger, M. Rubinstein, O. Krichevsky, and R. R. Netz, End-monomer dynamics in semiflexible polymers, *Macromolecules* **42**, 860–875 (2009).
46. M. Hinczewski and R. R. Netz, Global cross-over dynamics of single semiflexible polymers, *EPL* **88**, 18001 (2009).
47. D. A. Koster, A. Crut., S. Shuman, M.-A. Bjornsti, and N. H. Dekker, Cellular strategies for regulating DNA supercoiling: A single-molecule perspective, *Cell* **142**, 519–530 (2010).
48. V. Viglasky, F. Valle, J. Adamcik, I. Joab, D. Podhradsky, and G. Dietler, Anthracycline-dependent heat-induced transition from positive to negative supercoiled DNA, *Electrophoresis* **24**, 1703–1711 (2003).
49. T. R. Strick, J.-F. Allemand, D. Bensimon, A. Bensimon, and V. Croquetto, The elasticity of a single supercoiled DNA molecules by using magnetic beads, *Science* **271**, 1835–1837 (1996).
50. Z. Bryant, M. D. Stone, J. Gore, S. B. Smith, N. R. Cozzarelli, and C. Bustamante, Structural transitions and elasticity form torque measurements on DNA, *Nature* **424**, 338–341 (2003).
51. K. Besteman, S. Hage, N. H. Dekker, and S. G. Lemay, Role of tension and twist in single-molecule DNA condensation, *Phys. Rev. Lett.* **98**, 058103 (2007).
52. H. Brutzer, N. Luzzietti, D. Klaue, and R. Seidel, Energetics at the DNA supercoiling transition, *Biophys. J.* **98**, 1267–1276 (2010).
53. M. Wadati and H. Tsuru, Elastic model of looped DNA, *Physica D* **21**, 213–226 (1986).
54. H. Tsuru and M. Wadati, Elastic model of highly supercoiled DNA, *Biopolymers* **25**, 2083–2096 (1986).
55. F. Tanaka and H. Tanahashi, Elastic theory of supercoiled DNA, *J. Chem. Phys.* **83**, 6017–6026 (1986).
56. F. Julicher, Supercoiling transitions of closed DNA, *Phys. Rev. E* **49**, 2429 (1994).
57. J. F. Marko and E. D. Siggia, Statistical mechanics of supercoiled DNA, *Phys. Rev. E* **52**, 2912 (1995).
58. A. V. Vologodskii and J. F. Marko, Extension of torsionally stressed DNA by external force, *Biophys. J.* **73**, 123–132 (1997).
59. J. D. Moroz and P. Nelson, Torsional directed walks, entropic elasticity, and DNA twist stiffness, *Proc. Natl. Acad. Sci. USA* **94**, 14418–14422 (1997).

60. C. Bouchiat and M. Mezard, Elasticity model of a supercoiled DNA molecule, *Phys. Rev. Lett.* **80**, 1556 (1998).
61. S. Neukirch, Extracting DNA twist rigidity from experimental supercoiling data, *Phys. Rev. Lett.* **93**, 198107 (2004).
62. J. F. Marko and E. D. Siggia, Bending and twisting elasticity of DNA, *Macromolecules* **27**, 981–988 (1994).
63. R. D. Kamien, T. C. Lubensky, P. Nelson, and C. S. O'Hern, Direct determination of DNA twist-stretch coupling, *Europhys. Lett.* **38**, 237–242 (1997).
64. J. F. Marko, Stretching must twist DNA, *Europhys. Lett.* **38**, 183–188 (1997).
65. M. Y. Sheinin and M. D. Wang, Twist-stretch coupling and phase transition during DNA supercoiling, *Phys. Chem. Chem. Phys.* **11**, 4800–4803 (2009).
66. J. Langowski, Polymer chain models of DNA and chromatin, *Eur. Phys. J. E* **19**, 241–249 (2006).
67. M. D. Barkley and B. H. Zimm, Theory of twisting and bending of chain macromolecules: Analysis of the fluorescence depolarization of DNA, *J. Chem. Phys.* **70**, 2991–3007 (1979).
68. J. F. Marko, DNA under high tension: Overstretching, undertwisting, and relaxation dynamics, *Phys. Rev. E* **57**, 2134 (1998).
69. A. C. Maggs, Twist and writhe dynamics of stiff polymers, *Phys. Rev. Lett.* **85**, 5472 (2000).
70. S. V. Panyukov and Y. Rabin, Fluctuating elastic rings: Statics and dynamics, *Phys. Rev. E* **64**, 011909 (2001).
71. E. P. Petrov, T. Ohrt, R. G. Winkler, and P. Schwille, Diffusion and segmental dynamics of double-stranded DNA, *Phys. Rev. Lett.* **97**, 258101 (2006).
72. C. Bustamante, S. Smith, J. Liphardt, and D. Smith, Single-molecule studies of DNA mechanics, *Curr. Opin. Struct. Biol.* **10**, 279–285 (2000).
73. D. A. Koster, V. Croquette, C. Dekker, S. Shuman, and N. H. Dekker, Friction and torque govern the relaxation of DNA supercoils by eukaryotic topoisomerase IB, *Nature* **434**, 671–674 (2005).
74. C. Lavelle, Forces and torques in the nucleus: chromatin under mechanical constraints, *Biochem. Cell. Biol.* **87**, 307–322 (2009).
75. C. Levinthal and H. Crane, On the unwinding of DNA, *Proc. Natl. Acad. Sci. USA* **42**, 436–438 (1956).
76. L. F. Liu and J. D. Wang, Supercoiling of the DNA template during transcription, *Proc. Natl. Acad. Sci. USA* **84**, 7024–7027 (1987).
77. Y. Harada, O. Ohara, A. Takatsuki, H. Itoh, N. Shimamoto, and K. K. Jr, Direct observation of DNA rotation during transcription by Escherichia coli RNA polymerase, *Nature* **409**, 113 (2001).
78. P. Nelson, Transport of torsional stress in DNA, *Proc. Natl. Acad. Sci. USA* **96**, 14342–14347 (1999).
79. Y.-P. Tsao, H.-Y. Wu, and L. F. Liu, Transcription-driven supercoiling of DNA: Direct biochemical evidence from In vitro studies, *Cell* **56**, 111–118 (1989).
80. P. Droge and A. Nordheim, Transcription-induced conformational change

in a topologically closed DNA domain, *Nucleic Acids Res.* **19**, 2941–2946 (1991).
81. A. R. Rahmouni and R. D. Wells, Direct evidence for the effect of transcription on local DNA supercoiling *in vivo*, *J. Mol. Biol.* **223**, 131–144 (1992).
82. J. E. Krebs and M. Dunaway, DNA length is a critical parameter for eukaryotic transcription in vivo, *Mol. Cell. Biol.* **16**, 5821–5829 (1996).
83. F. Kouzine, J. Liu, S. Sanford, H.-J. Chung, and D. Levens, The dynamic response of upstream DNA to transcription-generated torsional stress, *Nat. Struct. Mol. Biol.* **11**, 1092–1100 (2004).
84. F. Kouzine, S. Sanford, Z. Elisha-Feil, and D. Levens, The functional response of upstream DNA to dynamic supercoiling *in vivo*, *Nat. Struct. Mol. Biol.* **15**, 146–154 (2008).
85. V. A. Stupina and J. C. Wang, DNA axial rotation and the merge of oppositely supercoiled DNA domains in Escherichia coli: Effects of DNA bends, *Proc. Natl. Acad. Sci. USA* **101**, 8608–8613 (2004).
86. P. Thomen, U. Bockelmann, and F. Heslot, Rotational drag on DNA: A single molecule experiment, *Phys. Rev. Lett.* **88**, 248102 (2002).
87. F. Leng and R. McMacken, Potent stimulation of transcription-coupled DNA supercoiling by sequence-specific DNA-binding proteins, *Proc. Natl. Acad. Sci. USA* **99**, 9139–9144 (2002).
88. A. Crut, D. A. Koster, R. Seidel, C. H. Wiggins, and N. H. Dekker, Fast dynamics of supercoiled DNA revealed by single-molecule experiments, *Proc. Natl. Acad. Sci. USA* **104**, 11957–11962 (2007).
89. C. W. Wolgemuth, T. R. Powers, and R. E. Goldstein, Twirling and whirling: Viscous dynamics of rotating elastic filaments, *Phys. Rev. Lett.* **84**, 1623 (2000).
90. S. Lim and C. S. Peskin, Simulations of the whirling instability by the immersed boundary method, *SIAM J. Sci. Comput.* **25**, 2066–2083 (2004).
91. H. Wada and R. R. Netz, Nonequilibrium hydrodynamics of a rotating filament, *Europhys. Lett.* **75**, 645–651 (2006).
92. H. Wada and R. R. Netz, Plectoneme creation reduces the rotational friction of a polymer, *EPL* **87**, 38001 (2009).
93. H. Schiessel, The physics of chromatin, *J. Phys. Condens. Matter* **15**, R699–R774 (2003).
94. R. R. Netz and J.-F. Joanny, Complexation between a semiflexible polyelectrolyte and an oppositely charged sphere, *Macromolecules* **32**, 9026–9040 (1999).
95. K.-K. Kunze and R. R. Netz, Complexes of semiflexible polyelectrolytes and charged spheres as models for salt-modulated nucleosomal structures, *Phys. Rev. E* **66**, 011918 (2000).
96. H. Boroudjerdi and R. R. Netz, Stability of strong polyelectrolyte-macroion complexes, *Europhys. Lett.* **71**, 1022–1028 (2005).
97. H. Boroudjerdi, Y.-W. Kim, R. R. Netz, X. Schlagberger, and A. Serr, Statics and dynamics of strongly charged soft matter, *Phys. Rep.* **416**, 129–199 (2005).

98. C. Holm, P. Kekicheff, and R. Podgornik, Eds., *Electrostatic effects in soft matter and biophysics*, NATO Science Series, (Kluwer Academic Publishers, Dordrecht, 2001).
99. J. Happel and H. Brenner, *Low Reynolds Number Hydrodynamics* (Noordhoff, Leyden, 1973).
100. M. Manghi, X. Schlagberger, Y.-W. Kim, and R. R. Netz, Hydrodynamic effects in driven soft matter, *Soft Matter* **2**, 653–668 (2006).
101. T. R. Powers, Dynamics of filaments and membranes in a viscous fluid, *Rev. Mod. Phys.* **82**, 1607 (2010).
102. A. E. H. Love, *A Treatise on the Mathematical Theory of Elasticity* (Dover, New York, 1944).
103. L. D. Landau and Lifshitz, *Theory of Elasticity* (Butterworth Heinemann, Boston, 1986).
104. A. Goriely and M. Tabor, The nonlinear dynamics of filaments, *Nonlin. Dynam.* **21**, 101–133 (2000).
105. S. V. Panyukov and Y. Rabin, Fluctuating filaments: Statistical mechanics of helices, *Phys. Rev. E* **62**, 7135 (2000).
106. G. Chirico, Torsional-bending infinitesimal dynamics of a DNA chain, *Biopolymers* **38**, 801–811 (1996).
107. H. Wada and R. R. Netz, Hydrodynamics of helical-shaped bacterial motility, *Phys. Rev. E* **80**, 021921 (2009).
108. R. E. Goldstein, T. R. Powers, and C. H. Wiggins, Viscous nonlinear dynamics of twist and writhe, *Phys. Rev. Lett.* **80**, 5232 (1998).
109. F. Mohammad-Rafiee and R. Golestanian, The effect of anisotropic bending elasticity on the structure of bent DNA, *J. Phys. Condens. Matter* **17**, S1165–S1170 (2005).
110. W. Kan and C. W. Wolgemuth, The shape and dynamics of the *Leptospiraceae*, *Biophys. J.* **93**, 54–61 (2007).
111. G. R. Kirchhoff, Über das gleichgewicht und die bewegung eines unendlich dünnen elastischen stabes, *J. Reine Anglew Math. (Crelle)* **56**, 285–343 (1859).
112. I. Klapper and M. Tabor, A new twist in the kinematics and elastic dynamics of thin filaments and ribbons, *J. Phys. A* **27**, 4919–4924 (1994).
113. I. Klapper, Biological applications of the dynamics of twisted elastic rods, *J. Comp. Phys.* **125**, 325–337 (1996).
114. A. Goriely and M. Tabor, New amplitude equations for thin elastic rods, *Phys. Rev. Lett.* **77**, 3537 (1996).
115. M. Nizette and A. Goriely, Towards a classification of Euler-Kirchhoff filaments, *J. Math. Phys.* **40**, 2830–2866 (1999).
116. S. A. Koehler and T. R. Powers, Twirling elastica: Kinks, viscous drag, and torsional stress, *Phys. Rev. Lett.* **85**, 4827 (2000).
117. J. M. T. Thompson and A. R. Champneys, From helix to localized writhing in the torsional post-buckling of elastic rods, *Proc. R. Soc. London A* **452**, 117–138 (1996).
118. B. Fain, J. Rudnick, and S. Östlund, Conformations of linear DNA, *Phys. Rev. E* **55**, 7364 (1997).

119. D. M. Stump, W. B. Fraser, and K. E. Gates, The writhing of circular cross-section rods: undersea cables to DNA supercoils, *Proc. R. Soc. London A* **454**, 2123–2156 (1998).
120. G. H. M. van der Heiden and J. M. T. Thompson, Helical and localised buckling in twisted rods: A unified analysis of the symmetric case, *Nonlin. Dynam.* **21**, 71–99 (2000).
121. A. Goriely and M. Tabor, Spontaneous helix hand reversal and tendril perversion in climbing plants, *Phys. Rev. Lett.* **80**, 1564 (1998).
122. R. E. Goldstein and A. Goriely, Dynamic buckling of morphoelastic filaments, *Phys. Rev. E* **74**, 010901(R) (2006).
123. C. W. Wolgemuth, R. E. Goldstein, and T. R. Powers, Dynamic supercoiling bifurcations of growing elastic filaments, *Physica D* **190**, 266–289 (2004).
124. G. Călugăreanu, L'intégrale de gauss et l'analyse des noeuds tridimensionnels, *Rev. Math. Pures Appl.* **4**, 5–20 (1959).
125. J. White, Self-linking and the gauss integral in higher dimensions, *Am. J. Math.* **91**, 693–728 (1969).
126. F. B. Fuller, Decomposition of the linking number of a closed ribbon: A problem from molecular biology, *Proc. Natl. Acad. Sci. USA* **75**, 3557–3561 (1978).
127. R. D. Kamien, Local writhing dynamics, *Eur. Phys. J. B* **1**, 1–4 (1998).
128. R. D. Kamien, The geometry of soft materials: a primer, *Rev. Mod. Phys.* **74**, 953 (2002).
129. S. Neukirch and E. Starostin, Writhe formulas and antipodal points in plectonemic DNA configurations, *Phys. Rev. E* **78**, 041912 (2008).
130. J. Aldinger, I. Klapper, and M. Tabor, Formulae for the calculation and estimation of writhe, *J. Knot Theory Ramifications* **4**, 343–372 (1995).
131. C. Bouchiat and M. Mezard, Reply to "Comment on 'Elasticity model of a supercoiled DNA molecule'", *Phys. Rev. Lett.* **88**, 089802 (2002).
132. C. Bouchiat and M. Mezard, Elastic rod model of a supercoiled DNA molecule, *Eur. Phys. J. E* **2**, 377 (2000).
133. J. Samuel, S. Sinha, and A. Ghosh, DNA elasticity: Topology of self-avoidance, *J. Phys.: Condens. Matter* **18**, S253–S268 (2006).
134. J. Samuel, S. Sinha, and A. Ghosh, Comment on "Writhe formulas and antipodal points in plectonemic DNA configurations", *Phys. Rev. E* **80**, 063901 (2009).
135. J. Samuel and S. Sinha, Tops and writhing DNA, *J. Stat. Phys.* **143**, 399–412 (2011).
136. V. Rossetto and A. C. Maggs, Comment on "Elasticity model of a supercoiled DNA molecule", *Phys. Rev. Lett.* **88**, 089801 (2002).
137. V. Rossetto and A. C. Maggs, Writhing geometry of open DNA, *J. Chem. Phys.* **118**, 9864–9874 (2003).
138. S. Neukirch and E. L. Starostin, Reply to "Comment on 'Writhe formulas and antipodal points in plectonemic DNA configurations'", *Phys. Rev. E* **80**, 063902 (2009).
139. J. D. Moroz and P. Nelson, Entropic elasticity of twist-storing polymers, *Macromolecules* **31**, 6333–6347, (1998).

140. E. L. Starostin. On the writhing number of a non-closed curve. In eds. J. A. Calvo, K. C. Millett, E. J. Rawdon, and A. Stasiak, *Physical and Numerical Models in Knot Theory Including Applications to the Life Sciences*, pp. 525–545. World Scientific (2005).
141. K. Klenin and J. Langowski, Computation of writhe in modeling of supercoiled DNA, *Biopolymers* **54**, 307–317 (2000).
142. S. Allison, R. Austin, and M. Hogan, Bending and twisting dynamics of short linear DNAs: Analysis of the triplet anisotropy decay of a 209 base pair fragment by Brownian simulation, *J. Chem. Phys.* **90**, 3843–3854 (1989).
143. G. Chirico and J. Langowski, Kinetics of DNA supercoiling studied by Brownian dynamics simulation, *Biopolymers* **34**, 415–433 (1994).
144. T. Schlick, Modeling superhelical DNA: recent analytical and dynamic approaches, *Curr. Opin. Struct. Biol.* **5**, 245–262 (1995).
145. K. Klenin, J. Merlitz, and J. Langowski, A Brownian dynamics program for the simulation of linear and circular DNA and other wormlike chain polyelectrolytes, *Biophys. J.* **74**, 780–788 (1998).
146. S. N. Goyal, N. C. Perkins, and C. L. Lee, Nonlinear dynamics and loop formation in Kirchhoff rods with implications to the mechanics of DNA and cables, *J. Comp. Phys.* **209**, 371–389 (2005).
147. D. L. Ermak and J. A. McCammon, Brownian dynamics with hydrodynamic interactions, *J. Chem. Phys.* **69**, 1352–1360 (1978).
148. J. Rotne and S. Prager, Variational treatment of hydrodynamic interaction in polymers, *J. Chem. Phys.* **50**, 4831–4837 (1969).
149. R. G. Larson, The rheology of dilute solutions of flexible polymers: Progress and problems, *J. Rheol. (N.Y.)* **49**, 1–70 (2005).
150. A. Goriely and M. Tabor, Nonlinear dynamics of filaments IV: Spontaneous looping of rods, *Proc. R. Soc. Lond.* **A454**, 3183–3202 (1998).
151. J. H. Michell, On the stability of a bent and twisted wire, *Messenger of Math.* **11**, 181–184 (1889-1990).
152. E. E. Zajac, Stability of two planar loop elasticas, *ASME J. Applied Math.* **29**, 136–142 (1962).
153. A. Goriely, Twisted elastic rings and the rediscoveries of Michell's instability, *J. Elasticity* **84**, 281–299 (2006).
154. T. Sakaue and H. Wada, *to be published* (2010).
155. H. Wada and R. R. Netz, Stretching helical nano-springs at finite temperature, *EPL* **77**, 68001 (2007).
156. J. Kaes, H. Strey, M. Baermann, and E. Sackmann, Direct measurement of the wave-vector-dependent bending stiffness of freely flickering actin filaments, *Europhys. Lett.* **21**, 865–870 (1993).
157. Y. Tsuda, H. Yashutake, A. Ishijima, and T. Yanagida, Torsional rigidity of single actin filaments and actin-actin bond breaking force under torsion measured directly by in vitro micromanipulation, *Proc. Natl. Acad. Sci. USA* **93**, 12937–12942 (1996).
158. R. P. Feynman and S. Weinberg, Eds., *Elementary Particles and the Laws of Physics: The 1986 Dirac Memorial Lectures* (Cambridge University Press, Cambridge, England, 1987).

159. A. C. Maggs, Writhing geometry at finite temperature: Random walks and geometric phases for still polymers, *J. Chem. Phys.* **114**, 5888–5896 (2001).
160. V. Rossetto and A. C. Maggs, Writhing geometry of stiff polymers and scattered light, *Eur. Phys. J. B* **29**, 323–326 (2002).
161. J. Samuel and R. Bhandari, General setting for Berry's phase, *Phys. Rev. Lett.* **60**, 2339 (1988).
162. F. B. Fuller, The writhing number of a space curve, *Proc. Natl. Acad. Sci. USA* **68**, 815–819 (1971).
163. S. R. de Groot and P. Mazur, *Non-equilibrium Thermodynamics* (North-Holland Publishing Co., Amsterdam, 1962).
164. P. Nelson, Comment on "Rotational Drag on DNA: A Single Molecule Experiment", *Phys. Rev. Lett.* **92**, 159801 (2004).
165. P. Thomen and F. Heslot, Reply to "Comment on 'Rotational Drag on DNA: A Single Molecule Experiment'", *Phys. Rev. Lett.* **92**, 159802 (2004).
166. J. Bednar, P. Furrer, V. Katritch, A. Stasiak, J. Dubochet, and A. Stasiak, Determination of DNA Persistence Length by Cryo-electron Microscopy. Separation of the Static and Dynamic Contributions to the Apparent Persistence Length of DNA, *J. Mol. Biol.* **254**, 579–594 (1995).
167. H. Yamakawa, *Modern Theory of Polymer Solutions* (Harper & Row, New York, 1971).
168. D. Garrivier and B. Fourcade, Twisting DNA with variable intrinsic curvature, *Europhys. Lett.* **49**, 390–395 (2000).
169. J. Rudnick and R. Bruinsma, DNA-protein cooperative binding through variable-range elastic coupling, *Biophys. J.* **76**, 1725–1733 (1999).
170. A. Balaeff, L. Mahadevan, and K. Shulten, Structural basis for cooperative DNA binding by CAP and *lac* repressor, *Structure* **12**, 123–132 (2004).
171. P. K. Purohit and P. C. Nelson, Effect of supercoiling on formation of protein-mediated DNA loops, *Phys. Rev. E* **74**, 061907 (2006).
172. D. Normanno, F. Vanzi, and F. S. Pavonne, Single-molecule manipulation reveals supercoiling-dependent modulation of *lac* repressor-mediated DNA looping, *Nucl. Acids Res.* **36**, 2505–2513 (2008).
173. I. D. Vlaminck, I. Vidic, M. T. J. van Loenhout, R. Kanaar, J. H. G. Lebbink, and C. Dekker, Torsional regulation of hRNA-induced unwinding of double-stranded DNA, *Nucleic Acids Res.* **38**, 4133–4142 (2010).
174. A. Sarkar and J. F. Marko, Removal of DNA-bound proteins by DNA twisting, *Phys. Rev. E* **64**, 061909 (2001).
175. G. Orphanides and D. Reinberg, RNA polymerase II elongation through chromatin, *Nature* **407**, 471–475 (2000).
176. G. Li, M. Levitus, C. Bustamante, and J. Widom, Rapid spontaneous accessibility of nucleosomal DNA, *Nat. Struct. Mol. Biol.* **12**, 46–53 (2005).
177. W. Möbius, R. A. Neher, and U. Gerland, Kinetic accessibility of buried DNA sites in nucleosomes, *Phys. Rev. Lett.* **97**, 208102 (2006).
178. C. Lavelle, DNA torsional stress propagates through chromatin fiber and participates in transcriptional regulation, *Nat. Struct. Mol. Biol.* **15**, 123–125 (2008).
179. A. Bancaud, N. C. e Silva, M. Barbi, G. Wagner, J.-F. Allemand, J. Mozzi-

conacci, L. C, V. Croquette, J.-M. Victor, A. Prunell, and J.-L. Viovy, Structural plasticity of single chromatin fibers revealed by torsional manipulation, *Nat. Struct. Mol. Biol.* **13**, 444–450 (2006).
180. A. Bancaud, G. Wagner, N. C. e Silva, C. Lavelle, H. Wong, J. Mozziconacci, M. Barbi, A. Sivolob, E. L. Cam, L. Mouawad, J.-L. Viovy, J.-M. Victor, and A. Prunell, Nucleosome chiral transition under positive torsional stress in single chromatin fibers, *Molecular Cell* **27**, 135–147 (2007).
181. P. Gupta, J. Zlatanova, and M. Tomschik, Nucleosome assembly depends on the torsion in the DNA molecule: A magnetic tweezers study, *Biophys. J.* **97**, 3150–3157 (2009).
182. J. Jin, L. Bai, D. S. Johnson, R. M. Fulbright, M. L. Kireeva, M. Kashlev, and M. D. Wang, Synergistic action of RNA polymerases in overcoming the nucleosomal barrier, *Nat. Struct. Mol. Biol.* **17**, 745–752 (2010).
183. B. D. Brower-Toland, C. L. Smith, R. C. Yeh, J. T. Lis, C. L. Peterson, and M. D. Wang, Mechanical disruption of individual nucleosomes reveals a reversible multistage release of DNA, *Proc. Natl. Acad. Sci. U.S.A.* **99**, 1960–1965 (2002).
184. I. M. Kulic and H. Schiessel, DNA spools under tension, *Phys. Rev. Lett.* **92**, 228101 (2004).
185. S. Mihardja, A. J. Spakowitz, Y. Zhang, and C. Bustamante, Effect of force on mononucleosomal dynamics, *Proc. Natl. Acad. Sci. USA* **103**, 15871–15876 (2006).
186. T. Wocjan, K. Klenin, and J. Langowski, Brownian dynamics simulation of DNA unrolling from the nucleosome, *J. Phys. Chem. B* **113**, 2639–2646 (2009).
187. B. Sudhanshu, S. Mihardja, E. F. Koslover, S. Mehraeen, C. Bustamante, and A. J. Spakowitz, Tension-dependent structural deformation alters single-molecule transition kinetics, *Proc. Natl. Acad. Sci. U.S.A.* **108**, 1885–1890 (2011).
188. T. Sakaue, K. Yoshikawa, S. H. Yoshimura, and K. Takeyasu, Histone core slips along DNA and prefers positioning at the chain end, *Phys. Rev. Lett.* **87**, 078105 (2001).
189. A. A. Zinchenko, K. Yoshikawa, and D. Baigl, Compaction of single-chain DNA by histone-inspired nanoparticles, *Phys. Rev. Lett.* **95**, 228101 (2005).
190. K. Luger, A. W. Maeder, R. K. Richmond, D. F. Sargent, and T. J. Rochmond, Crystal structure of the nucleosome core particle at 2.8 Å resolution, *Nature* **389**, 251–260 (1997).
191. C. A. Davey, D. F. Sargent, K. Luger, A. W. Maeder, and T. J. Rochmond, Solvent mediated interactions in the structure of the nucleosome core particle at 1.9 Å resolution, *J. Mol. Biol.* **319**, 1097–1113 (2002).
192. Y. Higuchi, T. Sakaue, and K. Yoshikawa, Torsional effect on the wrapping transition of a semiflexible polymer around a core as a model of nucleosome, *Phys. Rev. E* **82**, 031909 (2010).
193. I. V. Dobrovolskaia, M. Kenward, and G. Arya, Twist propagation in dinucleosome arrays, *Biophys. J.* **99**, 3355–3364 (2010).
194. J. Zlatanova, T. C. Bishop, J.-M. Victor, V. Jackson, and K. van Holde, The

nucleosome family: Dynamic and growing, *Structure* **17**, 160–171 (2009).

195. C. Becavin, M. Barbi, J.-M. Victor, and A. Lesne, Transcription within condensed chromatin: Steric hinderace facilitates elongation, *Biophys. J.* **98**, 824–833 (2010).
196. C. Hodges, L. Bintu, L. Lubkowska, M. Kashlev, and C. Bustamante, Nulceosomal fluctuations govern the transcription dynamics of RNA polymerase II, *Science* **325**, 626–628 (2009).
197. A. Sivolob and A. Prunell, Nucleosome conformational flexibility and implications for chromatin dynamics, *Phil. Trans. R. Soc. Lond. A* **362**, 1519–1547 (2004).

Chapter 7

Dynamics of Deformable Self-Propelled Particles: Relations with Cell Migration

Masaki Sano[1*], Miki Y. Matsuo[1†] and Takao Ohta[2‡]

[1]*Department of Physics, The University of Tokyo, Tokyo 113-0033, Japan*

[2]*Department of Physics, Kyoto University, Kyoto 606-8502, Japan*

Dynamics of deformable self-propelled particles are studied numerically and analytically. We introduce two theoretical models, a tensor model and a geometrical model. The tensor model is represented in the form of a coupled set of ordinary equations for the migration velocity and the modes of shape deformation. The geometrical model is a partial differential equation for the leading edge of a soft particle. We solve these equations to investigate the nonlinear coupling between migration and shape deformation. Experimental studies are also presented to measure the autocorrelation of the shape and the stress distribution in a migrating cell. The fundamental spatio-temporal patterns of translocation are identified from the data of autocorrelations.

1. Introduction

The active soft matter is defined such that it is an autonomous system accompanied with transport of materials and has attracted much attention recently from the view point of nonlinear dynamics and non-equilibrium statistical physics. Biological systems are one of the representative examples of the active soft matter.[1,2] Collective dynamics of interacting self-propelled objects have been modeled and studied extensively.[3–7] Various dynamical orders characteristic of pattern formation far from equilibrium have been investigated. The hydrodynamics of interacting swimming suspensions or

*`sano@phys.s.u-tokyo.ac.jp`
†`mikiy@jamstec.go.jp`
‡`takao@scphys.kyoto-u.ac.jp`

bacterias have been developed.[8–12] On the other hands, individual motions of swimming mocro-organisms have also been formulated by taking into account of the hydrodynamics effects.[13–15] These are an important subject of nonlinear dynamics to reveal fundamental mechanisms of efficient motions of micro-organisms.

Although there are a number of studies of self-propulsion as mentioned above, only a few theoretical models have been introduced for a *deformable* particle around its circular shape (spherical shape in three dimensions). More than twenty years ago, Shapere and Wiczek investigated the efficiency of swimming motion of a deformable object in two dimensions.[16] Recently, the correlation of the velocity fluctuations and the shape deformation in eucaryotic cells has been investigated experimentally.[17–21] Theoretical formulation of migration with deformation in specific living cells have been performed.[22,23] Self-propulsion of oily droplets due to Marangoni effect has also been investigated experimentally.[24–28] Our main aim is to study a deformable self-propelled particle from a general view point[29] without relying on any details of specific systems.

In this article, we shall present our recent studies of deformable self-propelled objects both theoretically and experimentally. The theoretical parts consist of analysis of two different models.[30] One is the tensor model in which shape deformation and the migration velocity are expressed by a coupled set of ordinary differential equations. The other is called the geometrical model where the motion of the leading edge of a soft particle is represented in two dimensions by a partial differential equation. We will show that the nonlinear coupling of shape deformation and migration produces a variety of dynamics even for an isolated single particle. It is interesting to see that the geometrical model exhibits a series of bifurcation of motions by changing only the radius of particle.

In the experimental parts, we focus ourselves on cell migration. In order to quantify the morphological dynamics of cell shape, the autocorrelation function is introduced and analyzed by varying the culture condition of Dictyosteluim discoidium. The measurement of mechanical stress distribution is also carried out for the understanding of the basic mechanism of shape deformations and cell migration.

The organization of the present article is as follows. In the next section (section 2), we describe the tensor model. This model is derived in section 3 by the singular perturbation method starting from an excitable reaction diffusion system in two dimensions. The set of ordinary differential equations is solved analytically and numerically in section 4 to reveal various dynamics of a self-propelled particle caused by the coupling between the migration velocity and the shape deformations. The relation between the tensor model in a special case in two dimensions and the dissipative wave model in one dimension is briefly remarked. The dynamics of the tensor model in three dimensions are shown in section 5. The experiments of cell migration is explained in section 6 where an autocorrelation function which characterizes several fundamental dynamic patterns of locomotion is introduced and analyzed. The distribution of mechani cal stress during cell migration is measured. In these experiments, the difference in dynamics between vegetative cells and starved cells is emphasized. In section 7, the time-evolution equation for a leading edge of migrating cell is considered. The motions are compared with those obtained from the tensor model. Summary and discussion are given in section 8.

2. Deformable Particle in Two Dimensions

We introduce a set of time-evolution equations for a deformable self-propelled particle.[29–31] To be specific, the theory is formulated in two dimensions but it is readily extended to three dimensions as shown in section 5. Weak deformations of a particle around a circular shape with radius R_0 can be represented as

$$R(\theta) = R_0 + \delta R(\theta, t) , \tag{1}$$

where

$$\delta R(\theta, t) = \sum_{n=-\infty}^{\infty} z_n(t) e^{in\theta} . \tag{2}$$

Since the translational motion of a particle will be incorporated in the velocity of the center of mass $\mathbf{v}$, the modes $n = \pm 1$ should be removed from the expansion (2). The mode with $n = 0$ is also excluded by assuming that the area (volume in three dimensions) is conserved. The modes $z_{\pm 2}$ represents an elliptical deformation of a particle. We introduce a second rank tensor as follows;

$$\begin{aligned} S_{11} &= -S_{22} = z_2 + z_{-2} , \\ S_{12} &= S_{21} = i(z_2 - z_{-2}) . \end{aligned} \tag{3}$$

This is a traceless and symmetric tensor. Similarly we introduce a third rank tensor from the modes $z_{\pm 3}$;

$$
\begin{aligned}
U_{111} &= z_3 + z_{-3} , \\
U_{222} &= -i(z_3 - z_{-3}) ,
\end{aligned}
\tag{4}
$$

and

$$
\begin{aligned}
U_{111} &= -U_{122} = -U_{212} = -U_{221} , \\
U_{222} &= -U_{112} = -U_{121} = -U_{211} .
\end{aligned}
\tag{5}
$$

It should be noted that the tensor S is equivalent with the nematic order parameter and U the order parameter of a banana-shaped liquid crystal[32,33]

The time-evolution equations of $\mathbf{v}$, S and U are derived by considering the possible couplings. Those take the following form up to cubic nonlinearity[30]

$$
\begin{aligned}
\frac{d}{dt}v_i &= \gamma v_i - \mathbf{v}^2 v_i - a_1 S_{ij} v_j - a_2 U_{ijk} v_j v_k - a_3 U_{ijk} S_{jk} \\
&\quad - a_4 (S_{mn} S_{mn}) v_i - a_5 (U_{\ell mn} U_{\ell mn}) v_i + a_6 S_{i\ell} S_{nm} U_{\ell nm} ,
\end{aligned}
\tag{6}
$$

$$
\begin{aligned}
\frac{d}{dt}S_{ij} &= -\kappa_2 S_{ij} + b_1 \left(v_i v_j - \frac{1}{2}\mathbf{v}^2 \delta_{ij} \right) + b_2 U_{ijk} v_k - b_3 (S_{mn} S_{mn}) S_{ij} \\
&\quad - b_4 \mathbf{v}^2 S_{ij} - b_5 (U_{\ell mn} U_{\ell mn}) S_{ij} + b_6 U_{ij\ell} S_{\ell m} v_m ,
\end{aligned}
\tag{7}
$$

$$
\begin{aligned}
\frac{d}{dt}U_{ijk} &= -\kappa_3 U_{ijk} - d_3 (U_{\ell mn} U_{\ell mn}) U_{ijk} \\
&\quad + d_1 \left[v_i v_j v_k - \frac{v_\ell v_\ell}{4} (\delta_{ij} v_k + \delta_{ik} v_j + \delta_{jk} v_i) \right] \\
&\quad + \frac{d_2}{3} \left[S_{ij} v_k + S_{ik} v_j + S_{jk} v_i - \frac{v_\ell}{2} (\delta_{ij} S_{k\ell} + \delta_{jk} S_{i\ell} + \delta_{ki} S_{j\ell}) \right] \\
&\quad - d_4 \mathbf{v}^2 U_{ijk} - d_5 (S_{mn} S_{mn}) U_{ijk} \\
&\quad + \frac{2d_6}{3} \left[S_{ij} S_{k\ell} v_\ell + S_{jk} S_{i\ell} v_\ell + S_{ki} S_{j\ell} v_\ell \right. \\
&\quad \left. - \frac{1}{2} (\delta_{ij} S_{nk} S_{n\ell} v_\ell + \delta_{jk} S_{ni} S_{n\ell} v_\ell + \delta_{ki} S_{nj} S_{n\ell} v_\ell) \right] .
\end{aligned}
\tag{8}
$$

The values of the coefficients are not determined by this phenomenological and symmetry argument. We shall derive, in section 3, a simplified version of the above set of equations starting from an excitable reaction-diffusion system. The coefficients are obtained explicitly as a function of the parameters of the excitable system.

We shall call the dynamics of Eqs. (6), (7) and (8) the tensor model. Propagation of a particle occurs from the first and the second terms in Eq. (6). That is, when γ is negative, a motionless state is stable whereas, for $\gamma > 0$, particle undergoes a translational motion. This is called a drift bifurcation.[34–39] When κ_2 and κ_3 are positive, a motionless particle takes a circular shape without deformation. However, when the particle migrates, it is deformed as the velocity is increased because of the couplings among v_i, S and U. We call this a migration-induced deformation. It should be noted that when κ_2 and κ_3 are negative, the circular shape is unstable even for a motionless state. These deformations can cause a drift motion, even when $\gamma < 0$, due to the couplings SU and SUU in Eq. (6). We call this motion a deformation-induced migration.

The simplified version of Eqs. (6) and (7) ignoring the variable $U_{\alpha\beta\gamma}$ has been introduced and studied in Ref. 29, which reads.

$$\frac{d}{dt}v_i = \gamma v_i - \mathbf{v}^2 v_i - aS_{ij}v_j \ , \tag{9}$$

$$\frac{d}{dt}S_{ij} = -\kappa S_{ij} + b\Big(v_i v_j - \frac{1}{2}\mathbf{v}^2\delta_{ij}\Big) \ . \tag{10}$$

Although this is quite simple, the solution is found to be highly nontrivial. By writing the velocity as $v_1 = v\cos\phi$, $v_2 = v\sin\phi$ and the amplitude of the second modes $z_{\pm 2}$ as $z_{\pm 2} = (s/4)e^{\mp 2i\theta}$, Eqs. (9) and (10) are rewritten as

$$\frac{d}{dt}v = v(\gamma - v^2) - \frac{1}{2}asv\cos 2(\theta - \phi) \tag{11}$$

$$\frac{d}{dt}\phi = -\frac{1}{2}as\sin 2(\theta - \phi) \tag{12}$$

$$\frac{d}{dt}s = -\kappa s + v^2 b\cos 2(\theta - \phi) \tag{13}$$

$$\frac{d}{dt}\theta = -\frac{v^2 b}{2s}\sin 2(\theta - \phi) \tag{14}$$

where v and s should be positive. It is noted from Eqs. (12) and (14) that only the relative angle $\psi = \theta - \phi$ enters into the time-evolution equations as

$$\frac{d}{dt}\psi = -\frac{1}{2}\left(-as + \frac{bv^2}{s}\right)\sin 2\psi. \tag{15}$$

This originates from the spatial isotropy.

The above set of equations has a stationary solution of a rectilinear motion for $\gamma > 0$. Without loss of generality, we may assume that the particle is moving along the x-axis, i.e., $\phi = 0$. There are two kinds of solution depending on the sign of the parameter b. When b is positive, the stationary solution is given by

$$\begin{aligned} \theta &= 0, \\ v^2 &= v_0^2 = \frac{\gamma}{1+B}, \\ s &= s_0 = \frac{b v_0^2}{\kappa}, \end{aligned} \tag{16}$$

where $B \equiv ab/(2\kappa)$. This means that elongation of the particle is along the migration direction. When b is negative, the solution is given by

$$\begin{aligned} \theta &= \pi/2, \\ v &= v_0, \\ s &= -s_0. \end{aligned} \tag{17}$$

In this case, the long axis of an elliptically deformed particle is perpendicular to the velocity. In order to avoid a singular behavior of v_0 and s_0 for $B \leq -1$ where a splitting of particle is expected to occur, we hereafter assume $ab > 0$ ($B > 0$).

The shape of real motile cells is discussed in section 6. For example, Keratocyte is elongated perpendicularly to the migration direction whereas Dicty cell is elongated parallel to the direction in a chemotaxic migration.

A remarkable property of the simplified version Eqs. (9) and (10) is that it exhibits a non-trivial bifurcation.[29] In fact, it is readily shown by a linear stability analysis of the stationary solutions that the rectilinear motion becomes unstable for $\gamma \geq \gamma_c$ where the threshold γ_c is given by

$$\gamma = \gamma_c = \frac{\kappa^2}{ab} + \frac{\kappa}{2}. \tag{18}$$

This instability occurs for both $\theta = 0$ and $\theta = \pi/2$. The bifurcation threshold is indicated on the $\gamma - \kappa$ plane in Fig. 1(a).

We make an ansatz that, when the straight motion is unstable, there occurs a circular motion along a trajectory of a closed circle. To this end, we put $v = v_r$, $s = s_r$ and $\theta = \omega t + \zeta/2$ and $\phi = \omega t$. Substituting these into Eqs. (11), (12), (13) and (14), one obtains after some algebra

$$v_r^2 = \gamma - \frac{\kappa}{2}, \tag{19}$$

and $s_r^2 = b v_r^2/a$, $\cos\zeta = \kappa/a s_r$ and $\omega^2 = (ab/4)(v_r^2 - v_c^2)$ where $v_c = \kappa/(ab)^{1/2}$. The frequency ω of rotation continuously increases from zero at

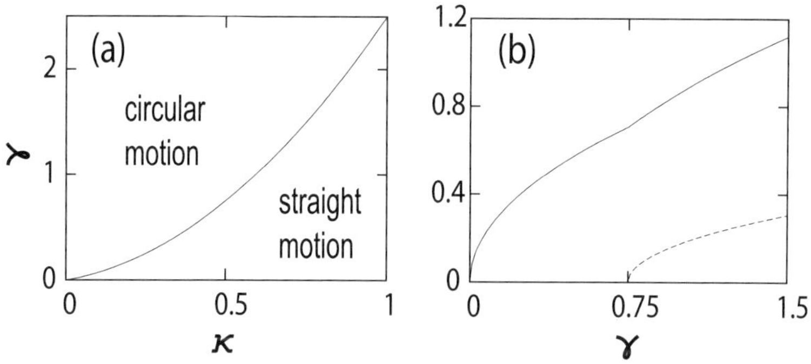

Fig. 1. (a) Stability diagram on the $\gamma - \kappa$ plane for $ab = 0.5$. (b) The velocity v (full line) and the frequency ω (broken line) as a function of γ for $a = -1.0, b = -0.5$ and $\kappa = 0.5$. The bifurcation occurs at $\gamma_c = 0.75$. These figures are reproduced from Ref. 29.

$\gamma = \gamma_c$. The velocity v is also continuous at the bifurcation point $\gamma = \gamma_c$. The γ−dependence of v and ω is displayed in Fig. 1(b).

In order to study the dynamics near the bifurcation at $\gamma = \gamma_c$ in more detail, we make a reduction of the variables. Substituting v_0 and s_0 into Eq. (15), one notes that the coefficient $-as + bv^2/s$ vanishes at the bifurcation point. Therefore, the angle variable ψ is found to be a slow variable near the stability threshold. This allows us to eliminate the other variables v and s by putting $dv/dt = ds/dt = 0$ in Eqs. (9) and (10). As a result, Eq. (15) is closed for ψ as

$$\begin{aligned}\frac{d}{dt}\psi &= F(\psi)\\ &= -\kappa\left[\frac{\gamma - \gamma_c + (\kappa/2 - \gamma)(1 - \cos^2 2\psi)}{-2\gamma_c + \kappa(1 - \cos^2 2\psi)}\right]\tan 2\psi, \qquad (20)\end{aligned}$$

where $F(\psi)$ is an odd function of ψ. This property comes from the parity symmetry that both clock-wise and counter clock-wise circular motions should be equally possible. It is readily verified that Eq. (20) for $-\pi/2 < \psi < \pi/2$ has only a stable solution $\psi = 0$ for $\gamma \leq \gamma_c$ whereas it has two stable solutions $\psi \neq 0$ for $\gamma \geq \gamma_c$. Therefore, an exchange of stability at $\gamma = \gamma_c$ occurs as a pitchfork bifurcation.

The circular motion predicted in Ref. 29 was observed later in Ref. 40 where numerical simulations of an excited domain were carried out in a three-component reaction diffusion system in two dimensions.

3. Particle Dynamics in an Excitable Reaction Diffusion System

In the previous section, we have derived the tensor model given by Eqs. (6), (7) and (8) for a deformable self-propelled particle phenomenologically based on a symmetry argument. In this section, we shall derive those equations analytically starting with an excitable reaction-diffusion system in two dimensions.[31]

3.1. *Excitable reaction-diffusion equations*

We start with a coupled set of reaction-diffusion equations for an activator u and an inhibitor v.

$$\tau\epsilon\frac{\partial u}{\partial t} = \epsilon^2\nabla^2 u + f\{u,v\} - v \ , \tag{21}$$

$$\frac{\partial v}{\partial t} = D\nabla^2 v + u - \gamma' v \ , \tag{22}$$

where

$$f\{u,v\} = -u + \theta(u - p'\{u,v\}) \ , \tag{23}$$

with $\theta(x) = 1$ for $x > 0$ and $\theta(x) = 0$ for $x < 0$. This piece-wise linear function is convenient to make all the calculations analytically possible. The functional $p'\{u,v\}$ contains a global coupling as

$$p' = p + \sigma[\int (u+v)d\mathbf{r} - W] \ , \tag{24}$$

where σ and W are positive constants, $0 < p < 1/2$ and the integral runs over the whole space. The constants τ and γ' are positive and chosen such that the system is excitable and that a localized stable pulse (particle) solution exists. Inside the particle, the variable u is positive whereas the outside of the particle is a rest state where u and v vanishes asymptotically away from the particle. The parameter ϵ stands for the width of the particle boundary. Hereafter we consider the limit $\epsilon \ll 1$.

First, let us consider the case that the global coupling is absent so that $f\{u,v\}$ is replaced by $f(u) = -u + \theta(u-p)$.[41,42] If $D = \gamma' = 0$, the set of equations (21) and (22) is known as the Fitz-Hugh-Nagumo equation for pulse propagation along nerve axion.[43] When the diffusion term of the inhibitor is present, $D \neq 0$, and is large enough, the system governed by Eqs. (21) and (22) exhibits a spatially periodic solutions, which contain a motionless isolated domain as a special case of infinite spatial period.[42] If the value of τ is decreased, this motionless solution becomes unstable and a breathing motion appears where the radius of a circular domain undergoes periodic oscillation while the center of mass is time-independent.[41,42,44]

The set of Eqs. (21) and (22) with (23) and (24) was introduced by Krischer and Mikhailov.[45] They investigated the collision of propagating particles in two dimensions by computer simulations and found a reflection of a pair of colliding particles when the propagating velocity is sufficiently small. The global coupling was introduced to avoid the breathing instability mentioned above. In fact, if one chooses sufficiently large values of σ in the global coupling (24), it becomes $p' = p$ with

$$\int d\mathbf{r}(u + v) = W \ , \tag{25}$$

and

$$f(u) = -u + \theta(u - p) \ . \tag{26}$$

In the limit $\epsilon \to 0$ and $\sigma \to \infty$ in (24), Eq. (21) becomes

$$-u + \theta(u - p) - v = 0 \ . \tag{27}$$

The interface position is defined through the condition $u = p$. It is shown from Eq. (27) that $u + v = 1$ inside the particle ($u > p$) whereas $u + v = 0$ outside the particle ($u < p$). Therefore the constraint (25) with a constant W means that the volume of an excited particle is independent of time and hence the breathing motion is prohibited. The bifurcation from a motionless particle to a propagating particle is supercritical in the limit $\sigma \to \infty$.[45]

Substituting Eq. (27) into Eq. (22) yields

$$\frac{\partial v}{\partial t} = D\nabla^2 v + \theta(u - p) - \beta v \ , \tag{28}$$

where $\beta = 1 + \gamma'$. The equilibrium solution of a motionless circular particle with radius R_0 in two dimensions is readily obtained.[42] The solution $v = \bar{v}$ is given from Eq. (28) for $0 < r < R_0$ by

$$\bar{v} = \frac{1}{\beta}[1 - R_0\kappa K_1(R_0\kappa)I_0(r\kappa)] \ , \tag{29}$$

and for $r > R_0$ by

$$\bar{v} = \frac{R_0\kappa}{\beta}I_1(R_0\kappa)K_0(r\kappa) \ , \tag{30}$$

where

$$\kappa = \sqrt{\frac{\beta}{D}}, \tag{31}$$

and I_n and K_n are the modified Bessel functions. The equilibrium profile of $\bar{u}$ is given by the relation $\bar{u} = \theta(R_0 - r) - \bar{v}$.

Equations (29) and (30) clearly indicate that the inhibitor v changes gradually in space with the characteristic length κ^{-1}. On the other hand, the activator u is discontinuous at $r = R_0$ in the limit $\epsilon \to 0$ and hence there is a sharp interface. In this situation, one can derive the interface equation of motion and the time-evolution equations for a deformed particle as described in the next subsection.

3.2. *Interface equation of motion*

In order to analyze the sharp spatial variation of u near $r = R_0$, one needs to rescale the space coordinate as $r' = r/\epsilon \sim O(1)$. In this magnified length scale, the spatial variation of v is negligible and the value of v in (21) can be replaced by the value at the interface $v(\mathbf{r}, t) = w$. As a result, Eq. (21) becomes the so-called time-dependent Gizburg-Landau equation. The interface equation of motion for this system is well known. See for example Ref. 46. For a given value of w, one can readily obtain the equation of motion for an arbitrarily deformed interface[42]

$$\tau V = \epsilon K + \tau c(w) + L, \tag{32}$$

where V is the normal component of the velocity directed from the inside to the outside of a particle and K is the mean curvature defined by $K = \nabla(\nabla u/|\nabla u|)$. The second term in Eq. (32) is the velocity for a flat interface and is related with w as[47]

$$\frac{c\tau}{\sqrt{(c\tau)^2 + 4}} = 1 - 2p - 2w. \tag{33}$$

The unknown constant w is determined by solving Eq. (28) where the term $\theta(u - p)$ is fixed if the interface configuration is given. The last term L in Eq. (32) is a Lagrange multiplier for the constraint of the particle area (volume in three dimensions) conservation (25) and is determined by the condition

$$\int d\omega V(\omega) = 0, \tag{34}$$

where $d\omega$ is the infinitesimal length (area in three dimensions) on the interface and the integral runs over the interface.

When the motion of the interface is slow compared with the relaxation rate of the inhibitor, one may employ a short time expansion for Eq. (28). In this approximation, the asymptotic solution of (28) can be written as

$$v(\mathbf{r},t) = G\theta - G^2\frac{\partial\theta}{\partial t} + G^3\frac{\partial^2\theta}{\partial t^2} - G^4\frac{\partial^3\theta}{\partial t^3} + \dots\,, \tag{35}$$

where G is defined through the relation

$$(-D\nabla^2 + \beta)G(\mathbf{r}-\mathbf{r}') = \delta(\mathbf{r}-\mathbf{r}')\,, \tag{36}$$

and the abbreviation such that $GA = \int d\mathbf{r}' G(\mathbf{r}-\mathbf{r}')A(\mathbf{r}')$ has been used. From Eq. (35), one can obtain the value of w. Substituting it into Eq. (32) with (33), one has a closed equation for an interface configuration.

The dynamics of self-propelled particles in one, two and three dimensions were developed based on the interface equation of motion (32).[34–39] It was shown that there is a drift bifurcation for some value of τ below which a motionless particle becomes unstable and it undergoes a drift motion. The migration velocity is arbitrarily small in the vicinity of the supercritical drift bifurcation. Therefore, the expansion (35) is validated. The interaction of particles was formulated in this way and the mechanism of reflection upon collisions of particles was clarified.[37] However, any deformations around a circular or spherical shape were not investigated. In the subsequent sections, we shall derive the equation of motion for a deformable excited particle for the reaction-diffusion equations (21) and (22).[31]

3.3. *Equation of motion for the center of mass $\boldsymbol{\rho}(t)$*

We consider a deformed particle with the center of mass $\boldsymbol{\rho}(t)$ in two dimensions. Its time-derivative $\dot{\boldsymbol{\rho}}(t)$ is given from a geometrical consideration by

$$\dot{\boldsymbol{\rho}} = \frac{1}{\Omega}\int d\omega V(\omega)\mathbf{R}(\omega)\,, \tag{37}$$

where the dot means the time derivative, Ω is the area (volume in three dimensions) of the particle and

$$\mathbf{R}(\theta) = R(\theta)\mathbf{e}_r\,, \tag{38}$$

with the radial unit vector $\mathbf{e}_r$. The distance from the center of mass to the interface is denoted by $R(\theta)$ with the angle θ with respect to the x axis. We assume that $R(\theta)$ is a single-valued function of θ for sufficiently weak deformations and expand as $R(\theta) = R_0 + \delta R(\theta, t)$ as Eq. (2).

The mean curvature and the normal component of the velocity are given up to the first order of the deviations, respectively, by

$$K(\theta,t) = -\frac{1}{R_0} - \frac{1}{R_0^2}\sum_{n=-\infty}^{\infty}(n^2-1)z_n(t)e^{in\theta} , \tag{39}$$

$$V(\theta,t) = \dot{\boldsymbol{\rho}}\cdot\mathbf{e}_r + \sum_{n=-\infty}^{\infty}\dot{z}_n(t)e^{in\theta} , \tag{40}$$

We shall derive a coupled set of the time-evolution equations for $\dot{\boldsymbol{\rho}}$, $z_{\pm 2}$ and $z_{\pm 3}$ under the assumption that the migration velocity is sufficiently small. Since the circular shape of a particle is assumed to be stable when it is motionless, the deformation of the particle is expected to be weak near the drift bifurcation where the propagating velocity is small.[37] In this condition, the truncation of the modes up to $n = \pm 3$ is justified.

The main steps of deriving the equation for $\dot{\boldsymbol{\rho}}$ are as follows. First of all, when the velocity c of the flat interface is small, one may expand Eq. (33) in powers of τc. Retaining up to the third order terms, one obtains from Eq. (32)

$$\begin{aligned}\tau V & -\epsilon K - L - \frac{(\tau V - \epsilon K - L)^3}{8} \\ & \approx \tau V - \epsilon K - L - \frac{\tau^3}{8} = 2(1-2p+2w) .\end{aligned} \tag{41}$$

The next step is to solve Eq. (35) with $\theta(u-p) = \theta(R_0 + \delta R - |\boldsymbol{r}-\boldsymbol{\rho}|)$ by means of the perturbation expansion in terms of δR to obtain w. The final step is to use the formulas Eqs. (37), (39) and (40). In this way, equation for the center of mass $\boldsymbol{v} = \dot{\boldsymbol{\rho}}$ is given up to the leading order by[31]

$$m\dot{v}_i + \frac{1}{2}(\tau - \tau_c)v_i + g v_i|\boldsymbol{v}|^2 = -a v_j S_{ji} , \tag{42}$$

where the tensor $S_{\beta\alpha}$ has been defined in Eq. (3) and

$$m = 2R_0\int_0^\infty dq q G_q^3 J_1(qR_0)^2 , \tag{43}$$

$$\tau_c = 4R_0\int_0^\infty dq q G_q^2 J_1(qR_0)^2 , \tag{44}$$

$$g = \frac{3R_0}{2}\int_0^\infty dq q^3 G_q^4 J_1(qR_0)^2 - \frac{3\tau^3}{64} , \tag{45}$$

and

$$a = 2(a_1 + a_2) , \tag{46}$$

with

$$a_1 = \frac{R_0}{2} \int_0^\infty dq q^2 G_q^2 J_1(qR_0) J_2(qR_0) \,, \tag{47}$$

$$a_2 = \frac{1}{4\pi R_0} \int_0^\infty dq q^2 \big[qJ_1(qR_0) - \frac{1}{R_0} J_2(qR_0)\big] \frac{\partial}{\partial q} (G_q^2 \theta_q^{(0)}) \,. \tag{48}$$

The functions G_q and $\theta_q^{(0)}$ are given, respectively, by

$$G_q \equiv \frac{1}{D} \int d\mathbf{r} (-\nabla^2 + \kappa^2)^{-1} \delta(\mathbf{r}) e^{-i\mathbf{q}\cdot\mathbf{r}} = \frac{1}{D(q^2 + \kappa^2)} \,, \tag{49}$$

$$\theta_q^{(0)} = \frac{2\pi R_0}{q} J_1(qR_0) \,. \tag{50}$$

The coefficient a is evaluated numerically and is found to be negative for $0 < \hat{R}_0 < \infty$ with $\hat{R}_0 \equiv R_0\kappa$. The left hand side of Eq. (42) has been obtained in Ref. 37. The coefficient m is positive and g is shown to be positive for $\tau \approx \tau_c$. Therefore, Eq. (42) indicates a drift bifurcation such that a motionless circular particle $\dot{\boldsymbol{\rho}} = 0$ is stable for $\tau > \tau_c$ whereas it looses stability for $\tau < \tau_c$ and undergoes a translational motion at the velocity $|\dot{\boldsymbol{\rho}}|^2 = (\tau_c - \tau)/(2g)$.

3.4. *Equations of motion for $S_{\alpha\beta}$ and $U_{\alpha\beta\gamma}$*

The time-evolution equations for $z_{\pm 2}(t)$ and $z_{\pm 3}(t)$ can also be obtained from the interface equation of motion (32). The procedure is essentially the same as that used for the center of mass $\dot{\boldsymbol{\rho}}$ as described in the preceding subsection. It turns out that, in order to take account of the coupling between z_n and $\dot{\boldsymbol{\rho}}$, we may consider only the linear order of the deviation δR. Linearizing Eq. (32) with respect to the deformation, one obtains

$$\tau \frac{\partial \delta R}{\partial t} = \frac{\epsilon}{R_0^2} \Big(\frac{\partial^2 \delta R}{\partial \phi^2} + \delta R \Big) - 4\delta w + L \,, \tag{51}$$

where $\delta w = w - w_0$ with w_0 the value of w for a circular particle and we have used the fact that $\tau dc(w_0)/dw_0 = -4$ as shown from Eq. (33). The most involved step is to evaluate δw for a deformed particle. However, we do not enter into the details. The final expression of equation for the tensor S which is related to the modes $z_{\pm 2}$ as Eq. (3) is given by[31]

$$\Gamma_2 \frac{dS_{ij}}{dt} = -K_2 S_{ij} + b[v_i v_j - \frac{\delta_{ij}}{2} \mathbf{v}^2] + b_1 U_{ij\ell} v_\ell \,, \tag{52}$$

where $\Gamma_2 = \tau - 4E_2$, $b = -4G_1$, $K_2 = (3\epsilon/R_0^2) - 4D_2$ and $b_1 = 2B_3$ and the following functions have been defined

$$B_n = R_0 \int_0^\infty dqq^2 G_q^2[-J_n(qR_0)J_{n-1}(qR_0) + J_1(qR_0)\frac{\partial}{\partial qR_0}J_1(qR_0)] , \tag{53}$$

$$D_n = R_0 \int_0^\infty dqqG_q[J_1(qR_0)^2 - J_n(qR_0)^2] , \tag{54}$$

$$E_n = R_0 \int_0^\infty dqqG_q^2 J_n(qR_0)^2 , \tag{55}$$

$$G_\ell = R_0 \int_0^\infty dqq^2 G_q^3 J_\ell(qR_0)J_{\ell+1}(qR_0) . \tag{56}$$

It is noted that the coefficient b is negative. In the vicinity of the bifurcation $\tau \sim \tau_c$ with τ_c given by (44), the coefficient Γ_2 is positive. The coefficient K_2 becomes negative for sufficiently large values of R_0 indicating an instability of a motionless circular particle.[42] We have evaluated the coefficient b_1 numerically and have found that it takes a finite positive value at $R_0 = 0$ and has a zero around at $\hat{R}_0 < 0.8$ and becomes negative vanishing asymptotically $R_0 \to \infty$.

Here we make a remark about the sign of the relevant coefficients. Since both a and b are negative, we may conclude that the particle is elongated perpendicularly to the migration velocity as shown in section 2. This is consistent with the numerical simulations.[45]

Similarly, the equation of motion for the third-rank tensor U_{ijk} defined by (4) is given by[31]

$$\Gamma_3 \frac{dU_{ijk}}{dt} = -K_3 U_{ijk} + 4d_1 \left[v_i v_j v_k - \frac{v_\ell v_\ell}{4}(\delta_{ij}v_k + \delta_{jk}v_i + \delta_{ki}v_j)\right] + \frac{2d_2}{3}\left[S_{ij}v_k + S_{jk}v_i + S_{ki}v_j - \frac{v_\ell}{2}(\delta_{ij}S_{k\ell} + \delta_{jk}S_{i\ell} + \delta_{ki}S_{j\ell})\right] , \tag{57}$$

where $\Gamma_3 = \tau - 4E_3$ and $K_3 = 8\epsilon/R_0^2 - 4D_3$ and the relation $S_{11} = -S_{22}$ has been used. The coefficients d_1 and d_2 are given, respectively, by

$$d_1 = R_0 \int_0^\infty dqq^3 G_q^4 J_1(qR_0)J_3(qR_0) , \tag{58}$$

$$d_2 = R_0 \int_0^\infty dqq^2 G_q^2 \left[J_2(qR_0)J_3(qR_0) + J_1(qR_0)\frac{\partial J_1(qR_0)}{\partial qR_0}\right]. \tag{59}$$

We have verified numerically that both d_1 and d_2 are positive.

We have derived the time-evolution equations for the center of mass, the second-rank tensor and the third rank tensor for deformations. Equations (42), (52) and (57) are indeed the same form as the tensor model given by Eqs. (6), (7) and (8), respectively. Some of the terms in the latter equations are omitted because the full derivation is very involved.

4. Dynamics of a Single Particle in Two Dimensions

4.1. *Numerical simulations I*

We have carried out numerical simulations for the tensor model, Eqs. (6), (7) and (8) in some different conditions.[30] First we show the results for the coupled set of equations for v_i and S_{ij} ignoring U_{ijk}.

$$\frac{d}{dt}v_i = \gamma v_i - \mathbf{v}^2 v_i - a_1 S_{ij} v_j \ , \tag{60}$$

$$\frac{d}{dt}S_{ij} = -\kappa_2 S_{ij} + b_1\left(v_i v_j - \frac{1}{2}\mathbf{v}^2\delta_{ij}\right) - b_3(S_{mn}S_{mn})S_{ij} \ . \tag{61}$$

This is justified when the relaxation of U is sufficiently rapid, i.e., κ_3 is large enough. We allow the case that κ_2 is negative in Eq. (61) and introduce the cubic term of S to study the deformation-induced migration. In this section, we put $b_3 = 1$ without loss of generality.

The set of equations (60) and (61) has been solved numerically for $a_1 = 1$ and $b_1 = 0.5$ and by changing the parameters κ_2 and γ. Note from Eqs. (16) and (17) that elongation of a particle for $\kappa_2 < 0$ is opposite to the case of $\kappa_2 > 0$. That is, a particle is elongated perpendicularly to the velocity of center of mass when b_1 is positive. The simple Euler scheme has been employed with the time increment $\Delta t = 10^{-3}$. We have checked numerical accuracy by using $\Delta t = 10^{-4}$. The phase diagram obtained in this way is displayed in Fig. 2.[30] In the region indicated by the cross symbols, the particle is motionless whereas it undergoes a rectilinear motion in the region of the open squares and a circular motion in the region of the open circles. The solid line is the boundary between the motionless state and the straight motion whereas the broken line is the phase boundary between the straight motion and the circular motion.

Now, we focus on the region for $\kappa_2 < 0$ and $\gamma < 0$ where the deformation-induced migration occurs. First of all, it should be noted that a rectilinear motion appears even in the deformation-induced migration as in the region

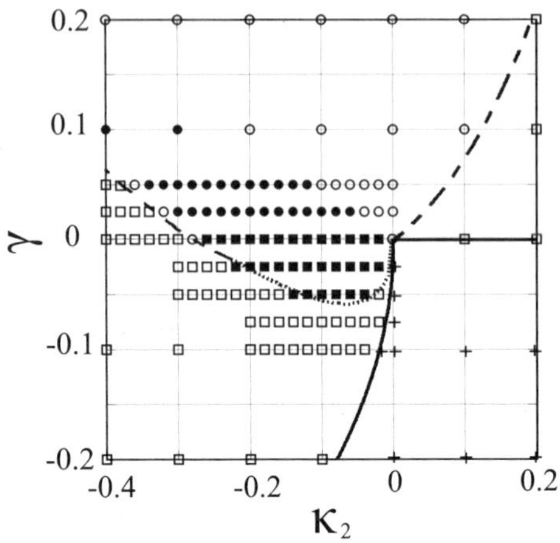

Fig. 2. Phase diagram obtained numerically for Eqs. (60) and (61) ignoring the U_{ijk} terms. The parameters are set to be $a_1 = 1.0$ and $b_1 = 0.5$. The meanings of the symbols and the lines are given in the text. This figure is reproduced from Ref. 30.

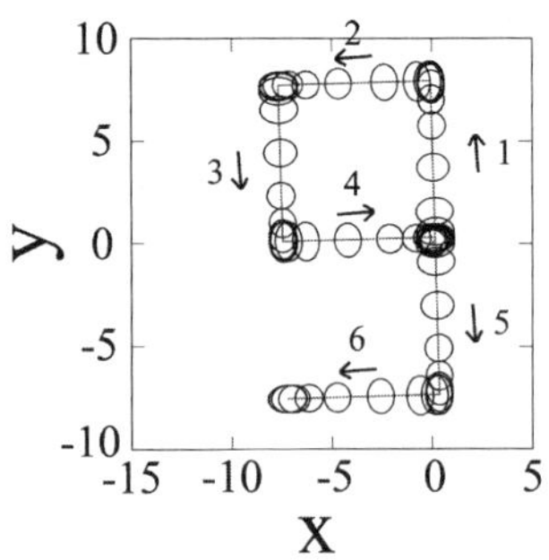

Fig. 3. Rectangular motion for $a_1 = 1.0$, $b_1 = 0.5$, $\kappa_2 = -0.1$ and $\gamma = -0.04$. The arrows and the digits indicate the direction of motion and the time-sequence of the motion respectively. This figure is reproduced from Ref. 30.

of the open squares. An interesting dynamics is observed in the region of the black squares in Fig. 2. The so-called rectangular motion appears which is displayed in Fig. 3 for $\gamma = -0.04$ and $\kappa_2 = -0.1$. In this motion, a propagating particle slows down and stop migrating. During the stopping interval the particle changes the shape and the propagation direction almost by 90°. Either clock-wise rotation or counter clock-wise rotation seem to occur at random and may depend on noises caused unavoidably in

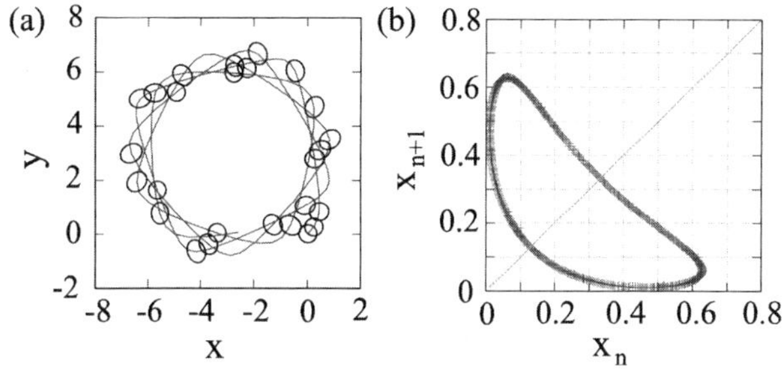

Fig. 4. (a) Quasi-periodic motion rotating in the counterclockwise direction for $a = 1.0$, $b = 0.5$, $\kappa_2 = -0.06$ and $\gamma = 0.02$. (b) Return map of the x coordinate. These figures are reproduced from Ref. 30.

the numerical computations. After changing the shape, the particle starts propagating again and repeats the process. The dotted line indicates the subcritical Hopf-bifurcation line between the straight motion and the rectangular motion. This is obtained analytically as the stability limit of the straight motion.

Another new dynamics has been found in the region $\kappa_2 < 0$ and $\gamma > 0$. A quasi-periodic motion is observed as shown in Fig. 4(a) for $\gamma = 0.02$ and $\kappa_2 = -0.06$. The return map of this motion is displayed in Fig. 4(b) where the values of the x-component of the location of the particle are plotted every time that the particle crosses the line $-3 < x < 3$ and $y = 3$.

4.2. *Numerical simulations II*

In this subsection, we show the results of numerical simulations of the full set of equations (6), (7) and (8).[30]

$$\frac{d}{dt} v_i = \gamma v_i - \mathbf{v}^2 v_i - a_1 S_{ij} v_j \,, \tag{62}$$

$$\begin{aligned} \frac{d}{dt} S_{ij} &= -\kappa_2 S_{ij} + b_1 \left(v_i v_j - \frac{1}{2}\mathbf{v}^2 \delta_{ij} \right) \\ &\quad + b_2 U_{ijk} v_k - b_3 (S_{mn} S_{mn}) S_{ij} \,, \end{aligned} \tag{63}$$

$$\begin{aligned} \frac{d}{dt} U_{ijk} &= -\kappa_3 U_{ijk} + d_1 \left[v_i v_j v_k - \frac{v_\ell v_\ell}{4} (\delta_{ij} v_k + \delta_{ik} v_j + \delta_{jk} v_i) \right] \\ &+ \frac{d_2}{3} \left[S_{ij} v_k + S_{ik} v_j + S_{jk} v_i - \frac{v_\ell}{2} (\delta_{ij} S_{k\ell} + \delta_{jk} S_{i\ell} + \delta_{ki} S_{j\ell}) \right] \,, \end{aligned} \tag{64}$$

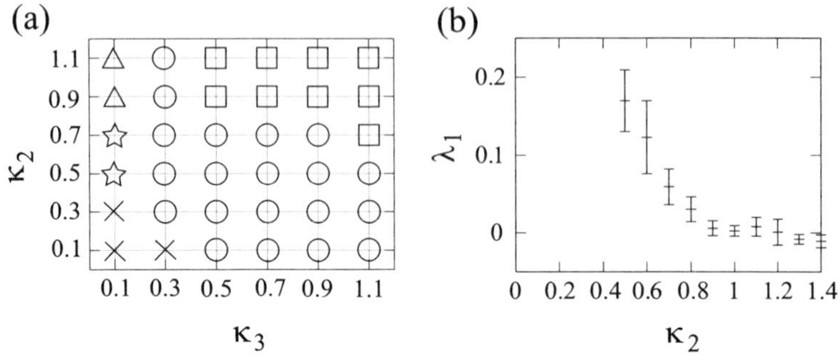

Fig. 5. (a) Phase diagram for Eqs. (62), (63) and (64) for $a_1 = -1.0$, $b_1 = -0.5$, $b_2 = 0.3$, $d_1 = 0.1$, $d_2 = 0.8$ and $\gamma = 1.0$. The meaning of the symbols is given in the text. In the region indicated by the symbol $\times$, we have no definite conclusion about the motion because of a numerical instability. (b) Maximum Lyapunov exponent obtained numerically for $\kappa_3 = 0.1$. These figures are reproduced from Ref. 30.

where the several terms have been omitted, for the sake of simplicity. For example, since the relaxation rates κ_2 and κ_3 are chosen to be positive in this subsection, the cubic term in Eqs. (7) and (8) are not considered, i.e., $b_3 = d_3 = 0$. That is, we are concerned with the migration-induced deformation in this subsection. The modified Euler method with the time increment either $\Delta t = 10^{-3}$ or $\Delta t = 10^{-4}$ has been employed. The relaxation rates κ_2 and κ_3 are varied to explore the possible dynamics.

The phase diagram is obtained as shown in Fig. 5(a).[30] The straight motion and the circular motion appear in the region indicated by the squares and by the circles respectively. A sequence of the snapshots of the particle shape in the circular motion is given in Fig. 6.

When κ_2 is large and κ_3 is small, a zig-zag motion is observed in the region of the triangles. The time-sequence of the snapshots for $\kappa_2 = 0.9$ and $\kappa_3 = 0.1$ is displayed in Fig. 7(a). The particle is traveling from the left to the right. The angle of the direction change in the zig-zag motion is about 60°. In the region indicated by the stars for the smaller values of κ_2 and for $\kappa_3 = 0.1$, the particle motion becomes chaotic. A chaotic trajectory is displayed in Fig. 7(b) for $\kappa_2 = 0.5$ and $\kappa_3 = 0.1$. We have evaluated the maximum Lyapunov exponent, λ_1, associated with the particle trajectory as depicted in Fig. 5(b). It is evident that λ_1 becomes positive for $\kappa_2 < 0.8$ and hence the dynamics is indeed a chaos. For smaller values of κ_2, any accurate value of the exponent is obtained because of a numerical instability.

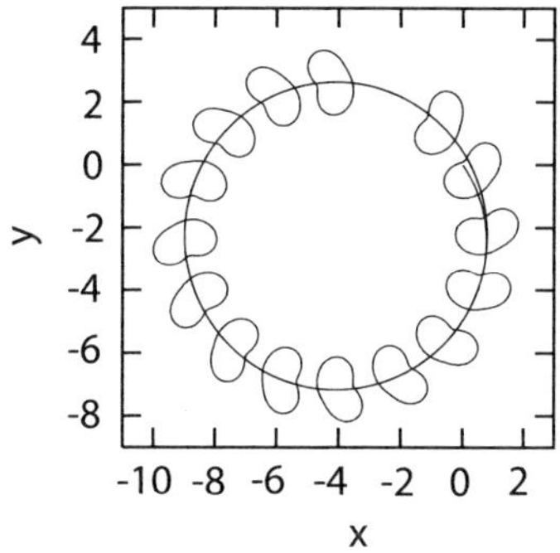

Fig. 6. Snapshots of a particle in a clock-wise circular motion for $a = -1.0$, $b = -0.5$, $b_1 = 0$, $d_1 = 0.25$, $d_2 = 0.15$, $\gamma = 1.0$, $\kappa_2 = 0.5$ and $\kappa_3 = 0.3$. The solid circle indicates the trajectory of the center of mass. This figure is reproduced from Ref. 31.

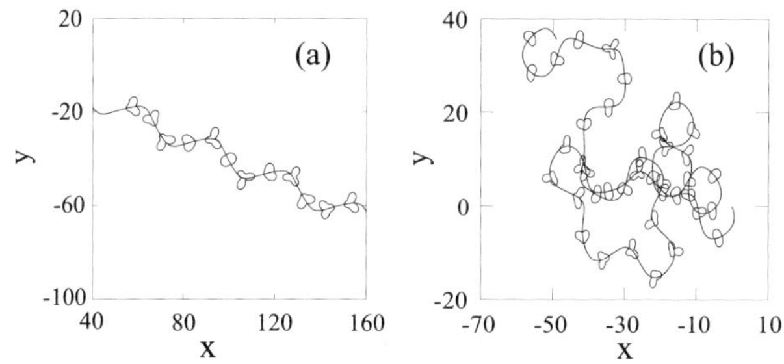

Fig. 7. (a) Zigzag motion for $\kappa_2 = 0.9$ and $\kappa_3 = 0.1$. The particle moves from the left to the right. (b) Chaotic motion for $\kappa_2 = 0.5$ and $\kappa_3 = 0.1$. These figures are reproduced from Ref. 30.

4.3. *Coupled set of equations for the Fourier amplitudes*

The set of equations (6), (7) and (8) in two dimensions can be written in terms of the Fourier components in Eq. (2). Here we define the complex variables as $z_1 = v_1 - iv_2$, $z_2 = \frac{1}{2}(S_{11} - iS_{12})$ and $z_3 = \frac{1}{2}(U_{111} + iU_{222})$. The time-evolution equations for z_1, z_2 and z_3 are given from Eqs. (6), (7) and (8) by

$$\dot{z_1} = (\gamma + d_{11}|z_1|^2 + d_{12}|z_2|^2 + d_{13}|z_3|^2)z_1 + e_{11}\bar{z_1}z_2 + e_{12}\bar{z_1}^2 z_3 + e_{13}\bar{z_2}z_3 + e_{14}\bar{z_3}z_2^2 \,, \tag{65}$$

$$\dot{z_2} = (-\kappa_2 + d_{21}|z_1|^2 + d_{22}|z_2|^2 + d_{23}|z_3|^2)z_2 + e_{21}z_1^2 + e_{22}\bar{z_1}z_3 + e_{23}\bar{z_2}z_3z_1 , \tag{66}$$

$$\dot{z_3} = (-\kappa_3 + d_{31}|z_1|^2 + d_{32}|z_2|^2 + d_{33}|z_3|^2)z_3 + e_{31}z_1^3 + e_{32}z_1z_2 + e_{33}\bar{z_1}z_2^2 , \tag{67}$$

where the bar indicates the complex conjugate. All the coefficients are real and are given by $d_{11} = -1$, $d_{12} = -8a_4$, $d_{13} = -16a_5$, $d_{21} = -b_4$, $d_{22} = -8b_3$, $d_{23} = -16b_5$, $d_{31} = -d_4$, $d_{32} = -8d_5$, $d_{33} = -16d_3$, $e_{11} = -2a_1$, $e_{12} = -2a_2$, $e_{13} = -8a_3$, $e_{14} = 16a_6$, $e_{21} = b_1/4$, $e_{22} = b_2$, $e_{23} = 2b_6$, $e_{31} = d_1/8$, $e_{32} = d_2/2$ and $e_{33} = 2d_6$. It is important to note that this set of equations is invariant under the transformation

$$(z_1, z_2, z_3) \to (e^{i\theta}z_1, e^{2i\theta}z_2, e^{3i\theta}z_3) , \tag{68}$$

for an arbitrary phase angle θ. This invariance arises from the isotropy of space.

When the variable z_3 is omitted, the set of equations (65) and (66) is the same as those considered by Armbruster et al.[48] They motivated to study some partial differential equation like the Kuramoto-Sivashinsky equation in one dimension under the periodic boundary condition[49] and obtained several bifurcations. We note the following correspondences between our results and theirs; the motionless circular shape particle ↔ the trivial solution, the deformed motionless particle ↔ the pure mode, the straight motion ↔ the standing wave, the rectangular motion ↔ the heteroclinic cycle, the circular motion ↔ the traveling wave and the quasi-periodic motion ↔ the modulated wave. The former is the motions obtained in our theory[30] whereas the latter is the terminology of Armbruster et al.[48]

5. Dynamics in Three Dimensions

Many of micro-organisms translocate in a three-dimensional space. A typical example is Listeria which causes locomotion together with spinning motion by polymerization of actin filaments.[50–52] Some of the experiments of chemically reacting oily droplets in a micrometer scale have also been carried out in three dimensions.[28] Therefore, it is necessary to extend our tensor model described in the preceding sections to three dimensions.[53,54] One of the advantageous features of our theory is that the representation of the time-evolution equations in terms of the tensor variables is independent of the dimensionality of space. For the sake of simplicity, however, we do not consider the third rank tensor U in this section.

The simplest and non-trivial version should read

$$\frac{d}{dt}v_i = \gamma v_i - \mathbf{v}^2 v_i - aS_{ij}v_j \tag{69}$$

$$\frac{d}{dt}S_{ij} = -\kappa S_{ij} + b\left(v_i v_j - \frac{1}{d}\mathbf{v}^2\delta_{ij}\right) , \tag{70}$$

where d is the dimensionality of space ($d = 3$ hereafter). The coefficients a and b are assumed to be $ab > 0$ in order to avoid a singular behavior as in two-dimensional case.[29] Other constants γ and κ are assumed to be positive throughout this section. Therefore, we restrict ourselves to the migration-induced deformation here. The above set of equations has been derived from an excitable reaction diffusion system in three dimensions by the method described in section 4.[54] Actually the coefficients a and b are found to be negative.

In three dimensions, the second rank symmetric tensor S_{ij} is given as follows. Let us put the distance between the surface and the center of mass of a particle as $R = R_0 + \delta R(\theta, \phi)$ with θ the azimuthal angle and ϕ the polar angle. The deviation $\delta R(\theta, \phi)$ is expanded in terms of the spherical harmonics as

$$\delta R(\theta, \phi) = \sum_{\ell=1}^{+\infty} \sum_{m=-\ell}^{+\ell} c_{\ell m} Y_{\ell m}(\theta, \phi) , \tag{71}$$

where the coefficients $c_{\ell m}$ are time-dependent. The tensor S_{ij} is given in terms of $c_{\ell m}$ as[54]

$$S_{11} = (3/2)^{1/2}(c_{22} + c_{2-2}) - c_{20} \tag{72}$$

$$S_{12} = S_{21} = i(3/2)^{1/2}(c_{22} - c_{2-2}) \tag{73}$$

$$S_{13} = S_{31} = (3/2)^{1/2}(c_{21} + c_{2-1}) \tag{74}$$

$$S_{22} = -(3/2)^{1/2}(c_{22} + c_{2-2}) - c_{20} \tag{75}$$

$$S_{23} = S_{32} = i(3/2)^{1/2}(c_{21} - c_{2-1}) \tag{76}$$

$$S_{33} = -(S_{11} + S_{22}) , \tag{77}$$

We have solved Eqs. (69) and (70) numerically. The 4th order Runge-Kutta method has been employed with time-increment $\Delta t = 10^{-3}$. The value of the parameters a and b is fixed, respectively, as $a = -1.0, b = -0.5$. Since the value of b is negative, the particle tends to elongate perpendicularly to the propagating direction. The results are summarized in the phase

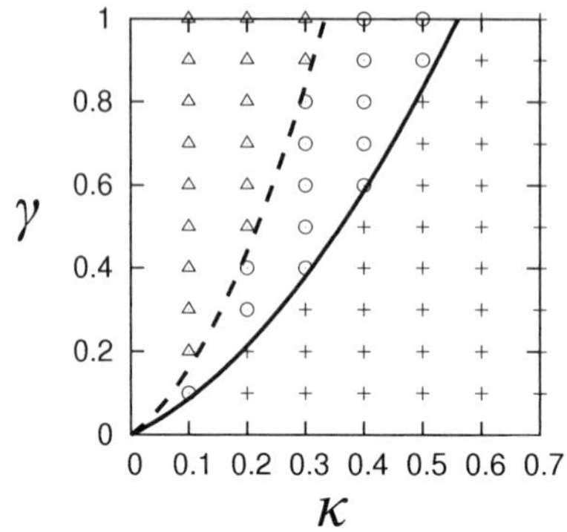

Fig. 8. Phase diagram for $a = -1.0$ and $b = -0.5$. In the region indicated by the cross symbols, the open circles and the triangles, a straight motion, a circular motion confined to a plane, and a helical motion are obtained, respectively. This figure is reproduced from Ref. 53.

diagram on the the $\gamma - \kappa$ plane as in Fig. 8. We have found three different self-propelled motions.[53] In the region indicated by the cross symbols, a straight motion is stable. In the region shown by the open circles and the triangles, a circular motion confined to a plane and a helical motion appear as shown in Fig. 9 and 10, respectively.

A straight motion takes an oblate spheroid as shown in Fig. 11(a) where the particle is propagating along the z-axis. For a circular motion or a helical motion, its shape takes a scalene ellipsoid as in Fig. 11(b) where the circular motion is on a plane parallel to the y-z plane.

In order to understand the numerical results, we solve Eqs. (69) and (70) in a stationary state and perform the linear stability analysis of the solutions. The solution of straight motion and circular motion can be obtained analytically. For instance, the velocity of a straight motion is given by $v_s^2 = 3\kappa\gamma/(3\kappa + 2ab)$. The velocity of a circular motion confined to a plane is given by

$$v_r^2 = \frac{\kappa\gamma}{\kappa + ab[(P_+^2 - 1/3)P_+^2 + \left(-P_+^2 + P_-^2 + 1/3\right) P_-^2/2]} \,. \tag{78}$$

where

$$P_\pm^2 = \frac{1}{2} \pm \frac{\kappa}{2}\left(\frac{6\kappa + ab}{ab(6\kappa\gamma - 3\kappa^2)}\right)^{1/2} \tag{79}$$

We have not succeeded in obtaining the analytical expression of the helical motion.

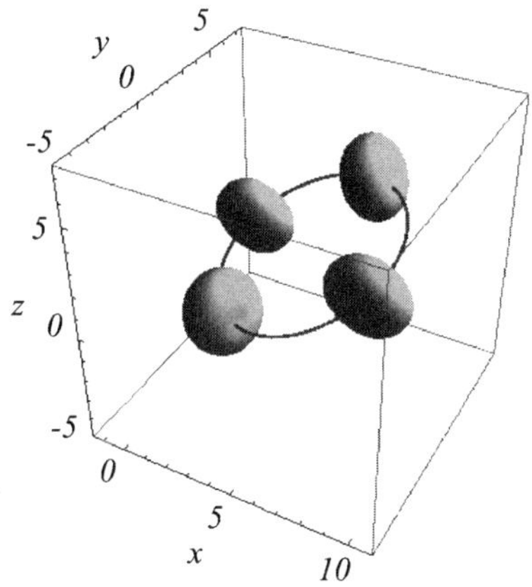

Fig. 9. Circular motion confined to a plane. The particle is undergoing a clock-wise rotation viewing from the top. This figure is reproduced from Ref. 53.

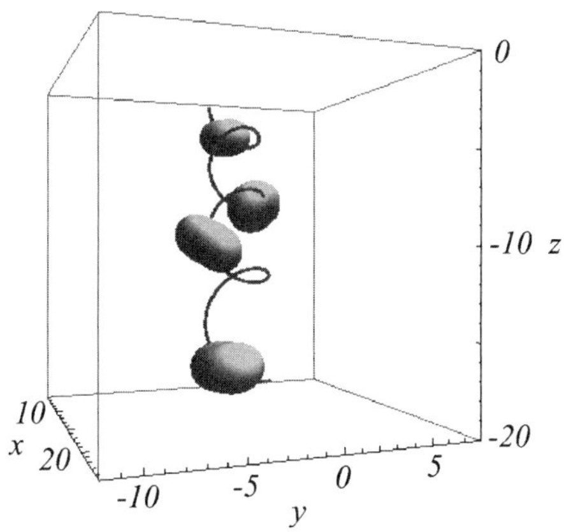

Fig. 10. Helical motion. The particle is moving from the top to the bottom. This figure is reproduced from Ref. 53.

The linear stability threshold of the straight motion is given by

$$\gamma = \frac{\kappa^2}{ab} + \frac{2}{3}\kappa \tag{80}$$

This is shown by the solid line in Fig. 8. The stability limit of the circular

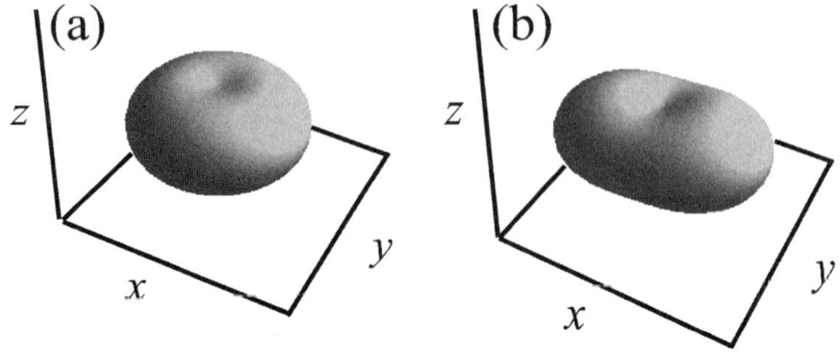

Fig. 11. (a) Oblate spheroid particle propagating along a straight line parallel to the z axis. The coefficients of the spherical harmonic expansion are chosen as $c_{00} = 1.0$ and $c_{20} = -0.2$. (b) Scalene ellipsoid undergoing a circular motion for $c_{00} = 1.0$ and $c_{22} = c_{2-2} = 0.5$. The motion is in a plane parallel to the $y - z$ plane. These figures are reproduced from Ref. 53.

motion is obtained as

$$\gamma = \frac{3\kappa^2}{ab} + \kappa \tag{81}$$

which is shown by the broken line in Fig. 8. The stability limits agree quantitatively with the numerical phase boundaries.

To be summarized, we have studied the dynamics of a deformable self-propelled particle in three dimensions based on Eqs. (69) and (70). We have found both numerically and theoretically a straight motion and a circular motion confined to a plane. A helical motion has also been obtained numerically. The linear stability analysis of the straight motion and the circular motion has been carried out, the results of which are found to be consistent with the phase diagram obtained numerically.

Before closing this section, we summarize the self-propelled motions predicted in our theory. We have shown various dynamics of self-propulsion in both two and three dimensions. In two dimensions, we have studied the case of deformation-induced migration as well as migration-induced deformation. Here we make some remarks on the results. First, a straight motion exists even for $\kappa_2 < 0$ and $\gamma < 0$ as shown in Fig. 2. Furthermore, this straight motion becomes unstable for smaller values of $|\gamma|$ and a rectangular motion appears. When $\kappa_2 > 0$ and $\gamma > 0$, but, the particle is sufficiently soft (i.e., κ_2 and κ_3 are small enough), zigzag motion and chaotic motion are obtained as shown in Fig. 7. We expect that these dynamical behaviors caused by the nonlinear coupling between shape defor-

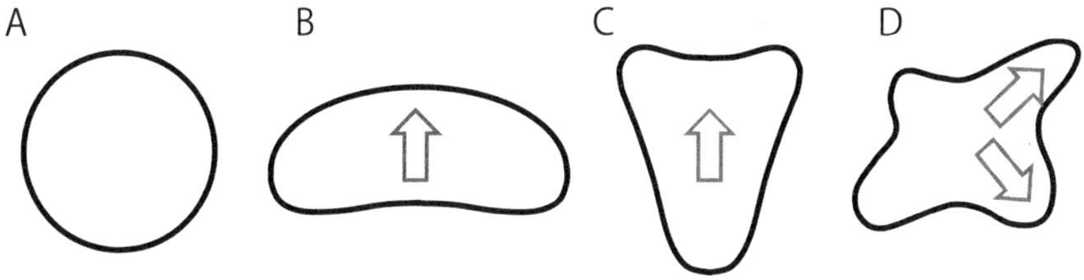

Fig. 12. Spontaneous symmetry breaking in cell shape. (A) Circular cell (Actin polymerization is prohibited by latrunculin A.) (B) Keratocyte cell. (C) Polarized cell in chemotaxis. (D) Amoeboid cell.

mations and migration will be observed experimentally, for example, in oily droplets undergoing chemical reactions.[27,28] Experiments in motile cells are also desired. Some of the recent related expe riments will be described in the next section.

6. Problems of Cell Migration

In the previous section, a general theory for a deformable self-propelled particle in two and three dimensions is explained. The theory can be adopted for the description and the prediction for the experimental systems exhibiting translation motion of domains accompanied with deformation. Such motions are often observed in the experiment of oily droplets[27,28] and closed loop of defects in liquid crystal.[55] Among the list of possible applications, the most challenging one might be the problem of cell migration.[18–20]

Cell migration is a highly complex process that integrates many spatial and temporal cellular events. Motile bacteria and most eukaryotic cells can move in a directed manner or seemingly random manner depending on the presence or absence of external cues. Directed cell migration toward a soluble ligand for chemotaxis is a general property of many motile eukaryotic cells. On the other hand, motile cells are able to migrate spontaneously even in the absence of external stimuli. Therefore it is interesting to consider how the spontaneous symmetry breaking in the cell shape is provoked in the absence of external stimuli. In order to address this question, a symmetric state in cell morphology, if it exists, will become an ideal starting point for the study.

In fact, a circular symmetric cell shape can be realized[56] when we treat the cell with latrunculin A which inhibits actin polymerization (Fig.12A). Here we project the cell body on to 2D plane and consider only 2D projected geometry for simplicity. This simplification will be better justified for crawling cells, because the cell enables its migration only through the control of adhesion, friction, and detachment from the substratum. Adhesion to the substratum, newly formed protrusions, and contraction at the rear are all detected by the 2D projected image of cell shape. It is worth asking a question what is the simplest symmetry breaking from the circular cell shape. There are two cases in the simplest symmetry braking types in eukaryotic cells. The first one is seen in straightly migrating fish keratocyte cells (Fig.12B). The second type can be seen in polarized cell during chemotactic motion (Fig. 12C) which is often seen in starved state of *Dictyostelium* cells. In both cases, cell shape is deformed from the circular shape, but the elongated direction of the cell shape is perpendicular to the moving direction for Keratocyte, while the elongated direction is parallel to the moving direction for Dicty cells. More complicated symmetry breaking are seen in amoeboid cells[57] in which the cell shape is far from circular shape as a result of multiple pseudpod extensions (Fig.12D). The variety of cell shapes and their remodeling are mainly controlled by actin polymerization dynamics. In this section, we investigate cell shape dynamics from a geometrical point of view. We will explain how the cell deforms its shape during migration and how the cell exerts force depending on its shape with the aid of localization in actin distribution.

6.1. *Experimental analysis of shape dynamics and migration*

We investigate the dynamics of both vegetative and starved *Dictyostelium discoideum* amoebae. We observed spontaneous cell locomotion of an individual amoeba in a homogeneous environment without any external stimuli by time-lapse videomicroscopy at a temporal resolution of 1 sec for 600 sec.[20] Circular maps were constructed around centroid of individual cell images (Fig.13). From binarized time-lapse images, we obtain the spatiotemporal profile of the cell shape $R(\theta, t)$ as follows. For each binarized image, we define the position of centroid as $\boldsymbol{M}(t) = (x_c(t), y_c(t))$. To calculate the dynamics of the cell shape in 2D cartesian system, we use circular mapping: the radial amplitude of extensions from centroid to the cell edge, $R(\theta, t)$, are calculated for each image. Where θ is the angle of the radial vector, and R is the distance from the centroid to the cell edge. We obtain 360 values ($\theta = 0 - 360$ [deg]) for $R(\theta, t)$ from one image data. $R(\theta, t)$ describes the spatiotemporal dynamics of the cell shape. (Figure 13).

Further we define velocity as the displacement of centroid,

$\boldsymbol{V}(t) = \dfrac{\boldsymbol{M}(t+\Delta t) - \boldsymbol{M}(t)}{\Delta t}$. The description by the field variable is convenient to recognize spatio-temporal patterns. We require two additional rules to avoid the difficulties in the calculations. First, if the function of cell contour becomes a multivalued function, we employ the closest value to the centroid. Second, if we obtain a picture in which centroid does not locate within a cell, we do not calculate any values but just take over the previous values. Although these rules are somewhat arbitrary, it does not affect our conclusions.

We investigate two different states of *Dictyostelium discoidium* (DD) cells in the development stage. Axenic type DD cells can take nutrient from the solution medium containing gulcose When DD cells is cultured in the medium containing gulcose, the cells are called in vegetative (VEG) state. In the VEG state, cells are less motile in regard to centroid motion and non-polarized compared with another state so- called the starved state. The starved state of the DD cells develops in 6 to 7 hours after the medium is changed into non-nutrient one. In the starved (STA) state, the cells look elongated and their centroid speed is faster than the VEG state.

Figure 14 (upper column) shows three characteristic examples of $R(\theta, t)$ in vegetative cells. As is clear from the figure, we can see the repeated extension and retraction of protrusions. However, it is difficult to find clear rules from these data because of abundant noise. To overcome this diffi-

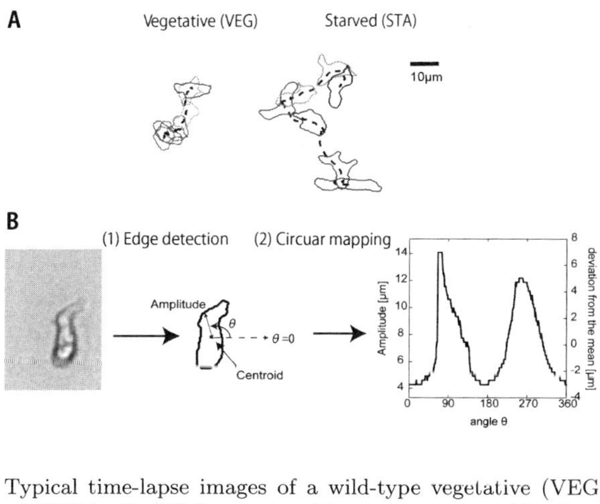

Fig. 13. (A) Typical time-lapse images of a wild-type vegetative (VEG) cells and a wild-type starved (STA) cell. (B) Summary of Image processing. (1) edge detection and (2) circular mapping of $R(\theta, t)$ from Ref. 20.

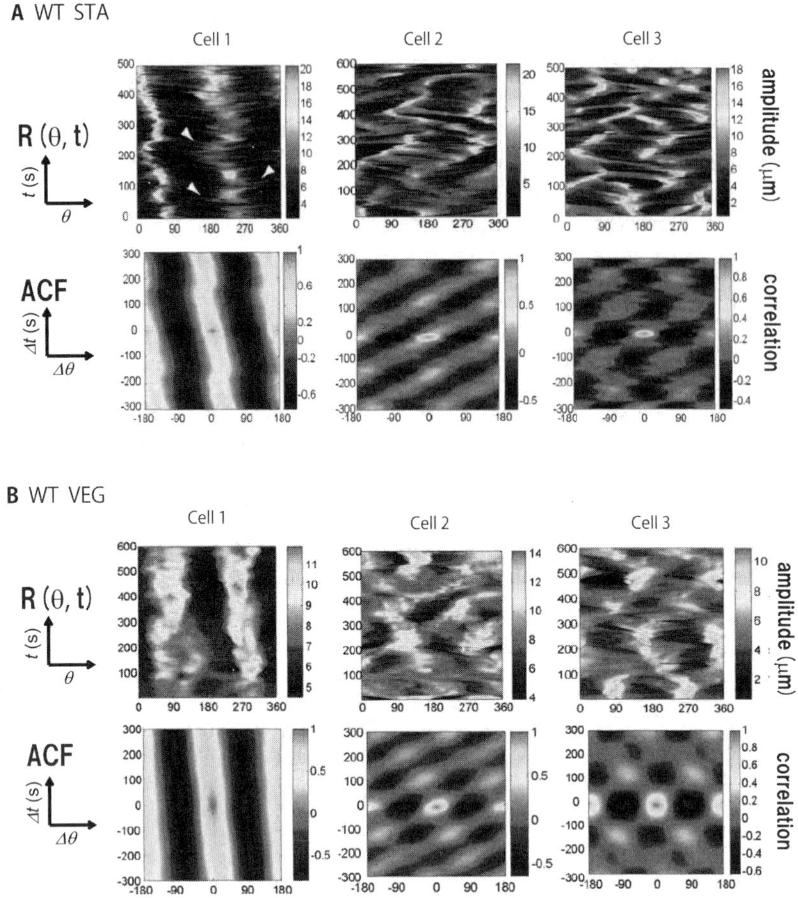

Fig. 14. Commonly observed orderly patterns of cell shape in wild-type cells. (A) $R(\theta, t)$ (Upper) and the corresponding autocorrelation function (Lower) are shown for the STA cells. (B) $R(\theta, t)$ (Upper) and the corresponding autocorrelation function (Lower) are shown for the WT VEG cells. Modified from Ref. 20.

culty, we calculated the autocorrelation function (ACF) of $R(\theta, t)$, which is $C(\Delta\theta, \Delta t)$ defined as,

$$C(\Delta\theta, \Delta t) = \frac{\langle \delta R(\theta + \Delta\theta, t + \Delta t)\delta R(\theta, t)\rangle_{\theta,t}}{\langle \delta R^2 \rangle_{\theta,t}} \tag{82}$$

where $\delta R(\theta, t) \equiv R(\theta, t) - \langle R(\theta, t)\rangle_\theta$.

The lower column in Fig.14a displays ACFs of the VEG state that we found clear patterns. The ACF of cell 1 (left) keeps higher correlation along the lines $\Delta\theta = 0$ and $\Delta\theta = 180$ degree with a little tilt, indicating that this cell keeps its shape stretching. The ACF of cell 2 (center) exhibits laterally propagating correlation, indicating that the protrusions of this cell laterally propagate. The ACF of cell 3 (Figure 1 (b) right) shows the periodic profile periodically increases and decreases that the correlation oscillates with 180 deg for angle and with 300 second for time, indicating that this vegetative cell periodically extends and retracts the protrusions. When a cell extends pseudopodia in a certain direction and then re-extending new pseudopodia perpendicular to the long axis of the cell, an oscillating pattern occurs. These results provide the evidence that the cell deforms by not random but the orderly patterns, stretching, rotating and oscillating even under the uniform environment.[20,58,59]

We next examined the morphological dynamics of the starved cell. After 7 hours starvation, a cell alters to move faster and have an elongated shape. Figure 14b display three examples of $R(\theta, t)$ (upper column) and $C(\Delta\theta, \Delta t)$ (lower column). We found the regular pattern from ACF; Cell 1(left), Cell2(center) and Cell3(right) deform their own shape by stretching, rotating and oscillating manners, respectively. Therefore, we conclude that both vegetative and starved cells can change their own shape not by random but by systematic manners even without any artificial confinement during spontaneous cell locomotion. We note that the membrane at the head (around 270°) shows larger fluctuations than that at the tail (around 90°) in the stretching cell (Fig.14b). Such an asymmetry was observed in neither the vegetative cell nor the rotating and oscillating starved cell.

We classified ACF of different cells by using the clustering analysis.[20] 70% of the vegetative cells exhibited these types of ordered pattern (n=53). And 66% of starved cells exhibited these types of pattern. The remaining cells exhibited a transient pattern between two of these types.

Commonly observed patterns raise a question; whether the transition of orderly patterns occurs or not in a single cell. To answer this question, we performed long-term observation of single WT cells for more than 30 minutes. We then calculated ACF of long-term data by averaging over moving windows of 10 min. We found a transition from elongating pattern to rotating patter, and transition from left-handed rotation to right-handed rotating pattern and *vice versa*. Moreover, 3.3 hour measurement of a WT vegetative cell revealed that a single cell can show three types of pattern as changing from rotating pattern to elongating one, and then to oscillating one (see fig.15). However, when we calculated ACF of this long-term

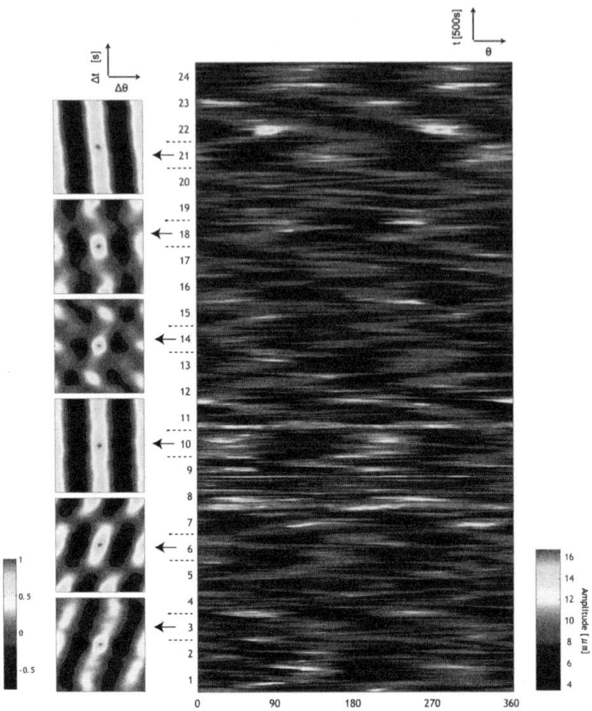

Fig. 15. Long-term measurement of the morphological dynamics of cell shape. We measured a single WT vegetative cell for 3.3 h and then calculated the autocorrelation function of $R(\theta, t)$ at each time window (500 s). Six examples of autocorrelation function are shown on the left side of $R(\theta, t)$. We found that the ordered pattern dynamically changes; for instance, from rotation to oscillation. From Ref. 20.

data by averaging over 3.3 hour, we no longer observed orderly patterns in ACF (data not shown). These results indicate that cells show orderly patterns lasting for 10 to 20 min and change them spontaneously although the transition among the orderly patterns is at random.

In the absence of external stimuli, individual *Eq.* cells spontaneously form actin filaments and extend 1 or 2 pseudopodia. In a sense, their shapes look like more or less random. The new pseudopodia are retracted or attached to the substrate. After the attachment of the pseudopodia to the substrate, the cell adopts a polarized morphology and retracts its rear edge and moves forward. Thus their motions also look like random. This kind of spontaneous cell migration allows the cells to forage and explore their surroundings by balancing random and directed migrations. To summarize, by analyzing the stochastic dynamics of cell shape and their coordination with migration, we found that *Dictyostelium* cells organize their own shape into three ordered patterns, elongation, rotation, and oscillation, in the absence of external stimuli. The cell switches among the different ordered patterns enabling overall random exploration for cells.[60,61]

6.2. *Experimental analysis of stress field*

Mechanical stress exerted by cell cytoskelton at the surface of cell-substratum and cell-cell interfaces is considered to be important in the regulation of many biological processes such as ameboid motions of isolated motile cells[62–65] and invasion of motile cells in the tissues. Since the stress is the cause of a mechanical motion and deformation, measurement of mechanical stress exerted by cell became an important task for the understanding of cell migration and development of biological tissues. In order to measure the stress quantitatively, the elastic substrate containing markers has been used by many researchers. But it is not yet a conventional technique. Here we explain the method of measuring stress exerted by cell and describe the results on ameboid cells.

The observation of the substrate deformation is generally performed by tracking fluorescent maicrobeads embedded inside polyacrylamide substrates. The Young's modulus E of the flexible substrates is usually adjusted in the range 200-1000 Pa by changing the proportion of acrylamide and *bis*-monomers for monitoring weak stresses by a single cell. A proper density of fluorescent beads, typically 0.2 μm in diameter and 0.22%v/v in concentration, is mixed in acrylamide and *bis*-acrylamide. After polymerization, we covalently coated the elastomer with type I collagen by using a cross-linker Sulfo-SANPAH, in order to provide a physiological surface for cells.

Traction force measurement by using elastomer requires subsequent force calculation from observed deformation. In the calculation, we use linear elasticity theory by assuming the linearity between deformation and stress. This is usually validated that the deformation is very small (0.1-1 μm) compared with the thickness (80μm) and the width ($\sim$ 20 mm) of the elastomer. If the stress distribution exerted by the cell is known, the deformation filed of the elastomer is easily calculated by convoluting the stress distribution with the appropriate Green's tensors. The observation gives the information of the deformation field, thus we need to deconvolve the deformation filed in order to estimate stress distribution exerted at the

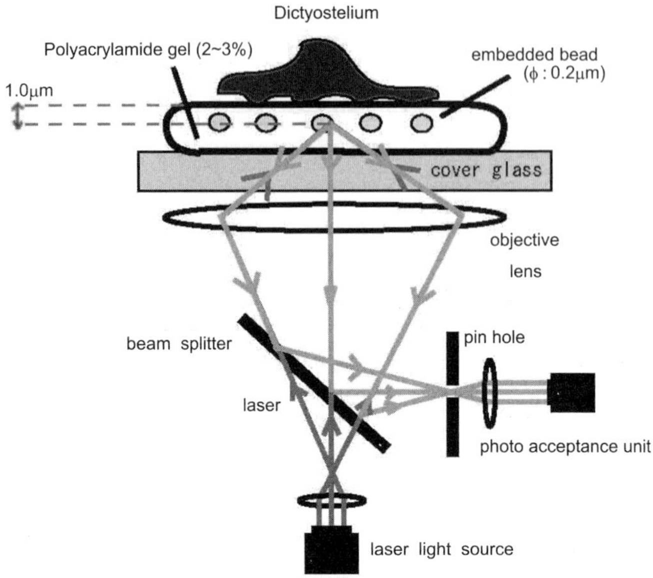

Fig. 16. Experimental setup for measuring cell traction force modified from Ref. 67.

cell-substratum surface.

In the framework of the linear elasticity theory, the Green's tensor is derived from the Boussinesq equations.[66] The deformation filed $\boldsymbol{u}(r)$ inside a semi-infinite elastic medium by a distribution of forces $\boldsymbol{F}(r')$ on the surface is described by

$$u_i(\boldsymbol{r}) = \int d\boldsymbol{r}' G_{ij}(\boldsymbol{r}-\boldsymbol{r}')F_j(\boldsymbol{r}'), \tag{83}$$

where i and j denote the x, y, z vectorial components. G_{ij} are Green functions which scale as $(r-r')^{-1}$. In the most of experiments, deformations are measured by using the confocal microscope along the plain with a fixed height , $r = (x, y, z_M)$, where the z_M is the distance from the surface. z_m is typically on the surface or close to the surface, e.g. 1 μm in the experiments. The fact that the Poisson ratio for the elastomer is close to 0.5 since the total volume of elastomer changes very little, simplifies the Green's function. As a result, the Green functions G_{xz}, G_{yz} can be neglected, and the problem is reduced to a 2D problem. We need to solve 2D force distribution (F_x, F_y) from the 2D displacement filed (u_x, u_y).

The integral equation of Eq.(83) becomes a set of linear equations,

$$u_{i\boldsymbol{r}_k} = \sum_j \sum_{\boldsymbol{r}'_\ell} G_{i\boldsymbol{r}_k\boldsymbol{r}'_\ell} F_{j\boldsymbol{r}'_\ell}. \tag{84}$$

We rewrite Eq. (84) as $\boldsymbol{u} = \boldsymbol{GF}$. We approximated the cellular force pattern into an ensemble of localized point-like forces under a triangular mesh, which is created within the individual cells by a Delaunay triangulation. The force at a single mesh point does not mean an exerted force at single focal adhesion contact. We need to solve the inverse problem to estimate force exerted at the cell-substratum interface from the information of deformation field. Inverse of the matrix G scales as $|\boldsymbol{r} - \boldsymbol{r}'|$, although most of area outside of cell attached surface are not subjected to external forces, thus the problem is generally an ill-posed problem. To overcome this problem, a regularization method is required.

$$F_\lambda = \text{argmin}||\boldsymbol{u} - \boldsymbol{GF}||_2 + \lambda||\boldsymbol{F}||_2 \tag{85}$$

We use Tikhonov regularization method for the calculations

A typical substrate deformation field exerted by a vegetative cell is shown in Fig.17. The length and direction of each vector indicate the magnitude and direction of deformation of the substrate caused by the cell contractility, respectively. The largest bead displacements were found in the range 200-500 nm for respectively aggregating and vegetative cells. Here they are directed inward in the rear part of the cell.

Figure 18 shows typical examples of traction force distribution of both vegetative cells and starved cells. Both cells show asymmetric traction force distribution during migration. The vegetative cell in Fig. 18A migrates downward. The vegetative cell gradually localize traction forces at the posterior part of the cell until t=12[sec]. And the cell abruptly proceeds and extends pseudopods at the anterior part of the cell at t=18[sec]. The starved cell in Fig. 18B has a polarized morphology. The starved cell migrates in the upper-left direction. Traction forces localized at the posterior part of the starved cell. The starved cell extends a pseudopod at every frame.

As the direct cause of pseudpod extension, the importance of myosin II localization was pointed out by many groups.[62,63] Because myosin II is one of the molecular motors controlling contractile forces, and localizes at the rear edge during chemotaxis. We observed the GFP-tagged myosin II and red-fluorescent beads simultaneously. At each measurement we take images with 2 different section. The one section is 1μm above the surface of the

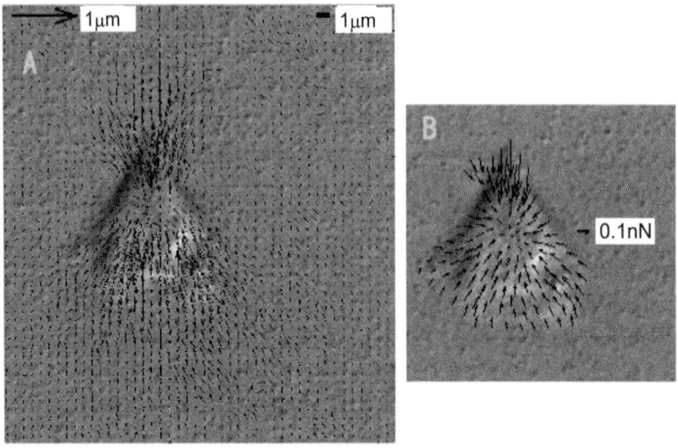

Fig. 17. (A) Displacement field of the elastomer estimated from beads displacement. (B) Calculated traction forces modified from Ref. 67.

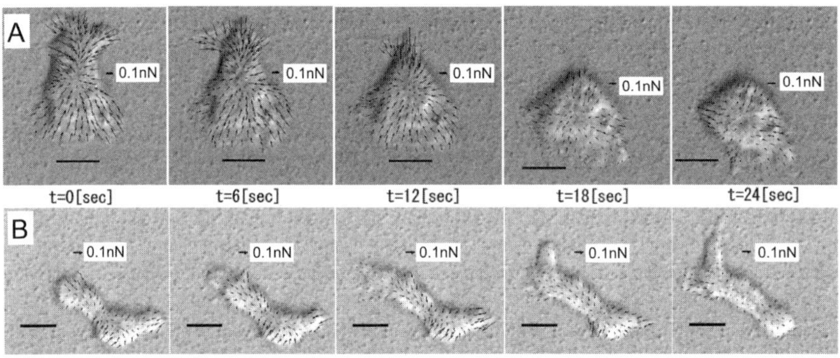

Fig. 18. Typical examples of traction forces field. (A) vegetative cells (B) starved cells. The scale bar represents 5μm. From Ref. 67.

elastomer and the other section is 1μm below the surface. The time interval between measurements is 20 seconds. Figure18 definitely shows that myosin II asymmetrically localize near the rear edge at which strong traction forces distribute. This indicates that the asymmetric force distribution is caused by the asymmetric myosin II distribution.

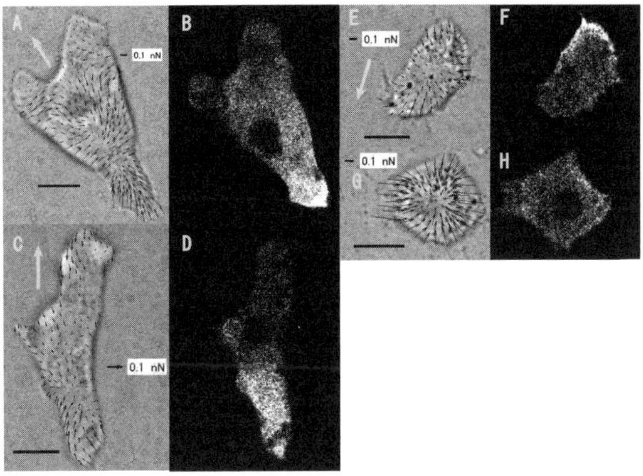

Fig. 19. Strength of the traction forces is highly correlated with myosin II localization. A, C: Traction force fields and B, D: fluorescent images of myosin II distributions for myosin-/GFP starved cells. E, G: Traction force fields and F, H: fluorescent images of myosin II distributions for myosin-/GFP vegetative cells. Scale bar is 5μm modified from Ref. 67.

7. Cell Migration Models

Clearly, shape is one of the fundamental degrees of freedom essential in motile cells.[68] The shape of motile cells is determined by many dynamic processes spanning several orders of magnitude in space and time, from local polymerization of actin monomers at subsecond timescales to global, cell-scale geometry that may persist for hours.[69] Therefore, understanding the mechanism of shape determination in cells is extremely challenging due to the numerous components involved and the complexity of their interactions. However, as long as we concentrate on the problems of mechanical determination of shape of motile cell, the mechanism of shape and motion might be understood by the techniques of reduction of degrees of freedom. Here, we give particular attention to a part of motile cell called leading edge, which is a characteristic periphery of active motile cell.[70,71] We show that the high correlation between shape and motility is explained by force balance on leading edge.

7.1. *Force balance on leading edge*

Many motile cells move on surfaces using flat motile appendages called lamellipodia as shown in Fig. 20(a). These appendages are made of a network of actin filaments (F-actin) enveloped by the cell membrane. The growth of filaments by polymerization at the lamellipodial periphery causes protrusion. Protorusion is one of the essential part of cell motility, which has been studied theoretically.[72–75] However, we must note that protrusion is not uniformly caused on the whole cell periphery.[21] This is known as a graded adhesion (firm at the front and weak at the rear) and contraction of the actin network, which lead to the forward translocation of the cell. Thus, the cell body at the rear of the motile cell is considered as a passive cargo; indeed, lamellipodial fragments without a nucleus are able to crawl with shapes and speeds similar to intact cells.[76,77] This fact justifies to focus on the front lamellipod without the cell body. The lamellipodium is a cytoskeletal protein actin projection on the mobile edge of the cell. It contains a quasi-two-dimensional actin mesh; the whole structure propels the cell across a substrate. Since the lamellipodium is moving front of cell, it is also called leading edge from the viewpoint of motility (Fig. 20(b)). Near the leading edge (lamellipodium), there are 3 characteristic phases: water, substrate, and cell. Therefore, the front motion might be treated as a problem of wetting property, which has been studied in soft-matter physics.[78]

Now, we consider the forces applied on one unit length of the leading edge. The core region, in which the details of contact force can be ignored, is sufficiently small compared with lamellipodium size. In this case, the force balance on the leading edge is simply satisfied by the balances between surface tensions of boundaries, where active force produced by a cytoplasmic matrix should be joined. Thus, force balance on the leading edge should satisfy the modified Young-Laplace equation

$$\gamma_{SL} - \gamma_{SC} - \gamma_{CL}\cos\theta_D + f_{\text{actin}}\sin\theta_D - f_{\text{fric}} = 0, \tag{86}$$

where γ_{SL} is the surface tension between substrate and water (cultured medium), γ_{SC} is the surface tension between substrate and cell, γ_{CL} is the surface tension between cell and water, f_{actin} is active force produced by actin filament, θ_D is dynamic contact angle, and f_{fric} is friction force (Fig. 20). The f_{actin} is a force produced by intracellular circumstances, which is characteristic in this dynamic balance equation. Among these forces, motile cell controls i) the active force f_{actin} and ii) contact force γ_{SL}, which is controlled by reactions about actin filament and contact proteins. However, it is not necessary to distinct these two actions for understanding the dynamics of cell motion, and, therefore, we define the active force by $f_{\text{active}} \equiv \gamma_{SL} - \gamma_{SC} - \gamma \cos\theta_D + f_{\text{actin}} \sin\theta_D$, with which the modified Young-Laplace equation becomes $f_{\text{active}} = f_{\text{fric}}$.

Leading edge is a one-dimensional structure connecting water, substrate, and cell. When the substrate is sufficiently stiff, the motion of the leading edge is restricted in a uniform two dimensional substrate surface. Thus, in the ideal situation, the motion of leading edge is modeled as a motion of line in two dimensional space. Because the force balance (86) is satisfied on each element of this one dimensional structure, the motion of leading edge specified by the position vector $\mathbf{r}$ in two dimensional space should be given by

$$\zeta \frac{d\mathbf{r}}{dt} = \mathbf{f}_{\text{active}} + \mathbf{f}_{\text{int}}, \tag{87}$$

where $\mathbf{f}_{\text{int}}$ is interaction between neighboring line elements. The parameter ζ is phenomenologically introduced dynamic friction constant, which is calculated from more detailed models. For example, when the contact force of cell is sufficiently strong, the cell motion accompanies the caterpillar motion of the intracellular fluid. In this case, the dynamic friction constant ζ is proportional to the viscosity of intracellular fluid.[78]

Thus, Eq. (87) is a fundamental model of leading edge of a motile cell. Within this consideration, we clarify the relationship between a purely geometric effect of the interface and the self-propelled motion. The assumption used in this consideration is that the leading edge is uniform, that is, parameters appeared in Eq. (86) are uniform all over the leading edge.

7.2. *Geometry and dynamics of leading edge*

We model a leading edge as an open interface having two free terminals with conserved arclength, whose dynamics is given by Eq. (87). This type of model is called geometrical model, which was first considered by Brower *et al.*;[79] they introduced a geometrical model whose dynamics is determined by local equations at the domain surface. Here, "local" implies that the velocity is locally determined by the curvature and its derivative. This is a drastic hypothesis because the motion of the interface is a nonlocal function of the interfacial coordinates in general. For example, in the case of spot motion in reaction diffusion systems, the complete motion might require the solution of a partial-differential equation throughout the entire d-dimensional space.

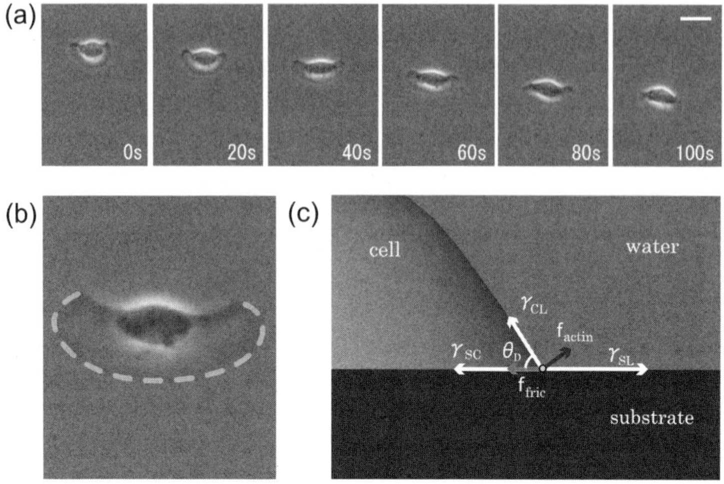

Fig. 20. Concept of leading edge. (a) Sequential phase-contrast images of a fish keratocyte cell indicating the geometry of leading edge. The scale bar implies 20 μ m. (b) Forces applied to the leading edge of a motile cell. (c) Force balance at the leading edge.

Nevertheless, Brower *et al.* clarified that some features of the motion are reproduced by the kinematics of a locality assumption.[79] Thus, with the aid of Brower's description, we examine whether only geometric constraints can cause self-propelled motion. Within this framework, we will conclude that the geometrical model can describe the occurrence of self-propelled motion, while the breakdown of locality must be assumed.

Now, we give a description of one-dimensional interfacial surface with local dynamics in a two-dimensional space. We rewrite Eq. (87) as

$$\frac{d\mathbf{r}}{dt} = \hat{L}_0\mathbf{r}, \tag{88}$$

where we assumed $\zeta = 1$. The nonlinear operator $\hat{L}_0$ is generally described by

$$\hat{L}_0\mathbf{r} = U\mathbf{n} + V\mathbf{t} \tag{89}$$

in two-dimensional space, where $\mathbf{n}$ and $\mathbf{t}$ are the normal and tangential vectors, respectively, and U and V are the associated forces. In general interface growth dynamics, the operator $\hat{L}_0$ can have a complicated functional form of the interfacial position. Here, we use the locality assumption in which $\hat{L}_0$ includes only the local quantity of interface position. That is, U and V are the functionals of $\mathbf{r}$ and its derivatives: $U \simeq U(\mathbf{r}, \nabla\mathbf{r}, \cdots)$.

Equation (88) must be gauge invariant implying that we can arbitrarily use the extrinsic coordinate α as a parametrization. Here, we specifically select the parameter $-\pi \leq \alpha \leq \pi$ satisfying $\dot{\alpha} = 0$ as an extrinsic representation. This parametrization is called orthogonal gauge by Brower *et al.*[79] On the other hand, it is advisable to use a gauge invariant intrinsic coordinate s for the description of the equation in which the normal and tangential vectors become unit vectors. The Frenet-Serret formula $\mathbf{r}_s = \mathbf{t}, \mathbf{t}_s = -\kappa\mathbf{n}$, and $\mathbf{n}_s = \kappa\mathbf{t}$ are applied for the relationship among $\mathbf{r}, \mathbf{t}$, and $\mathbf{n}$, where κ is the local curvature of the interface. The metric g is defined by $\sqrt{g} \equiv \partial s/\partial\alpha$, and the arclength s is given by $s(\alpha) = \int^{\alpha} \sqrt{g} d\alpha'$.

In order to append the spontaneous curvature into the model, we assume the phenomenological relation

$$U = 1 - \kappa, \tag{90}$$

where κ is the curvature. The first term of the r.h.s, 1, means that active force f_{active} is uniformly applied to the leading edge from the intracellular circumstance, that is, $f_{\text{active}} = \text{const}$, while const $= 1$ is assumed for simplicity. The second term is the effect of the curvature. V is treated as a multiplier to preserve the arclength under the boundary condition $V = 0$ at $s = -\pi, \pi$. Note that both of U and V are nonequilibrium active forces, which is not determined by a variational principle. Equations (88), (89), and (90) have a trivial stable uniform solution with $\kappa = 1$. When the interface is specified by $-\pi < s \leq \pi$, the trivial solution is given by

$$\mathbf{r} = \mathbf{r}(0) + \int_0^s dse^{i(s+\theta_0)}, \tag{91}$$

where θ_0 is the phase of the origin. We used the conventional notation for vector and complex number describing the curve by the complex position $\mathbf{r} = x+iy$. In this article, we continue to use this notation for the description of vectors. Hereafter, we investigate the dynamics in the vicinity of the trivial solution.

Now, we analyze Eq. (88), from which we extract the dynamics of shape and centroid. We request that the metric is invariant over the time-evolution, which is necessary to preserve the arclength. Under this conditions, we derive the dynamical equation of shape by

$$\partial_t\theta = \theta_{ss} - \theta_s \int^s \theta_s(1 - \theta_s)ds, \tag{92}$$

where θ is defined by $\kappa = \theta_s$, whose geometrical meaning is depicted in Fig.(21). Next, we consider the centroid of the interface. When the interface

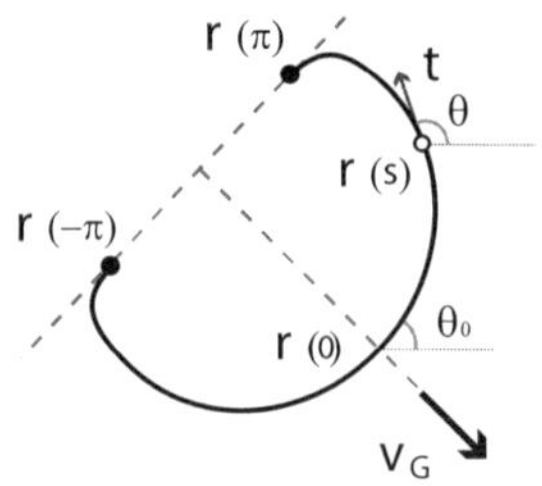

Fig. 21. Motion of broken interface. The interface moves to the direction perpendicular to the line connected between the terminal points.

is assumed to be intrinsically uniform, the centroid is described by

$$\mathbf{r}_G = \int_{-\pi}^{\pi} \mathbf{r} ds. \tag{93}$$

To understand the relationship between centroid motion and shape, we assume that some applied perturbation imparted the deformation including only the first Fourier mode to the trivial solution; $\theta = s - a_1 \sin s + \theta_0$. In this case, clearly, the mode $n = 1$ is strictly related to the distance between the terminal points, and thus the total centroid velocity is given by

$$v_G = -2\pi i a_1 e^{i\theta_0} + \mathcal{O}(a_1^2). \tag{94}$$

Thus, we obtained two fundamental equations (92) and (94) from which we understand several facts applicable to the geometrical model with the locality assumption. Equation (92) shows that the locality assumption always leads the neutral stability for $n = 1$ mode of deformation. With Eq. (94), this fact further implies that the locality assumption gives the neutral stability of the centroid motion. This means that the property of the solution is qualitatively changed by an arbitrary small perturbation, which breaks the symmetry. For example, even when we consider the static solution of the interface, the solution might start to self-propel under the noisy circumstances. To understand the robust and universal description of the interface, we should perform singular perturbation to unfold this singular point.[80]

The locality assumption imparts simplicity to the interface dynamics, but at the same times, it restricts the possible variety of interface dynamics. Here, we relax the restriction for locality and assume that the breakdown of locality is sufficiently weak to be treated as a perturbation. We consider the dynamics of interface

$$\frac{d\mathbf{r}}{dt} = \hat{L}_0\mathbf{r} + \epsilon\hat{L}\mathbf{r}, \tag{95}$$

where $\hat{L}$ is a nonlocal perturbation operator that need not satisfy the locality assumption.

Our aim is to obtain a simple perturbation to break the neutral stability. Here, we show one possible candidate of the simple perturbation. The protocol for the nonlocal perturbation is based on two hypotheses; the first one is called "the locality retrieval condition" which means that the locality is maintained by at least one point on the curve. We assume that the locality is recovered at the origin $s = 0$, hence the operator becomes zero at the recovery point, $\hat{L}\mathbf{r}(0) = 0$. The second hypothesis is that the interface interacts with some virtual interface satisfying the locality retrieval condition. For the virtual interface, we especially consider a precise circle $\mathbf{r}_0$ whose arclength and phase at the origin are identical to the neutrally stable closed interface. The perturbation term $\hat{L}\mathbf{r}$ is assumed to impart harmonic interaction between the broken interface and the virtual interface. From these conditions, $\hat{L}\mathbf{r}$ is determined as

$$\hat{L}\mathbf{r} = \mathbf{r} - \mathbf{r}_0, \tag{96}$$

where

$$\mathbf{r}_0(s) = \mathbf{r}(0) + \int_0^s ds e^{i(s+\theta_0)}. \tag{97}$$

From Eq. (96), we get the motion of shape as

$$\begin{aligned}\frac{\partial\theta}{\partial t} &= -\frac{\partial}{\partial s}U - \frac{\partial\theta}{\partial s}\int^s \theta_s U ds \\ &\quad +\epsilon\sin\phi - \epsilon\frac{\partial\theta}{\partial s}\int^s (1-\cos\phi)ds. \end{aligned} \tag{98}$$

When ϵ is small but finite, the $n = 1$ mode is the sole slow critical mode effective in long-term dynamics, hence it is possible to derive the effective description including only the $n = 1$ mode.

Our construction of the nonlocal perturbation in which we introduced a virtual potential circle might seem unusual to some readers. We could have selected any other type of nonlocal perturbation; however, this type is heuristically introduced as a simple example of a possible nonlocal perturbation. It is important to note that the local description given by Eqs. (88) and (89) is structurally unstable. Hence, almost any type of nonlocal perturbation unfolds the neutral stability of Brower's dynamics. Therefore, we conclude that the slow mode equation that will be derived here after is valid within the $\mathcal{O}(\epsilon)$ neighbor of the system constructed under the locality assumption.

Now, we perform singular perturbation to derive the coarse-grained dynamical equation, assuming that ϵ is sufficiently small. Equation (92) is approximated and rewritten as

$$\hat{\ell}_0 \phi = \frac{\partial \phi}{\partial t} - \epsilon \phi + N_2(\phi) + N_3(\phi), \tag{99}$$

where the variable ϕ is a phase deviation from the exact circle defined by $\theta = s + \phi + \theta_0$. and $\hat{\ell}_0$ is the linear operator

$$\hat{\ell}_0 = \frac{\partial^2}{\partial s^2} + 1. \tag{100}$$

N_2 and N_3 are the second and third order nonlinear functions of ϕ given by $N_2(\phi) = -\phi\phi_s - \int \phi_s^2 ds$, and $N_3(\phi) = -\phi_s \int \phi_s^2 ds$. We define the small parameter ε as $\epsilon = R\varepsilon$, with which we give the naïve perturbation series $\phi = \phi_0 + \varepsilon\phi_1 + \varepsilon^2\phi_2 + \cdots$. Further introducing the multiple slow time scales $t = \varepsilon^{-1}\tau + \varepsilon^{-2}\tau_2$, we obtain the perturbation series. The $\mathcal{O}(\varepsilon)$ equation gives the zero eigenvalue function $\phi_1 = -a_1(t)\sin s$. From the solvability condition for the $\mathcal{O}(\varepsilon^2)$ equation,[81] we get the slow mode equation by

$$\frac{da_1}{d\tau} = Ra_1 - a_1^2 + \mathcal{O}(a_1^3), \tag{101}$$

which is the normal-form equation of transcritical bifurcation. Here, we simultaneously consider the centroid velocity in the lowest order approximation. Although we introduced the nonlocal perturbation, the perturbation is $\mathcal{O}(\epsilon)$; hence, the effect of the perturbation on the centroid is significantly less than that of the locality assumption. Thus, it is sufficient to consider Eq. (94) as the lowest order velocity relation. Considering Eq. (94) with Eq. (101), we finally conclude that Eq. (101) is the drift bifurcation equation of a self-propelled interface. Figure 22 shows an example of the spontaneous drift dynamics of the broken interface obtained numerically by solving Eq. (95). We see that the initially circular interface becomes an arc accompanying a spontaneous drift motion. This interface motion is quite similar to the motion by Keratocyte cell in Fig. 20(a). Thus, we conclude that the transcritical drift bifurcation robustly appears with the unfolding of Brower's description of interface.

Finally, we give a brief explanation on the origin of the transcriticality. Our theory is constructed by assuming that the curve has spontaneous cur-

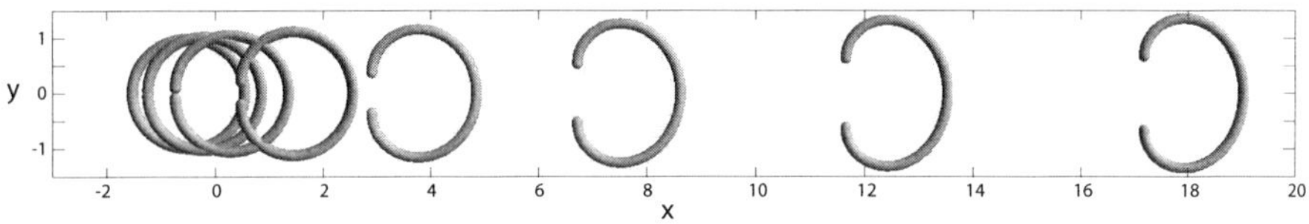

Fig. 22. Spontaneous drift motion of broken interface. The value of the parameter is set to $R = 0.5$. The initial state is selected to a complete circular shape $a_1 = 0$ and a small perturbation $\delta a_1 = 10^{-4}$ is appended at $t = t_{\text{on}}$. The resultant drift motions obtained at $t - t_{\text{on}} = 0, 12, 14, 16, 18, 20, 22, 24$ are depicted.

vature, which implies that the convex and concave bendings have different effects. With this bending asymmetry, Eq. (99) is not invariant against the operation $\phi \to -\phi$, from which the transcriticality originates. If we assumed the other simple case $U \sim -\kappa$, instead of Eq. (90), the bending symmetry is recovered, hence the drift bifurcation should become supercritical.

7.3. *Complex active motion of leading edge*

We have described the steady movement of the leading edge by the geometrical model. The steady moving leading edge is considered as a model of keratocyte cell, which shows simply straight forward motion. On the other hand, we showed that *Dictyostelium* cell shows a variety of motions, which include rotation and oscillation. Is our geometrical type model extendable to show Dicty type rich motion? Here, we show that this is really extendable by using symmetry argument and reduction method.

General geometrical model is given by $\frac{d\mathbf{r}}{dt} = \hat{L}\mathbf{r}$, where the operator $\hat{L}$ includes local and nonlocal interactions. This equation is transformed into the equations of shape and centroid location expressed by the phase variable ϕ,

$$\partial_t \phi = F(\phi), \tag{102}$$

where the function F has a complicated form depending on $\hat{L}$. The centroid motion is given by Eq. (94) as long as the operator $\hat{L}$ is close to the operator $\hat{L}_0$ that includes only local interactions. Equation (102) is assumed to have a trivial solution $\phi = 0$. Now, we consider what is the simplest form of F that shows complex dynamics in ϕ. For that, we consider a symmetry that the equation of ϕ must satisfy. Because the 2-dimensional space is uniform, the dynamics of ϕ should be invariant against the following transformations: (i) $\phi \to \phi + \phi_0$, (ii)$s \to s + s_0$ with ϕ_0 and s_0 constants and (iii)$s \to -s$

and $\phi \to -\phi$. Keeping these in mind, we write down the equation for ϕ up to bilinear order of ϕ as

$$\partial_t \phi = \epsilon_2 \partial_s^2 \phi + \epsilon_4 \partial_s^4 \phi + \epsilon_6 \partial_s^6 \phi + g_1 \partial_s \phi \partial_s^2 \phi + g_2 \partial_s^2 \phi \partial_s^3 \phi + g_3 \partial_s \phi \partial_s^4 \phi + \cdots . \quad (103)$$

Considering these terms, we conclude that the minimal non-trivial equation for ϕ is given by[30]

$$\partial_t \phi = \partial_s^4 \phi + \partial_s^6 \phi - \partial_s^2 \phi \partial_s^3 \phi, \quad (104)$$

where the coefficients ϵ_4, ϵ_6, and g_2 are eliminated by redefining t, s and ϕ and the sign of third term of the r.h.s. is chosen to be negative without loss of generality. Note that this equation contains one parameter, the system size L. When we consider Eqs. (104) and Eq. (94) as a model of leading edge, we call these equations active cell model.

The active cell model shows several types of deformation-centroid coupled motion depending on the value of L, which includes motionless, straight motion, circular motion, rectangular motion, quasi periodic motion, zig-zag motion, and chaotic motion.[30] Direct numerical calculation shows that the active cell model makes a steady stable deformed domain in a broad range of the parameter except for the three characteristic windows; $W_1 \simeq \{L|L \in [12.0, 14.4]\}, W_2 \simeq \{L|L \in [21.0, 21.9]\}$, and $W_3 \simeq \{L|L \in [29.0, 29.9]\}$. Within these windows, the domain shows a complex deformation-centroid coupled motion; circular, quasi-periodic, and rectangular motions in W_1, circular and chaotic motions in W_2, and zigzag motions in W_3. The straight motion is obtained when the value of parameter is lower than W_1 (when $L = 6.2 \sim 12.0$). In Fig. 23(a), we show the trajectory obtained within W_2, where the model seems to give a chaotic trajectory. Fig. 23(b) also shows a trajectory obtained within W_2, but the trajectory in this case seems more regulated than (a). This trajectory is a kind of heteroclinic cycle, where the orbit within one metastable cycle almost periodically jumps to the other metastable cycle via the saddle point. Fig. 23(c) and (d) show the two kinds of trajectory seen in W_3, in which the effective deformation modes are mode-3 and mode-4 here. In our calculations, the mode-2 and mode-3 effective zigzag motion has not been found in the active cell model. However, the two kinds of transient mode-2,3 zigzag motions are seen in the way of the chaotic trajectory (Fig. 23(a)).

In conclusion, we proposed a geometrical model of motile cell. In our geometrical model, we assumed that the motile cell is retracted by the

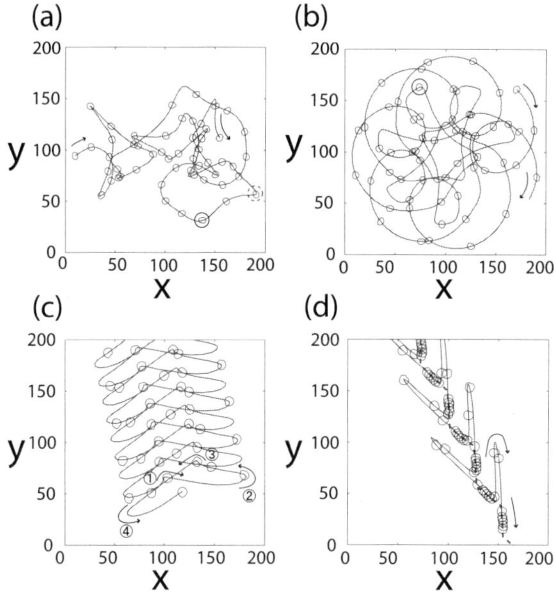

Fig. 23. Motion trajectories of active cell model. The values of system size parameter used here are (a)$L = 21.6$, (b)$L = 21.8$, (c)$L = 29.2$, and (b)$L = 29.6$. (a) Chaotic motion. The circles indicates characteristic turns; the solid circle: $\pm 60°$ turn, the broken circle: $\pm 120°$ turn. (b)Quasi-periodic heteroclinic motion. The orbit periodically jumps from one metastable cycle to the other metastable cycle through the saddle point (solid circle). (c)(d)Mode 3-4 effective zigzag motions. These mode 3-4 zigzag motions have different turn angles: (c)$180° \pm 45°$ and (d) $\pm 45°$. In order to clarify the motion, we appended the velocity parameter v_0 defined by $\boldsymbol{v}_G = -v_0 i a_1$. The velocity parameters used here are (a)(b)(d) $v_0 = 5.0$ and (c) $v_0 = 20.0$. From Ref. 30.

active uniform contact line called leading edge. Because the leading edge is assumed to be uniform, the dynamics of the leading edge is crucially determined by its geometrical shape. The correlation between dynamics and geometry in our geometrical model give a possible interpretation of motion of keratocyte and Dictyostelim. We consider that the geometrical type model is useful to study the motility of living cell.

8. Summary and Discussion

We have described our recent theoretical and experimental studies of self-propelled dynamics. The theory is based on the two model equations which take account of the coupling between shape deformations and migration velocity. One is the tensor model which is a set of ordinary differential equations for the deformation modes up to $n = 3$. The other is the geometrical model in the form of a partial differential equation for the leading edge of a cell. Both models exhibit commonly, a motionless state, straight motion, circular motion, rectangular motion and chaotic motion. Zigzag motion and quasi-periodic motion are also obtained in the two models. However the details are different as can be seen in comparison of Fig. 7(a) and Fig. 23(c) and (d) for the zigzag motions, and Fig. 4 and Fig. 23(b) for the quasi-periodic motions. These similarity and difference can be understood as follows. The Fourier expansion of the geometrical model takes the same form of the mode equations as Eqs. (65), (66) and (67). However, since the geometrical model contains higher modes, it exhibits more complicated dynamics as shown in Fig. 23.

In the experimental part, dynamics of Dictyostelium discoideum amoebae have been explained. In order to investigate fundamental patterns of cell motion, the auto-correlation function of shape deformations and the distribution of local distortion (local stress) inside a motile cell have been measured. In these procedures, three basic dynamical patterns, stretching, rotating, oscillating are identified.

When we compare these experiments with our theories, we may expect that stretching and oscillating would be explained by a straight motion and a zigzag motion, or, their modified versions. However, any motion corresponding to rotating has not been obtained in our models. It is remarked that rotating (or, spinning) should be distinguished from the circular motion in Fig. 6 where the trajectory of center of mass takes a circular loop. Spinning motion is often observed in living cells in three dimensions like Listeria.[22,50–52] Therefore, we admit that to incorporate spinning degrees of freedom in our models is an important remaining problem. Another problem is to extend the tensor model to take account of the hydrodynamic effects. This is necessary when we deal with droplets self-propelled by Marangoni effect in three dimensions and swimming micro-organisms as the elaborated theories have been developed in specific biological systems.[13–15] Finally, we emphasize that stochasticity in deformable self-propelled objects,[82] in particular, the cross correlations of shape fluctuations and velocity fluctuations would be an interesting future subject. This study is of importance not only for our understanding of efficiency of the active transport in living systems but also for developing statistical physics far from equilibrium.

Acknowledgement

This work was supported by the JSPS Core-to-Core Program "International research network for non-equilibrium dynamics of soft matter" and the Grant-in-Aid for the priority area "Soft Matter Physics" both from the Ministry of Education, Culture, Sports, Science and Technology (MEXT) of Japan.

References

1. G. Taylor, The action of waving cylindrical tails in propelling microscopic organisms, *Proc. R. Soc. Lond. A* **211**, 225 (1952).
2. M. Percell, Life at low reynolds number, *Am. J. Phys.* **45**, 3 (1977).
3. T. Vicsek et. al, Novel type of phase transition in a system of self-driven particles, *Phys. Rev. Lett.* **75**, 1226 (1995).
4. N. Shimoyama, K. Sugawara, T. Mizuguchi, Y. Hayakawa, and M Sano, Collective motion in a system of motile elements, *Phys. Rev. Lett.* **76**, 3870 (1996).
5. J. Toner and Y. Tu, Flocks, herds, and schools: a quantitative theory of flocking, *Phys. Rev. E* **58**, 4828 (1998).
6. G. Grégoire and H. Chaté, Onset of collective and cohesive motion, *Phys. Rev. Lett.* **92**, 025702 (2004).
7. F. Ginelli, F. Peruani, M. Bär, and H. Chaté, Large-scale collective properties of self-propelled rods, *Phys. Rev. Lett.* **104**, 184502 (2010).
8. Y. Hatwalne, S. Ramaswamy, M. Rao and R. A. Simha, Rheology of active-particle suspensions, *Phys. Rev. Lett.* **92**, 118101 (2004).
9. A. W. C. Lau and T. C. Lubensky, Fluctuating hydrodynamics and microrheology of a dilute suspension of swimming bacteria, *Phys. Rev. E* **80**, 011917 (2009).
10. B. M. Haines and A. Sokolov, I. S. Aranson, L. Berlyand and D. A. Karpeev, Three-dimensional model for the effective viscosity of bacterial suspensions, *Phys. Rev. E* **80**, 041922 (2009)
11. A. Baskaran and M. C. Marchetti, Statistical mechanics and hydrodynamics of bacterial suspensions, *Proc. Nat. Acad. Sci.* **106** 15567 (2009).
12. L. Giomi, T. B. Liverpool and M. C. Marchetti, Sheared active fluids: thickening, thinning, and vanishing viscosity, *Phys. Rev. E* **81**, 051908 (2010).
13. T. Ishikawa and T. J. Pedely, The rheology of a semi-dilute suspension of swimming model micro-organisms, *J. Fluid Mech.* **588**, 399 (2007) .

14. H. Wada and R. R. Netz, Hydrodynamics of helical-shaped bacterial motility, *Phys. Rev. E* **80**, 021921 (2009).
15. T. Ishikawa, Suspension biomechanics of swimming microbes, *J. R. Soc. Interface* 2009, online 12 August.
16. A. Shapere and F. Wilczek, Self-propulsion at low Reynolds number, *Phys. Rev. Lett.* **20**, 2051 (1987).
17. K. Keren, Z. Pincus, G. M. Allen, E. L. Barnhart, G. Marriott, A. Mogilner and J. A. Theriot, Mechanism of shape determination in motile cells, *Nature* **453**, 475 (2008).
18. L. Bosgraaf, P. J. M. Van Haastert, The Ordered extension of pseudopodia by amoeboid cells in the absence of external cues, *PLoS one* **4**, e5253 (2009).
19. L. Li, S. F. Norrelykke and E. C. Cox, Persistent cell motion in the absence of external signals: a search strategy for eukaryotic cells, *PLoS one* **3**, e2093 (2008).
20. Y. T. Maeda, J. Inose, M. Y. Matsuo, S. Iwaya, and M. Sano, Ordered patterns of cell shape and orientational correlation during spontaneous cell migration, *PLoS one* **3**, e3734 (2008).
21. K. Keren, P. T. Yam, A. Kinkhabwala, A. Mogilner, and J. A. Theriot, Intracellular fluid flow in rapidly moving cells, *Nature Cell Biol.* **10**, 1219 (2009).
22. H. Boukellal, O. Campas, J-F. Joanny, J. Prost and C. Sykes, Soft listeria: actin-based propulsion of liquid drops, *Phys. Rev. E* **69**, 061906 (2004).
23. S. I. Nishimura, M. Ueda and M. Sasai, Cortical factor feedback model for cellular locomotion and cytofission, *PLoS Comp. Bio.* **5**, e1000310 (2009).
24. Y. Sumino, N. Magome, T. Hamada, and K. Yoshikawa, Self-running droplet: emergence of regular motion from nonequilibrium noise, *Phys. Rev. Lett.* **94**, 068301 (2005).
25. K. Nagai, Y. Sumino, H. Kitahata, and K. Yoshikawa, Mode selection in the spontaneous motion of an alcohol droplet, *Phys. Rev. E* **71**, 065301(R) (2005).
26. K. Suzuki *et al.*, Sparkling morphological changes and spontaneous movements of self-assemblies in water induced by chemical reactions, *Chemistry Letters* **38**, 1010 (2009).
27. T. Toyota, K. Takakura, Y. Kageyama, K. Kurihara, N. Maru, K. Ohnuma, K. Kaneko and T. Sugawara, Population study of sizes and components of self-reproducing giant multilamellar vesicles, *Langmuir* **24**, 3037 (2008).
28. T. Toyota, N. Maru, M. M. Hanczyc, T. Ikegami and T. Sugawara, Self-propelled oil droplets consuming ĄgfuelĄh surfactant, *J. Am. Chem. Soc.* **131**, 5012 (2009).
29. T. Ohta and T. Ohkuma, Deformable self-propelled particles, *Phys. Rev. Lett.* **102**, 154101 (2009).
30. T. Hiraiwa, Y. Matsuo, T. Ohkuma, T. Ohta and M. Sano, Dynamics of a deformable self-propelled domain, *Europhys. Lett.* **91**, 20001 (2010).
31. T. Ohta, T. Ohkuma and K. Shitara, Deformation of a self-propelled domain in an excitable reaction-diffusion system, *Phys. Rev. E* **80**, 056203 (2009).
32. L. G. Fel, Tetrahedral symmetry in nematic liquid crystals, *Phys. Rev. E* **52**,

702 (1995).
33. H. R. Brand, H. Pleiner and P. E. Cladis, Tetrahedratic cross-couplings: novel physics for banana liquid crystals, *Physica A* **351**, 189 (2005).
34. T. Ohta, J. Kiyose and M. Mimura, Collision of propagating pulses in a reaction-diffusion system, *J. Phys. Soc. Jpn.* **66**, 1551 (1997).
35. C. P. Schenk, M. Or-Guil, M. Bode and H.-G. Purwins, Interacting pulses in three-component reaction-diffusion systems on two-dimensional domains, *Phys. Rev. Lett.* **78**, 3781 (1997).
36. M. Or-Guil, M. Bode, C. P. Schenk, and H.-G. Purwins, Spot bifurcations in three-component reaction-diffusion systems: the onset of propagation, *Phys. Rev. E* **57**, 6432 (1998).
37. T. Ohta, Pulse dynamics in a reaction-diffusion system, *Physica D* **151**, 61 (2001).
38. S.-I. Ei, M. Mimura and M. Nagayama, Pulse-pulse interaction in reaction-diffusion systems, *Physica D* **165**, 176 (2002).
39. Y. Nishiura, T. Teramoto, and K. Ueda, Scattering of traveling spots in dissipative systems, *Chaos* **15**, 047509 (2005).
40. T. Teramoto, K. Suzuki and Y. Nishiura, Rotational motion of traveling spots in dissipative systems, *Phys. Rev. E* **80**, 046208 (2009).
41. S. Koga and Y. Kuramoto, Localized Patterns in Reaction-Diffusion Systems, *Prog. Theor. Phys.* **63**, 106 (1980).
42. T. Ohta, M. Mimura and R. Kobayashi, Higher-dimensional localized patterns in excitable media, *Physica D* **34**, 115 (1989).
43. R. FitzHugh, Impulses and physiological states in theoretical models of nerve membrane, *Biophys. J.* **1**, 445 (1961); J. Nagumo, S. Arimoto, and S. Yoshizawa, An active pulse transmission line simulating nerve axon, *Proc. IRE* **50**, 2061 (1962).
44. B. Kerner and V. V. Osipov, Autosolitons, *Sov. Phys. Usp.* **32**, 101 (1989).
45. K. Krischer and A. Mikhailov, Bifurcation to traveling spots in reaction-diffusion systems, *Phys. Rev. Lett.* **73**, 3165 (1994).
46. L. M. Pismen, *Patterns and Interfaces in Dissipative Dynamics* (Springer, Berlin, 2006).
47. J. Rinzel and D. Terman, Propagation phenomena in a bistable reaction-diffusion system, *SIAM J. Appl. Math.* **42**, 1111 (1982).
48. D. Armbruster, J. Guckenheimer and P. Holmes, Heteroclinic cycles and modulated travelling waves in systems with O(2) symmetry, *Physica D* **29**, 257 (1988).
49. D. Armbruster, J. Guckenheimer and P. Holmes, Kuramoto-Sivashinsky dynamics on the center-unstable manifold, *SIAM J. Appl. Math.* **49**, 676 (1989).
50. W. L. Zeile, F. Zhang, R. B. Dickinson and D. L. Purich, Listeria's right handed helical rocket-tail trajectories: mechanistic implications for force generation in actin-based motility, *Cell Motil. Cytoskeleton* **60**, 121 (2005).
51. H. C. Crenshaw, A new look at locomotion in microorganisms: rotating and translating, *Amer. Zool.* **36**, 608 (1996).
52. V. B. Shenoy, D. T. Tambe, A. Prasad and J. A. Theriot, A kinematic description of the trajectories of Listeria monocytogenes propelled by actin

comet tails, *Proc. Natl. Acad. Sci. USA* **104**, 8229 (2007).
53. T. Hiraiwa, K. Shitara and T. Ohta, Dynamics of a deformable self-propelled particle in three dimensions, *Soft Matter* **7**, 3083 (2011).
54. K. Shitara, T. Hiraiwa and T. Ohta, Deformable self-propelled domain in an excitable reaction-diffusion system in three dimensions, *Phys. Rev. E* **83**, 066208 (2011).
55. T. Kawagishi, T. Mizuguchi, and M. Sano, Points, walls, and loops in resonant oscillatory media, *Phys. Rev. Lett.* **75**, 3768 (1995)
56. C. A. Parent, B. J. Blacklock, W. M. Froehlich, D. B. Murphy, and P. N. Devreotes, G protein signaling events are activated at the leading edge of chemotactic cells, *Cell* **95**, 81-91 (1998).
57. N. Andrew and R. H. Insall, Chemotaxis in shallow gradients is mediated independently of PtdIns 3-kinase by biased choices between random protrusions, *Nat. Cell Biol.* **9**, 193-200 (2007).
58. T. Killich, P. J. Plath, X. Wei, H. Bultmann, L. Rensing, et al., The locomotion, shape and pseudopodial dynamics of unstimulated dictyostelium cells are not random, *J. Cell Sci.* **106**, 1005-1013 (1993).
59. H. G. Dobereiner, B. J. Dubin-Thaler, J. M. Hofman, H. S. Xenias, T. N. Sims, et al., Lateral membrane waves constitute a universal dynamic pattern of motile cells, *Phys. Rev. Lett.* **97**, 038102 (2006).
60. G. Gerisch, T. Bretschneider, A. Muller-Taubenberger, E. Simmeth, M. Ecke, et al., Mobile actin clusters and traveling waves in cells recovering from actin depolymerization, *Biophys. J.* **87**, 3493 (2004).
61. O. D. Weiner, W. A. Marganski, L. F. Wu, S. J. Altschuler and M. W. Kirschner, An actin-based wave generator organizes cell motility, *PLoS Biology* **5**, e221 (2007).
62. M. L. Lombardi, D. A. Knecht, M. Dembo, and J. Lee, Traction force microscopy in dictyostelium reveals distinct roles for myosin II motor and actin-crosslinking activity in polarized cell movement, *J Cell Sci.* **120**, 1624 (2007).
63. K. S. K. Uchida, T, Kitanishi-Yumura and S. Yumura, Myosin II contributes to the posterior contraction and the anterior extension during the retraction phase in migrating dictyostelium cells, *J Cell Sci* **116**, 51 (2003).
64. H. Delanoe-Ayari, S. Iwaya, Y. T. Maeda, J. Inose, C. Riviere, M. Sano, and J.-P. Rieu, Changes in the magnitude and distribution of forces at different dictyostelium developmental stages, *Cell Motility and the Cytoskeleton* **65**, 314 (2008).
65. H. Delanoe-Ayari, J. P. Rieu, and M. Sano, 4D traction force microscopy reveals asymmetric cortical forces in migrating dictyostelium cells, *Phys. Rev. Lett.***105**, 248103 (2010).
66. L. D. Landau, and E. M. Lifshitz. *Theory of Elasticity*, 3rd Ed. J. B. Sykes and W. H. Reid, translators. Pergamon Press, Oxford, UK (1986).
67. S. Iwaya, *Masters Thesis*, Department of Physics, The University of Tokyo (2008).
68. X. Jiang, D. A. Bruzewicz, A. P. Wong, M. Piel, and G. M. Whitesides, Directing cell migration with asymmetric micropatterns, *Proc. Natl. Acad. Sci.* **102**, 975 (2005).

69. P. Lenz, ed., *Cell Motility (Biological and Medical Physics, Biomedical Engineering)* (Springer, New York, 2008).
70. J. Lee, A. Ishihara, and K. Jacobson, How do cells move along surfaces? *Trends Cell Biol.* **3**, 366 (1993).
71. C. W. Wolgemuth, J. Stajic, and A. Mogilner, Redundant mechanisms for stable cell locomotion revealed by minimal models, *Biophys. J.* **101**, 545 (2011).
72. C. W. Wolgemuth, Lamellipodial contractions during crawling and spreading, *Biophys. J.* **89**, 1643 (2005).
73. M. Buenemann, H. Levine, W-J. Rappel, and L. Sander, The role of cell contraction and adhesion in dictyostelium motility, *Biophys. J.* **99**, 50 (2010).
74. K. John, P. Peyla, K. Kassner, J. Prost, and C. Misbah, Nonlinear study of symmetry breaking in actin gels: implications for cellular motility, *Phys. Rev. Lett.* **100**, 068101 (2008).
75. A. C. Callen-Jones, J-F. Joanny, and J. Prost, Viscous-fingering-like instability of cell fragments, *Phys. Rev. Lett.* **100**, 258106 (2008).
76. S. Schaub, S. Bohnet, V. M. Laurent, J-J. Meister, and A. B. Verkhovsky, Comparative maps of motion and assembly of filamentous actin and myosin II in migrating cells, *Mol. Biol. Cell* **18**, 3723 (2007).
77. B. Rubinstein, M. F. Fournier, K. Jacobson, A. B. Verkhovsky, and A. Mogilner, Actin-myosin viscoelastic flow in the keratocyte lamellipod, *Biophys. J.* **97**, 1853 (2009).
78. P-G. de Gennes D. Quere, and F. Brochard-Wyart, *Capillarity and Wetting Phenomena: Drops, Bubbles, Pearls, Waves.*(Springer, New York, 2003).
79. R. C. Brower, D. A. Kessler, J. Koplik and H. Levine, Geometrical models of interface evolution, *Phys. Rev. A* **29** 1335 (1984).
80. S. Wiggins, *Introduction to Applied Nonlinear Dynamical Systems and Chaos.* (Springer, Berlin, 2003).
81. Y. Kuramoto, *Chemical Oscillations, Waves, and Turbulence.* (Springer, Berlin, 1984).
82. B. ten Hagen, S. van Teeffelen and H. Loewen, Brownian motion of a self-propelled particle, *J. Phys.: Condens. Matter* **23** 194119 (2011).

Index

3D, 93
  image, 97, 112, 121
  observation, 93
  structure, 93
7CB, 59
8CB, 61

actin filament, 315
actin polymerization, 389, 390
active
  cell model, 408
  force, 400
  motile cell, 399
  soft matter, 365
  transport, 411
atomic force microscopy, 277
Avrami model, 100
azobenzene, 146, 147, 174
  H-aggregation, 148, 157, 178
  photostationary mixture, 175

bend
  persistence length, 276
bending
  elasticity, 277
birefringence, 147
bottleneck effcct
  bottleneck effect , 180
breakup, 89

Brewster angle microscopy, 147, 175
Brownian coagulation, 220
Brownian motion, 14
buckling, 307

carbon black suspension, 69
CCR, *see* convective constraint release
cell migration, 366, 395, 399
characteristic
  length, 137
  time, 105
chemotaxis, 389, 397
chirality, 173
  enantiomeric excess, 177
  homochiral terrestrial life, 179
  orientational, 162, 165, 167, 175
  racemic mixture, 174
  selection, 177
chromatin, 338
circular dichroism, 175
coalescence, 89, 96, 97, 111, 121, 162, 165
  rate, 99
Cole expression, 75
column, 99
columnar structure, 110, 113

complex modulus, 112
concentration fluctuations, 207, 209
conductivity, 91, 95, 97, 120, 133
confocal scanning laser microscopy, 93
convective constraint release, 53
correlated diffusion, 226, 259
correlation length, 100, 207, 211
coupled modulated bilayers, 217
coupling diffusion coefficients, 226, 259
CSLM, 93
curvature, 288

deformable particle, 366, 367
deformation-induced migration, 369, 379, 388
depolarization factor, 129
diagonal element, 119
Dicty cell, 370, 390
Dictyostelium, 366, 390, 395, 407, 411
dielectric constant, 91, 95, 97, 111, 120, 133
dielectric intensity, 44
dimensional analysis, 101, 131
dimethylsiloxane, 91
dipole moment, 127
dipole-dipole interaction, 96
DMS, 91
DNA
  replication, 278
  transcription, 278
DNA supercoiling, 277
DNA-protein complex, 338
Doi-Ohta
  model, 91, 139
  scaling relation, 134

domain drag coefficient, 233
domain growth exponent, 222
drift bifurcation, 368, 375–377, 406
droplet deformation, 101
droplet-dispersed structure, 112
dynamic structure factor, 245, 249, 252

effective diffusion coefficient, 212, 218
effective polarizability, 104
elastic constants, 158
elastic energy, 289
electric interfacial tension, 130
electric torque, 126
electrode polarization, 64
electrorheological effect, 91
electrostatic interaction, 104
ellipsoid, 119, 123
elongation, 111, 121
emulsion, 89
energy dissipation function, 4
ER fluid, 91
Euler angles, 302
excluded volume effects, 255

first mode, 134
Fitz-Hugh-Nagumo equation, 373
fluctuation-dissipation, 158
fluorescent dye, 96
free-energy functional, 157
Frenet-Serret equations, 288
friction coefficient, 5, 105
Fuller's theorem, 297

Gauss-Bonnet theorem, 296
Gaussian curvature, 297
Gaussian polymer chain, 242

generalized Frenet basis, 286
geometric phase, 320
geometrical model, 366, 401, 407, 408, 410
glassy relaxation, 44
Green–Kubo expression, 74
Green–Kubo theorem, 43
growth rate, 103

helical motion, 386
helix, 288
hierarchical model, 96
histon octamer, 338
hydrodynamic coupling, 257
hydrodynamic screening length, 206
hydrodynamic torque, 126
hydrodynamics interactions, 303

immiscible
  blend, 91, 118
    ER fluid, 92
  fluid blend, 89
  polymer blend, 97, 118
induced dipole-dipole interaction, 121
interface, 89, 374–377, 395, 401, 403, 404, 406
  tensor, 90
interfacial
  stress, 119
  tension, 90

keratocyte, 370, 389, 401, 406–408
Kirchhoff kinetic analogy, 295
Kirchhoff rod equations, 294
Kratky-Porod wormlike chain model, 277
Kuramoto-Sivashinsky equation, 384

LacI protein, 338
Lagrange multiplier, 289
Langevin equations, 303
Langmuir monolayers, 146, 151, 173
  domain textures, 150
  axisymmetric textures, 153
  bend distortion, 148
  boojum, 166
  collective precession, 159
  compressibility, 182
  condensates, 175
  diffusion coefficient, 179
  flow, 180
  flow contraction, 181
  flow profile, 182
  hexatic phases, 161
  orientational order, 147
  permeability constant, 182
  splay distortion, 148, 152
Langmuir trough, 151
  microchannel, 180
Langmuir-Blodgett monolayers, 146
LAOS, *see* large-amplitude shear oscillation
large-amplitude shear oscillation, 78
LCP, 91, 95, 97, 111, 119, 133
leading edge, 366, 399–401, 403, 407, 408
Leslie-Ericksen's equation, 32
Li^+-O-Li^+ bonding, 66
Lifshitz point, 217
Lifshitz-Slyozov, 220
linking number, 295

liquid crystal, 58
liquid crystalline polymer, 91
locality retrieval condition, 405
loss modulus, 113

Macdonald theory, 78
major axis, 100
Marangoni effect, 366, 411
Mason number, 133
Maxwell stress, 92, 110, 118, 119
  tensor, 113
Maxwell-Wagner model, 137
mean curvature, 374
mesogen, 95
methyl phenyl silicone oil, 133
microemulsion, 216
microfluidics, two-dimensional, 179
  network of microchannels, 186
migration-induced deformation, 369, 382, 385, 388
minor axis, 100
mobility tensors, 206
modified Bessel function, 116
moment, 292
momentum decay model, 231
morphology, 89, 118
Morse potential, 340
MPS, 133
mutual equilibration, 52

nematic order parameter, 368
network structure, 121
Newtonian, 90, 131
non-
  equilibrium statistical physics, 365
nonlocal perturbation, 405, 406
nucleosome, 338

off-diagonal element, 119
Onsager's kinetic equaiton, 7
optical tweezers, 277
oscillation measurement, 110
osmotic
  bulk modulus, 28
  pressure, 24
  stress tensor, 29
Ostwald ripening, 200

pattern formation, 365
photoalignment, 151, 157
photoexcitation, 157
photoisomerization, 147
PIB, 95, 97, 111, 119, 133
plectoneme, 321
polyelectrolyte, 283
polyisobutylene, 95
polymer relaxation time, 245, 246, 252
Possion's ratio, 285
preaveraging approximation, 243

raft, 198
Rayleighian, 4
reaction diffusion, 366, 371, 372, 375, 385, 402
  excitable, 368
reciprocal relation, 5
relaxation rate, 211
reorientational dynamics, 152
Reynolds number, 284
rheo-dielectric behavior, 37
rheology, 118
rheometer, 93
RNA polymerase, 279, 338
Rotne-Prager mobility tensor, 303
Rouse dynamics, 293

rubbery relaxation, 44

salt/PEO composite, 63
scaling
  function, 101, 132
  property, 131
  relation, 102
screening length, 203
second mode, 134
self-assembled monolayer, 187
self-organization, 146, 151
self-propelled
  dynamics, 410
  interface, 406
  motions, 385, 388, 401, 402
  objects, 365
  particle, 366, 375
    deformable, 367, 372, 388, 389
self-propulsion, 366, 388
semiflexible polymer, 275
singular perturbation, 366, 404, 406
Smectic-C, 151, 161, 171, 178
soft lithography, 187
soft matter, 145
spatial correlation function, 100
sphere model, 105
spheroid, 99
spheroid model, 108
stacked fluid membranes, 257
static elastic modulus, 117
steady shear flow, 119
step electric field, 121, 130
Stokes equation, 204
Stokes formula, 303
Stokes-Einstein relation, 203
storage modulus, 112
stress tensor, 90
structural
  persistence length, 335
structural anisotropy, 122
superhydrophobic environment, 186
surface tension, 90, 110

tensor model, 366, 368, 379, 384, 410
ternary lipid mixture, 199
tilt angle, 124
time-correlation function, 210
Time-dependent
  Ginzburg–Landau model, 209
topological defects, 146, 161, 178
  annihilation, 163, 165, 172, 178
  asymmetric motion, 168
  charge, 162
  dynamics, 161
  formation, 162, 165
torsion, 288
transcription dynamics, 345
transcriticality, 406
transient
  process, 119, 131
  shear stress, 120
traveling waves, 156
twirling, 312
twist, 288
twist
  elasticity, 277
  persistence length, 276
type-A dipole, 48
type-B dipole, 46

variational principle, 8, 10
variational principle
  hydrodynamic, 7
  Onsager's, 7

viscosity, 183
viscous stress, 118, 119
vortex, 175
  stirring, 175

wetting contrast, 186
whirling, 313
writhe, 294, 295

Young's modulus, 284